AF606198

VOLUME FIVE HUNDRED AND SIXTY SEVEN

METHODS IN ENZYMOLOGY

Calorimetry

METHODS IN ENZYMOLOGY

VOLUME FIVE HUNDRED AND SIXTY SEVEN

METHODS IN ENZYMOLOGY

Calorimetry

Edited by

ANDREW L. FEIG
Department of Chemistry
Wayne State University
Detroit, MI
USA

AMSTERDAM • BOSTON • HEIDELBERG • LONDON
NEW YORK • OXFORD • PARIS • SAN DIEGO
SAN FRANCISCO • SINGAPORE • SYDNEY • TOKYO
Academic Press is an imprint of Elsevier

Academic Press is an imprint of Elsevier
50 Hampshire Street, 5th Floor, Cambridge, MA 02139, USA
525 B Street, Suite 1800, San Diego, CA 92101–4495, USA
The Boulevard, Langford Lane, Kidlington, Oxford OX5 1GB, UK
125 London Wall, London, EC2Y 5AS, UK

First edition 2016

Notices
Knowledge and best practice in this field are constantly changing. As new research and experience broaden our understanding, changes in research methods, professional practices, or medical treatment may become necessary.

Practitioners and researchers must always rely on their own experience and knowledge in evaluating and using any information, methods, compounds, or experiments described herein. In using such information or methods they should be mindful of their own safety and the safety of others, including parties for whom they have a professional responsibility.

To the fullest extent of the law, neither the Publisher nor the authors, contributors, or editors, assume any liability for any injury and/or damage to persons or property as a matter of products liability, negligence or otherwise, or from any use or operation of any methods, products, instructions, or ideas contained in the material herein.

ISBN: 978-0-12-802906-0
ISSN: 0076-6879

For information on all Academic Press publications visit our website at http://store.elsevier.com/

CONTENTS

CONTRIBUTORS

Monique Bastidas
Department of Chemistry, The Pennsylvania State University, University Park, Pennsylvania, USA

Guillaume Bec
Biophysics & Structural Biology Team, IBMC, UPR9002-CNRS, University of Strasbourg, Strasbourg, France

Philip C. Bevilacqua
Huck Institutes of the Life Sciences; Department of Chemistry; Department of Biochemistry and Molecular Biology, and Center for RNA Molecular Biology, The Pennsylvania State University, University Park, Pennsylvania, USA

M. Lucia Bianconi
Instituto de Bioquímica Médica Leopoldo de Meis, Universidade Federal do Rio de Janeiro, Rio de Janeiro, Brazil

Delta K. Boyles
Department of Chemistry, Mississippi State University, Mississippi State, Mississippi, USA

Mark L. Brader
Protein Pharmaceutical Development, Biogen, Cambridge, Massachusetts, USA

Curtis F. Brewer
Department of Molecular Pharmacology, Albert Einstein College of Medicine, Bronx, New York, USA

Christie G. Brouillette
Center for Structural Biology, and Department of Chemistry, University of Alabama at Birmingham, Birmingham, Alabama, USA

Samuel E. Butcher
Department of Biochemistry, University of Wisconsin–Madison, Madison, Wisconsin, USA

Margaret C. Carpenter
Department of Chemistry, Dartmouth College, Hanover, New Hampshire, USA

Stefano Ciurli
Laboratory of Bioinorganic Chemistry, Department of Pharmacy and Biotechnology, University of Bologna, Bologna, Italy

Molly L. Croteau
Department of Chemistry, Dartmouth College, Hanover, New Hampshire, USA

Cyrielle Da Veiga
Biophysics & Structural Biology Team, IBMC, UPR9002-CNRS, University of Strasbourg, Strasbourg, France

Tarun K. Dam
Laboratory of Mechanistic Glycobiology, Department of Chemistry, and Biotechnology Research Center, Michigan Technological University, Houghton, Michigan, USA

Carmelo Di Primo
University of Bordeaux, and INSERM U869, Laboratoire ARNA, Pessac, France

Philippe Dumas
Biophysics & Structural Biology Team, IBMC, UPR9002-CNRS, University of Strasbourg, Strasbourg, France

Joseph P. Emerson
Department of Chemistry, Mississippi State University, Mississippi State, Mississippi, USA

Eric Ennifar
Biophysics & Structural Biology Team, IBMC, UPR9002-CNRS, University of Strasbourg, Strasbourg, France

Kareem Fakhfakh
Michael Smith Laboratories, and Department of Chemical and Biological Engineering, University of British Columbia, Vancouver, British Columbia, Canada

Ni Fan
Laboratory of Mechanistic Glycobiology, Department of Chemistry, Michigan Technological University, Houghton, Michigan, USA

Julia A. Fecko
Huck Institutes of the Life Sciences, The Pennsylvania State University, University Park, Pennsylvania, USA

Charles Haynes
Michael Smith Laboratories, and RES'EAU Water Research Network, Department of Chemical and Biological Engineering, University of British Columbia, Vancouver, British Columbia, Canada

Kate L. Henderson
Department of Chemistry, Mississippi State University, Mississippi State, Mississippi, USA

Scott Horowitz
Howard Hughes Medical Institute and Department of Molecular, Cellular, and Developmental Biology, University of Michigan, Ann Arbor, Michigan, USA

Curtis B. Hughesman
Michael Smith Laboratories, University of British Columbia, Vancouver, British Columbia, Canada

Beatriz Ibarra-Molero
Facultad de Ciencias, Departamento de Química-Física, Universidad de Granada, Granada, Spain

Vincent Kao
Department of Chemical and Biological Engineering, University of British Columbia, Vancouver, British Columbia, Canada

Sandro Keller
Molecular Biophysics, University of Kaiserslautern, Kaiserslautern, Germany

Shun-ichi Kidokoro
Department of Bioengineering, Nagaoka University of Technology, Nagaoka, Japan

Pavel Landsman
Protein Pharmaceutical Development, Biogen, Cambridge, Massachusetts, USA

Chad W. Lawrence
Department of Chemistry, The Pennsylvania State University, University Park, Pennsylvania, USA

Vu H. Le
Department of Chemistry, Mississippi State University, Mississippi State, Mississippi, USA

Edwin A. Lewis
Department of Chemistry, Mississippi State University, Mississippi State, Mississippi, USA

A. Louise Creagh
Michael Smith Laboratories, and Department of Chemical and Biological Engineering, University of British Columbia, Vancouver, British Columbia, Canada

Luis A. Marky
Department of Pharmaceutical Sciences, University of Nebraska Medical Center, Omaha, Nebraska, USA

Luca Mazzei
Laboratory of Bioinorganic Chemistry, Department of Pharmacy and Biotechnology, University of Bologna, Bologna, Italy

Eva Muñoz
AFFINImeter Scientific & Development Team, Software 4 Science Developments, S. L. Ed. Emprendia, Campus Vida, Santiago de Compostela, A Coruña, Spain

Victor Muñoz
Centro Nacional de Biotecnología, Consejo Superior de Investigaciones Científicas, Madrid, Spain, and School of Engineering, University of California, Merced, California, USA

Athi N. Naganathan
Department of Biotechnology, Bhupat & Jyoti Mehta School of Biosciences, Indian Institute of Technology Madras, Chennai, India

Shigeyoshi Nakamura
Department of Bioengineering, Nagaoka University of Technology, Nagaoka, Japan

Vicki Nienaber
Zenobia Therapeutics, San Diego, California, USA

William G. Noid
Department of Chemistry, The Pennsylvania State University, University Park, Pennsylvania, USA

William Palau
University of Bordeaux, and INSERM U869, Laboratoire ARNA, Pessac, France

Angel Piñeiro
AFFINImeter Scientific & Development Team, Software 4 Science Developments, S. L. Ed. Emprendia, and Department of Applied Physics, Faculty of Physics, University of Santiago de Compostela, Campus Vida, Santiago de Compostela, A Coruña, Spain

Colette F. Quinn
Department of Chemistry, Dartmouth College, Hanover, New Hampshire, USA

Michael I. Recht
Palo Alto Research Center, Palo Alto, California, USA

Calliste Reiling-Steffensmeier
Department of Pharmaceutical Sciences, University of Nebraska Medical Center, Omaha, Nebraska, USA

Javier Rial
AFFINImeter Scientific & Development Team, Software 4 Science Developments, S. L. Ed. Emprendia, Campus Vida, Santiago de Compostela, A Coruña, Spain

Juan Sabin
AFFINImeter Scientific & Development Team, Software 4 Science Developments, S. L. Ed. Emprendia, Campus Vida, Santiago de Compostela, A Coruña, Spain

Debashish Sahu
Department of Chemistry, The Pennsylvania State University, University Park, Pennsylvania, USA

Jose M. Sanchez-Ruiz
Facultad de Ciencias, Departamento de Química-Física, Universidad de Granada, Granada, Spain

Scott A. Showalter
Huck Institutes of the Life Sciences; Department of Chemistry; Department of Biochemistry and Molecular Biology, and Center for RNA Molecular Biology, The Pennsylvania State University, University Park, Pennsylvania, USA

Melanie L. Talaga
Laboratory of Mechanistic Glycobiology, Department of Chemistry, Michigan Technological University, Houghton, Michigan, USA

Deniz B. Temel
Protein Pharmaceutical Development, Biogen, Cambridge, Massachusetts, USA

Martin Textor
Molecular Biophysics, University of Kaiserslautern, Kaiserslautern, Germany

Francisco E. Torres
Palo Alto Research Center, Palo Alto, California, USA

Raymond C. Trievel
Department of Biological Chemistry, University of Michigan, Ann Arbor, Michigan, USA

Kirk A. Vander Meulen
Department of Biochemistry, University of Wisconsin–Madison, Madison, Wisconsin, USA

Dean E. Wilcox
Department of Chemistry, Dartmouth College, Hanover, New Hampshire, USA

Zhengrong Yang
Center for Structural Biology, University of Alabama at Birmingham, Birmingham, Alabama, USA

Neela H. Yennawar
Huck Institutes of the Life Sciences, The Pennsylvania State University, University Park, Pennsylvania, USA

Barbara Zambelli
Laboratory of Bioinorganic Chemistry, Department of Pharmacy and Biotechnology, University of Bologna, Bologna, Italy

PREFACE

Reports using modern biological calorimetry began to appear in the literature in the late 1960s and early 1970s. The first generation of microcalorimeters were home built, required large samples, and had sensitivity issues such that large heats of reaction were required for analysis to be successful. As calorimeters evolved and became more sophisticated; however, their utility grew tremendously. In the 1990s, the instruments crossed a significant threshold. They became sufficiently sensitive to be broadly applicable in biology. Shortly thereafter, modest-priced commercial instruments became available such that individual investigators and core facilities could afford to own a microcalorimeter. This led to a rapid rise in their usage. A simple search of PubMed shows that in 1990, approximately 500 cited calorimetry as a key word (Fig. 1). By 2010, 20 years later, this number had increased five-fold, with more that 2500 papers listing calorimetry as a key word. Calorimetry has become a standard method for the biophysical characterization of biological molecules and reactions.

Since 1972, the *Methods in Enzymology* Series has published 26 reviews on the topic of biological calorimetry, starting with a groundbreaking review by one of the founders of the field John Sturtevant (Sturtevant, 1972). Many of these chapters were in volumes focused on specific biological macromolecule (e.g., RNA, GPCRs) or on specific applications of calorimetry to a molecule or structure central to that volume (e.g., membranes). Here in Volume 567, we provide for the first time an entire volume dedicated to applications of microcalorimetry across biology. While not every permutation of biocalorimetry is represented, the collected methods provide a sampling of the ways in which calorimeters are used to understand biological systems and how they are sometimes combined with other types of analysis to dissect the complex behavior of biomolecules.

The calorimetry literature features two dominant classes of experiments: isothermal titration calorimetry (ITC) and differential scanning calorimetry (DSC). The first section of this volume looks at ITC experiments. Here, small increments of a titrant are injected into the titrand, while feedback circuitry maintains a constant temperature between a sample cell and a reference cell. This experiment is exquisitely sensitive for the measurement of binding, allowing complete thermodynamic characterization of a reaction (ΔH, ΔS, ΔG and K_d, and n) from a single titration if measured under the

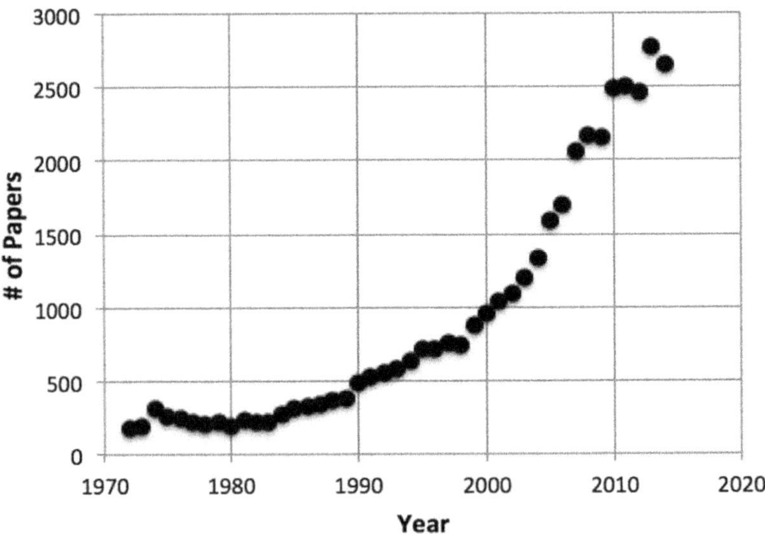

Figure 1 Number of papers indexed per year in PubMed for which "Calorimetry" is listed as a key word.

proper conditions. The experiment measures all heat sources during the reaction, however, and that can cause some practical problems. For instance, if the buffer of the titrant and titrand do not match exactly, then a heat of dilution is observed superimposed upon the binding data. In addition, since most biological reactions are carried out in buffered solutions, biochemists often choose to ignore the protons gained or lost as part of a reaction. However, buffer protonation can contribute significantly to the heats of reaction obscuring the fundamental result of an experiment unless proper corrections are applied. Buffers, even Good's buffers (Good, 1966), are also not always innocent. As a result of their metal binding behavior, buffers sometimes act as enzyme inhibitors or interact with the metal ions critical for activity. These behaviors can be assessed through carefully designed ITC analysis.

Most scientists view ITC as a method to measure the thermodynamics of binding reactions, but as the reviews in this volume show, ITC today is being used for a much wider set of applications. Heat is a very sensitive readout on reactions and is a superb way to measure the rates of enzymatic reactions using natural substrates without resorting to coupled-enzyme assays. ITC allows the user to incrementally increase the substrate concentration to measure Michaelis–Menten kinetics or raise inhibitor concentrations to assess enzyme inactivation. By using enthalpy arrays or high-throughput ITC instruments, the method is even suitable to screening compound libraries for bioactive molecules.

The second section of this volume is devoted to scanning calorimetry. In the DSC instrument, sample and reference cells are gradually heated maintaining the same temperature in each. If molecules within the sample cell undergo phase transitions or unfold, the endo- or exothermic processes can be measured. The heat capacity of the solution changes during the transition, resulting in a deflection in the power function. When measured relative to a comparable sample lacking the macromolecule, one can obtain the enthalpy associated (ΔH) with the transition as well as the transition temperature. Through modeling of the thermogram, a full thermodynamic description of the transition energetics (including ΔG, ΔH, ΔS, and ΔC_P) can be obtained as long as the transition is reversible.

DSC is used for biophysical characterization of many macromolecules. It can be used alone or in conjunction with other methods, including ITC. The reviews collected in this volume describe a few of these applications to illustrate the importance of DSC toward understanding the physical properties of proteins, nucleic acids, and lipids. One particularly enlightening result comes from the use of DSC in pharmaceutical formulation work. Dogma has been that the most thermally stable molecules make for the best pharmaceutical preparations. An illustrative example presents work on therapeutic antibodies and how DSC together with other biophysical methods dissects the multidomain unfolding behavior that impacts their biological activity.

Finally, a somewhat unconventional but hopefully highly useful article describes a biophysical instrumentation facility within an academic department. Because so many investigators want access to calorimetry but do not necessarily need an instrument of their own, the shared-instrument model is quite attractive. This paper articulates some of the best practices that allow shared calorimetry facilities to be successful scientific resources while keeping costs reasonable for the users. It describes the economics required to acquire and maintain the instruments as well as issues related to training new users and the critical mass of investigators necessary to make the investment pay off. This article should be useful for departments or institutions interested in establishing shared facilities or updating the business practices of existing ones.

Andrew L. Feig

REFERENCES

Good, N. (1966). Hydrogen ion buffers for biological research. *Biochemistry*, *5*, 467–477.
Sturtevant, J. M. (1972). Calorimetry. *Methods in Enzymology*, *26*, 227–253.

SECTION I

ITC

CHAPTER ONE

Isothermal Titration Calorimetry Measurements of Metal Ions Binding to Proteins

Colette F. Quinn[1], Margaret C. Carpenter[2], Molly L. Croteau, Dean E. Wilcox[3]
Department of Chemistry, Dartmouth College, Hanover, New Hampshire, USA
[3]Corresponding author: e-mail address: dean.wilcox@dartmouth.edu

Contents

Abstract

ITC measurements involving metal ions are susceptible to a number of competing reactions (oxidation, precipitation, and hydrolysis) and coupled reactions involving the buffer and protons. Stabilization and delivery of the metal ion as a well-defined and well-characterized complex with the buffer, or a specific ligand, can suppress undesired

[1] Current address: TA Instruments, Lindon, UT.
[2] Current address: Department of Chemistry and Biochemistry, University of Colorado, Boulder, CO.

Methods in Enzymology, Volume 567
ISSN 0076-6879
http://dx.doi.org/10.1016/bs.mie.2015.08.021

solution chemistry and, depending on the stability of the metal complex, allow accurate measurements of higher affinity protein-binding sites. This requires, however, knowledge of the thermodynamics of formation of the metal complex and accounting for its contribution to the experimentally measured values (K_{ITC} and ΔH_{ITC}) through a *post hoc* analysis that provides the condition-independent binding thermodynamics (K, $\Delta G°$, ΔH, ΔS, and ΔC_P). This analysis also quantifies the number of protons that are displaced when the metal ion binds to the protein.

ABBREVIATIONS

DTT dithiothreitol
EDTA ethylenediaminetetraacetic acid
ITC isothermal titration calorimetry
NIST National Institute of Science and Technology
TCEP tris(2-carboxyethyl)phosphine

1. INTRODUCTION

Certain metal ions play important roles in biology, while others are disruptive to biological pathways and thus toxic. In most of these essential and deleterious roles, the metal is bound to a protein or other biological macromolecule. Since binding of the metal to the protein is requisite for function (e.g., metalloenzyme, metal–protein complex), the thermodynamics of binding is fundamental to the biology. While several methods are capable of measuring the equilibrium that quantifies the affinity of a protein for a metal ion (K_d), it is necessary to quantify the change in enthalpy (ΔH), either indirectly by van't Hoff analysis or directly by calorimetric measurements, to gain insight about specific binding interactions and determine the binding thermodynamics that result in this affinity.

The technique of isothermal titration calorimetry (ITC), which became commercially available nearly 30 years ago (Freire, Mayorga, & Straume, 1990; Wiseman, Williston, Brandts, & Lin, 1989), has proven to be very successful and is widely used for quantitative measurements of binding thermodynamics. The basis of this technique is the quantitative measure of heat flow associated with a binding event that, when conducted as a titration of one species into the other, allows the binding equilibrium (K or K_d), and thus the binding free energy ($\Delta G° = -RT \ln K$), to be quantified. Since the heat flow associated with binding at constant pressure is the change in enthalpy ($\Delta H = q_P$), one measurement is, in theory, capable of providing the binding free energy, the binding enthalpy and the binding entropy (ΔS) from the fundamental relationship, $\Delta G^\circ = \Delta H - T\Delta S$. However, since ITC directly measures the total heat flow, all processes that occur upon addition of the titrant will contribute to the measured enthalpy, whether or not they are the binding reaction of interest.

Development of the ITC method and instrumentation was largely driven by a need to quantify the binding thermodynamics of biological macromolecules (e.g., protein–protein, protein–DNA, and receptor–ligand). When ITC measurements involve metal ions, there are a number of special issues that must be considered, many of which were first identified in a pair of papers by Brandts and coworkers on Fe^{3+} binding to transferrin (Lin, Mason, Woodworth, & Brandts, 1991, 1993). This chapter will outline these issues and provide guidance for the design of ITC measurements and the analysis of ITC data that is needed to obtain accurate thermodynamic values for metal ions binding to proteins.

2. METAL ION PROPERTIES RELEVANT TO ITC MEASUREMENTS

Since each metal ion, including different oxidation states of the same metal, has its own unique chemical properties, it is important to be as informed as possible about these properties, specifically those that are relevant in aqueous solution.

2.1 Aqueous Solution Chemistry of Metal Ions

Three key properties of metal ions can compromise the ability to accurately measure their binding thermodynamics with ITC. These will vary with the metal and should be considered for each case, as they can lead to competing reactions that contribute heat to the experimental enthalpy and affect the binding equilibrium.

First is the susceptibility of the metal ion to *oxidation* or, less commonly, reduction. Since the calorimeter cell is not sealed and the stirring ensures rapid mixing of any O_2 that is introduced, aerobic oxidation is a concern. While oxygen-scavenging agents, such as sodium dithionite, can be included, their oxidation will contribute to the heat that is measured. Therefore, when the metal ion or the protein is susceptible to oxidation, this competing reaction should be suppressed by placing the calorimeter under an inert atmosphere. While modern titration calorimeters are small enough to fit into commercial glove boxes, Fig. 1 shows a custom Plexiglas® glove box that was designed and built for a MicroCal VP-ITC.

Second is the issue of *solubility*, as the solution chemistry of certain metal ions can lead to precipitation. Phosphate forms insoluble species with many transition metal ions and, for this reason, phosphate should generally be avoided as the buffer with metal ions. Certain metal ions may have low solubility in other buffer solutions, which should be identified before hand and avoided. Another example is the low solubility of the halide salts of Ag^+, requiring the elimination of chloride from ITC solutions.

A third issue, which is known as *hydrolysis* (Nakon & Krishnamoorthy, 1983) is the deprotonation of a metal-bound water (Eq. 1), resulting in a hydroxide ligand.

$$[LM^{n+} - H_2O]^{+\,n} + H_2O \rightleftarrows [LM^{n+} - OH]^{+\,(n-1)} + H_3O^+ \tag{1}$$

Such metal-hydroxo species are susceptible to oligomerization and reduced solubility. Since this process typically depends on concentration, hydrolysis can lead to an unintended (and unidentified) heat upon dilution of a metal titrant solution into the cell.

Finally, there may be other unique solution chemistry of the metal ion of interest. One example is the favorable ($K \sim 10^6$) disproportionation of Cu^+ in aqueous solution (Eq. 2) (Latimer, 1952),

Figure 1 Custom glove box for a MicroCal VP-ITC that allows ITC measurements under an inert atmosphere.

$$2Cu^{+}_{(aq)} \rightleftarrows Cu^{0}_{(s)} + Cu^{2+}_{(aq)} \tag{2}$$

which is independent of oxidation and must be suppressed to stabilize copper in this oxidation state for accurate thermodynamic measurements (Johnson et al., 2015).

2.2 Buffers

Since metal ions are Lewis acids (electron pair acceptors) and buffers are Lewis bases (electron pair donors), there will always be an interaction between a metal ion and the basic form of the buffer (Table 1 provides protonation values for several common buffers). Since the buffer is normally present in large excess, this interaction cannot be ignored, but can sometimes be useful. There are no innocent buffers, including those, like

Table 1 Buffer Protonation Values at 25 °C[a]

Buffer	Log K_{HB}	ΔH kJ mol^{-1} (kcal mol^{-1})
ACES[b]	6.79	−39.1 (−9.34)
BES	7.11	−25.4 (−6.07)
bis-Tris	6.54	−28.7 (−6.86)
Cacodylic acid[b]	6.19	2.1 (0.5)
Cholamine^{+}	6.89[c]	−48.1 (−11.5)[d]
DIPSO	7.58	−30.2 (−7.22)
HEPES	7.52	−21.8 (−5.22)
HEPPSO	7.98	−23.7 (−5.67)
Imidazole	7.0	−36.9 (−8.81)
MES[b]	6.24	−14.8 (−3.54)
MOPS[b]	7.18	−21.0 (−5.01)
PIPES	6.93	−11.6 (−2.78)
TAPSO	7.55	−39.1 (−9.34)
TES	7.6	−32.2 (−7.7)
Tris	8.10	−47.5 (−11.36)

[a]Values from NIST (2004), except where indicated; $\mu = 0.10$, except where indicated.
[b]$\mu = 0.00$.
[c]Zawisza, Croteau, Wilcox, and Bal (2015).
[d]Blasie and Berg (2002).

the positively charged amine buffer cholamine chloride (2-(aminoethyl) trimethylammonium chloride), that would seem to have a weak interaction with metal ions (Blasie & Berg, 2003). There are also cases of a unique interaction between a metal ion and a buffer, such as the redox reaction of Cu^{2+} with 4-(2-hydroxyethyl)piperazine-1-ethanesulfonate (HEPES) (Hegetschweiler & Saltman, 1986).

Because metal–buffer interactions are unavoidable, it is important to know the predominant metal–buffer species that is present in the buffered solution. This will allow its formation constant and heat of formation to be subtracted from the experimental values to obtain the condition-independent binding constant and enthalpy. While several metal–buffer combinations have been investigated thoroughly, the best values for the formation constants are often available from the National Institute of Science and Technology (NIST) (NIST Standard Reference Database, 2004) or the International Union of Pure and Applied Chemistry (Petitt & Powell, 1993), although earlier (Magyar & Godwin, 2003) and recent (Ferreira, Pinto, Soares, & Soares, 2015) reviews have compiled useful metal–buffer stabilities. However, the formation enthalpies of these metal–buffer species, which are needed for the analysis of ITC data (vide infra), are not as readily available.

Nevertheless, the interaction of buffer with metal ions can be beneficial when it suppresses competing precipitation and hydrolysis reactions. Further, buffers that form a more stable complex with the metal may be useful to quantify binding to higher affinity sites, as this may put the competition between the buffer and the protein for the metal ion within the stability range of the ITC measurement (so-called c window).

2.3 Protons

Since the proton is also a Lewis acid, metal ions will often compete with protons when binding to a protein, and this competition needs to be considered. The pH of the solution and relevant pK_a values will dictate (a) the concentration of the basic form of the buffer that interacts with the metal, (b) the protons that the metal will displace from certain amino acids (e.g., Cys, His, and Tyr), and (c) any other (de)protonations that are coupled to metal binding. These will have an impact on the experimentally measured binding equilibrium and particularly the binding enthalpy when there is a coupled (de)protonation of the buffer, and need to be included in a *post hoc* analysis to obtain condition-independent binding thermodynamics.

Nevertheless, a series of ITC measurements with buffers that have a range of protonation enthalpies can be used to determine how many protons are displaced, or less commonly bind, when a metal ion binds to a protein at a given pH (Grossoehme, Akilesh, Guerinot, & Wilcox, 2006). This can provide valuable insight about the binding residues and the coordination site of the metal ion. Alternatively, a similar analysis of ITC data at two or more pH values can be used to determine the pK_a associated with the (de) protonation.

3. ITC MEASUREMENTS INVOLVING METAL IONS

Direct addition of a metal ion into a solution of the protein ($M \rightarrow P$) is the most obvious ITC measurement of metal binding. However, there may be cases when the reverse titration ($P \rightarrow M$) is useful, such as when the metal ion has limited solubility that can only be achieved at the lower concentration in the cell or when stirring in the cell compromises the integrity of the metal-free protein (Carpenter & Wilcox, 2014). Further, when more than one metal ion binds to the protein, additional insight may be obtained by comparing titration data obtained in both directions, since the isotherm for the reverse titration should have reciprocal stoichiometries (Spuches, Kruszyna, Rich, & Wilcox, 2005). An example is formation of the 1:2 complex of Ni^{2+} and His (Fig. 2), where the two equilibria for each direction and their relationships are indicated in Scheme 1.

Another potentially useful ITC measurement is the displacement of one metal ion by another (Eq. 3) to obtain the difference in their binding thermodynamics.

$$M^{n+} - \text{protein} + M'^{n+} \rightleftarrows M'^{n+} - \text{protein} + M^{n+} \tag{3}$$

If the thermodynamics of the displaced metal binding to the protein are known, or can be measured independently, then the binding thermodynamics of the displacing metal can be determined from these measurements (Carpenter et al., 2015).

A third type of ITC measurement that has proven useful for certain cases is the titration of a chelating ligand into a solution of the metalloprotein (Eq. 4) (Carpenter & Wilcox, 2014; Carpenter et al., 2015; Grossoehme et al., 2006; Pedroso et al., 2014).

$$\text{chelate} + M^{n+} - \text{protein} \rightleftarrows \text{protein} + M^{n+} - \text{chelate} \tag{4}$$

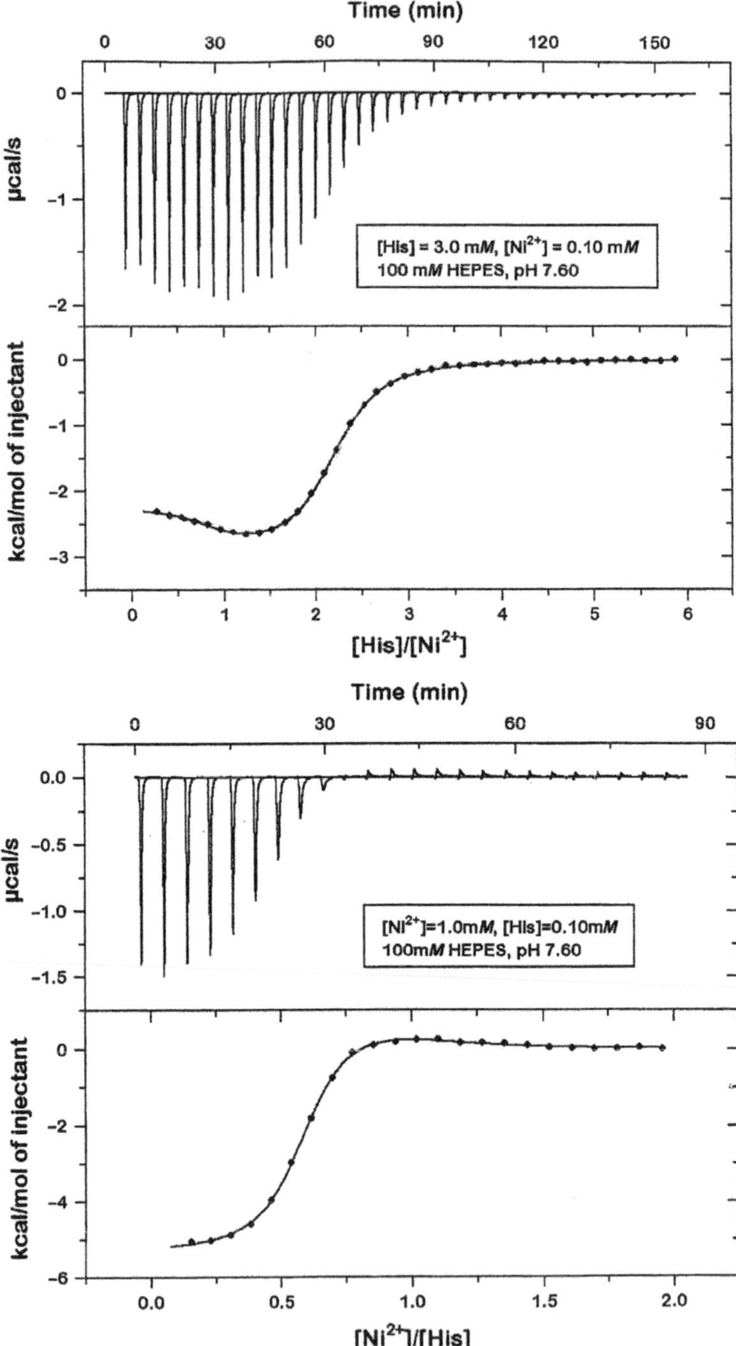

Figure 2 ITC thermograms of the amino acid histidine binding to Ni^{2+} in 100 m*M* HEPES buffer, pH 7.6; upper: forward titration, His $\rightarrow$ Ni^{2+}; lower: reverse titration, Ni^{2+} $\rightarrow$ His.

Forward direction: His→Ni^{2+}

$$His + Ni^{2+} \rightleftarrows Ni(His)^{+} + H^{+} \qquad K_1$$

$$His + Ni(His)^{+} \rightleftarrows Ni(His)_2 + H^{+} \qquad K_2$$

Reverse direction: Ni^{2+}→His

$$Ni^{2+} + 2\,His \rightleftarrows Ni(His)_2 + 2\,H^{+} \qquad \beta_2 = K_1 \cdot K_2$$

$$Ni^{2+} + Ni(His)_2 \rightleftarrows 2\,Ni(His)^{+} \qquad K' = K_1 / K_2$$

Scheme 1 Relevant equilibria for titrations of Ni^{2+} and His.

Based on the principle of microscopic reversibility, the chelation data can provide the binding thermodynamics, after accounting for the stronger metal–chelate interaction.

Since there is a heat from dilution of the titrant solution when it is added to the cell, this dilution heat must be subtracted from the binding isotherm to obtain accurate experimental enthalpies. While a heat of dilution can be measured from a control titration (e.g., metal into the cell without the protein), the heat associated with the last few injections of an ITC measurement, when there is no longer any binding, provides the dilution heat for the exact experimental conditions and is preferred. However, a metal control titration can provide evidence for metal-based reactions (e.g., hydrolysis) that may unknowingly be contributing to the ITC measurement. A significant change in the dilution heat during a control titration indicates a concentration-dependent background that needs to be carefully subtracted or eliminated with different experimental conditions (e.g., buffer).

Protein solutions often include other species, such as a salt (NaCl), reductant (tris(2-carboxyethyl)phosphine (TCEP), dithiothreitol (DTT)), preservative (NaN_3), and/or detergent (sodium dodecyl sulfate). As long as their concentrations are matched between the titrant and cell solutions and they are not involved in other reactions during the ITC measurement, their presence may not significantly affect the ITC results. As indicated above, aerobic oxidation of reducing agents is a concern, particularly the exothermic oxidation of DTT, which should be avoided. However, it is crucial to know if any of these species have an interaction with the metal ion so they can be removed or this interaction can be included in the data analysis. For example, Bal and coworkers have quantified the stabilities of metal complexes with the common reducing agents DTT (Krezel et al., 2001), TCEP (Krezel, Latajka, Bujacz, & Bal, 2003), and dithiobutylamine

(Adamczyk, Bal, & Krezel, 2015). Finally, in any situation where there is a competition between the protein and some other ligand, maintaining the ligand at a high and essentially constant concentration will simplify the data analysis; otherwise a competition-binding model (Sigurskjold, 2000) that accounts for significant changes in concentration (activity) must be used.

It is often not difficult to obtain ITC data, but fitting and analyzing the experimental data to obtain accurate thermodynamic values can be a real challenge. Thus, it is critical to evaluate the initial data and possibly to modify the experimental conditions to obtain optimal data. This may involve adjusting the concentrations or conditions to increase the signal-to-noise or to shift the binding isotherm to accentuate certain portions, such as the initial substoichiometric region. For complicated binding data, a higher resolution isotherm may be obtained by concatenating the data from a second, and even third, filling of the syringe with the titration solution and linking the two, or three, sequential sets of binding data (Grossoehme et al., 2006; Spuches et al., 2005). Finally, it is crucial to repeat ITC measurements to ensure their reproducibility and to determine the error associated with the experimental values.

4. FITTING ITC DATA INVOLVING METAL IONS

Several models are widely available to extract binding values from ITC data. These include (a) the one-site model with one set of binding parameters (n, K, and ΔH); (b) the two-sites model with two sets of binding sites that fill independently, each with their own set of binding parameters; (c) the sequential binding model with integer binding stoichiometry and sequential filling of the binding sites, each with their own values of K and ΔH, and (d) the competition-binding model that accounts for the changing concentration (activity) of the displaced species (Sigurskjold, 2000). While other models can be developed to analyze ITC binding isotherms (e.g., three-sites model (Bou-Abdallah, Woodhall, Velazquez-Campoy, Andrews, & Chasteen, 2005)), most binding data can be analyzed with a judicious use of these models (Grossoehme, Mulrooney, Hausinger, & Wilcox, 2007).

In certain cases, it may be useful to consider the fit values obtained from an analysis of the data with different models. For example, when there is a competition for the metal ion, the one-site or two-sites models can provide accurate values for the binding enthalpy but the competition model is needed to obtain an accurate binding constant (Carpenter & Wilcox, 2014; Carpenter et al., 2015; Pedroso et al., 2014). When the isotherm contains two binding events, a comparison of the fit values with the one-site, two-sites and sequential binding models may provide evidence for an interaction between the two sites. Finally, a comparison of the goodness of fit with different models needs to account for the degrees of freedom (independent fit parameters) with the different models. Therefore, fit comparisons should be based on the residual sum of squares divided by the degrees of freedom (Johnson et al., 2015).

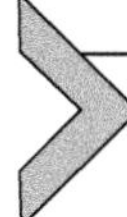

5. *POST HOC* ANALYSIS OF ITC DATA INVOLVING METAL IONS

ITC data can be fit to provide the binding stoichiometry (n), binding (equilibrium) constant (K), and binding enthalpy (ΔH), from which it is possible to calculate the binding entropy $\Delta S = (\Delta H + RT \ln K)/T$ for each binding event. However, because of the properties of metal ions outlined above, there are inevitably one or more coupled equilibria that are associated with metal binding. The contributions of these equilibria to the experimental values (K_{ITC} and ΔH_{ITC}) need to be subtracted to obtain accurate K, $\Delta G°$, ΔH, and ΔS values for the binding of interest (i.e., metal ion binding to the protein).

5.1 Binding Enthalpy (ΔH)

The experimental enthalpy from ITC measurements of metal ions binding to proteins (ΔH_{ITC}) typically contains multiple contributions, including the heat associated with (a) dissociation of the buffer from the metal ion; (b) the metal ion binding to the protein, which often displaces one or more protons; and (c) any displaced protons binding to the buffer (Scheme 2).

Since the second of these is desired, it is crucial to account for contributions from the first and third. The first equilibrium can be determined independently and a growing compilation of metal–buffer enthalpies (Table 2) provides these values. The third equilibrium can be determined from the protonation enthalpy of the buffer (Table 1), as long as the number of

$$\text{metal–buffer} \rightleftarrows \text{metal} + \text{buffer} \qquad \Delta H_{\text{M-buffer}}$$

$$\text{metal} + \text{protein} \rightleftarrows \text{metal–protein} + n\,\text{H}^+ \qquad \Delta H_{\text{M-protein}}$$

$$\underline{n\,\text{H}^+ + n\,\text{buffer} \rightleftarrows n\,\text{bufferH}^+ \qquad \Delta H_{\text{bufferH+}}}$$

$$\text{protein} + \text{metal–buffer} + (n-1)\,\text{buffer} \rightleftarrows \text{metal–protein} + n\,\text{bufferH}^+ \qquad \Delta H_{\text{ITC}}$$

Scheme 2 Individual and net equilibria, and their enthalpies, for a metal ion binding to a protein in buffered aqueous solution.

Table 2 Metal–Buffer Thermodynamic Values at 25 °C[a]

Metal	Buffer	pH	Major Species[b] in Excess Buffer at the Indicated pH	Log *K* or log *β*[c] of Major Species	ΔH_{MB}[d] kJ mol^{-1} (kcal mol^{-1})
Ag^+	bis-Tris	7.4	MB	2.00	0.075 (0.018)
	HEPES	7.4	MB	6.0[e]	22.2 (5.3)
	Tris	7.0	MB_2	6.57[e]	−62.8 (−15)
Cd^{2+}	ACES	7.25	nd	nd	−13.4 (−3.2)[f]
	bis-Tris	7.0	MB	2.47[g]	−2.5 (−0.6)
	HEPES	7.4; ***7.4***	MBH^+	4.16	3.1 (0.73); ***1.7*** (***0.41***)[h]
	TAPSO	7.4	MB	4.2	17.2 (4.1)
	TES	7.4	nd	nd	−6.7 (−1.6)
	Tris	7.4; ***7.4***	MB	1.88	−6.8 (−1.62); ***−7.6*** (***−1.81***)[h]
Co^{2+}	ACES	7.4	MB	3.49	−11.7 (−2.8)
	bis-Tris	7.25	MB	1.78[g]	−5.1 (−1.23)
	Cacodylic acid	6.0	MB	2.30[i]	5.4 (1.3)[i]
	DIPSO	7.4	MB	3.56	8.4 (2.0)
	MES	6.0	MB	2.04	1.3 (0.3)[i]
	PIPES	7.25; ***6.0***	MB	3.12	−0.02 (−0.08)[f]; ***2.4*** (***0.58***)[i]
	TAPSO	7.4	MB	3.53	−6.3 (−1.5)
	TES	7.4	MB	2.07	3.3 (0.8)
	Tris	7.4	MB	1.73	−12.4 (−2.97)
Cu^{2+}	ACES	7.25; ***7.0***	MB_2	8.04	−48.1 (−11.5)[f]; ***−48.1*** (***−11.5***)[j]
	BES	7.0	MB	3.24	−23.8 (−5.7)[k]
	bis-Tris	7.0; ***7.0***	MB	5.13	−26.4 (−6.3); ***−30.5*** (***−7.3***)[k]
	Cholamine$^+$	7.0	nd	nd	−76.1 (−18.2)
	DIPSO	7.0	MB_2	8.1	−28.0 (−6.7)

Table 2 Metal–Buffer Thermodynamic Values at 25 °C—cont'd

Metal	Buffer	pH	Major Species in Excess Buffer at the Indicated pH	Log *K* or log *β* of Major Species	ΔH_{MB} kJ mol^{-1} (kcal mol^{-1})
	HEPES	7.4	MB	3.22	6.3 (1.5)[l]
	HEPPSO	7.0	MB	2.05	−2.5 (−0.6)
	Imidazole	7.4	MB_4	12.6[e]	−54.0 (−12.9)[l]
	MOPS	7.0	MB	4.0	−14.6 (−3.5)
	TAPSO	7.0	MB_2	8.2	−39.3 (−9.4)
	TES	7.0; ***7.0***	MB	3.90[m]	−18.0 (−4.3); ***−20.0*** (***−4.8***)[k]
	Tris	7.0; ***7.4***	MB^+; $\boldsymbol{MB_2^+}$	−2.22[n]; (***1.47***)[n]	−61.5 (−14.7); ***−70.7*** (***−16.9***)[l]
Fe^{2+}	MOPS	7.25	MB	1.14[f]	−6.7 (−1.61)
Ni^{2+}	ACES	7.25	MB	3.61	−33.1 (−7.9)[f]
	bis-Tris	7.0	MB	3.59[g]	−10.0 (−2.4)
	Cacodylic acid	6.0	MB	2.33[i]	5.0 (1.2)[i]
	MES	6.0	MB	2.06	1.7 (0.40)[i]
	PIPES	6.0	MB	3.20	2.6 (0.63)[i]
	TES	7.45	MB	3.35	0.13 (0.032)
	Tris	8.22	MB_2	4.6	−17.2 (−4.1)
Pb^{2+}	HEPES	7.4	MB	3.14	3.1 (0.73)
	MES	7.4	nd	nd	5.2 (1.24)
	Tris	7.4	MB	1.96	−6.8 (−1.62)
Zn^{2+}	ACES	7.25; ***7.4***	MB_2	3.74	−19.7 (−4.7)[f]; ***−21.8*** (***−5.2***)
	bis-Tris	7.0; ***7.25***	MB	2.38	−1.8 (−0.4); ***−1.3*** (***−0.3***)
	HEPES	7.4; ***7.4***	MB	3.27	1.3 (0.3); ***1.5*** (***0.35***)[h]
	MES	6.5	nd	nd	2.9 (0.7)
	MOPS	7.4	nd	nd	0.9 (0.2)

Continued

Table 2 Metal–Buffer Thermodynamic Values at 25 °C—cont'd

Metal	Buffer	pH	Major Species in Excess Buffer at the Indicated pH	Log K or log β of Major Species	ΔH_{MB} kJ mol^{-1} (kcal mol^{-1})
	PIPES	7.25; ***7.4***	nd	nd	0.9 (0.2); ***1.9 (0.5)***
	Tris	7.4; ***7.4***	MB	2.27	−8.4 (−2.0); ***−6.4 (−1.52)***[h]

[a]$\mu=0.1$, except where indicated; values in bold italics are from an independent measurement and may be at a different pH.
[b]M = metal, B = buffer, nd = not determined.
[c]All values are from Ferreira et al. (2015), unless otherwise indicated; nd = not determined.
[d]Values were determined with metal-binding measurements using a known ligand, such as EDTA, trien, or TEPA, in a large excess of buffer.
[e]NIST (2004).
[f]Grossoehme et al. (2006).
[g]$\mu=1.0$.
[h]Chekmeneva, Prohens, Díaz-Cruz, Ariño, and Esteban (2008).
[i]Wyrzykowski, Pilarski, Jacewicz, and Chmurzyński (2013).
[j]Song, Weitz, Hendrich, Lewis, and Emerson (2013).
[k]Nettles, Song, Farquhar, Fitzkee, and Emerson (2015).
[l]Quinn (2010).
[m]Nakon and Krishnamoorthy (1983).
[n]Nagaj et al. (2013).

displaced protons (n_{H+}) is known. This can be found either from other experimental information (e.g., metal coordination to the protein and the residue pK_a's) or determined from a simple analysis of data from ITC measurements in different buffers (Grossoehme et al., 2006), where the enthalpy contribution from buffer protonation gives the number of displaced protons.

A procedure to determine the condition-independent metal-binding enthalpy, $\Delta H_{M\text{-protein}}$, at a given pH would involve the following steps.

1. Find experimental conditions (pH, buffer) that give a good binding isotherm and binding enthalpy (ΔH_{ITC}).
2. At this pH, collect data in at least one more buffer ($\Delta H'_{ITC}$) that has a different protonation enthalpy ($\Delta H_{bufferH+}$).
3. Find (Table 2), or determine, the metal interaction with the buffers ($\Delta H_{M\text{-buffer}}$).
4. Use these values (ΔH_{ITC}, $\Delta H_{bufferH+}$, $\Delta H_{M\text{-buffer}}$) for each of the buffers to determine the number of protons (n_{H+}) that are displaced, or bind, upon metal binding.
5. Subtract the metal–buffer enthalpy ($\Delta H_{M\text{-buffer}}$) and the buffer protonation enthalpy ($n_{H+}\cdot\Delta H_{bufferH+}$) from the experimental enthalpy (ΔH_{ITC}) for each buffer to obtain condition-independent metal-binding enthalpies ($\Delta H_{M\text{-protein}}$).

$$\text{chelateH}^+_m \rightleftarrows \text{chelate} + m\,\text{H}^+ \qquad \Delta H_{\text{chelateH+}}$$

$$\text{metal–protein} \rightleftarrows \text{metal} + \text{protein} \qquad \Delta H_{\text{M-protein}}$$

$$\text{chelate} + \text{metal} \rightleftarrows \text{chelate–metal} \qquad \Delta H_{\text{M-chelate}}$$

$$m\,\text{H}^+ + m\,\text{buffer} \rightleftarrows m\,\text{bufferH}^+ \qquad \Delta H_{\text{bufferH+}}$$

$$\text{chelateH}^+_m + \text{metal–protein} + m\,\text{buffer} \rightleftarrows \text{chelate–metal} + \text{protein} + m\,\text{bufferH}^+ \qquad \Delta H_{\text{ITC}}$$

Scheme 3 Individual and net equilibria, and their enthalpies, for the chelation of a metal ion bound to a protein in buffered aqueous solution.

6. Compare the resulting condition-independent binding enthalpies for consistency.

There may be other contributions to the experimental binding enthalpy that need to be subtracted in more complex cases. For example, the chelate in a chelation titration experiment is often protonated in the absence of the metal ion, and its deprotonation enthalpy must be included in the analysis (Scheme 3).

5.2 Equilibrium Constant (*K*)

The experimental equilibrium constant reflects not only the binding equilibrium of interest but the contributions from any competing equilibria. This includes equilibria associated with the metal–buffer interaction (the buffer is competing with the protein for the metal ion) and any pH-dependent competition between the metal ion and protons. Competition from other species is included in the analysis by competition factors, Q (Eq. 5; X=buffer, protons, etc.) (Grossoehme, Spuches, & Wilcox, 2010),

$$Q = \sum (\beta_n [X]_n) \tag{5}$$

where

$$K_{\text{M-protein}} = K_{\text{ITC}} Q \tag{6}$$

For example, Eq. (7) indicates Q when the buffer forms a 1:1 complex with the metal ion,

$$Q_{\text{buffer}} = (1 + K_{\text{M-buffer}}[\text{buffer}]). \tag{7}$$

If ITC measurements are done in a well-buffered solution, then the concentrations of the acidic and basic forms of the buffer are relatively constant and any coupled (de)protonation of the buffer does not contribute significantly to the binding constant K.

5.3 Binding Entropy (ΔS) and Heat Capacity (ΔC_P)

The fit of an experimental binding isotherm provides values for K_{ITC} and ΔH_{ITC}, from which a value of ΔS_{ITC} can be calculated. These values, however, are unique for the experimental conditions. When there are coupled equilibria, as there invariably are when metal ions bind to a protein, the condition-independent values for K and ΔH are required to determine the condition-independent value for ΔS from the standard relationship, $\Delta G° = \Delta H - T\Delta S$. For the same reason, the condition-independent value for the change in heat capacity upon binding, ΔC_P, requires the temperature dependence of the condition-independent binding enthalpy in the analysis, $\Delta C_P = \Delta\Delta H/\Delta T$.

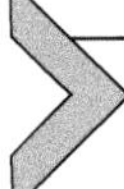

6. ADDITIONAL CONSIDERATIONS WITH METAL IONS

6.1 Comparison with Other Methods

Several methods, including equilibrium dialysis and various spectroscopic techniques, can be used to determine the equilibrium constant for metal ions binding to proteins. However, two methods commonly used in protein-binding studies, surface plasmon resonance and fluorescence anisotropy, may not be able to detect a metal ion binding to a protein, unless the binding results in a significant change in the refractive index of the protein or a change in its hydrodynamic properties, respectively. Since heat flow (change in enthalpy) accompanies essentially all binding events, ITC is well suited for quantifying metal-binding equilibria.

Direct measure of the binding enthalpy requires a calorimetric method, and ITC allows the complete binding thermodynamics to be quantified. A van't Hoff analysis of the temperature dependence of the binding constant can indirectly provide a binding enthalpy, but this analysis assumes a negligible change in the heat capacity over the experimental temperature range, which is often not the case with proteins.

6.2 Sources of Error

Modern titration calorimeters are remarkably sensitive and able to measure the small amounts of heat associated with binding in dilute aqueous solution. However, any value that is reported from ITC experimental data must include the error that is associated with the value. Typically, the variability among measurements is larger than the error associated with fitting a single data set, and a statistically significant value requires enough replicates to report the error associated with the value.

6.3 Tips

Titration calorimetry measurements can be compromised by contaminants, such as metal precipitates. Careful cleaning with a chelating agent (i.e., ethylenediaminetetraacetic acid (EDTA), nitrilotriacetic acid) is needed to ensure that the cell and syringe are free of metal contaminants.

Since the thermodynamics of metal ions binding to common ligands (e.g., EDTA) are well known, simple control titrations (e.g., metal → EDTA) can not only ensure that the instrument is responding properly but can also verify that the metal titrant solution is well behaved and accurately delivering the desired metal ion.

When little net heat is found with an expected binding event, it may be possible to adjust the conditions to amplify the magnitude of the signal. While this can be achieved by increasing the sample concentration, this may not be experimentally feasible. However, if any protonation or deprotonation is coupled to the binding, then changing the buffer to one with a higher or lower enthalpy of protonation would modulate this contribution to the net experimental heat. Alternatively, since the binding enthalpies of proteins in aqueous solution are generally sensitive to temperature ($\Delta C_P \neq 0$), the net enthalpy will likely vary with temperature.

Individual injection peaks of an ITC thermogram may provide useful insight about the binding kinetics and mechanism. Multiphasic peaks indicate a multistep binding mechanism, such as rapid contact binding followed by a slower rearrangement. When metal ions bind to different sites, they may have a different injection peak shape, in addition to a different binding enthalpy, which is quantified by integration of the peak area.

7. CONCLUSION

To obtain accurate thermodynamic values for metal ions binding to proteins, all possible competing and coupled equilibria that may be associated with metal ions in ITC measurements need to considered. The former should be suppressed and the latter must be included in a *post hoc* analysis of the data to obtain condition-independent binding thermodynamics. However, the judicious choice of buffer can be useful in ITC measurements involving metal ions. By ensuring the delivery of a well-defined metal–buffer, or metal–ligand, complex of appropriate stability, competing reactions can be suppressed and binding to higher affinity protein sites can be accurately determined. In addition, ITC measurements in different buffers can be used to quantify the number of protons that are displaced when the metal ion binds to the protein at a given pH.

ACKNOWLEDGMENTS

We thank NSF (CHE-1308598; CHE-0910746) and NIH (P42-ES07373) for support of our calorimetric measurements of metal ions binding to proteins, Yi Zhang for Figure 2, and Anne Spuches, Nicholas Grossoehme, Michael Stevenson, and Mizuho Horioka for their contributions to Table 2.

REFERENCES

Adamczyk, J., Bal, W., & Krezel, A. (2015). Coordination properties of dithiobutylamine (DTBA), a newly introduced protein disulfide reducing agent. *Inorganic Chemistry, 54*, 596–606.

Blasie, C. A., & Berg, J. M. (2002). Structure-based thermodynamic analysis of a coupled metal binding-protein folding reaction involving a zinc finger peptide. *Biochemistry, 41*, 15068–15073.

Blasie, C. A., & Berg, J. M. (2003). Kinetics and thermodynamics of copper(II) binding to apoazurin. *Journal of the American Chemical Society, 125*, 6866–6867.

Bou-Abdallah, F., Woodhall, M. R., Velazquez-Campoy, A., Andrews, S. C., & Chasteen, N. D. (2005). Thermodynamic analysis of ferrous ion binding to *Escherichia coli* ferritin EcFtnA. *Biochemistry, 44*, 13837–13846.

Carpenter, M. C., Shami Shah, A., DeSilva, S., Gleaton, A., Su, A., Goundie, B., et al. (2015). Thermodynamics of Pb(II) and Zn(II) binding to MT-3, a neurologically important metallothionein. *Metallomics*. under review.

Carpenter, M. C., & Wilcox, D. E. (2014). Thermodynamics of formation of the insulin hexamer: Metal-stabilized proton-coupled assembly of quaternary structure. *Biochemistry, 53*, 1296–1301.

Chekmeneva, E., Prohens, R., Díaz-Cruz, J. M., Ariño, C., & Esteban, M. (2008). Thermodynamics of Cd^{2+} and Zn^{2+} binding by the phytochelatin (y-Glu-Cys)$_4$-Gly and its precursor glutathione. *Analytical Biochemistry, 375*, 82–89.

Ferreira, C. M. H., Pinto, I. S. S., Soares, E. V., & Soares, H. M. V. M. (2015). (Un)suitability of the use of pH buffers in biological, biochemical and environmental studies and their interaction with metal ions—A review. *RSC Advances, 5*, 30989–31003.

Freire, E., Mayorga, O. L., & Straume, M. (1990). Isothermal titration calorimetry. *Analytical Chemistry, 62*, 950A–959A.

Grossoehme, N. E., Akilesh, S., Guerinot, M., & Wilcox, D. E. (2006). Metal-binding thermodynamics of the histidine-rich sequence from the metal-transport protein IRT1 of *Arabidopsis thaliana. Inorganic Chemistry, 45*, 8500–8508.

Grossoehme, N. E., Mulrooney, S. B., Hausinger, H. P., & Wilcox, D. E. (2007). Thermodynamics of Ni^{2+}, Cu^{2+}, and Zn^{2+} binding to the urease metallochaperone UreE. *Biochemistry, 46*, 10506–10516.

Grossoehme, N. E., Spuches, A. M., & Wilcox, D. E. (2010). Application of isothermal titration calorimetry in bioinorganic chemistry. *Journal of Biological Inorganic Chemistry, 15*, 1183–1191.

Hegetschweiler, K., & Saltman, P. (1986). Interaction of copper(II) with N-(2-hydroxyethyl)piperazine-N′-ethanesulfonic acid (HEPES). *Inorganic Chemistry, 25*, 107–109.

Johnson, D., Stevenson, M. J., Almadidy, Z., Jenkins, S., Wilcox, D. E., & Grossoehme, N. E. (2015). Stabilization of Cu(I) for binding and calorimetric measurements in aqueous solution. *Dalton Transactions*, *44*, 16494–16505.

Krezel, A., Latajka, R., Bujacz, G. D., & Bal, W. (2003). Coordination properties of tris(2-carboxyethyl)phosphine, a newly introduced thiol reductant, and its oxide. *Inorganic Chemistry*, *42*, 1994–2003.

Krezel, A., Lesniak, W., Jezowska-Bojczuk, M., Mlynarz, P., Brasun, J., Kozlowski, H., et al. (2001). Coordination of heavy metals by dithiothreitol, a commonly used thiol group protectant. *Journal of Inorganic Biochemistry*, *84*, 77–88.

Latimer, W. M. (1952). *Oxidation potentials* (2nd ed.). Englewood Cliffs: Prentice-Hall, Inc.

Lin, L. N., Mason, A. B., Woodworth, R. C., & Brandts, J. F. (1991). Calorimetric studies of the binding of ferric ions to ovotransferrin and interactions between binding sites. *Biochemistry*, *30*, 11660–11669.

Lin, L. N., Mason, A. B., Woodworth, R. C., & Brandts, J. F. (1993). Calorimetric studies of the binding of ferric ions to human serum transferrin. *Biochemistry*, *32*, 9398–9406.

Magyar, J. S., & Godwin, H. A. (2003). Spectropotentiometric analysis of metal binding to structural zinc-binding sites: Accounting quantitatively for pH and metal ion buffering effects. *Analytical Biochemistry*, *320*, 39–54.

Nagaj, J., Stokowa-Soltys, K., Kurowska, E., Frączyk, T., Jeżowska-Bojczuk, M., & Bal, W. (2013). Revised coordination model and stability constants of Cu(II) complexes of tris buffer. *Inorganic Chemistry*, *52*, 13927–13933.

Nakon, R., & Krishnamoorthy, C. R. (1983). Free-metal ion depletion by "Good's" buffers. *Science*, *221*, 749–750.

Nettles, W. L., Song, H., Farquhar, E. R., Fitzkee, N. C., & Emerson, J. P. (2015). Characterization of copper(II) binding sites in human carbonic anhydrase II. *Inorganic Chemistry*, *54*, 5671–5680.

NIST Standard Reference Database 46 version 8.0. (2004). *NIST critically selected stability constants of metal complexes database*. Gaithersburg, MD: U.S. Department of Commerce; National Institute of Standards and Technology.

Pedroso, M. M., Ely, F., Mitic, N., Carpenter, M. C., Gahan, L. R., Wilcox, D. E., et al. (2014). Comparative investigation of the reaction mechanisms of the organophosphate-degrading phosphotriesterases from *Agrobacterium radiobacter* (OpdA) and *Pseudomonas diminuta* (OPH). *Journal of Biological Inorganic Chemistry*, *19*, 1263–1275.

Petitt, L. D., & Powell, H. K. J. (1993). *IUPAC Stability Constants Database*. West Yorkshire, UK: Academic Software.

Quinn, C. F. (2010). *Analyzing ITC data for the enthalpy of binding metal ions to ligands*. http://www.tainstruments.com/pdf/MCAPN-2010-02%20ITC%20Metal%20Ions%20Binding%20to%20Ligands.pdf. accessed on 03/08/2015.

Sigurskjold, B. W. (2000). Exact analysis of competition ligand binding by displacement isothermal titration calorimetry. *Analytical Biochemistry*, *277*, 260–266.

Song, H., Weitz, A. C., Hendrich, M. P., Lewis, E. A., & Emerson, J. P. (2013). Building reactive copper centers in human carbonic anhydrase II. *Journal of Biological Inorganic Chemistry*, *18*, 595–598.

Spuches, A. M., Kruszyna, H. G., Rich, A. M., & Wilcox, D. E. (2005). Thermodynamics of the As(III)-thiol interaction: Arsenite and monomethylarsenite complexes with glutathione, dihydrolipoic acid, and other thiol ligands. *Inorganic Chemistry*, *44*, 2964–2972.

Wiseman, T., Williston, S., Brandts, J. F., & Lin, L. (1989). Rapid measurement of binding constants and heats of binding using a new titration calorimeter. *Analytical Biochemistry*, *179*, 131–137.

Wyrzykowski, D., Pilarski, B., Jacewicz, D., & Chmurzyński, L. (2013). Investigation of metal-buffer interactions using isothermal titration calorimetry. *Journal of Thermal Analysis and Calorimetry*, *111*, 1829–1836.

Zawisza, I., Croteau, M.L., Wilcox, D.E., & Bal, W. (2015). Unpublished results.

CHAPTER TWO

Assessing Coupled Protein Folding and Binding Through Temperature-Dependent Isothermal Titration Calorimetry

Debashish Sahu*, Monique Bastidas*, Chad W. Lawrence*[,1], William G. Noid*, Scott A. Showalter*[,†,2]

*Department of Chemistry, The Pennsylvania State University, University Park, Pennsylvania, USA
†Department of Biochemistry and Molecular Biology, The Pennsylvania State University, University Park, Pennsylvania, USA
[2]Corresponding author: e-mail address: sas76@psu.edu

Contents

Abstract

Broad interest in the thermodynamic driving forces of coupled macromolecular folding and binding is motivated by the prevalence of disorder-to-order transitions observed when intrinsically disordered proteins (IDPs) bind to their partners. Isothermal titration calorimetry (ITC) is one of the few methods available for completely evaluating the thermodynamic parameters describing a protein–ligand binding event. Significantly, when the effective ΔH° for the coupled folding and binding process is determined by ITC in a

[1] Present address: Environmental Molecular Sciences Laboratory, Pacific Northwest National Laboratory, Richland, WA 99354.

Methods in Enzymology, Volume 567
ISSN 0076-6879
http://dx.doi.org/10.1016/bs.mie.2015.07.032

temperature series, the constant-pressure heat capacity change (ΔC_p) associated with these coupled equilibria is experimentally accessible, offering a unique opportunity to investigate the driving forces behind them. Notably, each of these molecular-scale events is often accompanied by strongly temperature-dependent enthalpy changes, even over the narrow temperature range experimentally accessible for biomolecules, making single temperature determinations of $\Delta H°$ less informative than typically assumed. Here, we will document the procedures we have adopted in our laboratory for designing, executing, and globally analyzing temperature-dependent ITC studies of coupled folding and binding in IDP interactions. As a biologically significant example, our recent evaluation of temperature-dependent interactions between the disordered tail of FCP1 and the winged-helix domain from Rap74 will be presented. Emphasis will be placed on the use of publically available analysis programs written in MATLAB that facilitate quantification of the thermodynamic forces governing IDP interactions. Although motivated from the perspective of IDPs, the experimental design principles and data fitting procedures presented here are general to the study of most noncooperative ligand binding equilibria.

1. INTRODUCTION

Intrinsically disordered proteins (IDPs) lack a significant bias toward any particular spatially and temporally stable equilibrium structure under native conditions (Tompa, 2002; van der Lee et al., 2014). In contrast to IDP folding, their interactions appear to be more harmonious with the expectation of a funneled free energy surface that guides the IDP conformational ensemble toward a folded structure in the bound state (Csermely, Palotai, & Nussinov, 2010; Papoian & Wolynes, 2003). This result suggests there may be other parallels between folding of globular proteins and IDP interactions. Certain 3D folds are repeated in proteins with specific functions; similarly, some biological processes are optimally performed by IDPs, providing a strong incentive for better understanding the energetics of IDP interactions (Dunker, Silman, Uversky, & Sussman, 2008; Fuxreiter et al., 2008). The prevalence of IDP disorder-to-order transitions induced by interactions (Dyson & Wright, 2002; Tompa & Fuxreiter, 2008; Wright & Dyson, 2009), which often involve regions of the IDP possessing partially preformed structure (Mohan et al., 2006), has stimulated research into the thermodynamic driving forces coupling folding and binding (Ferreon & Hilser, 2004). The hypothesis emerging from this work is that native disorder provides a molecular pathway to high fidelity, yet reversible, association in processes including cellular signaling and gene transcription (Ganguly et al., 2012; Wright & Dyson, 2009). Accordingly, there is considerable interest in quantification of the thermodynamic forces that govern IDP interactions.

When applicable, isothermal titration calorimetry (ITC) is one of the most robust methods available for thermodynamic studies (Ghai, Falconer, & Collins, 2012). ITC provides a direct measurement of the binding enthalpy ($\Delta H°$) and equilibrium association constant (K_A, more commonly reported as $K_D = K_A^{-1}$), enabling the calculation of the Gibbs free energy ($\Delta G°$) through the known temperature and parametric determination of the binding entropy ($\Delta S°$). Significantly, when $\Delta H°$ is determined by ITC in a temperature series, the constant-pressure heat capacity change (ΔC_p) associated with the binding process is also experimentally accessible. As a result, several temperature-dependent ITC data sets have been reported for IDPs binding to macromolecular partners (Cervantes et al., 2011; Cho et al., 2010; Drobnak et al., 2013; Ferreon & Hilser, 2004; Lacy et al., 2004; Spolar & Record, 1994; Xu, Oruganti, Gopalan, & Foster, 2012). For both unfolding and ligand dissociation events, large changes in $\Delta H°$ are often seen, even to the extent of producing a transition from endothermic to exothermic binding over narrow and near-room temperature ranges. Thus, measuring $\Delta H°$ at a single temperature is often less informative than typically assumed (Cooper, 2010) and the significance of experimentally determining ΔC_p when investigating coupled IDP folding and binding should not be underestimated. In particular, the influence of ΔC_p on complex stability and the insight it provides into, e.g., the role of forces including the hydrophobic effect in their binding mechanism is of special significance.

While this argument suggests that temperature-dependent studies are essential to an optimal ITC-based investigation of IDP folding and binding, the resulting data must be interpreted with caution. Specifically, it is clear that attributing too much mechanistic significance to entropy–enthalpy compensation is dangerous, because systematic under-reporting of ITC errors enriches the natural correlation that is already expected between these parameters when they are reported over narrow (experimentally accessible) temperature ranges (Chodera & Mobley, 2013). Perhaps more important than the simple presence of entropy–enthalpy compensation is that this phenomenon suggests interactions involving molecular disorder-to-order transitions should be relatively robust to temperature fluctuations, because the Gibbs free energy will experience relatively small temperature-induced changes over biologically tolerated temperature ranges as a result. This may be particularly significant where the interaction under investigation is formed through the accumulation of multiple cooperative and weak interactions, which will give rise to the large observed ΔC_p (Cooper, 2010), as is nearly always the case for protein folding or coupled folding and binding.

As a model system to investigate the vital role of disordered proteins for regulating eukaryotic gene transcription (Tantos, Han, & Tompa, 2012), our laboratories have collaborated to study interactions between the intrinsically disordered C-terminal region of the TFIIF-associating CTD phosphatase (FCP1) and the cooperatively folded winged-helix domain from Rap74 (Kumar, Showalter, & Noid, 2013; Lawrence, Bonny, & Showalter, 2011; Lawrence, Kumar, Noid, & Showalter, 2014; Lawrence & Showalter, 2012; Wostenberg, Kumar, Noid, & Showalter, 2011). Briefly, FCP1 performs an essential function in transcription by dephosphorylating the hyperphosphorylated C-terminal domain of RNA Polymerase II in preparation for subsequent rounds of transcription. Biochemical studies suggest that the interaction between Rap74 and FCP1 facilitates this function. As illustrated in Fig. 1, our work has shown that the Rap74 binding surface of FCP1, which forms an amphipathic α-helix in complex (Kamada, Roeder, & Burley, 2003; Nguyen et al., 2003), is only partially α-helical in the unbound state (Lawrence et al., 2011). Given this structural knowledge, we set out to derive an easily generalized framework for investigating coupled folding and binding of IDPs. As described above, temperature-dependent ITC provides a thorough characterization of the effective thermodynamic properties of coupled folding and binding processes. Therefore, we have evaluated the temperature dependence of the FCP1–Rap74 interaction by ITC, using a general thermodynamic model in which the large observed change in ΔC_p is directly fit (Lawrence et al., 2014).

Here, we will document the procedures we have adopted in our laboratory for designing, executing, and globally analyzing temperature-dependent ITC studies of coupled folding and binding in IDP interactions.

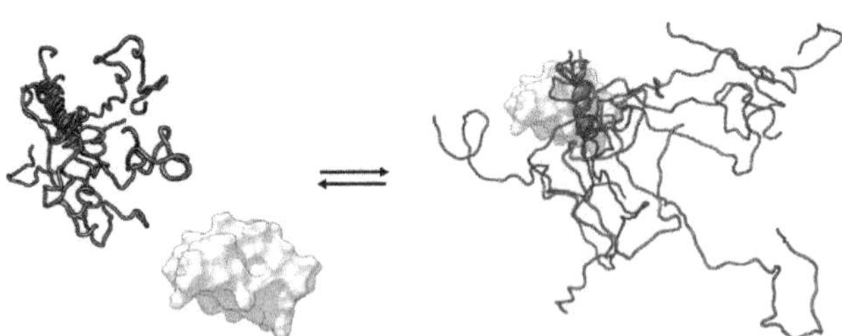

Figure 1 An illustration of the interaction between FCP1 (dark gray ribbon) and Rap74 (white surface model), which is the object of the temperature-dependent isothermal titration calorimetry study used to generate examples throughout the text.

Due to the nature of this volume, we will assume the reader is generally familiar with the ITC technique and therefore will not focus on a detailed protocol for conducting the individual titrations. Instead, our focus will be on avoiding the common pitfalls in analyzing temperature-dependent ITC data sets. Emphasis will be placed on the use of analysis programs we have written in MATLAB that are made publically available by the authors and which facilitate quantification of the thermodynamic forces that govern IDP interactions.

2. SAMPLE PREPARATION

In general, preparation of IDPs for ITC assays is no different from the preparation of cooperatively folding proteins for the same purpose. However, there are two general trends in designing purification strategies for high yield and high purity when working with IDPs that bear mentioning. Perhaps surprisingly, many IDPs seem to express well in *Escherichia coli*, without significant problems due to proteolysis. On the other hand, the yield of very small proteins from *E. coli* overexpression is often poor, which can make it necessary to express intrinsically disordered regions as fusion proteins incorporating a large domain, such as GST or maltose binding protein, in order to improve yield. Note, however, that this strategy was not necessary for the C-terminal region of FCP1, for which a representative purification protocol is presented below. Finally, caution must be exercised when selecting molecular weight cutoffs (MWCOs) for membranes used in dialysis or centrifugal concentration, because the nonglobular nature of IDP structure allows them to pass through membranes that are nominally restrictive enough to retain a protein with a particular IDP's mass. In general, selecting a MWCO that is less than half of the IDP mass is a good initial strategy that can be relaxed based on experience to improve membrane performance or purification speed and efficiency.

2.1 Plasmid Construction

A synthetic gene with *E. coli* optimized codon usage and also encoding a nonnative N-terminal tryptophan was purchased from Geneart to encode FCP1 (residues 878–961 of the human polypeptide sequence). The DNA encoding this FCP1 construct was cloned into the pET-47b (Novagen) expression vector to create the plasmid pET-47b-FCP1W, which was transformed into *E. coli* BL21 (DE3) competent cells. The same procedure was repeated with an *E. coli* optimized synthetic gene encoding residues 434–517 of the human Rap74 polypeptide sequence, producing the plasmid pET-47b-Rap74.

2.2 Purification of FCP1 and Rap74

1. Inoculate LB media with *E. coli* bearing either the pET-47b-FCP1W or the pET-47b-Rap74 plasmid and then grow to an optical density between 0.8 and 0.9 at 600 nm. Induce expression through the addition of IPTG to a final concentration of 1.0 m*M*, followed by an additional 3–4 h of growth.
2. This step is performed at 4 °C. Harvest the cells by centrifugation for 20 min at 3300 times gravity. Suspend the cell pellet in lysis buffer (50 m*M* Tris, adjusted to pH 7.5 before use, 500 m*M* NaCl, 20 m*M* imidazole, 5 m*M* β-mercaptoethanol), 1 × protease inhibiting cocktail set V (EMD Millipore), and 1 × DNase (Promega). Lyse by sonication and then clarify the cell lysates by centrifugation for at least 30 min at 14,000 × *g*.
3. Equilibrate an immobilized metal affinity chromatography column containing a 5 mL settled bed volume of Ni^{2+}-NTA resin (Thermo Fisher), using 40 mL of lysis buffer. Pass the clarified cell lysate over the column, followed by passage of 40 mL wash buffer (lysis buffer plus 0.1% Triton X-100), and a final rinse with an additional 20 mL of lysis buffer. All steps may be performed using gravity flow.
4. Elute the His-tagged fusion protein in 20 mL of elution buffer (50 m*M* Tris, adjusted to pH 7.5 before use, 500 m*M* NaCl, 200 m*M* imidazole, 5 m*M* β-mercaptoethanol) and transfer the collected material to pres-oaked 1000 Da MWCO dialysis tubing. Add His-tagged 3C Protease to the tubing prior to closing and dialyze against 1.6 L of dialysis buffer (50 m*M* Tris, adjusted to pH 7.5 before use, 300 m*M* NaCl, 5 m*M* β-mercaptoethanol) overnight at 4 °C.
5. Achieve removal of the recombinant His-tag and His-tagged 3C Protease by passing the contents of the dialysis tubing back over a reequilibrated Ni^{2+}-NTA column using gravity flow. To ensure complete elution of FCP1 or Rap74, wash the column with an additional 20 mL of dialysis buffer.
6. Concentrate the protein using a 3000 Da MWCO centrifugal concentrator with a PES membrane (EMD Millipore) and store in lysis buffer until ready for ITC.

7. The day before ITC experiments are to be initiated, transfer FCP1 and Rap74 to separate pieces of 1000 Da MWCO dialysis tubing; codialyze FCP1 and Rap74 overnight against a shared reservoir of 50 m*M* cacodylate, pH 6.5, 50 m*M* KCl. Determine final concentrations of FCP1 and Rap74, either by absorbance at 280 nm or using a Direct Detect FT-IR (EMD Millipore).

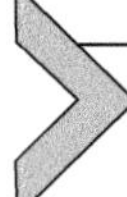

3. SPECIAL CONSIDERATIONS FOR ITC OF INTRINSICALLY DISORDERED PROTEINS

3.1 Experimental Design

In general, the ITC experiment is set up by loading the ligand at high concentration into the injection syringe, which is then inserted into the sample cell that has been filled with a less concentrated solution containing the macromolecule. For one-to-one interactions, the assignment of the labels "ligand" and "macromolecule" is essentially semantic; thus, the investigator is at liberty to select which macromolecular species to inject from the syringe based on practical considerations. It must be stressed that all processes accompanied by the evolution or uptake of nonnegligible heat will be detected by the ITC experiment and so it is important to minimize contributions from off-target processes. Of relevance to the present discussion, many IDPs undergo reversible assembly state changes over the concentration ranges typically accessed upon loading into the syringe. Therefore, it is generally preferred that the IDP be loaded into the sample cell and the well-folded binding partner into the syringe, even if the IDP is the less massive species.

Locally, we have access to both a MicroCal, Inc. (now a part of Malvern) VP-ITC and a MicroCal AutoITC$_{200}$ for our work. While IDP binding interactions display just as wide an array of associated affinities as those measured between pairs of well-folded partners (see table 1 in Gibbs & Showalter, 2015), IDP interactions most commonly are accompanied by dissociation constants in the micromolar range. Therefore, the protocols discussed here will be presented under the assumption that a VP-ITC (or equivalent) is being used.

3.2 Calorimetric Parameters

1. Total injections—this is dependent on the volume of each titration, but typically on the order of 30–40 injections. The user can overestimate this value, without adversely affecting the titration; the instrument will report insufficient volume in the syringe and save the data. In addition,

the first injection is always set to a lower volume to account for inconsistencies in syringe filling (Pierce, Raman, & Nall, 1999) and to eliminate the known "backlash" anomaly (Mizoue & Tellinghuisen, 2004); but this initial point is not included in the analysis.

2. Cell temperature—the typical range of temperatures for these studies is 5–40 °C. If possible, determine the thermal denaturation temperature (e.g., by differential scanning calorimetry) for the cooperatively folded partner of the IDP, which will enable the final temperature range to be selected such that thermal denaturation of the folded protein can be avoided.
3. Reference power—the reference power is the constant power applied to equilibrate the temperatures between the reference and sample cells. Typically the value is set to 21 μJ/s (5 μCal/s in the MicroCal software), but for interactions yielding large heats, the reference power should be set to 42 μJ/s (10 μCal/s) or greater.
4. Initial delay—set to the minimum value of 60 s to establish a baseline.
5. Concentrations—the concentrations of the samples should be entered here, but if necessary, minor corrections can be made during analysis.
6. Injection parameters—typical settings include:
   a. Volume—the volume of each injection.
   b. Duration—the duration in seconds of each injection. As a practical value to establish a starting point, this parameter is usually set to produce an injection rate of 0.5 μL/s. For example, if the injection volume is 10 μL, the default setting for this parameter is 20 s.
   c. Spacing—after each injection, the differential heat signal should return completely to baseline. The necessary spacing between each injection will need to be determined based on the system, but typically 240 or 360 s is a good starting point.
   d. Filter period—this parameter alters the period in which a single data point is collected. For reactions occurring rapidly, 2 s is sufficient to represent a peak, which can be properly integrated to produce an accurate enthalpy.
7. Feedback mode/gain—The feedback mode dictates the response time of the instrument. For increased sensitivity when small heats of injection are expected, the passive (none) mode will provide the best data. For the quickest response time, a high response time is selected. This mode is recommended in the manual.
8. ITC equilibration options—typically, Fast Equil. and Auto are selected. This will ensure that the temperatures of the cell and sample are within

specified tolerances for the starting temperature. Note that these settings will dictate an automatic period of equilibration with stirring and final baseline equilibration prior to initiating the injection schedule.

4. DATA FITTING AND ANALYSIS

There are many commercial or open source software packages available for the analysis of ITC data, including Origin® (bundled with the MicroCal calorimeters we have used for this work and thus described here), SEDPHAT (Schuck, 2000), Affinimeter (Burnouf et al., 2012), or NanoAnalyze (TA Instruments). In our laboratory, we generally perform all of our ITC data fitting and analysis using in-house MATLAB macros that have features we will also discuss in this section. All MATLAB macros used for the analysis presented in this article are freely available through the Penn State ScholarSphere Data Repository at the following stable URL: https://scholarsphere.psu.edu/downloads/5712mr21f.

4.1 Standard Data Fitting in Origin

Standard ITC data analysis, as originally proposed by Wiseman et al., begins with fitting to the single class of sites model in Origin®, yielding apparent values for $\Delta H°$, K_A, and the stoichiometry (n), as presented elsewhere in this volume (Wiseman, Williston, Brandts, & Lin, 1989). However, collecting calorimetric data at a single temperature can be problematic if the apparent binding enthalpy happens to be near zero at the temperature selected. For example, the raw ITC data generated for the present study of FCP1 binding to Rap74 (Fig. 2) clearly show a near-zero differential heat for each injection at 20 °C (Fig. 2B). Origin® analysis of nearly isoenthalpic data typically yields unreliable stoichiometries and association constants. Further, the uncertainty in the measured $\Delta H°$ is likewise increased, although this is rarely detrimental as determination that $\Delta H°$ was near zero is qualitatively insightful on its own.

Often, during the early design stages of the experiment, titrations will be performed at several temperatures in order to find the "optimal" temperature that yields both strong heats to measure and a robust determination of the equilibrium constant. As is clear from Fig. 3A, this search often reveals that the binding enthalpy varies dramatically with temperature and, in particular, demonstrates a large ΔC_p, which suggests that the binding event is driven by many, weak cooperative interactions. The conventional method for quantitative calculation of the ΔC_p associated with an interaction

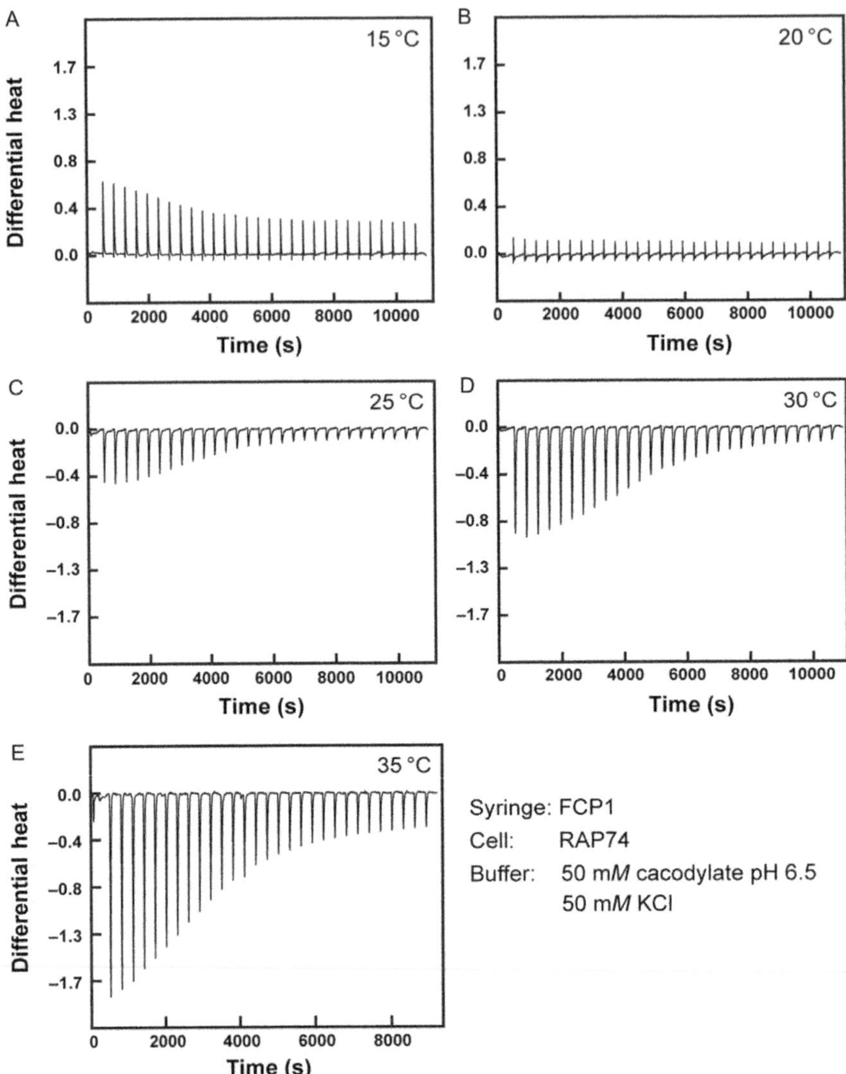

Figure 2 Isotherms generated by titration of FCP1 from the syringe into Rap74 in the cell of a VP-ITC at (A) 283 K, (B) 288 K, (C) 303 K, (D) 308 K, and (E) 308 K. The buffer conditions used for the ITC experiments were 50 m*M* cacodylate pH 6.5 and 50 m*M* KCl.

between ligand and macromolecule is to repeat the binding titration at different temperatures and obtain the apparent binding enthalpy as a function of temperature. A linear fit of the resulting $\Delta H^{\circ}(T)$ data series yields ΔC_{p} as the slope of the fit. For most macromolecular interactions, this procedure works well; there is a linear dependence of the enthalpic contribution to

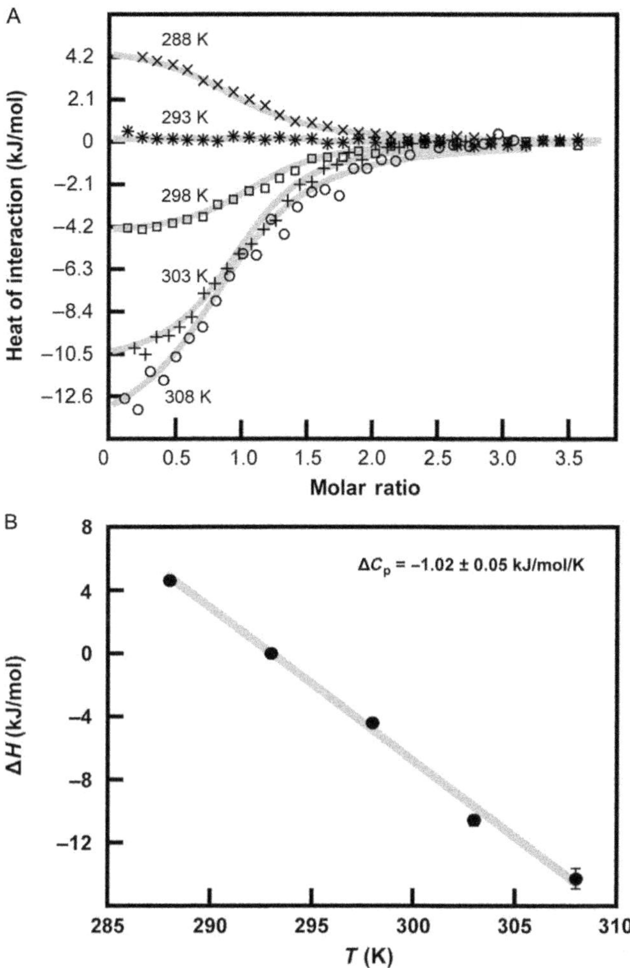

Figure 3 Serial analysis of temperature-dependent ITC data in Origin® yields an estimate of the constant-pressure heat capacity change (ΔC_p) from a linear fit to the apparent binding enthalpy (ΔH). (A) Overlay of individual fits (gray line) for each temperature, with the experimentally observed integrated heats of injection displayed at 288 K (×), 293 K (*), 298 K (□), 303 K (+), and 308 K (○). (B) Determination of ΔC_p from a linear fit (gray line) of the fitted ΔH (black circles) at each temperature, with the best-fit value provided as an inset.

ligand binding on temperature over the narrow temperature ranges experimentally accessed, unless a thermal unfolding transition for one or both of the biomolecules is approached. For example, the binding of FCP1 by Rap74 is accompanied by a linear temperature dependence to ΔH° over the range from 288 to 308 K (Fig. 3A). When the individual data sets are fit in Origin®, ΔC_p can also be computed by plotting the fitted $\Delta H°$ as a function of temperature and performing a linear fit, yielding the heat capacity change $\Delta C_p = -1.02 \pm 0.05$ kJ/mol/K as the slope (Fig. 3B). Large negative ΔC_p values are typically interpreted to arise from changes in the amount of solvent exposed surface area upon binding (Dill, 1990), although there are many possible causes (Cooper, Johnson, Lakey, & Nollmann, 2001).

Protocol for data fitting

1. After each experiment, the data are manually integrated to minimize the error in the heat of injection.
2. Prior to fitting the data in Origin®, the baseline is corrected for heats of dilution of the buffer and/or the ligand. The baseline correction for the dilution of ligand into buffer is accounted for in one of two ways: a separate titration of ligand into buffer is collected or a 5-point baseline average is computed. For the analysis of FCP1–Rap74 interaction data, the raw data are manually integrated and a 4-point baseline average is used for baseline subtraction of the heats of dilution of Rap74 using the simple math function in Origin®.
3. Baseline corrected data are fit to a one-site binding model, where 100 iterations of fitting are completed until χ^2 is no longer reduced.
4. Repeat this procedure for each titration performed at the various temperatures.

4.2 Data Fitting in MATLAB

Generating accurate individual fits of ITC data to a single set of sites binding model in the Origin® software requires that the upper baseline approach nearly to zero without a sign inversion. Ideally, no heat should be released or absorbed during the final ~5 injections, but often this is not the case; various factors including buffer mismatch, dilution effects, or improper baseline subtraction can cause significantly nonzero heats for the final points or even transitions crossing from, e.g., exothermic to endothermic heats. This is problematic because the fitting function used in Origin® does not allow the best-fit line to cross zero (i.e., the fit is constrained to be semidefinite). The result of an improperly treated upper baseline will be artifacts in the fitted binding affinity, enthalpy, and number of binding sites. In order to circumvent this fitting issue, we relax the upper baseline constraints using in-house written MATLAB fitting macros that add an offset to the single class of sites binding model. This new method allows the user to directly take the integrated heat data from the ITC with less extensive intervention and preprocessing.

The code we provide for fitting temperature-dependent ITC data is written to implement the Levenberg–Marquardt (LM) algorithm (MATLAB function fit.m with LM options) for large-scale fitting of nonlinear functions. An alternative method that appears attractive is to implement the standard nonlinear numerical fitting algorithm present in the base MATLAB library (MATLAB function nlinfit.m), but in our experience, this algorithm tends to become trapped in local minima during the fitting of ITC data. Initial estimates of the apparent binding association equilibrium constant (K_A) and number of binding sites (n) must be provided, noting that quality initial guesses can improve fitting efficiency dramatically. It is not necessary for the user to provide initial guesses for $\Delta H°$ and the baseline offset, because the code is written to extract high-quality initial guesses for these parameters directly from the integrated heats provided as input. Finally, we note that as distributed our macro is designed to perform 10^6 LM iterations, which generally should be sufficient with reasonable initial estimates for the fitting parameters.

Our modified algorithm adds an additional fitting parameter (the upper baseline) compared to the simpler model implemented in Origin®, but in general the best-fit values for the thermodynamic parameters generated when our modified method is applied to well-collected data do not vary significantly with respect to those generated by the conventional method (Table 1). For example, when the binding of FCP1 by Rap74 is fit in MATLAB using ITCFit.m (Fig. 4), the fitted $\Delta C_p = -1.16 \pm 0.07$ kJ/mol/K, which is in good agreement with $\Delta C_p = -1.02 \pm 0.05$ cal/mol/K fitted in Section 4.1 using Origin® derived fits to the raw ITC data. Note that there are data sets for which the agreement between these two methods will be reduced. Specifically, a poorly defined upper or lower baseline in the data reduces the quality of the fit in both algorithms and differences are expected to arise. Notably, in cases where the upper baseline is poorly defined, Origin® will fix the upper baseline to zero and force the fitting of the observed heat released to conform with this choice, resulting in a poor convergence with the experimental data as well. In these cases, it is always advisable to repeat the experiments with increased concentration of macromolecule in the cell such that the "*c* value" (Wiseman et al., 1989) is improved.

Finally, we note that the addition of an extra parameter in the fitting algorithm does lead to increased values for the estimated uncertainties of the fitted parameters. This is of little concern because it has been noted that

Table 1 Summary of Best-Fit Parameters for FCP1 Binding to Rap74

T (K)	N	ΔH° (kJ/mol)	K_D (μM)
Origin			
288	1.02 ± 0.02	4.8 ± 0.1	2.2 ± 0.2
293	N.B.D.	N.B.D.	N.B.D.
298	1.15 ± 0.02	-4.6 ± 0.1	1.4 ± 0.2
303	1.04 ± 0.02	-11.1 ± 0.3	1.9 ± 0.3
308	0.96 ± 0.03	-14.9 ± 0.7	3.1 ± 0.5
MATLAB			
288	1.03 ± 0.02	5.1 ± 0.2	2.7 ± 0.3
293	N.B.D.	N.B.D.	N.B.D.
298	1.18 ± 0.02	-4.9 ± 0.2	1.9 ± 0.3
303	1.08 ± 0.02	-12.7 ± 0.5	3.1 ± 0.5
308	1.01 ± 0.04	-17.7 ± 1.4	5.3 ± 1.2

These parameters produce the gray best-fit lines in Figs. 3A and 4A. For the 293 K data heats were not sufficient to constrain the fit and the results are reported as no binding detected (N.B.D.)

the true error in ITC fitting parameters is probably systematically underestimated (Chodera & Mobley, 2013). It is our view that the simpler data handling in our MATLAB-driven fitting procedure minimizes user bias and removes potential artifacts, making an inflation of the error estimates a price well worth paying.

Protocol for data fitting

1. At the end of an ITC experiment, the user will initially integrate the heat released or absorbed in each injection, as outlined in Section 4.1.
2. Following integration, the data series containing injection volume (v_i), ligand concentration (X_T), macromolecule concentration (M_T), molar ratio (X_T/M_T), and heat of injection (ndh) in tabular form can be accessed in Origin® through the "Windows" menu tab. These data are saved in a standard ASCII text format that is directly used by our provided MATLAB scripts as inputs. Macromolecule and ligand refer here to the contents in the cell and the syringe, respectively.
3. The file generated in step 2, along with the active cell volume (V_o) and syringe concentration of ligand (X_o), are used as inputs into the MATLAB fitting macro (ITCFit.m), which fits the data to a modified

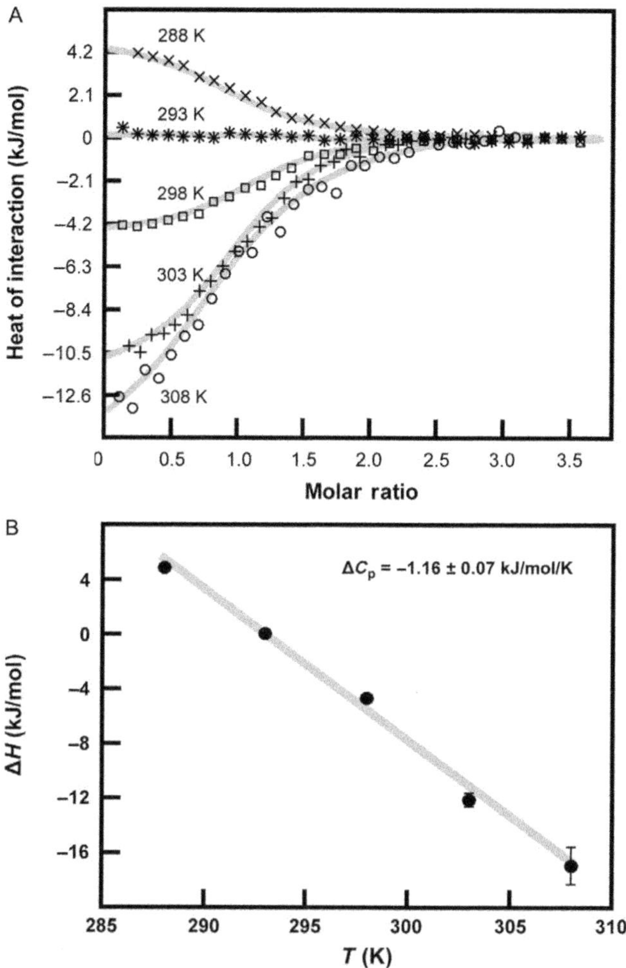

Figure 4 Serial analysis of temperature-dependent ITC data in MATLAB yields an estimate of the constant-pressure heat capacity change (ΔC_p) from a linear fit to the apparent binding enthalpy (ΔH). (A) Overlay of individual fits (gray line) for each temperature, with the experimentally observed integrated heats of injection displayed at 288 K (×), 293 K (*), 298 K (□), 303 K (+), and 308 K (○). (B) Determination of ΔC_p from a linear fit (gray line) of the fitted ΔH (black circles) at each temperature, with the best-fit value provided as an inset.

single class of sites binding model as discussed above. Initial estimates of the apparent binding association equilibrium constant (K_A) and number of binding sites (n) must be provided. If the fit does not converge, it is recommended to repeat the fitting procedure using different initial estimates of K_A and n.

4.3 Global Analysis of Temperature-Dependent Data Sets

The procedure outlined in the previous two sections is robust and effective, but the elegance of ITC data is that it directly measures heats, granting unique access to fundamental thermodynamic quantities capable of producing direct mechanistic insights. In our view, it is advantageous to consider global fitting of temperature-dependent ITC series, as we will describe in this final section. One clear advantage of a global data analysis is that it minimizes the impact of uncertainty inherent to data sets that happen to have been collected near the isoenthalpic temperature for the process under investigation. A second advantage of the method is that it directly estimates ΔC_p. Global data analysis inherently requires model building, noting that the single set of sites model discussed above will typically need to be discarded, because the equilibrium association constant need not be independent of temperature. The greatest strength of this approach is that the models used can rely entirely on fundamental thermodynamic equations, with only the interpretation depending on system-specific assumptions, making the results reasonably model independent.

A global model must determine both ΔH° and K_A at each temperature T for which ITC experiments were performed. Our central assumption is that ΔC_p is constant over the relevant temperature range. The standard enthalpy and entropy then vary according to

$$\Delta H^\circ(T) = \Delta C_p(T - T_H) + \Delta H_0 \quad (1)$$

$$\Delta S^\circ(T) = \Delta C_p \ln(T/T_S) + \Delta S_0 \quad (2)$$

where $\Delta H_0 = \Delta H^\circ(T_H)$ and $\Delta S_0 = \Delta S^\circ(T_S)$. From these expressions, the association constant is determined from $K_A(T) = \exp[-\Delta G^\circ(T)/RT]$ with $\Delta G^\circ(T) = \Delta H^\circ(T) - T\Delta S^\circ(T)$. Note that T_H and T_S are not necessarily the same and that these expressions are exact over the range for which ΔC_p is constant.

Previous studies have demonstrated that the interaction between FCP1 and Rap74 is mediated by a large nonpolar interface with only one direct salt-bridge interaction (Kamada et al., 2003). This extensive nonpolar interface, as well as the large, negative heat capacity that is inferred from independently fitting the ITC data at each temperature, suggests that hydrophobic forces may drive the coupled folding–binding interaction between FCP1 and Rap74. Moreover, these considerations suggest that the large, negative heat capacity primarily reflects the desolvation of nonpolar surface area. Accordingly, we interpreted the measured thermodynamic parameters with a simple hydrophobic model described by Baldwin (1986). In this model, $T_H = 295$ K is the temperature at which the enthalpy of hydrophobic interactions vanishes, while $T_S = 386$ is the temperature at which the entropy of hydrophobic interactions vanishes. Within the context of this model, ΔC_p is assumed to describe hydrophobic effects. In contrast, ΔH_0 and ΔS_0 are interpreted as temperature-independent contributions to $\Delta H^\circ(T)$ and $\Delta S^\circ(T)$ that are *not* due to hydrophobic effects; e.g., ΔH_0 and ΔS_0 are interpreted to describe contributions from electrostatic interactions.

When $T_H = 295$ K and $T_S = 386$ K are predetermined by the Baldwin model, the global fit depends upon three parameters. Using the MATLAB script Fcn_Fit_Cp_Ndil.m (distributed through the Penn State ScholarSphere as described elsewhere; see *Protocol for data fitting* below for details), we determined $\Delta C_p = -1.01 \pm 0.04$ kJ/mol/K, $\Delta H_0 = -2.24 \pm 0.23$ kJ/mol, and $\Delta S_0 = -0.17 \pm 0.01$ kJ/mol/K (Fig. 5). Indeed, these parameters are quite consistent with the interpretation of the Baldwin model. The best-fit ΔC_p is in good agreement with the relation proposed by Spolar and Record (1994) for relating solvated surface areas to heat

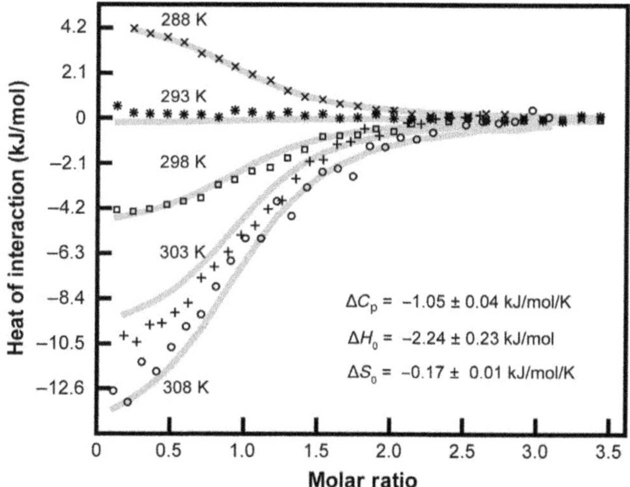

Figure 5 Global fit of temperature-dependent ITC titrations in MATLAB yields a direct estimate of the constant-pressure heat capacity change (ΔC_p), as well as temperature-independent contributions to the overall enthalpy change (ΔH_0) and entropy change (ΔS_0). The output best-fit parameters are displayed as an inset. Calculated isotherms resulting from the global fit are displayed as a gray line for each temperature, while the experimentally observed integrated heats of injection are displayed at 288 K (×), 293 K (*), 298 K (□), 303 K (+), and 308 K (○).

capacity changes. The best-fit ΔH_0 is consistent with the presence of a small number of favorable polar interactions at the protein–protein interface. Finally, the best-fit ΔS_0 is consistent with the expectation of the rotational and translational entropy loss due to the association of the two proteins, although ΔS_0 is somewhat smaller than that estimated by Spolar and Record. We previously suggested that this difference may be due to the "fuzziness" (Tompa & Fuxreiter, 2008) that we have previously observed in the Rap74–FCP1 protein complex (Lawrence & Showalter, 2012; Wostenberg et al., 2011).

In closing, we note that although the results of the global fit are quite consistent with the Baldwin model, the fitting procedure does not fundamentally depend upon this model. Rather, the Baldwin model provides a molecular interpretation of the parameters that are determined by the global analysis. In particular, the Baldwin model relates ΔC_p to hydrophobic effects, while attributing ΔH_0 and ΔS_0 to other sources. However, the only essential assumption of the global fitting procedure we describe is that ΔC_p is constant over the experimental temperature range, which is identical to the assumption made for serial fitting as described in Sections 4.1 and 4.2. The definition of $T_H = 295$ K and $T_S = 386$ K only results in trivial shifts in ΔH_0 and ΔS_0, while having no effect on ΔC_p. Thus, our thermodynamic model can be readily applied for interactions that are not dominated by hydrophobic interactions.

Protocol for data fitting

1. At the end of an ITC experiment, the user will initially integrate the heat released or absorbed in each injection, as outlined in Section 4.1.
2. Following integration, the data series containing injection volume (v_i), ligand concentration (X_T), macromolecule concentration (M_T), molar ratio (X_T/M_T), and heat of injection (*n*dh) tabular form can be accessed in Origin® through the "Windows" menu tab. These data are saved in a standard ASCII text format that is directly used by our provided MATLAB scripts as inputs after the following changes are made. The first line must be commented out if headers are saved through Origin®; the last line must also be commented out in order to discard the last point. Macromolecule and ligand here refer to the contents in the cell and the syringe, respectively.
3. The ITC data files generated in step 2, along with the active cell volume (V_0), syringe concentration of ligand (X_0), experiment temperature (T) in K, and number of initial points to exclude (npd, which will typically be at least 1), are provided as inputs into the MATLAB fitting macro

Table 2 An Example of the Master Input File That Is Provided to the Macro Fcn_Fit_Cp_Ndil.m for Global Fitting of Temperature-Dependent ITC Data Sets

File_name.DAT	T	V_0	X_0	npd
15C.DAT	288	1427.47	0.3	2
20C.DAT	293	1427.47	0.3	1
25C.DAT	298	1427.47	0.3	1
30C.DAT	303	1427.47	0.3	2
35C.DAT	308	1427.47	0.3	1

(Fcn_Fit_Cp_Ndil.m) in a separate user written, tab delimited text file. A sample of what this file will look like is provided as Table 2. Running this protocol will produce best-fit values for ΔC_p, ΔH_0, and ΔS_0.

5. SUMMARY

There is significant interest in establishing general theories to describe IDP interactions (Levy, Cho, Onuchic, & Wolynes, 2005; Liu, Faeder, & Camacho, 2009; Whitford, Sanbonmatsu, & Onuchic, 2012). Quantitative analysis of temperature-dependent ITC data sets offers an unequaled opportunity to evaluate structure–function relationships for IDPs. Through these studies, it is possible to establish the energetic consequences of intrinsic disorder and to define the driving forces for disordered protein interactions, which is especially interesting when binding is coupled to a folding transition in the IDP. This chapter provides a framework for designing, executing, and analyzing temperature-dependent ITC studies in IDP interactions. The analysis stage of the research protocol documented here is largely achieved using MATLAB macros to facilitate either serial or global fitting of the temperature-dependent data series. All of the MATLAB scripts needed for this protocol are freely available through the stable URL: https://scholarsphere.psu.edu/downloads/5712mr21f.

In the closing section of this chapter, we have provided a specific example of a procedure implemented in MATLAB for global fitting of temperature-dependent ITC data that has the benefit of directly estimating the apparent ΔC_p for the transition(s) being monitored in the titration. Although our case study involved coupled folding and binding, the thermodynamic framework presented is not limited to describing interactions of this nature; however, we do note that the functional form derived is only rigorously applicable to one-to-one binding. With this caveat in mind, it is our general finding that the global fitting protocol we have established here is preferable to serial fitting of temperature-dependent data. The principle strength of this method is its reliance on directly fitting to fundamental thermodynamic equations, from which any desired equation of state may be calculated. Extension of the model to incorporate additional thermodynamic observables (e.g., pressure effects, the influence of added osmolytes, or macromolecular crowding) is straightforward. Of practical significance, this fitting method reliably incorporates data from (nearly) isoenthalpic titrations, making these data sets valuable contributors to the overall constraint of the parameters. In order to optimally utilize our protocol, we recommend balancing data collection over a wide temperature range and the need to sample where ΔC_p for the transition(s) is constant. In practice, this means that the upper temperature limit should be ~10 °C below the thermal denaturation midpoint of any folded biomolecules. For best results, we find that collecting data at a minimum of 4–6 temperatures provides a robust fit, with passage through the isoenthalpic point preferred if possible. Although we have only shown a single titration at each temperature for clarity in the figures presented here (Fig. 5), we generally find fitting of triplicate titrations at each selected temperature to provide a more robust sampling of the errors associated with the measurements.

For the case study we present, our experiments demonstrate that the FCP1–Rap74 interaction is readily reversed and characterized by a modest equilibrium dissociation constant of 2–5 μM under nearly all assayed conditions. This suggests that intrinsic disorder may provide an important functional advantage by making IDP interactions more robust with respect to certain local perturbations, although this appears somewhat in contrast to the notion that IDPs function as molecular rheostats. Moreover, our ITC protocol reveals a large and negative ΔC_p associated with this interaction. Without making any assumptions regarding the chemical details of the binding interface, this large and negative ΔC_p suggests that the FCP1–Rap74 interaction is driven by hydrophobic forces resulting from desolvating the binding surface. Furthermore, as a consequence of this large and negative ΔC_p, the binding interaction switches from entropically driven to enthalpically driven near room temperature. This emphasizes the importance of ITC experiments at multiple temperatures because ITC experiments performed at a single temperature would fail to reveal this transition. This transition is particularly important, as it is generally believed that the entropic penalty associated with folding IDPs plays a critical role in regulating their moderate binding affinity. Consequently, we assert that conducting variable-temperature ITC experiments provides a powerful framework for elucidating the thermodynamic forces that drive IDP interactions, as well as for revealing the unique biophysical properties and functional advantages associated with intrinsic disorder.

ACKNOWLEDGMENTS

We would like to thank the Penn State Automated Biomolecular Calorimetry Facility and its staff for the vital role they play in our research program and for providing helpful discussion during the preparation of this manuscript. We also thank Dr. Sushant Kumar for assistance validating the MATLAB scripts presented. This work was funded in part by an NSF CAREER award (MCB-0953918) to S.A.S. and a NIH predoctoral fellowship (F31GM101936) to M.B. This work was also supported by fellowships from the Sloan Foundation and the Penn State Institute for Cyberscience awarded to WGN. We thank the Penn State ScholarSphere team (http://scholarsphere.psu.edu) for its help in establishing permanent and stable web hosting for the products of our research program.

REFERENCES

Baldwin, R. L. (1986). Temperature dependence of the hydrophobic interaction in protein folding. *Proceedings of the National Academy of Sciences of the United States of America, 83*, 8069–8072.

Burnouf, D., Ennifar, E., Guedich, S., Puffer, B., Hoffmann, G., Bec, G., et al. (2012). kinITC: A new method for obtaining joint thermodynamic and kinetic data by isothermal titration calorimetry. *Journal of the American Chemical Society, 134*, 559–565.

Cervantes, C. F., Bergqvist, S., Kjaergaard, M., Kroon, G., Sue, S. C., Dyson, H. J., et al. (2011). The RelA nuclear localization signal folds upon binding to IkappaBalpha. *Journal of Molecular Biology, 405*, 754–764.

Cho, S., Swaminathan, C. P., Bonsor, D. A., Kerzic, M. C., Guan, R., Yang, J., et al. (2010). Assessing energetic contributions to binding from a disordered region in a protein–protein interaction. *Biochemistry, 49*, 9256–9268.

Chodera, J. D., & Mobley, D. L. (2013). Entropy–enthalpy compensation: Role and ramifications in biomolecular ligand recognition and design. *Annual Review of Biophysics, 42*, 121–142.

Cooper, A. (2010). Protein heat capacity: An anomaly that maybe never was. *Journal of Physical Chemistry Letters, 1*, 3298–3304.

Cooper, A., Johnson, C. M., Lakey, J. H., & Nollmann, M. (2001). Heat does not come in different colours: Entropy–enthalpy compensation, free energy windows, quantum confinement, pressure perturbation calorimetry, solvation and the multiple causes of heat capacity effects in biomolecular interactions. *Biophysical Chemistry, 93*, 215–230.

Csermely, P., Palotai, R., & Nussinov, R. (2010). Induced fit, conformational selection and independent dynamic segments: An extended view of binding events. *Trends in Biochemical Sciences, 35*, 539–546.

Dill, K. A. (1990). Dominant forces in protein folding. *Biochemistry, 29*, 7133–7155.

Drobnak, I., De Jonge, N., Haesaerts, S., Vesnaver, G., Loris, R., & Lah, J. (2013). Energetic basis of uncoupling folding from binding for an intrinsically disordered protein. *Journal of the American Chemical Society, 135*, 1288–1294.

Dunker, A. K., Silman, I., Uversky, V. N., & Sussman, J. L. (2008). Function and structure of inherently disordered proteins. *Current Opinion in Structural Biology*, *18*, 756–764.

Dyson, H. J., & Wright, P. E. (2002). Coupling of folding and binding for unstructured proteins. *Current Opinion in Structural Biology*, *12*, 54–60.

Ferreon, J. C., & Hilser, V. J. (2004). Thermodynamics of binding to SH3 domains: The energetic impact of polyproline II (PII) helix formation. *Biochemistry*, *43*, 7787–7797.

Fuxreiter, M., Tompa, P., Simon, I., Uversky, V. N., Hansen, J. C., & Asturias, F. J. (2008). Malleable machines take shape in eukaryotic transcriptional regulation. *Nature Chemical Biology*, *4*, 728–737.

Ganguly, D., Otieno, S., Waddell, B., Iconaru, L., Kriwacki, R. W., & Chen, J. (2012). Electrostatically accelerated coupled binding and folding of intrinsically disordered proteins. *Journal of Molecular Biology*, *422*, 674–684.

Ghai, R., Falconer, R. J., & Collins, B. M. (2012). Applications of isothermal titration calorimetry in pure and applied research—Survey of the literature from 2010. *Journal of Molecular Recognition*, *25*, 32–52.

Gibbs, E. B., & Showalter, S. A. (2015). Quantitative biophysical characterization of intrinsically disordered proteins. *Biochemistry*, *54*, 1314–1326.

Kamada, K., Roeder, R. G., & Burley, S. K. (2003). Molecular mechanism of recruitment of TFIIF-associating RNA polymerase C-terminal domain phosphatase (FCP1) by transcription factor IIF. *Proceedings of the National Academy of Sciences of the United States of America*, *100*, 2296–2299.

Kumar, S., Showalter, S. A., & Noid, W. G. (2013). Native-based simulations of the binding interaction between RAP74 and the disordered FCP1 peptide. *Journal of Physical Chemistry B*, *117*, 3074–3085.

Lacy, E. R., Filippov, I., Lewis, W. S., Otieno, S., Xiao, L. M., Weiss, S., et al. (2004). p27 binds cyclin-CDK complexes through a sequential mechanism involving binding-induced protein folding. *Nature Structural & Molecular Biology*, *11*, 358–364.

Lawrence, C. W., Bonny, A., & Showalter, S. A. (2011). The disordered C-terminus of the RNA Polymerase II phosphatase FCP1 is partially helical in the unbound state. *Biochemical and Biophysical Research Communications*, *410*, 461–465.

Lawrence, C. W., Kumar, S., Noid, W. G., & Showalter, S. A. (2014). Role of ordered proteins in the folding-upon-binding of intrinsically disordered proteins. *Journal of Physical Chemistry Letters*, *5*, 833–838.

Lawrence, C. W., & Showalter, S. A. (2012). Carbon-detected N-15 NMR spin relaxation of an intrinsically disordered protein: FCP1 dynamics unbound and in complex with RAP74. *Journal of Physical Chemistry Letters*, *3*, 1409–1413.

Levy, Y., Cho, S. S., Onuchic, J. N., & Wolynes, P. G. (2005). A survey of flexible protein binding mechanisms and their transition states using native topology based energy landscapes. *Journal of Molecular Biology*, *346*, 1121–1145.

Liu, J., Faeder, J. R., & Camacho, C. J. (2009). Toward a quantitative theory of intrinsically disordered proteins and their function. *Proceedings of the National Academy of Sciences of the United States of America*, *106*, 19819–19823.

Mizoue, L. S., & Tellinghuisen, J. (2004). The role of backlash in the "first injection anomaly" in isothermal titration calorimetry. *Analytical Biochemistry*, *326*, 125–127.

Mohan, A., Oldfield, C. J., Radivojac, P., Vacic, V., Cortese, M. S., Dunker, A. K., et al. (2006). Analysis of molecular recognition features (MoRFs). *Journal of Molecular Biology*, *362*, 1043–1059.

Nguyen, B. D., Abbott, K. L., Potempa, K., Kobort, M. S., Archambault, J., Greenblatt, J., et al. (2003). NMR structure of a complex containing the TFIIF subunit RAP74 and the RNA polymerase II carboxyl-terminal domain phosphatase FCP1. *Proceedings of the National Academy of Sciences of the United States of America*, *100*, 5688–5693.

Papoian, G. A., & Wolynes, P. G. (2003). The physics and bioinformatics of binding and folding-an energy landscape perspective. *Biopolymers*, *68*, 333–349.

Pierce, M. M., Raman, C. S., & Nall, B. T. (1999). Isothermal titration calorimetry of protein–protein interactions. *Methods*, *19*, 213–221.

Schuck, P. (2000). Size-distribution analysis of macromolecules by sedimentation velocity ultracentrifugation and lamm equation modeling. *Biophysical Journal*, *78*, 1606–1619.

Spolar, R. S., & Record, M. T., Jr. (1994). Coupling of local folding to site-specific binding of proteins to DNA. *Science*, *263*, 777–784.

Tantos, A., Han, K. H., & Tompa, P. (2012). Intrinsic disorder in cell signaling and gene transcription. *Journal of Molecular and Cellular Endocrinology*, *348*, 457–465.

Tompa, P. (2002). Intrinsically unstructured proteins. *Trends in Biochemical Sciences*, *27*, 527–533.

Tompa, P., & Fuxreiter, M. (2008). Fuzzy complexes: Polymorphism and structural disorder in protein–protein interactions. *Trends in Biochemical Sciences*, *33*, 2–8.

van der Lee, R., Buljan, M., Lang, B., Weatheritt, R. J., Daughdrill, G. W., Dunker, A. K., et al. (2014). Classification of intrinsically disordered regions and proteins. *Chemical Reviews*, *114*, 6589–6631.

Whitford, P. C., Sanbonmatsu, K. Y., & Onuchic, J. N. (2012). Biomolecular dynamics: Order–disorder transitions and energy landscapes. *Reports on Progress in Physics*, *75*, 076601.

Wiseman, T., Williston, S., Brandts, J. F., & Lin, L. N. (1989). Rapid measurement of binding constants and heats of binding using a new titration calorimeter. *Analytical Biochemistry*, *179*, 131–137.

Wostenberg, C., Kumar, S., Noid, W. G., & Showalter, S. A. (2011). Atomistic simulations reveal structural disorder in the RAP74-FCP1 complex. *Journal of Physical Chemistry B*, *115*, 13731–13739.

Wright, P. E., & Dyson, H. J. (2009). Linking folding and binding. *Current Opinion in Structural Biology*, *19*, 31–38.

Xu, Y., Oruganti, S. V., Gopalan, V., & Foster, M. P. (2012). Thermodynamics of coupled folding in the interaction of archaeal RNase P proteins RPP21 and RPP29. *Biochemistry*, *51*, 926–935.

CHAPTER THREE

Fragment-Based Screening for Enzyme Inhibitors Using Calorimetry

Michael I. Recht*[,1], Vicki Nienaber[†], Francisco E. Torres*
*Palo Alto Research Center, Palo Alto, California, USA
[†]Zenobia Therapeutics, San Diego, California, USA
[1]Corresponding author: e-mail address: michael.recht@parc.com

Contents

Abstract

Isothermal titration calorimetry (ITC) provides a sensitive and accurate means by which to study the thermodynamics of binding reactions. In addition, it enables label-free measurement of enzymatic reactions. The advent of extremely sensitive microcalorimeters

Methods in Enzymology, Volume 567
ISSN 0076-6879
http://dx.doi.org/10.1016/bs.mie.2015.07.023

have made it increasingly valuable as a tool for hit validation and characterization, but its use in primary screening is hampered by requiring large quantities of reagents and long measurement times. Nanocalorimeters can overcome these limitations of conventional ITC, particularly for screening libraries of 500–1000 compounds such as those encountered in fragment-based lead discovery. This chapter describes how nanocalorimetry and conventional microcalorimetry can be used to screen compound libraries for enzyme inhibitors.

1. INTRODUCTION

1.1 Nanocalorimeter Arrays

Isothermal titration calorimetry (ITC) directly measures the heat evolved during a reaction, enabling determination of ΔH, ΔG, ΔS, and stoichiometry in a single experiment (Peters, Frasca, & Brown, 2009; Wiseman, Williston, Brandts, & Lin, 1989). In addition to binding reactions, ITC has been used as a label-free method to measure enzymatic reactions (Bianconi, 2003; Olsen, 2006; Todd & Gomez, 2001). Because the measurement is label-free and requires no immobilization of either the target or ligand, calorimetry can be used to perform a direct enzyme activity assay without modifying the substrate and, unlike conventional colorimetric or fluorescence-based enzyme assays, is unaffected by the spectroscopic properties of the sample. When used in screening applications, each calorimetric measurement yields full enzyme kinetic parameters for reactions performed in the presence of each compound (Recht et al., 2012, 2014). This allows discrimination of inhibition mechanism (competitive, noncompetitive, or mixed) and, for competitive inhibitors, determination of K_I for each compound from the primary screening data.

Conventional microcalorimetry is hampered by requiring large quantities of reagents (0.2–1.4 mL) and long measurement times (10–30 min). Nanocalorimeters can overcome these limitations of conventional ITC. Enthalpy arrays, which are nanocalorimeters in a 96-detector array format, enable measurement of the thermodynamics and kinetics of molecular interactions using small sample volumes (500 nL) and short measurement times (typically 5–10 min; Table 1; Torres et al., 2004).

Each detector cell on the array consists of two identical thermistor sensing elements connected in a Wheatstone bridge configuration that provides a differential temperature measurement between the sample specimen and reference specimen (Fig. 1; Recht et al., 2008; Torres et al., 2004). Each site

Table 1 Compound Screening Capabilities of Calorimeters

Calorimeter	Kinetic Characterization			Throughput (Per hour)	[Enzyme] (*M*)	Moles of Enzyme Per Reaction	Moles of Substrate Per Reaction
	k_{cat}	K_M	K_I				
Enthalpy array (500 nL) (Recht et al., 2008, 2009; Torres et al., 2004)	$k_{cat} > 0.5\ s^{-1}$	10^{-6} to $10^{-3}\ M$	10^{-7} to $10^{-3}\ M$	12	1.5×10^{-5}	7.5×10^{-12}	7.5×10^{-10}
iTC200/PEAQ-ITC, Nano ITC (200 μL) (Cai, Cao, & Lai, 2001; Freyer & Lewis, 2008; Peters et al., 2009)	$k_{cat} > 0.02\ s^{-1}$	10^{-8} to $10^{-3}\ M$	10^{-8} to $10^{-3}\ M$	3	5×10^{-7}	1×10^{-10}	5×10^{-8}

k_{cat} is the turnover number, K_M is the Michaelis constant, and K_I is the competitive inhibition constant. Enzyme concentration and moles of enzyme and substrate are calculated assuming $k_{cat} = 2\ s^{-1}$, $K_M = 5\ \mu M$, $[S]_0$ (enthalpy array) $= 300 \times K_M$, $[S]_0$ (ITC) $= 50 \times K_M$, $\Delta H = -5000$ cal/mol, and throughput as indicated in table.

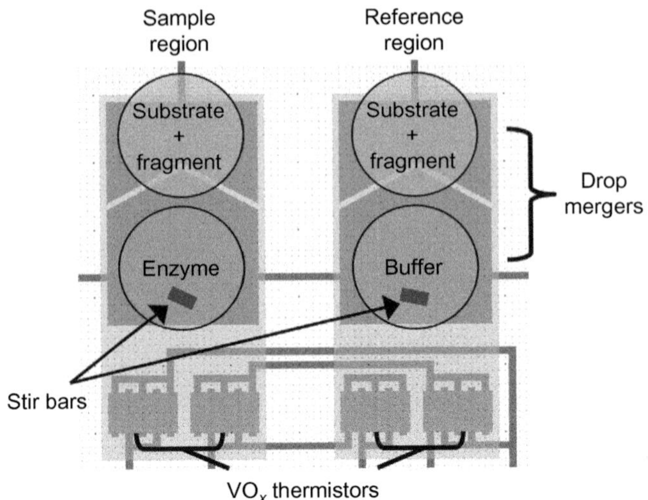

Figure 1 Schematic of a single enthalpy array detector. Sample and reference regions are designated based on the material (substrate (alone or with fragment), enzyme, or buffer) in the drops deposited on each region. PEGylated magnetic stir bars (400 μm × 200 μm × 15 μm) are manually placed on array drop merger elements using an *X–Y–Z* micromanipulator. A single stir bar is placed at the location where the enzyme solution or buffer solution will be deposited on the array. *Adapted with permission from Recht et al. (2009). Copyright, Elsevier, Inc.* (See the color plate.)

incorporates a pair of electrostatic drop mergers, enabling simultaneous isothermal merging of the sample and reference specimens. While electrostatically driven merging coupled with the surface tension of the drops does cause some mixing, the time constant is large because diffusive mixing still dominates. Rapid mixing is achieved via magnetic micro stir bars placed in one drop of both the sample and reference pair (De Bruyker et al., 2011). As shown in Fig. 1 for an enzymatic reaction, the sample specimen consists of a pair of 250 nL droplets: substrate (with or without fragment) and enzyme. The reference specimen consists of a 250-nLdroplet of the same substrate (with or without fragment) solution used in the sample specimen and a second 250-nL droplet containing buffer only. Upon merging, the final reaction volumes of the sample and reference specimens are 500 nL each.

1.2 Fragment-Based Lead Discovery

Fragment-based lead discovery (FBLD) is an approach in which small, low complexity chemical fragments of 6–15 heavy atoms are screened for binding to, or inhibiting the activity of, a target. The hits from the screen are then linked and/or elaborated, ideally yielding early lead compounds for drug discovery (Carr, Congreve, Murray, & Rees, 2005; Erlanson, McDowell, & O'Brien, 2004; Jahnke & Erlanson, 2006).

Hits arising from FBLD usually bind with low affinity; therefore, sensitive detection methods are required (Fattori, 2004). A common approach to library screening has been to use biophysical methods such as X-ray crystallography (Nienaber et al., 2000) or NMR (Shuker, Hajduk, Meadows, & Fesik, 1996). Since these structure-based approaches are expensive in terms of time and materials, alternative screening approaches using surface plasmon resonance (Giannetti, 2011; Hämäläinen et al., 2008) and high-concentration screening using HTS assays (Boettcher et al., 2010; Godemann et al., 2009) have become common.

The calorimetry methods presented here are ideally suited for FBLD because they are highly sensitive to weakly binding ligands. These calorimetry methods are sensitive to K_I in the mM range (Recht et al., 2012, 2014). So, in addition to detection of ligand binding, the methods provide an accurate read-out of functional activity. Hence, the methods may also be used to guide medicinal chemistry optimization of fragments during the fragment-to-lead phase. Because the enzymatic substrate can be used in its native form, one may assay a range of biologically relevant substrates, including other macromolecules. This can facilitate detection of ligand binding not only at the active site but also at other protein–protein interaction and/or allosteric sites.

Fragment screening has a number of advantages over other screening methods that can be fully realized using a calorimetric assay. First, one need not screen a very large library to find a hit, especially if the fragments are kept small (low molecular weight) and simple. Hann computationally demonstrated that small, simple molecules have a much higher probability of binding to a complex site (such as an enzyme active site) than complex molecules (Hann, Leach, & Harper, 2001). Depending on the druggability of the target, one may find ample hits from a fragment library of even a few hundred compounds, in particular if accurate K_I are obtained in the mM range. Second, fragments have been shown to detect allosteric binding sites (Heinrich et al., 2013; Jahnke et al., 2010; Saalau-Bethell et al., 2012) or even create a new binding pocket (Murray et al., 2010). Higher molecular weight, more complex molecules may be sterically excluded from these sites. Because the calorimetric approach outlined here measures enzyme activity, an accurate functional read-out can be obtained in the initial screening stage, permitting identification of functionally relevant allosteric/remote sites. Third, fragment to lead with a low molecular weight starting point (average $M_W \approx 150$ Da), provides sufficient room to grow the molecule while staying within successfully launched drug chemical space ($M_W < \approx 350$–400 Da) (Wenlock, Austin, Barton, Davis, & Leeson, 2003). Careful monitoring of ligand efficiency (Hopkins, Groom, & Alex, 2004) and other properties (e.g., selectivity) can provide multiple chemically diverse starting points for lead optimization. The ability of calorimetry to provide accurate K_I for low molecular weight, weakly binding ligands during the screening and lead optimization phases allows for more accurate structure-activity relationship development early in the fragment-to-lead stage of the discovery process.

We have demonstrated that an enthalpy array enzyme activity assay can be used in FBLD to identify low-affinity competitive inhibitors of the phosphodiesterases PDE4A (Recht et al., 2012) and PDE10A (Recht et al., 2014) prior to moving into X-ray crystallography studies. With the knowledge that the compounds act as competitive inhibitors, one could consider methods other than X-ray crystallography to obtain information about the specific interactions between the compound and the protein (Recht et al., 2012).

The first method below describes the use of the Palo Alto Research Center enthalpy array system consisting of a custom-built measurement chamber and arrays of nanocalorimeters (Recht et al., 2008; Torres et al., 2004). As this proprietary system is not commercially available, a second method adapted for use with commercially available power-compensation ITC instruments (MicroCal/Malvern Instruments, TA Instruments) is also included.

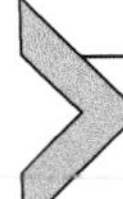

2. REQUIRED MATERIALS

2.1 Equipment

- Nanocalorimeter (Torres et al., 2004)
- Enthalpy arrays; fabrication of the arrays is described in Torres et al. (2004) with improved thermistors described in De Bruyker et al. (2007) and Recht et al. (2008).
- Micro stir bars (De Bruyker et al., 2011)
- *X*–*Y*–*Z* micromanipulator model 7372E or equivalent (West Bond, Anaheim, CA)
- Polymer cap
- 384 well plates for enzyme, substrate, and buffer samples
- Optional: commercial ITC (MicroCal ITC_{200}, PEAQ-ITC, or equivalent).

2.2 Reagents

- Enzyme solution (in reaction buffer plus 2% (v/v) DMSO)
- Substrate solution (in reaction buffer)
- Fragment library (see Section 4; solutions of 200 m*M* compound in anhydrous DMSO)
- Enzyme reaction buffer—nearly any aqueous buffer is acceptable. For enzymes requiring the presence of a reducing agent, better results are often obtained using tris(2-carboxyethyl)phosphine instead of dithiothreitol (Thomson & Ladbury, 2004).
- Anhydrous DMSO
- ddH_2O.

2.3 Software

- MATLAB
- Origin

3. INSTRUMENTATION

As shown in Fig. 2, there is a cap above the detector during measurements to isolate the drops and detector from the environment. There is also a

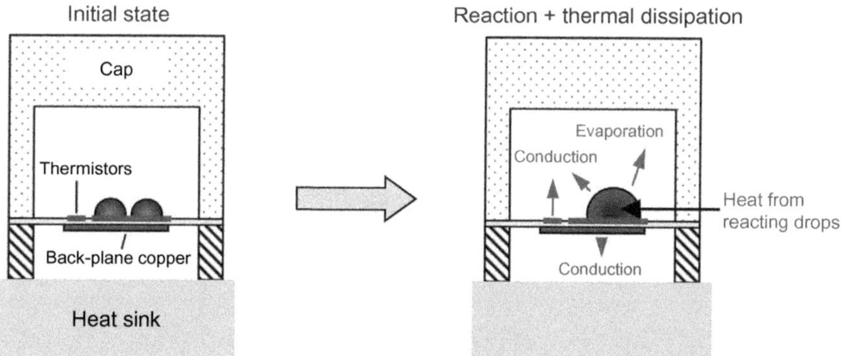

Figure 2 Side view of one detector of an array before and after drop merging. The drops and detector are isolated from the environment by a cap above and a heat sink below. Pogo pins through the cap provide electrical connections from the array to the electronics on a printed circuit board (not shown). The copper heat sink has a large mass × specific heat, which serves to lessen thermal fluctuations. The back-plane copper underneath the drops and thermistors evens out the temperature. Upon merging, drops heat (or cool for endothermic reactions) and the heat dissipates by conductive and evaporative heat transport. (See the color plate.)

copper block below the detector that has a substantial thermal mass to provide the necessary thermal stability during measurements. When the drops merge and a biochemical reaction occurs, heat is released (exothermic) or absorbed (endothermic), changing the temperature of the sample drop relative to the reference drop. Both conductive and evaporative heat transport occur through the air around the drops. The insulation provided by the quiescent air around the drop and detector limits the rate of heat transport to the cap and heat sink, allowing the temperature rise to persist long enough to measure it. The characteristic time for heat dissipation is $\approx$2.3 s.

The nanocalorimeter system (Fig. 3A) consists of an electromagnetic field shielded, temperature-regulated housing surrounding a noncontact liquid dispenser, a robot, and data acquisition hardware. In the dispense chamber (left), reagents are deposited on the array using a Deerac eight channel spot-on™ original equipment manufacturer (OEM) module (Deerac/Labcyte). For the volumes we are dispensing (250 nL), Deerac specifies a dispense accuracy of ±5% and precision of <7% CV. For quality control, every time a set of Equator tips is installed in our spot-on™ OEM module, the diameters for a set of nine water drops on a hydrophobic surface are measured for each tip. If necessary, the mean size is adjusted through a software interface and the tip is retested. For drops dispensed from each of the eight tips, we have observed that the diameters after adjustment are almost always 850 ± 10 μm, which corresponds to <5% CV by volume. In those rare instances where the drop diameters fall outside of this range, the tip in question is discarded.

In the measurement chamber (Fig. 3B), the array sits atop the aluminum or copper block. Within the block are individually addressable magnetic stirrers. In addition, the chamber houses a temperature-controlled custom data acquisition printed circuit board (PCB). Electrical contact between the array and the PCB is made via pogo pins. The detector electronics on the PCB include an instrumentation amplifier, as well as drive electronics to distribute the sine wave drive voltage from the lock-in amplifier to the Wheatstone bridge devices. The instrumentation amplifier is placed on the PCB near the array so that the signal gets amplified close to where it is generated on the array, reducing the pickup of noise before amplification. Additional circuitry actuates the electrostatic merging. The PCB is attached to a metal block, which has cooling water passing through it to remove the heat of the PCB. The metal block has two sections separated by a thermoelectric cooler to accurately maintain the correct temperature. A polymer cap minimizes evaporation of the droplets during the measurements. In addition,

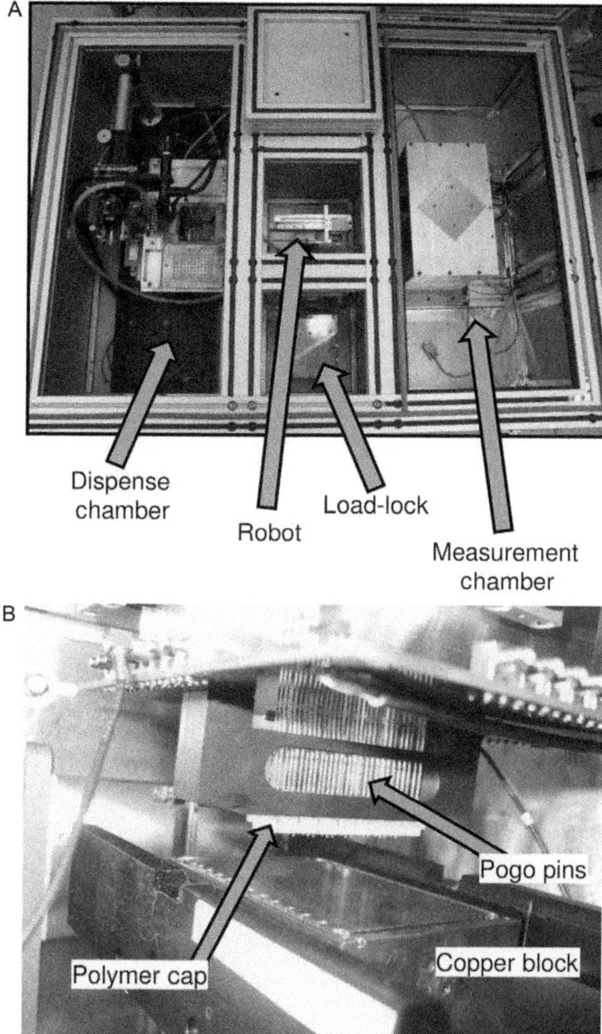

Figure 3 Photographs of PARC enthalpy array nanocalorimeter system. (A) Top view of complete system showing the dispense chamber (left), the robot which moves the array between the dispense chamber and the measurement chamber (center), and the measurement chamber (right). Not shown in this image are electronics external to the device chamber including a lock-in amplifier and function generator, as well as the *X–Y–Z* micromanipulator for depositing stir bars on the array. (B) View inside the measurement chamber showing the copper block on which the array sits, the polymer cap which controls evaporation of the reaction droplets, the pogo pins which make electrical contact with the thermistor bridge and drop mergers on the array.

external to the chamber are a lock-in amplifier (Stanford Research Systems model SR850), function/arbitrary waveform generator (Agilent model 33120A), and a high-voltage power amplifier/supply (TREK model 677B).

4. FRAGMENT LIBRARY SELECTION

The ideal chemical properties for a fragment library are consistent with calorimetric screening. The fragments should be low molecular weight and simple to increase probability of binding and facilitate fragment to lead. Other chemical properties should also be consistent with optimization into a successful drug candidate. Wenlock et al. (2003) showed that compounds that drop out of the clinic tend to be high molecular weight and high clogP. During optimization, compound molecular weight typically increases by about 100 Da. Lipophilicity, measured by clogP, also increases during optimization (Hann, 2011; Leach, Hann, Burrows, & Griffen, 2006). An observed requirement for permeability is to keep the number of hydrogen bond donors as low as possible, average ≈ 2 (Wenlock et al., 2003; Winiwarter et al., 1998, 2003). As discussed above, calorimetry is sensitive to low-affinity ligands. It also requires a relatively high solubility (≈ 2 mM in 2% (v/v) DMSO) which is consistent with a more hydrophilic starting point (clogP < 2). As most commercial fragment libraries are designed based upon the rule-of-three where molecular weight is less than 300 Da, clogP < 3 and h-bond donors < 3 (Congreve, Carr, Murray, & Jhoti, 2003), we prepared a new library for calorimetric screening. The library was assembled based upon Zenobia's fragment library collection (http://zenobiafragments.com). Compounds were selected based upon Zenobia's guide of two (Go2) where molecular weight is <200 Da, clogP < 2 and h-bond donors ≤ 2 (Nienabar, personal communication). In general, additional libraries may be assembled from in house or other commercial compound collections based upon Go2 chemical properties. The chemical properties of Zenobia's fragment collection are summarized in Table 2.

The number of compounds that are required to identify a hit depends on the druggability of the target and diversity of the fragment collection (the primary limiting reagent is protein supply). For a low molecular weight fragment collection, we measure diversity at the chemical core (e.g., Murcko core) level and secondarily, at the functional group level. For a "difficult target," a screen of 1000 compounds that are chemically diverse should be sufficient. If no hits are found with this screen, it is a sign that the program will likely be very challenging. For more druggable active sites, even a few

Table 2 Properties of Zenobia's Fragment Collection

980 Compounds Total	**Mean**	**Min**	**Max**
Flexibility			
% rotatable bonds[a]	26	0	100
Properties of core[b]			
% hetero-atom in core	16	0	80
% aromatic atom in core	73	0	100
% aliphatic atom in core	27	0	100
Complexity and 3D character			
Fragment Complexity[c]	301	19	1294
Plane of Best Fit[d]	0.18	0.00	0.85
SP3 Character[e]	0.09	0.00	0.37
Chemical properties			
Molecular weight	154	82	294
Heavy atoms count	11	6	19
XLogP	0.95	−4.44	4.52
tPSA	48	0	133
Hydrogen bond acceptors	2	0	7
Hydrogen bond donors	1	0	5

[a]Number of rotatable bonds/total number of bonds.
[b]All are number of described atoms/total number of atoms in the Murcko Core (Bemis & Murcko, 1996).
[c]Fragment complexity = $|\#\ \text{bonds}^2 - \#\ \text{heavy atoms}^2 + \#\ \text{heavy atoms}| + (\#\ \text{heteroatoms}/100)$ (Nilakantan et al., 2006).
[d]Deviation of atoms from an imaginary plane through the molecule (Firth, Brown, & Blagg, 2012).
[e]Number of sp3 atoms/total number of atoms.

hundred carefully chosen compounds may be sufficient to identify a set of diverse starting points for fragment to lead. For the PDE4/PDE10 examples cited above (Recht et al., 2012, 2014), a diverse set of 245 compounds were selected from the Zenobia fragment collection for chemical core diversity with inclusion of additional fragments that could be chemical analogues of known PDE substrates. Thirteen hits were found for PDE4 and 24 for PDE10. Two of the PDE10 hits were successfully optimized to low and submicromolar potency.

4.1 Materials Needed

- Standard laboratory liquid handling tools; multichannel pipettes are sufficient.
- SpeedVac vacuum concentrator (SPD1010, SPD2010, or equivalent) for assembling an internal preformatted library in 96 well plates.
- Soluble (4 m*M* in aqueous (2% v/v DMSO) solution), fragment collection (we use the Go2 to increase probability of hits).
- Organic solvents, typically DMSO.

4.2 Software Needed

- Software for viewing chemical structures either as SMILES or in SDF format (Hyleos ChemFileBrowser is a free program to view SDF files; many online translators may be used to translate SMILES strings).
- Database to store data (could be Excel or more sophisticated package).

4.3 Protocol to Prepare of Fragment Library for Screening

1. Determine screening capacity.
2. Purchase preformatted commercial library or assemble fragment library. To assemble an internal preformatted fragment library:
   a. Identify compounds that will make up the library. We suggest that the compounds meet the Go2 and sample sufficient chemical space to identify a hit.
   b. Purchase compounds.
   c. Receive and check compounds into internal database.
   d. Test solubility of compounds in DMSO (ideally compounds will be soluble at 200 m*M*).
   e. Test solubility of compounds in a 50% acetonitrile solution at 100 m*M*; if compounds do not dissolve in this solution, they may be dissolved in DMSO or DMSO–HCl. Solvent is dependent on the equipment available for drying the compounds in plates.
   f. Test solubility of compound at 1 m*M* aqueous solution (1–2% (v/v) DMSO).
   g. For compounds that are soluble in steps d–f, prepare a 100 m*M* stock of each compound in 50% acetonitrile (or DMSO if applicable). These stocks may be stored in deep well blocks to facilitate sample manipulation.
   h. Pipet fixed amount of each 50% acetonitrile solution in one or more V-bottom 96 well plates (e.g., 100 μL).

  i. Dry plates in Savant SpeedVac plate centrifuge overnight.
  j. Plates may be stored desiccated at room temperature.
3. In advance of screening, resuspend each fragment sample to 200 mM in DMSO (or other organic solvent).

5. CALORIMETRIC ENZYME ASSAY: NANOCALORIMETER

5.1 Sample Considerations and Preparation

It is important to ensure that the enzyme and substrate solutions are made in identical buffer and that the final DMSO concentration in all solutions is matched. Any mismatch in DMSO concentration or buffer composition may produce thermal artifacts in the data. Since the heat associated with these effects evolves quickly, this is less of a problem in an extended (2–3 min) enzymatic reaction compared to an instantaneous binding reaction, but higher quality data will be obtained by ensuring that buffer/DMSO concentrations are matched in each solution.

The concentration of enzyme (and substrate) should be selected to yield a good signal-to-noise ratio and a sufficiently long reaction in comparison to initial transients (such as substrate, fragment, or DMSO dilution heat). For an enzyme-catalyzed reaction, the temperature signal is proportional to [E], so the enzyme concentration needs to be high enough to yield a sufficient signal (500 μK or 0.5 μW) at maximum reaction velocity (V_{max}) to give a signal-to-noise ratio of at least 5:1. The ratio of enzyme to substrate should be large enough to achieve a steady-state enzymatic reaction that is long in comparison to initial transients. With magnetic mixing, these transients typically last less than 10 s, so a 60–90 s enzymatic reaction in the absence of inhibitor is usually sufficient (Recht et al., 2009). The substrate concentration should also be adjusted so that at the start of the reaction, $[S]_0 \approx 50 \times K_M$. By keeping [S] high at the start of the reaction, K_I for stronger competitive inhibitors can be determined in the initial screen.

The DMSO concentration in the reaction should typically be kept below 5% (v/v) to avoid adverse effects on enzyme activity. We have generally used a final DMSO concentration of 2% (v/v), which yields a fragment concentration of 2 mM in the final reaction when starting from 200 mM stocks (Recht et al., 2012, 2014).

As shown in Fig. 4A, 30 s of data are captured prior to initiation of the reaction at $t=0$ s. In screening mode, reaction data are typically captured for 4 min following initiation of the reaction to ensure that all substrate has been converted to product even in the presence of an inhibitor. In cases where the

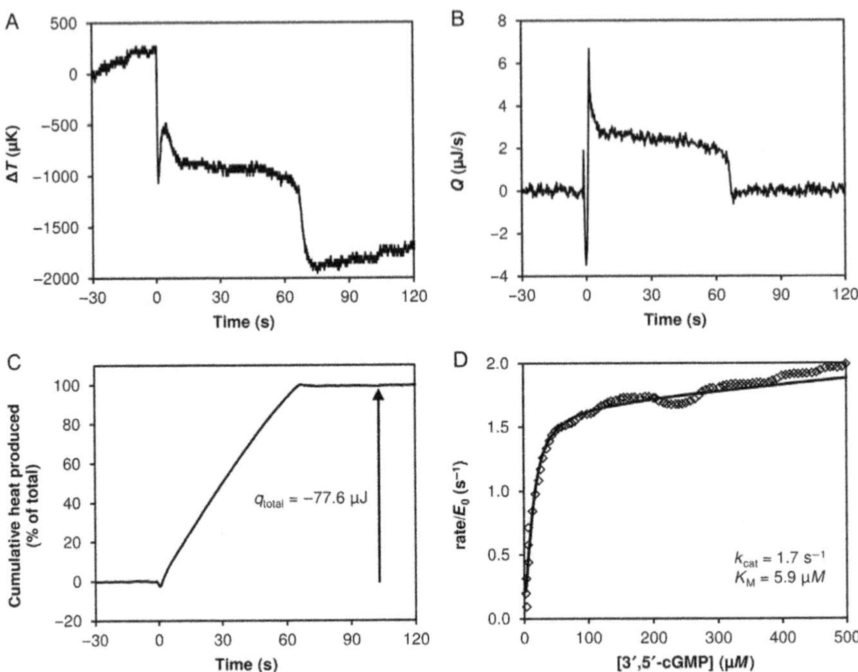

Figure 4 Enzymatic reaction measured by enthalpy arrays. (A) Plot of temperature difference versus time. The reaction is initiated at $t=0$ s. (B) Plot of power (Q) versus time for the reaction shown in (A), following baseline correction. Q is integrated relative to the corrected baseline. (C) Integral of $Q(\tau)$ over time, yielding total enthalpy of the reaction. (D) PDE10A hydrolysis of 3′,5′-cGMP. Reaction contained 15 μ*M* PDE10A and 2 m*M* 3′,5′-cGMP. Solid curve is fit of data to Eq. (2). Fitting the open black points yielded $k_{cat}=1.7\ s^{-1}$ and $K_M=5.9\ \mu M$.

inhibitor is highly potent, the compound is retested either at a lower concentration or by recording data for a longer period of time.

5.2 Data Analysis

As shown in Fig. 4A, we record the differential temperature as a function of time. This is converted to an enthalpy by deconvoluting $Q(\tau)$ from the equation

$$T(t) = \int_{\text{merge}}^{t} Q(\tau)\mathfrak{I}(t-\tau)\,d\tau. \tag{1}$$

In Eq. (1), $T(t)$ is the temperature change (T) as a function of time (t) relative to the baseline temperature, $Q(\tau)$ is the rate of heat generation as a function of time, and $\mathfrak{I}(t-\tau)$, the "impulse response," is the temperature rise that would occur at a time t given a Dirac delta function impulse of heat at time τ (Recht et al., 2008). To obtain $Q(\tau)$ from $T(t)$, we used COMSOL® (COMSOL, Inc.) simulations of our nanocalorimeters to determine $\mathfrak{I}(t-\tau)$ (Torres et al., 2006). To deconvolute $Q(\tau)$, we approximate $Q(\tau)$ as a piecewise linear continuous function and use a least-squares regression to determine the fitting coefficients. Following calculation of $Q(\tau)$ (Fig. 4B), we integrate $Q(\tau)$ over time to obtain the total enthalpy (Fig. 4C). The data is subsequently transformed into the reaction velocity as a function of substrate (Fig. 4D) (Recht et al., 2009) and k_{cat} and K_M for experiments with no inhibitor are determined by fitting the data for $Q/([E]\Delta H)$ versus [S] to a modified form of the general rate equation:

$$\frac{Q}{[E](-\Delta H)} = \text{rate} = k_{cat}[ES] = k_{cat}\frac{(E_0 + [S] + K_M) - \sqrt{(E_0 + [S] + K_M)^2 - 4 \times E_0 \times [S]}}{2}(1 + \beta[S]) \tag{2}$$

In Eq. (2), Q is the rate of heat generation, k_{cat} is the turnover number, E_0 is the total enzyme concentration, [S] is the substrate concentration, ΔH is the enthalpy change per mole of substrate reacted, and β is the slope for Q versus [S] at [S] greater than K_M (Recht et al., 2009). The $(1+\beta[S])$ term modifies the general rate equation and is clearly distinguishable from the effect of K_M. For standard Michaelis–Menten kinetics and ideal solution enthalpies, β will be zero, but we have observed reactions for which there is a clear nonzero slope for Q versus [S] at [S] greater than the observed K_M (Recht et al., 2009). The general rate equation is used because if the concentration of enzyme in the reaction is high relative to K_M ($E_0/K_M > 0.1$), depletion of substrate from solution becomes significant (Cha, 1970; Goldstein, 1944).

For competitive inhibitors, one is able to observe a change in the apparent K_M, allowing determination of K_I. Upon examination of the Michaelis–Menten equation modified for competitive inhibition (Fersht, 1999, eq. 3.32):

$$\text{rate} = \frac{k_{cat}E_0[S]}{[S] + K_M(1 + [I]/K_I)} \tag{3}$$

in which [I] is the concentration of free inhibitor, the apparent K_M is seen to be the true K_M multiplied by $(1+[I]/K_I)$. At inhibitor concentrations $[I] > K_I$, the shift in apparent K_M becomes significant. When $[I] \gg E_0$ holds, as it will in screening fragments, the concentration of free inhibitor [I] is close to the total concentration of inhibitor I_0, making it reasonable to use I_0 in the above equation in place of [I], the standard practice in enzymology.

Examples of data from the screening of a fragment library are shown in Fig. 5. Figure 5A and B shows a comparison of the reaction of PDE10A with 3′,5′-cGMP in the absence of inhibitor and in the presence of a fragment which is a competitive inhibitor. In contrast, most compounds in the fragment library yield data like that shown in Fig. 5C and D. The reaction is the presence of the compound is indistinguishable from the control reaction.

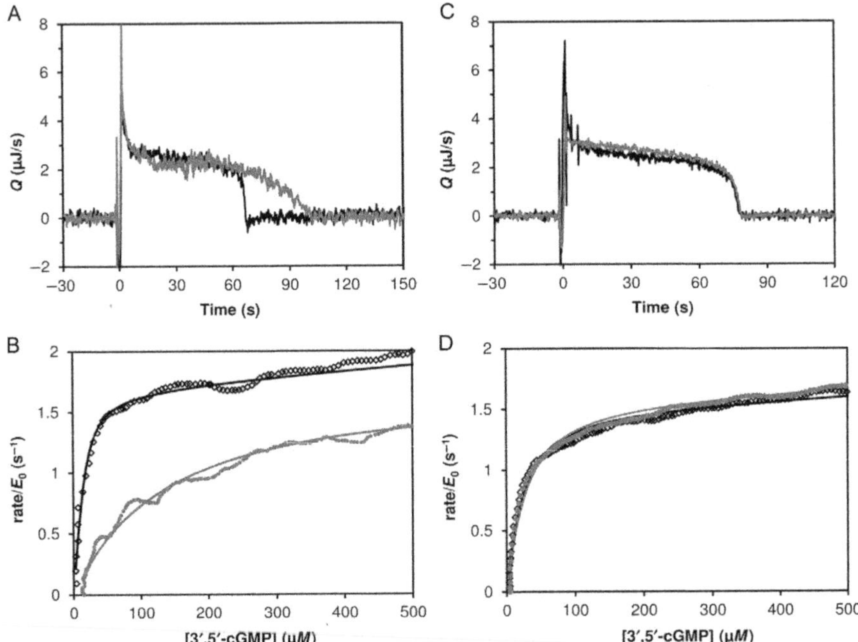

Figure 5 PDE10A hydrolysis of 3′,5′-cGMP in the absence (black, open) and presence (maroon, solid) of fragments. Reactions contained 15 μ*M* PDE10A and 2 m*M* 3′,5′-cGMP. Solid curves are fit of data to Eq. (2). (A) Plot of power (*Q*) versus time for the reaction in absence and presence of 2 m*M* fragment ZT0429. Note the signal returns to original levels as substrate is completely hydrolyzed and in particular more slowly than the reaction in the absence of inhibitor. This difference reflects a change in apparent K_M for 3′,5′-cGMP in the presence of ZT0429. (B) Rate versus remaining 3′,5′-cGMP concentration in the absence and presence of 2 m*M* ZT0429. Solid curves are fit of data to Eq. (2); K_I for ZT0429 is 300 μ*M*. (C) Plot of power (*Q*) versus time for the reaction in absence and presence of 2 m*M* fragment ZT0417. Note the signal for the reaction in the presence of fragment is nearly identical to the control reaction. (D) Rate versus remaining 3′,5′-cGMP concentration in the absence and presence of 2 m*M* ZT0417. Solid curves are fit of data to Eq. (2); K_I for ZT0417 is $\gg$2000 μ*M*. (See the color plate.)

5.3 Protocol

1. Install polymer cap in the measurement chamber and close the chamber lid.
2. Populate array with micro stir bars. Using an *X–Y–Z* micromanipulator, place a single bar on each region of array as indicated in Fig. 1. The bars will remain in place unless the array is sharply bumped or inverted.
3. Place the array in the dispense chamber
4. Prepare enzyme solution at twice the final reaction concentration. If the stock solution of enzyme is at least 100-fold higher concentration than the desired concentration in the reaction, or if it is stored in reaction buffer, it can be diluted directly into the reaction buffer (including 2% DMSO), otherwise it should be dialyzed or buffer exchanged (using an Amicon® Ultra centrifugal filter or equivalent) against the reaction buffer.
5. Prepare substrate solution in reaction buffer (with 2% (v/v) DMSO) at twice the final reaction concentration. If the substrate is originally in a crystalline or lyophilized powder form, it can be dissolved directly in the reaction buffer. If the substrate is already in solution and large enough molecular weight, it should be dialyzed (or buffer exchanged) against the reaction buffer.
6. Prepare substrate with fragment solutions. A separate solution of substrate with fragment is required for each fragment to be tested. It is only necessary to prepare 150 μL of each solution. In place of the DMSO added to the substrate solution above (step 2), add 3 μL 200 m*M* fragment stock (in anhydrous DMSO). This will yield a solution containing 4 m*M* fragment, which will result in 2 m*M* fragment in the combined reaction droplet.
7. Prepare buffer solution by adding anhydrous DMSO to a final concentration of 2% (v/v).
8. Load well plate: Eight tips are used for each aspirate, which is sufficient to measure a control reaction and test five different fragments. Load 100 μL of solution in the wells:
   Tip 1: Enzyme
   Tip 2: Substrate
   Tip 3: Buffer
   Tip 4: Substrate plus fragment 1
   Tip 5: Substrate plus fragment 2
   Tip 6: Substrate plus fragment 3

Tip 7: Substrate plus fragment 4

Tip 8: Substrate plus fragment 5

9. Load 30–50 μL of reagent solutions in tips. Following aspiration of the reagents, the tips are prepared by dispensing 20–30 droplets (50 nL each).

10. Reactions are performed in groups of eight (one row). Reactions are typically performed in triplicate. Drops are dispensed on the array, an image of each site is recorded, the robot moves the array into the measurement chamber, and the array equilibrates. Following a 2-min equilibration, the first reaction is recorded. The remaining seven reactions are automatically recorded sequentially upon completion of the first reaction. The array is returned to the dispense chamber, an image of each site is again captured, and the system is ready to proceed with the next set of reactions. An entire array of measurements can be programmed to run automatically, but usually a total of 18 reactions (control + 5 fragments in triplicate) are recorded before the reagents are purged, tips are cleaned, and a new set of fragments is loaded.

11. Analyze data as described in Section 5.2.

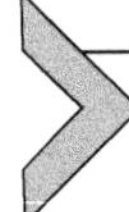

6. CALORIMETRIC ENZYME ASSAY: COMMERCIAL MICROCALORIMETERS

Many of the considerations outlined above for nanocalorimeters hold true for microcalorimeters, but there are several modifications arising due to differences in the reaction volumes of the two types of calorimeters which affect the fluidics and instrument response time. While enthalpy arrays combine droplets of equal volumes in a reaction, microcalorimeters inject small volumes (0.2–10 μL) of reactant into a larger sample cell (200 μL). This has several implications for experimental design. First, in a microcalorimeter it is prudent to combine the enzyme and fragment in the sample cell and inject the substrate solution rather than trying to inject a substrate plus fragment solution. The heat from the large dilution factor (20- to 1000-fold) of the injected solution results in a large transient signal that can exceed the available power-compensation range. It is also more difficult to achieve the desired fragment concentration in the reaction, as it must be present in the syringe at a concentration 20–1000 × the desired value. This can also be an issue for substrate when trying to achieve $[S]_0 > 50 \times K_M$ at the start of the reaction. If large dilution heats are encountered, one must reduce the concentration of substrate in the syringe and try to achieve the desired $[S]_0$

by increasing the injection volume. If $[S]_0$ is $<50\times K_M$, it may be necessary to retest more potent inhibitors at a lower concentration to obtain an accurate determination of K_I. The incubation of enzyme with fragment in the sample cell could reveal inhibitors that have slow binding kinetics which would be missed in the standard enthalpy array protocol above. Second, because the response time of a commercial ITC is approximately 10 s, the reaction time should be extended so that it is at least an order of magnitude greater than the instrument response time constant (Morin & Freire, 1991). This translates to the uninhibited control reaction lasting at least 100 s and preferably 3–4 min. If a large injection volume (>10 μL) is required, the reaction time should be extended so that it is at least an order of magnitude greater than the injection time. The data capture time is increased to 10–15 min when screening compounds.

The throughput of a microcalorimeter for screening a compound library for enzyme inhibitors is on the order of three reactions per hour (Table 1). It is advisable to run, at minimum, a control (no inhibitor) reaction at the start and end of each day and each time a new batch of enzyme is introduced. Our standard practice is to test each compound in duplicate, meaning that only ≈10 compounds per day can be tested. As shown in Table 1, each reaction performed in a microcalorimeter uses about 10 × the amount of enzyme and 100 × the amount of substrate as a reaction performed using calorimetry, although the concentration of each is lower in the microcalorimeter.

While achieving sufficient signal-to-noise is rarely a problem with microcalorimeters, power-compensation calorimeters have a differential power limit which may be exceeded when increasing the enzyme concentration so that the high starting concentration of substrate (e.g., $[S]_0>50\times K_M$) is converted to product in a reasonable amount of time. Therefore, one must balance the time per reaction with the maximum signal measurable by the instrument.

6.1 Required Materials

- MicroCal ITC_{200} (Malvern Instruments)
- Loading syringe
- Enzyme solution
- Substrate solution
- Fragment solution (200 m*M* in DMSO)
- 5% solution of Contrad 70 (for cell cleaning)
- ddH_2O

6.2 Cleaning Before and After Runs

Standard cleaning before a run involves filling the sample cell with 5% Contrad 70 and then flushing exhaustively with ddH_2O. At the conclusion of a run, use the filling syringe to remove the sample, fill the cell with 5% Contrad 70, then flush exhaustively with ddH_2O. Remove residual water prior to starting next run. If the sample cell is particularly dirty, this procedure can be run overnight as well, soaking the instrument in 5% Contrad 70 followed by flushing the instrument with water the next day. The titration syringe should also be cleaned thoroughly prior to the first run of the day, whenever the substrate solution is changed, and at the conclusion of the day's experiments.

6.3 Protocol 2: Microcalorimeter

1. Prepare 300 μL of enzyme solution in reaction buffer (either with 2% (v/v) DMSO or 2 m*M* fragment plus enough DMSO to bring final concentration to 2% (v/v)).
2. Prepare 100 μL of substrate solution in reaction buffer (with 2% (v/v) DMSO).
3. Rinse the cell with reaction buffer and empty completely.
4. Fill the sample cell with enzyme solution (≈200 μL) being careful not to introduce bubbles into the cell. Then load the syringe with the substrate solution (≈40 μL). Carefully lower the titration syringe into the top of the sample cell and prepare the instrument by setting the data collection parameters: reference power (12 μcal/s); initial delay (60 s); single injection, stirring speed (1000 RPM); 240–1200 s between injections. The time between injections can be optimized by taking the length of reaction observed in the absence of inhibitor and doubling it for reactions in the presence of a compound from the library.

6.4 Data Processing and Analysis

Although data analysis can be performed using the Origin software supplied by the manufacturer, the automated routines for enzyme assay analysis assume trace enzyme concentration relative to [I] and K_M. Therefore, we usually export the data as a text file and analyze it in MATLAB using Eqs. (2) and (3) as above.

7. SUMMARY

The calorimetric enzyme assay has several advantages over other enzyme assay techniques: it allows the use of natural substrates, it is unaffected by the spectroscopic properties of the sample, and it yields full enzyme kinetic parameters for reactions performed in the absence or presence of an inhibitor. These properties make calorimetry, in particular nanocalorimetry, a powerful technique to screen libraries of compounds to identify enzyme inhibitors.

ACKNOWLEDGMENTS

This work was supported by the grant R01EB009191 from the National Institute of Biomedical Imaging and Bioengineering, National Institutes of Health (NIH). Its contents are solely the responsibility of the authors and do not necessarily represent the official views of the NIH.

REFERENCES

Bemis, G. W., & Murcko, M. A. (1996). The properties of known drugs. 1. Molecular frameworks. *Journal of Medicinal Chemistry, 39*, 2887–2893.

Bianconi, M. L. (2003). Calorimetric determination of thermodynamic parameters of reaction reveals different enthalpic compensations of the yeast hexokinase isozymes. *The Journal of Biological Chemistry, 278*(21), 18709–18713.

Boettcher, A., Ruedisser, S., Erbel, P., Vinzenz, D., Schiering, N., Hassiepen, U., et al. (2010). Fragment-based screening by biochemical assays: Systematic feasibility studies with trypsin and MMP12. *Journal of Biomolecular Screening, 15*(9), 1029–1041.

Cai, L., Cao, A., & Lai, L. (2001). An isothermal titration calorimetric method to determine the kinetic parameters of enzyme catalytic reaction by employing the product inhibition as probe. *Analytical Biochemistry, 299*(1), 19–23.

Carr, R. A., Congreve, M., Murray, C. W., & Rees, D. C. (2005). Fragment-based lead discovery: Leads by design. *Drug Discovery Today, 10*(14), 987–992.

Cha, S. (1970). Kinetic behavior at high enzyme concentrations. Magnitude of errors of Michelis-Menten and other approximations. *The Journal of Biological Chemistry, 245*(18), 4814–4818.

Congreve, M., Carr, R., Murray, C., & Jhoti, H. (2003). A "rule of three" for fragment-based lead discovery? *Drug Discovery Today, 8*(19), 876–877.

De Bruyker, D., Recht, M. I., Bhagat, A. A., Torres, F. E., Bell, A. G., & Bruce, R. H. (2011). Rapid mixing of sub-microlitre drops by magnetic micro-stirring. *Lab on a Chip, 11*(19), 3313–3319.

De Bruyker, D., Wolkin, M. V., Recht, M. I., Torres, F. E., Bell, A. G., Anderson, G. B., et al. (2007). MEMS-based enthalpy arrays. In *Paper presented at the Transducers '07, Lyon & Eurosensors XXI: The 14th international conference on solid-state sensors, actuators, and microsystems*.

Erlanson, D. A., McDowell, R. S., & O'Brien, T. (2004). Fragment-based drug discovery. *Journal of Medicinal Chemistry, 47*(14), 3463–3482.

Fattori, D. (2004). Molecular recognition: The fragment approach in lead generation. *Drug Discovery Today, 9*(5), 229–238.

Fersht, A. (1999). *Structure and mechanism in protein science: A guide to enzyme catalysis and protein folding*. New York, NY: W.H. Freeman.

Firth, N. C., Brown, N., & Blagg, J. (2012). Plane of best fit: A novel method to characterize the three-dimensionality of molecules. *Journal of Chemical Information and Modeling, 52*(10), 2516–2525.

Freyer, M. W., & Lewis, E. A. (2008). Isothermal titration calorimetry: Experimental design, data analysis, and probing macromolecule/ligand binding and kinetic interactions. *Methods in Cell Biology*, *84*, 79–113.

Giannetti, A. M. (2011). From experimental design to validated hits a comprehensive walk-through of fragment lead identification using surface plasmon resonance. *Methods in Enzymology*, *493*, 169–218.

Godemann, R., Madden, J., Kramer, J., Smith, M., Fritz, U., Hesterkamp, T., et al. (2009). Fragment-based discovery of BACE1 inhibitors using functional assays. *Biochemistry*, *48*(45), 10743–10751.

Goldstein, A. (1944). The mechanism of enzyme-inhibitor-substrate reactions: Illustrated by the cholinesterase-physostigmine-acetylcholine system. *The Journal of General Physiology*, *27*(6), 529–580.

Hämäläinen, M. D., Zhukov, A., Ivarsson, M., Fex, T., Gottfries, J., Karlsson, R., et al. (2008). Label-free primary screening and affinity ranking of fragment libraries using parallel analysis of protein panels. *Journal of Biomolecular Screening*, *13*(3), 202–209.

Hann, M. M. (2011). Molecular obesity, potency and other addictions in drug discovery. *Medicinal Chemistry Communications*, *2*(5), 349–355.

Hann, M. M., Leach, A. R., & Harper, G. (2001). Molecular complexity and its impact on the probability of finding leads for drug discovery. *Journal of Chemical Information and Computer Sciences*, *41*(3), 856–864.

Heinrich, T., Seenisamy, J., Emmanuvel, L., Kulkarni, S. S., Bomke, J., Rohdich, F., et al. (2013). Fragment-based discovery of new highly substituted 1H-pyrrolo[2,3-b]- and 3H-imidazolo[4,5-b]-pyridines as focal adhesion kinase inhibitors. *Journal of Medicinal Chemistry*, *56*(3), 1160–1170.

Hopkins, A. L., Groom, C. R., & Alex, A. (2004). Ligand efficiency: A useful metric for lead selection. *Drug Discovery Today*, *9*(10), 430–431.

Jahnke, W., & Erlanson, D. A. (2006). *Fragment-based approaches in drug discovery*. Weinheim: Wiley-VCH. [Chichester: John Wiley, distributor].

Jahnke, W., Rondeau, J. M., Cotesta, S., Marzinzik, A., Pelle, X., Geiser, M., et al. (2010). Allosteric non-bisphosphonate FPPS inhibitors identified by fragment-based discovery. *Nature Chemical Biology*, *6*(9), 660–666.

Leach, A. R., Hann, M. M., Burrows, J. N., & Griffen, E. J. (2006). Fragment screening: An introduction. *Molecular BioSystems*, *2*(9), 430–446.

Morin, P. E., & Freire, E. (1991). Direct calorimetric analysis of the enzymatic activity of yeast cytochrome c oxidase. *Biochemistry*, *30*(34), 8494–8500.

Murray, C. W., Carr, M. G., Callaghan, O., Chessari, G., Congreve, M., Cowan, S., et al. (2010). Fragment-based drug discovery applied to hsp90. Discovery of two lead series with high ligand efficiency. *Journal of Medicinal Chemistry*, *53*(16), 5942–5955.

Nienaber, V. L., Richardson, P. L., Klighofer, V., Bouska, J. J., Giranda, V. L., & Greer, J. (2000). Discovering novel ligands for macromolecules using X-ray crystallographic screening. *Nature Biotechnology*, *18*(10), 1105–1108.

Nilakantan, R., Nunn, D. S., Greenblatt, L., Walker, G., Haraki, K., & Mobilio, D. (2006). A family of ring system-based structural fragments for use in structure-activity studies: Database mining and recursive partitioning. *Journal of Chemical Information and Modeling*, *46*(3), 1069–1077.

Olsen, S. N. (2006). Applications of isothermal titration calorimetry to measure enzyme kinetics and activity in complex solutions. *Thermochimica Acta*, *448*, 12–18.

Peters, W. B., Frasca, V., & Brown, R. K. (2009). Recent developments in isothermal titration calorimetry label free screening. *Combinatorial Chemistry & High Throughput Screening*, *12*(8), 772–790.

Recht, M. I., De Bruyker, D., Bell, A. G., Wolkin, M. V., Peeters, E., Anderson, G. B., et al. (2008). Enthalpy array analysis of enzymatic and binding reactions. *Analytical Biochemistry, 377*(1), 33–39.

Recht, M. I., Sridhar, V., Badger, J., Bounaud, P. Y., Logan, C., Chie-Leon, B., et al. (2014). Identification and optimization of PDE10A inhibitors using fragment-based screening by nanocalorimetry and X-ray crystallography. *Journal of Biomolecular Screening, 19*(4), 497–507.

Recht, M. I., Sridhar, V., Badger, J., Hernandez, L., Chie-Leon, B., Nienaber, V., et al. (2012). Fragment-based screening for inhibitors of PDE4A using enthalpy arrays and X-ray crystallography. *Journal of Biomolecular Screening, 17*(4), 469–480.

Recht, M. I., Torres, F. E., De Bruyker, D., Bell, A. G., Klumpp, M., & Bruce, R. H. (2009). Measurement of enzyme kinetics and inhibitor constants using enthalpy arrays. *Analytical Biochemistry, 388*, 204–212.

Saalau-Bethell, S. M., Woodhead, A. J., Chessari, G., Carr, M. G., Coyle, J., Graham, B., et al. (2012). Discovery of an allosteric mechanism for the regulation of HCV NS3 protein function. *Nature Chemical Biology, 8*(11), 920–925.

Shuker, S. B., Hajduk, P. J., Meadows, R. P., & Fesik, S. W. (1996). Discovering high-affinity ligands for proteins: SAR by NMR. *Science, 274*(5292), 1531–1534.

Thomson, J. A., & Ladbury, J. E. (2004). Isothermal titration calorimetry: A tutorial. In J. E. Ladbury & M. L. Doyle (Eds.), *Biocalorimetry 2: Applications of calorimetry in the biological sciences* (pp. 37–58). Chichester: Wiley.

Todd, M. J., & Gomez, J. (2001). Enzyme kinetics determined using calorimetry: A general assay for enzyme activity? *Analytical Biochemistry, 296*(2), 179–187.

Torres, F. E., Kuhn, P., De Bruyker, D., Bell, A. G., Wolkin, M. V., Peeters, E., et al. (2004). Enthalpy arrays. *Proceedings of the National Academy of Sciences of the United States of America, 101*(26), 9517–9522.

Torres, F. E., Recht, M. I., Bell, A. G., De Bruyker, D., Wolkin, M. V., Peeters, E., et al. (2006). Modeling the PARC nanocalorimeter using COMSOL. In *Paper presented at the COMSOL user's conference, Las Vegas, NV.*

Wenlock, M. C., Austin, R. P., Barton, P., Davis, A. M., & Leeson, P. D. (2003). A comparison of physiochemical property profiles of development and marketed oral drugs. *Journal of Medicinal Chemistry, 46*(7), 1250–1256.

Winiwarter, S., Ax, F., Lennernas, H., Hallberg, A., Pettersson, C., & Karlen, A. (2003). Hydrogen bonding descriptors in the prediction of human in vivo intestinal permeability. *Journal of Molecular Graphics & Modelling, 21*(4), 273–287.

Winiwarter, S., Bonham, N. M., Ax, F., Hallberg, A., Lennernas, H., & Karlen, A. (1998). Correlation of human jejunal permeability (in vivo) of drugs with experimentally and theoretically derived parameters. A multivariate data analysis approach. *Journal of Medicinal Chemistry, 41*(25), 4939–4949.

Wiseman, T., Williston, S., Brandts, J. F., & Lin, L. N. (1989). Rapid measurement of binding constants and heats of binding using a new titration calorimeter. *Analytical Biochemistry, 179*(1), 131–137.

CHAPTER FOUR

Measuring Multivalent Binding Interactions by Isothermal Titration Calorimetry

Tarun K. Dam*,†,1, Melanie L. Talaga*, Ni Fan*, Curtis F. Brewer‡,1

*Laboratory of Mechanistic Glycobiology, Department of Chemistry, Michigan Technological University, Houghton, Michigan, USA

†Biotechnology Research Center, Michigan Technological University, Houghton, Michigan, USA

‡Department of Molecular Pharmacology, Albert Einstein College of Medicine, Bronx, New York, USA

[1]Corresponding authors: e-mail address: tkdam@mtu.edu; fred.brewer@einstein.yu.edu

Contents

Abstract

Multivalent glycoconjugate–protein interactions are central to many important biological processes. Isothermal titration calorimetry (ITC) can potentially reveal the molecular and thermodynamic basis of such interactions. However, calorimetric investigation of multivalency is challenging. Binding of multivalent glycoconjugates to proteins (lectins)

Methods in Enzymology, Volume 567
ISSN 0076-6879
http://dx.doi.org/10.1016/bs.mie.2015.08.013

often leads to a stoichiometry-dependent precipitation process due to noncovalent cross-linking between the reactants. Precipitation during ITC titration severely affects the quality of the baseline as well as the signals. Hence, the resulting thermodynamic data are not dependable. We have made some modifications to address this problem and successfully studied multivalent glycoconjugate binding to lectins. We have also modified the Hill plot equation to analyze high quality ITC raw data obtained from multivalent binding. As described in this chapter, ITC-driven thermodynamic parameters and Hill plot analysis of ITC raw data can provide valuable information about the molecular mechanism of multivalent lectin–glycoconjugate interactions. The methods described herein revealed (i) the importance of functional valence of multivalent glycoconjugates, (ii) that favorable entropic effects contribute to the enhanced affinities associated with multivalent binding, (iii) that with the progression of lectin binding, the microscopic affinities of the glycan epitopes of a multivalent glycoconjugate decrease (negative cooperativity), (iv) that lectin binding to multivalent glycoconjugates, especially to mucins, involves internal diffusion jumps, (bind and jump) and (v) that scaffolds of glycoconjugates influence their entropy of binding.

1. INTRODUCTION

Glycan (carbohydrate)–protein interactions play important roles in a variety of biological processes including cellular recognition, adhesion, inflammation, cancer metastasis, growth, pathogen recognition, and immune defense (Lis & Sharon, 1998; Varki, 1993; Varki et al., 2009). The majority of the glycan-mediated interactions are multivalent in nature, where glycoconjugates with multiple glycan epitopes bind to their protein receptors or lectins. The valance of a lectin is often more than one. Multivalent binding may lead to the formation of lectin–glycan cross-linked complexes (Dam, Roy, Das, Oscarson, & Brewer, 2000). Multivalent binding enhances the intrinsic weak affinity associated with monovalent glycan–lectin interaction. However, the thermodynamic basis of this increased affinity is not properly understood. Isothermal titration calorimetry (ITC), a technique that directly determines thermodynamic-binding parameters, could be a useful tool for defining the thermodynamics of multivalent-binding interaction. In a single experiment, ITC measures the value of n (the number of binding sites of the protein), ΔH (the enthalpy of binding), and K_a (the association constant). From the K_a value, the free energy of binding (ΔG), and the entropy of binding (ΔS) can be calculated. However, it is difficult to study multivalent glycan–lectin interactions by ITC because cross-linking interactions between the reactants (lectins and their ligands) often lead to the formation of insoluble precipitates. Such precipitation during an ITC titration adversely affects the baseline as well as the overall integrity of the experiments. To overcome the challenges associated with multivalent binding during ITC experiments, we have made a few simple modifications in order to obtain unambiguous thermodynamic-binding data. We have successfully employed ITC to study binding interactions of lectins with different multivalent ligands, namely, synthetic dendrimers (Dam et al., 2000), globular glycoproteins (Dam et al., 2005; Talaga et al., 2014), and mucins (Dam et al., 2007). As described in this chapter, ITC is capable of providing more than just thermodynamic data. ITC-driven-binding data and their further analysis (Dam, Roy, Page, & Brewer, 2002a, 2002b) can provide important information regarding the molecular mechanism of multivalent lectin–glycan interactions.

2. DETERMINING THE THERMODYNAMICS OF LECTIN BINDING TO MULTIVALENT SYNTHETIC ANALOGS

In our initial studies, we selected mannose/glucose-specific lectin concanavalin A (ConA) and the seed lectin from *Dioclea grandiflora* (DGL) to investigate if it is feasible to employ ITC for studying multivalent lectin-carbohydrate interactions. Binding of monovalent ligands to ConA and DGL has been investigated extensively using ITC measurements (Chervenak & Toone, 1995, 1996; Dam, Oscarson, & Brewer, 1998; Mandal, Kishore, & Brewer, 1994). X-ray crystal structures of these two tetrameric lectins are known, and their binding sites are well defined. Therefore, we chose these two well-characterized lectins and studied their binding interactions with synthetic multivalent carbohydrates **1–3** (Fig. 1).

2.1 Methods

2.1.1 Isothermal Titration Calorimetry

ITC experiments were performed with MCS-ITC and VP-ITC instruments from Microcal, Inc., (Northampton, MA). Injections of 4 μL of a carbohydrate or glycoprotein solution were added from a computer-controlled microsyringe at an interval of 4 min into the sample solution of lectin (cell volume of 1.34 mL) with 350 rpm stirring. Control experiments were performed by making identical injections of respective glycoconjugate into a cell, containing buffer without lectin, showed insignificant heats of dilution. The concentrations of lectins were from 10 to 100 μ*M* and those of the glycans and glycoconjugates from 0.06 to 4.0 m*M*. The experimental data were

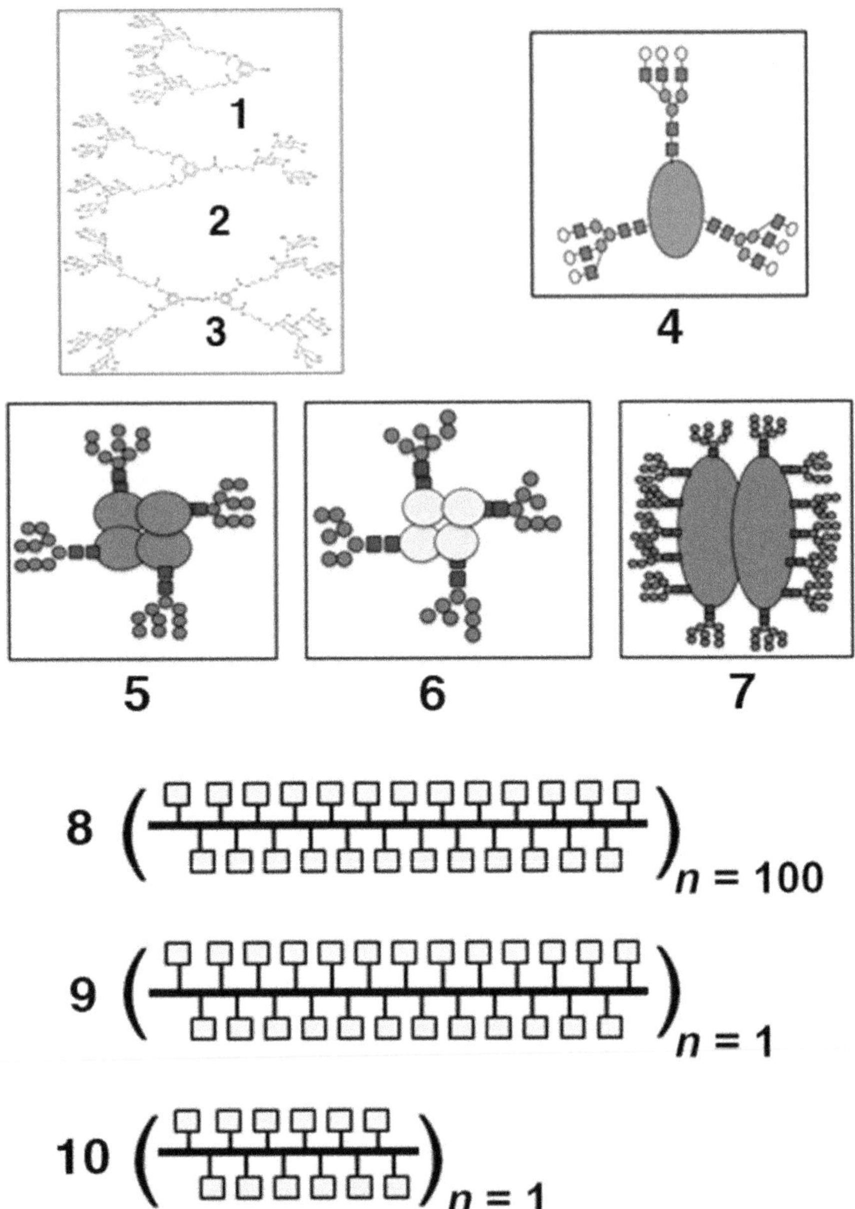

Figure 1 Multivalent glycoconjugates used in the present studies: synthetic multivalent analogs (**1–3**); globular glycoproteins including asialofetuin (ASF) (**4**), SBA (soybean agglutinin) (**5**), avidin (**6**), and invertase (**7**); porcine submaxillary mucins (PSM) including the 100-repeat 81-residue polypeptide O-glycosylation domain of PSM containing only peptide-linked-GalNAc residues (Tn-PSM) (**8**), the single 81-residue polypeptide O-glycosylation domain of PSM containing peptide-linked α-GalNAc residues (81-mer Tn-PSM) (**9**), and the 38/40-residue polypeptide(s) derived from the 81-residue polypeptide O-glycosylation domain of PSM containing peptide-linked α-GalNAc residues (38/40-mer Tn-PSM) (**10**). The number of glycan chains in **8** (Tn-PSM) is ~2300. The number of α-GalNAc residues in **9** (81-mer Tn-PSM) is ~23, while the number of α-GalNAc residues in **10** (38/40-mer Tn-PSM) is ~11–12. (See the color plate.)

fitted to a theoretical titration curve using software supplied by Microcal, with ΔH (enthalpy change in kJ/mol), K_a (association constant in M^{-1}) and n (number of binding sites per lectin monomer), as adjustable parameters. The parameter $c = K_a\, M_t(0)$, where $M_t(0)$ is the initial macromolecule concentration. This parameter is important in titration microcalorimetry (Wiseman, Williston, Brandts, & Lin, 1989). All experiments were performed with c values $1 < c < 200$. The instrument was calibrated using the calibration kit containing ribonuclease A and cytidine 29-monophosphate supplied by Microcal. Thermodynamic-binding parameters were calculated from the equation:

$$\Delta G = \Delta H - T\Delta S = -RT\, \ln K_a \tag{1}$$

where ΔG, ΔH, and ΔS are the changes in free energy, enthalpy, and entropy of binding. T is the absolute temperature and $R = 1.98$ cal/mol/K.

2.1.2 Determination of Functional Valence of Multivalent Ligands

In order to determine the exact number of participating epitopes of a multivalent ligand that are involved in lectin binding (functional valence), the following equation was used (Dam et al., 2000, 2007; Talaga et al., 2014):

$$\text{Functional valence} = 1/n, \tag{2}$$

where n is the n value directly determined by ITC.

It is important to determine the functional valence of a multivalent ligand because the structural valence (the actual number of epitopes) of a multivalent ligand may or may not be equal to its functional valence. ITC can directly and accurately measure functional valence if the experiments are done with sufficient care.

Note: At certain stoichiometric conditions, binding of multivalent ligands to ConA and DGL results in the formation of insoluble cross-linked complexes at pH 7.2. Insoluble complex formation (or precipitation) during ITC measurement affects the quality of the thermodynamic data. Such precipitation was prevented by doing one or more of the following: (i) using lectins and ligands in reduced concentrations, (ii) using a buffer with a low pH (pH 5.0), and (iii) using a buffer with low salt concentration. Under such conditions, ConA and DGL exist as a dimeric protein, and thus, insoluble complex formation was significantly prevented. We have also found that just by using a lower concentration of lectins (10–25 μM) and the ligands, multivalent interactions could be measured

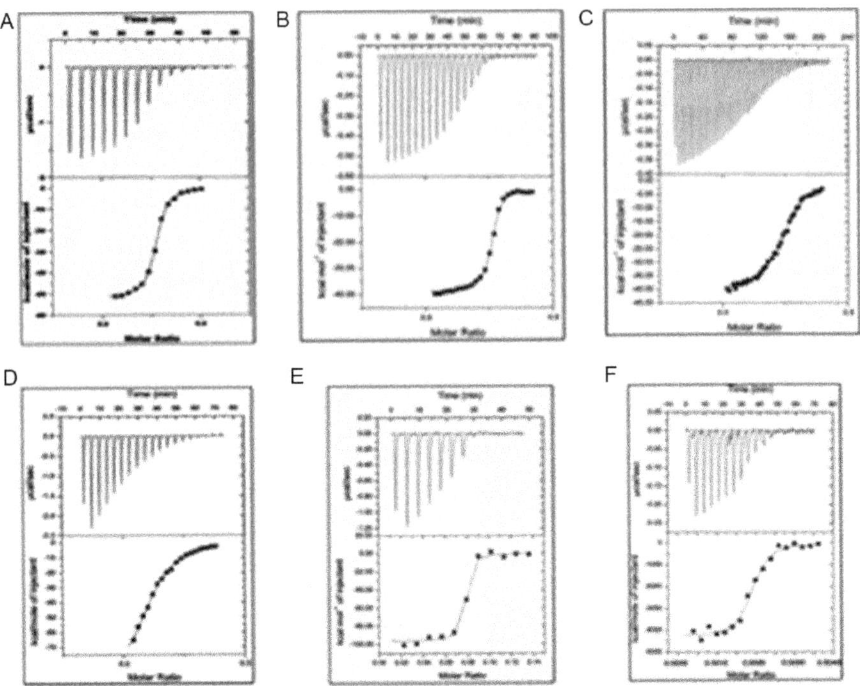

Figure 2 ITC profiles of ConA with **3** (**A**), ConA with SBA (**B**), ConA with avidin (**C**), galectin-3 with ASF (**D**), ConA with invertase (**E**), SBA with **8** (Tn-PSM) (**F**).

by ITC even at pH 7.2, if the binding affinity is high. Thus, the ITC experiments reported here were performed under conditions, where the formation of insoluble cross-linked complexes was arrested or significantly slowed. As a result, the quality of the ITC-driven thermodynamic data was not compromised. Representative ITC profiles are shown in Fig. 2.

It should be mentioned that precipitation formation is not common to all multivalent interactions. Precipitation occurs only when the reactants (ligands and receptors) noncovalently cross-link each other after binding. The extent of cross-linking is stoichiometry dependent. If the ratio of the reactants is kept off their stoichiometric ratio, the extent of cross-linking, and thus the amount of precipitation reduces significantly. It is often helpful to use an alternate technique (such as quantitative precipitation) to determine (i) if the reactants precipitate after they interact with each other and (ii) the stoichiometry. Such alternate experiments help design reliable ITC experiments of multivalent interactions.

Reliable thermodynamic data can only be obtained when the quality of an ITC profile is high and when the correct fitting program is chosen and exact concentrations of the reactants are used. An unperturbed baseline, symmetrical (minimum scattering) and sufficient data points (titrations should last a reasonable time before saturation) and maximum possible saturation are the hallmark of a high quality ITC profile.

Concentrations of the reactants (receptors and their ligands) must be determined with extreme care before using them for fitting the data. We used molar concentrations of multivalent carbohydrate analogs, not their epitope equivalence.

Incorrect reactant concentrations severely affect the quality of thermodynamic parameters determined by ITC. Therefore, extra attention should be paid when using partially soluble and partially active reactants. A partially active protein (receptor) could be titrated against a known monovalent ligand. If the protein is partially active, the *n* value will be fractional. The inactive protein molecules can be excluded through repeated chromatography till the *n* value of the protein with a known monovalent ligand becomes ~1.0. In our experiments, certain preparations of recombinant galectins showed fractional *n* values (0.6–0.7, indicating that 40–30% of galectin molecules were inactive in those preparations) with the monovalent ligand LacNAc. After rechromatography, the *n* values with LacNAc reached 1.0. We used those rechromatographed samples for ITC experiments.

Partially soluble or partially active ligands (especially synthetic ligands) should be checked carefully. If their solubility and activity cannot be enhanced and their concentrations are not determined correctly, they should be excluded from ITC experiments.

Finally, a reference titration (injecting the reactant into buffer) must be done with each ligand.

2.2 Data Analysis and Interpretation

2.2.1 *ITC can Reliably Measure the Enhanced Affinities of Multivalent Carbohydrates*

The trisaccharide trimannoside or TriMan [3,6-di-*O*-(α-D-mannopyranosyl)-α-D-mannopyranoside] is a high-affinity monovalent ligand of ConA and DGL. Synthetic analogs **1**, **2**, and **3** (in Fig. 1) possess two, three, and four terminal TriMan residues, respectively. The ITC data in Table 1 for analog **1** show 6-fold higher K_a values for ConA relative to monovalent TriMan. The same analog shows 5-fold greater K_a values for DGL (Table 1) relative to TriMan. Analogs **2** and **3** show 11- and

Table 1 Thermodynamic-Binding Data of ConA and DGL Obtained with Synthetic Multivalent Analogs

	K_a ($M^{-1} \times 10^{-4}$)	$-\Delta G$ (kJ/mol)	$-\Delta H$ (kJ/mol)	$-T\Delta S$ (kJ/mol)	*n* (sites/ monomer)
ConA					
TriMan	39	31.8	61.5	29.7	1.0
1	250	36.4	109.6	73.2	0.53
2	420	37.7	121.3	83.7	0.51
3	1350	40.6	221.8	181.2	0.26
DGL					
TriMan	122	34.7	67.8	33.1	1.0
1	590	38.5	115.1	76.6	0.51
2	1000	40.2	134.7	94.6	0.40
3	6500	44.4	245.6	201.3	0.25

ITC experiments were performed at 27 °C.
Errors in K_a range from 1% to 7%. Errors in ΔG are less than 1%. Errors in ΔH are 1–4%.
Errors in $T\Delta S$ are 1–7%. Errors in n are less than 2%.

35-fold higher K_a values for ConA, respectively, and 8- and 53-fold higher K_a values for DGL relative to the TriMan. These results are consistent with hemagglutination inhibition data.

2.2.2 Information Obtained from ITC-Driven n Values

ITC-driven *n* value could be used to determine (i) the functional valence (1/*n*) of a multivalent ligand and (ii) the nature of lectin–ligand cross-linking. The *n* value for **1** binding to ConA is 0.53 and for binding to DGL is 0.51, respectively. These values are close to the theoretical value of 0.50 for binding of a divalent ligand (1/0.5 = 2) to the two lectins. Therefore, the functional valence (1/*n*) of **1** is 2, which is equal to its structural valence. This indicates that **1** is primarily involved in divalent crossing-linking interactions with the lectins. *n* values for **3** binding to ConA and DGL are 0.26 and 0.25, respectively. This indicates that the functional valence of **3** is 1/0.25 = 4. Thus, all four TriMan moieties of **3** bind to ConA and DGL.

2.2.3 Structural Valence of a Multivalent Ligand May Differ from Its Functional Valence

The above data show that the values of *n* for binding of **1** and **3** to ConA and DGL are inversely proportional to the number of binding epitopes in the analogs. However, the value of *n* for the binding of **2** to ConA is 0.5 instead of the predicted value of 0.33 based on the structural valence of triantennary analog which possesses three TriMan residues. Thus, **2** is functionally bivalent in binding to ConA as indicated by its ITC-derived value of *n*. On the other hand, the value for **2** binding to DGL is 0.40 which is less than 0.50 for bivalent binding but higher than 0.33 for trivalent binding. The *n* value of 0.40 suggests that greater than half of the molecules of **2** are involved in trivalent binding to DGL and less than half are involved in bivalent binding to the lectin. Thus, the ITC results show that analog **2** is bivalent in binding to ConA, and bivalent and trivalent in binding to DGL. This difference in functional valence of **2** for ConA and DGL is also interesting in light of the similarity in overall structures of the two lectins (Rozwarski, Swami, Brewer, & Sacchettini, 1998). The results for **2** indicate the importance of ITC measurements in determining the functional valence of a multivalent carbohydrate for specific lectins, which may differ from the structural valence of the carbohydrate. As mentioned in the subsequent sections, differences between structural and functional valence were observed in multivalent glycoproteins, especially in mucins.

The value of *n* and thus the functional valence is sensitive to the reactant concentrations. Therefore, it is mandatory to check the purity, activity, and proper concentrations of the reactants before doing ITC experiments (described in Section 2.1.2). Otherwise, the value of functional valence will be flawed. Functional valence could be confirmed by quantitative precipitation assays.

2.2.4 Macroscopic-Binding Enthalpy of Multivalent Carbohydrate Analogs Is the sum of the Microscopic Enthalpy of the Epitopes

Calorimetric data show that for higher affinity multivalent analogs, the observed value of ΔH per mole of the analog is approximately the sum of the ΔH values of the individual epitopes. Similar observations have been made for the binding of a trivalent system of receptor and ligand derived from vancomycin and D-Ala-D-Ala (Rao, Lahiri, Isaacs, Weis, & Whitesides, 1998). **1** has ΔH value of −109.6 kJ/mol for binding to ConA which is a little less than twice the ΔH value of −61.5 kJ/mol for TriMan (Table 1). The same is true for **1** binding to DGL (Table 1). The ΔH value for tetravalent analog **3** binding to ConA is −221.8 kJ/mol, which is approximately four times as great as the ΔH value of the free TriMan (−61.5 kJ/mol). The ΔH value for **3** binding to DGL is −245.6 kJ/mol, which is also nearly four times as great as the ΔH value of the free TriMan (−67.8 kJ/mol). The ΔH values of analog **2** are more complicated since the analog appears to be close to bivalent for ConA ($n=0.51$), and partially trivalent for DGL ($n=0.40$).

2.2.5 Macroscopic TΔS of High-Affinity Carbohydrate Analogs Is More Than the Sum of Microscopic TΔS

The largest increases in K_a values are for tetravalent analog **3** binding to ConA (35-fold) and DGL (53-fold) (Table 1). However, these enhancements are small compared to the very large increases in affinity that occur when a multivalent ligand binds to a single receptor molecule possessing multiple binding sites. For example, the affinity of a trivalent derivative of vancomycin for a trivalent derivative of D-Ala-D-Ala is $\sim 10^{17}/M$, whereas the affinity of the corresponding monovalent analogs is $\sim 10^{6}/M$ (Rao et al., 1998). In the latter study, thermodynamic measurements showed that both ΔH and $T\Delta S$ scaled proportionally to the number of binding epitopes in the molecules. These thermodynamic findings are characteristics of the binding of a multivalent ligand to a single multivalent receptor molecule. However, in the present study, ΔH scales proportionally to the number of binding epitopes in the higher affinity multivalent carbohydrates in Table 1, but $T\Delta S$ does not. Instead, $T\Delta S$ is much more negative than if it proportionally scaled to the number of epitopes in the carbohydrates. For example, **3** contains four trimannosyl-binding epitopes, and to a first approximation, the ΔH value for ConA of −221.8 kJ/mol is four times the ΔH of −61.5 kJ/mol for TriMan (Table 1). However, the observed $T\Delta S$ value for **3** is −181.2 kJ/mol, not −118.8 kJ/mol as it would be if it scaled with the $T\Delta S$ value of −29.7 kJ/mol for TriMan (Table 1). The resulting ΔG value of **3** would also be much greater if $T\Delta S$ scaled with valence since the difference between ΔH and $T\Delta S$ would be greater. However, the observed ΔG value(s) for **3** are much smaller. The same is true for the other multivalent carbohydrates in Table 1. Hence, the finding in the present study that ΔH values scale but not $T\Delta S$ values is characteristic of the binding of a multivalent ligand to separate and unconnected receptor molecules. In the present case, binding of a single molecule of **3** to four separate ConA or DGL molecules.

2.2.6 Favorable TΔS Contributes to the Enhanced Affinities of Multivalent Carbohydrates

The enhancements in affinity of **3** for ConA (35-fold) and DGL (53-fold) relative to TriMan are not due to **3** binding to single molecules of ConA or DGL that possesses four binding sites. Rather, the increases in affinity of **3** are due to binding of four separate lectin molecules to **3**. This means that the observed (macroscopic) K_a value for **3** is actually the average of the four microscopic K_a values at each of its four epitopes, since each epitope is involved in binding to a separate lectin molecule. It therefore follows that if ΔH is constant at each epitope, as determined above, and is approximately the same as that of TriMan, then increases in the overall microscopic K_a values at the four epitopes require more favorable $T\Delta S$ contributions than that of TriMan.

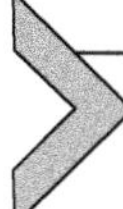

3. DETERMINATION THAT THE EPITOPES OF MULTIVALENT ANALOGS POSSESS A GRADIENT OF DECREASING MICROSCOPIC-BINDING AFFINITIES (K_A)

The fact that analogues **1–3** must have microscopic K_a values associated with their multiple epitopes leads to a prediction regarding their binding interactions with lectins such as ConA and DGL. Sequential binding of each epitope in analogues **1–3** results in diminished valence for each molecule; therefore, binding of the analogs should occur with decreasing microscopic affinity (negative cooperativity) as their effective valence decreases. This expected decrease in functional valence of the multivalent analogues and concomitant decrease of microscopic affinities (negative-binding cooperativity) can be verified with Hill plot analysis (Di Cera, 1995; Stryer, 1998) of the ITC raw data.

The Hill plot has the advantage of assigning numerical values to the degree of cooperativity of the system. The Hill plot also has the advantage of being a logarithmic representation which allows plotting of all obtainable data, unlike double-reciprocal or half-reciprocal plots that often have open upper limits on the abscissa and ordinate (Dam et al., 2002a).

3.1 Methods

3.1.1 Hill Plot Analysis of ITC Data

The total concentration of ligand $X_t(i)$ as well as lectin $M_t(i)$ after the ith injection and the heat evolved on the ith injection $Q(i)$ are readily available from the ITC raw data file after each experiment. The concentration correction is automatically done by the Origin software. The concentration of bound ligand $X_b(i)$ after the ith injection is

$$X_b(i) = [Q(i)/(\Delta H V_o)] + X_b(i-1) \tag{3}$$

where $Q(i)$ (μcal) is the heat evolved on ith injection, ΔH(cal/mol) is the enthalpy change, V_o (mL) is the active cell volume, and X_b (mM) is the concentration of bound ligand. X_b is equal to M_b, the concentration of bound protein, and in the present study of multivalent ligands, the more general expression is $M_b = (X_b)$ (functional valence of ligand). The concentration of free ligand (X_f) after the *i*th injection was determined as

$$X_f(i) = X_t(i) - X_b(i) \quad (4)$$

Hill plots were constructed by plotting log[$Y(i)/[1 - Y(i)]$] versus log [$X_f(i)$], where $Y(i)$ is [$X_b(i)$](functional valency of ligand)/$M_t(i)$, which are modified versions of the Hill plots (Di Cera, 1995) that take into account the functional valence of the ligand. The functional valence of **1**, **2**, and **3** binding to ConA and DGL was obtained from our previous study (Table 1; Dam et al., 2000). A program was created using Microsoft Excel for construction of Hill plots. Work sheet data of total ligand as well as total lectin concentration and the amount of heat evolved were copied from the ITC raw data files and pasted on the appropriate columns of the program. After calculation, the program shows the profiles of Hill plots. Delta Graph was then used for further analysis of the plots. The validity of the information obtained from the Hill plot [log[$Y/(1 - Y)$] versus log(X_f)] was tested by directly fitting the binding data of monovalent TriMan to the Hill equation. The Hill slope was found to be the same by direct fitting or plotting the Hill equation data. Similar attempts at direct fitting of the ITC data for the multivalent analogues failed since the Hill *n* values change throughout the binding process.

Note: Hill plot analysis of ITC raw should only be done when the quality of the ITC run is high (described in Section 2.1.2). Multiplication of (X_b/M_t) with the correct functional valence of the ligand is mandatory to obtain an unambiguous Hill plot. Otherwise, the Hill plot will not dispose around the zero point on the ordinate. Therefore, Hill plot analysis of ITC data should not be attempted without determining the correct functional valence. As mentioned before (in Section 2.2.3), functional valence could be double checked by quantitative precipitation assays.

Curved Hill plots are indeed difficult to interpret. We fitted progressive three data points, beginning with the first injection point (1, 2, 3; then 2, 3, 4; then 3, 4, 5; etc.) at a time and determined the slope values. As stated below (in Section 4), "reverse ITC" data validated our interpretation of curved Hill plots. Therefore, combining "reverse ITC" with Hill plot analysis (whenever possible) would be an appropriate approach.

3.2 Data Analysis and Interpretation

3.2.1 Microscopic Affinities of the Epitopes of Multivalent Analogs Decrease with the Progression of Lectin Binding

The Hill plot of ITC data for TriMan binding to ConA is shown in Fig. 3A. The plot is essentially a straight line with a slope of 0.94, which is close to a

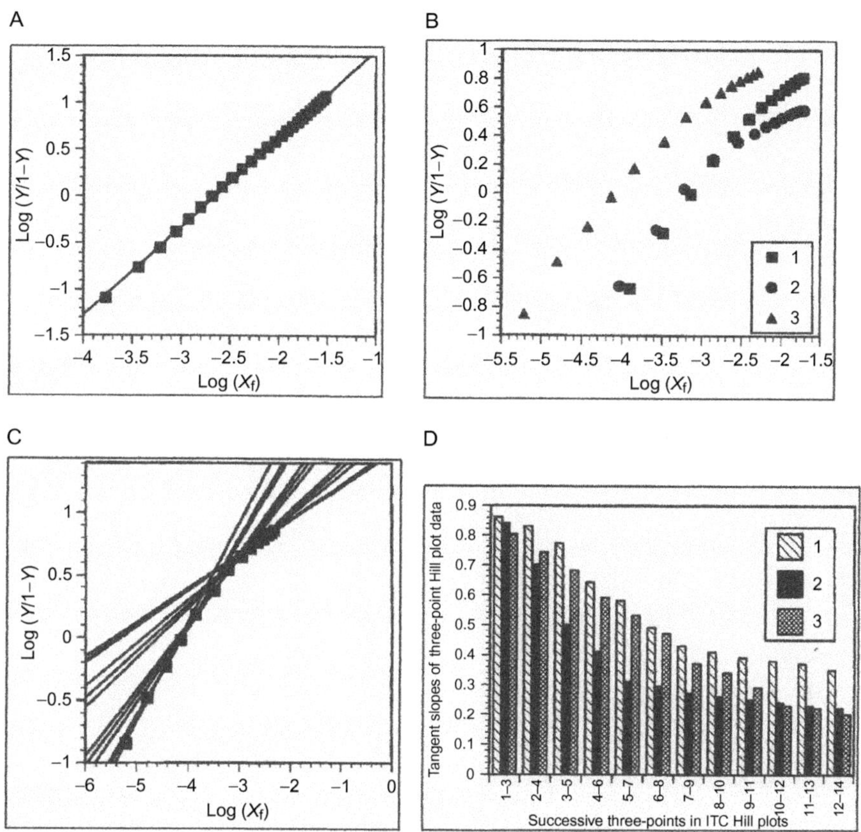

Figure 3 (A) Hill plot of the ITC data of TriMan (800 μ*M*) binding to ConA (20 μ*M*). In the plot, $Y=(X_b)$ (functional valence of sugar)/M_t, where X_b is the concentration of bound ligand, M_t is the concentration of total lectin, and X_f is the concentration of free sugar. The functional valence of TriMan is 1 (Table 1). (B) Hill plots of the ITC data for analogues **1–3** binding to ConA (20 μ*M*). The concentration of **1** was 650 μ*M*, **2** was 670 μ*M*, and **3** was 240 μ*M*. The definition of *Y* is given in the text. The functional valence of **1** and **2** is 2, and the functional valence of **3** is 4 (Table 1). (C) Tangent slopes of progressive three-point intervals of the Hill plot for analogue **3** with ConA. (D) Bar graphs of the three-point tangent slopes of the ITC data Hill plots of analogues **1–3** binding to ConA. *Reprinted from Dam et al. (2002a) with permission from the American Chemical Society.*

value of 1.0 for noncooperative-binding interactions (Di Cera, 1995; Stryer, 1998). A similar plot with a slope of 0.95 is observed for DGL (not shown). Importantly, the absence of allosteric interactions in ConA and DGL upon binding TriMan allows application of Hill plot analysis to the ITC-binding data for analogues **1–3**. This can be done because the incremental heats measured upon sugar addition and binding are proportional to the number of moles of ligand bound and not due to allosteric transitions in the lectins. These general conditions are required for Hill plot analysis of ITC data (Indyk & Fisher, 1998). Therefore, Hill plots of the ITC-binding data for **1–3** to both lectins were examined. Hill plots of the ITC data for analogues **1–3** binding to ConA are shown in Fig. 3B. All three plots are curvilinear rather than linear. The plots are also disposed around the zero point on the ordinate, as observed for monovalent TriMan (Fig. 3A), only after correction for the functional valence of the multivalent sugars. This provides confirmation of the ITC-derived functional valence of **1**, **2**, and **3** for ConA in our previous study (Dam et al., 2000; Table 1). The tangent slopes of progressive three-point intervals of the *x*-axis of the Hill plot for **3** are shown in Fig. 3C, which shows decreasing tangent slopes along the binding curve. Figure 3D shows a bar graph comparison of the three-point tangent slopes of all three analogues binding to ConA. All three analogues possess initial tangent slope values between 0.8 and 0.9. The final tangent slope value of **1** is 0.35, while the final values for **2** and **3** are 0.24 and 0.22, respectively. These results indicate increasing negative cooperativity (decreasing microscopic-binding affinity) in the binding of ConA to analogs **1–3**. Similar results were obtained when Hill plots for **1–3** binding to DGL were analyzed. Importantly, the increasing negative-binding cooperativity of observed during the binding of lectins to **1–3** is due to the multivalency of **1–3** and not the lectins since TriMan shows no cooperativity effects.

The physical basis for the increasing negative-binding cooperativity of **1–3** can be understood, in part, by the reduction in functional valence of the analogues as they bind to an increasing number of lectin molecules. For example, Fig. 4 shows the various microequilibria constants for **3** as it sequentially binds one, two, three, and four molecules of ConA (or DGL). The functional valence of unbound **3** (species A) is tetravalent, the functional valence of **3** with one-bound lectin molecule (species B) is trivalent, the functional valence of **3** with two-bound lectin molecules (species C) is divalent, and the functional valence of **3** with three-bound lectin molecules (species D) is monovalent. The increasingly curvilinear Hill plots in Fig. 3 for **3** are consistent with the decreasing valence and increasing

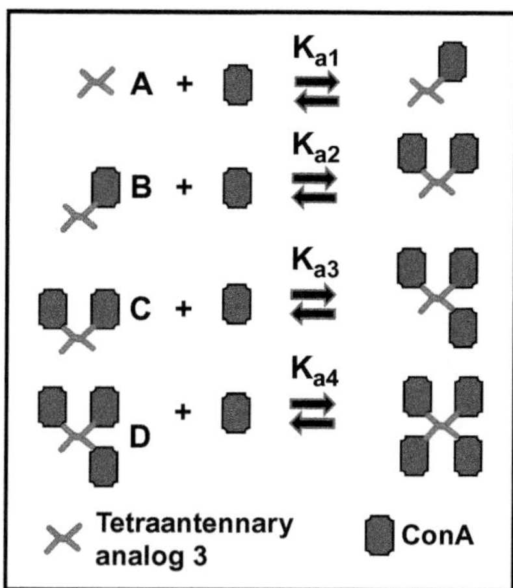

Figure 4 Schematic representation of ConA binding to analog **3**. Four microequilibrium constants of the tetravalent analog **3** are represented by K_{a1}, K_{a2}, K_{a3}, and K_{a4} for binding of ConA to the first epitope of **3** (species A), to the second epitope of **3** (species B), to the third epitope of **3** (species C), and to the fourth epitope of **3** (species D) to give the fully bound complex.

negative-binding cooperativity of **3** with increasing sequential occupancy of the four epitopes of the analog. The same is true for analogs **1** and **2** with ConA and DGL. Another physical factor which may play a role in the curvilinear Hill plots of **1–3** with ConA and DGL is the formation of noncovalent cross-linked complexes between lectin molecules and multivalent carbohydrates (Dam et al., 2000). Such cross-linking interactions along with the decreasing functional valence of **1–3** upon binding sequential lectin molecules play a role in the curvilinear Hill plots of the analogs.

3.2.2 Microscopic K_a Values and Kinetic Effects of Multivalency

The observed (macroscopic) K_a values for multivalent analogues **1–3** in Table 1 are the average of the microscopic-binding-free energy terms ($-\Delta G$) of the different epitopes of the analogs. For example, analogue **3** possesses four TriMan epitopes, and ITC data in Table 1 indicate that all four arms of **3** are involved in binding to separate molecules of ConA and DGL, respectively. As shown in Fig. 4, the four microequilibrium constants of **3** can be represented by K_{a1}, K_{a2}, K_{a3}, and K_{a4} for binding of a ConA or DGL molecule to the first arm of **3** (species **A**), to the second arm of **3** (species **B**), etc. Therefore, the observed macroscopic ΔG values of **3** (ΔG_{obs}) for ConA and DGL in Table 1 are the averages of the four microscopic ΔG.

$$\Delta G_{obs}(3) = (\Delta G_1 + \Delta G_2 + \Delta G_3 + \Delta G_4)/4 \qquad (5)$$

The relative values of ΔG_1, ΔG_2, ΔG_3, and ΔG_4 must decrease, based on the decreasing ΔG values (Table 1) of **2**, **1**, and TriMan, respectively, which have the same valence as species **B**, **C**, and **D** of **3**, respectively. Thus, it is expected that $K_{a1} > K_{a2} > K_{a3} > K_{a4}$ for **3** binding to ConA and DGL as shown in Fig. 4.

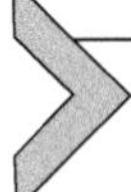

4. DIRECT CALORIMETRIC MEASUREMENTS OF MICROSCOPIC THERMODYNAMIC BINDING PARAMETERS BY REVERSE ITC

Binding thermodynamics of the epitopes of multivalent analogs was directly measured by "reverse" ITC. In this modified approach, the lectin (which is placed in the ITC cell during a "normal" ITC) was placed in the syringe. The multivalent analogs (which are placed in the ITC syringe during a "normal" ITC) were place in the ITC cell. This was done to study the microscopic-binding thermodynamics of individual epitopes of a multivalent analog.

4.1 Methods

4.1.1 Reverse ITC

"Reverse" ITC experiments were performed with ConA titrated into solutions of **2** and **3**. The data were fitted using Origin version 5.0 software from Microcal. The number of binding epitopes on the carbohydrates was selected from the "normal" ITC data previously obtained in Table 1. Importantly, selection of other values for the number of epitopes failed to provide a fit of the data. As an important control, ConA was titrated into a solution of monovalent TriMan. The data were fitted using a two-site model since data in Table 1 from the normal ITC experiment demonstrated two binding sites for analogs **1** and **2**.

Note: To obtain reliable data from reverse ITC, one must (i) have a good control run with a monovalent analog, (ii) know right functional valence from "normal" ITC to choose the correct fitting model, and (iii) take extra precaution when using more than two-site binding model to fit the data. Reverse ITC with tetravalent **3** and ConA produced high quality ITC profiles. However, a four-site binding model could not produce unambiguous thermodynamic data.

4.1.1.1 Why and When Use "Reverse ITC"?

While regular ITC determines macroscopic-binding thermodynamics, "reverse ITC" measures the microscopic-binding parameters of individual epitopes of a multivalent ligand. "Reverse ITC" could reveal and compare the K_a (thus the ΔG), ΔH and $T\Delta S$ values of individual epitopes of multivalent ligands. From this comparison, one can conclude whether the epitopes show a negative cooperativity (decreasing affinities), positive cooperativity (increasing affinities), or no cooperativity (similar affinities). "Reverse ITC" also discloses any change in the relative contributions of enthalpy and entropy from epitope to epitope. When possible, "reverse ITC" could be used as a complementary approach of Hill plot analysis. "Reverse ITC" is suitable for any system, where the correct functional valence of the multivalent ligand is known, and the macroscopic affinity is in the lower micromolar range.

4.2 Data Analysis and Interpretation

4.2.1 Microscopic ΔG *and* TΔS *Values of the Epitopes Are Dissimilar*

The reverse ITC measurements show two microscopic K_a values for **1**: K_{a1} is $1.6 \times 10^7/M$ and K_{a2} is $8.8 \times 10^5/M$. Hence, the microscopic affinity constant of the first epitope of **1** is 18 times greater than that of the second epitope. The latter value, in turn, is close to the affinity constant of TriMan (Table 1). These findings indicate that binding of the first molecule of ConA to **1** occurs with much higher affinity than binding of the second molecule of ConA. These results are consistent with the increasing negative cooperativity of **1** binding to ConA and DGL. Importantly, Table 1 shows that the observed macroscopic ΔG value for **1** binding to ConA in the normal ITC experiment is −36.4 kJ/mol. Reverse ITC data show that ΔG_1 is −41.0 kJ/mol and ΔG_2 is −33.1 kJ/mol for the two epitopes of **1**. The average value of ΔG_1 and ΔG_2 is −37.7 kJ/mol, which is very similar to −36.4 kJ/mol observed in Table 1. Thus, the observed macroscopic K_a for **1** in Table 1 agrees with the average of the two microscopic K_a values for **1** obtained by reverse ITC. This is an important result confirming the validity of the two microscopic K_a values determined in the reverse ITC measurement (Dam et al., 2002b).

The microscopic enthalpies of binding ΔH_1 is −52.3 kJ/mol and ΔH_2 is −51.5 kJ/mol. This indicates that the enthalpies of binding of the two epitopes of **1** are essentially the same (Dam et al., 2002b).

These values are very similar to that for TriMan (−54.8 kJ/mol) obtained from reverse ITC run (Dam et al., 2002b), indicating constant ΔH values for the trimannosyl moieties of **1** in binding to ConA. These findings agree with our previous results using normal ITC measurements (Table 1). Knowing the microscopic ΔG_1 and ΔG_2 values for the two epitopes of **1** along with the microscopic ΔH_1 and ΔH_2 values allows calculation of the corresponding microscopic entropy of binding ($T\Delta S$) values of the two epitopes. The calculated $T\Delta S_1$ value is -11.3 kJ/mol and $T\Delta S_2$ is -17.6 kJ/mol. Hence, there is a 6.3 kJ/mol more favorable entropy of binding for the first epitope of **1** relative to the second epitope.

As described above, the functional valence of **2** with ConA was 2 rather than 3 (Table 1). Hence, **2** is functionally bivalent with ConA as is **1**. Importantly, the reverse ITC data of ConA with **2** could only be fit with a two binding site model ($n=2$ per molecule of **2**) but not a three-site model (Dam et al., 2002b). The calculated $T\Delta S_1$ value of **2** is -12.1 kJ/mol and $T\Delta S_2$ is -17.2 kJ/mol. Hence, there is a 5.0 kJ/mol more favorable entropy of binding value for the first epitope of **2** relative to the second epitope of **2**. These results provide a direct demonstration of (i) the favorable entropy effect for binding of the first epitope of a functionally bivalent carbohydrate to a lectin and (ii) decreasing microscopic-binding affinity (negative cooperativity) with the progression of binding.

5. ITC STUDIES OF LECTIN BINDING TO MULTIVALENT GLYCOPROTEINS

As stated below, calorimetric investigation provided important information about lectin–glycoprotein interactions.

5.1 Methods

5.1.1 ITC Titration and Hill Plot Analysis

Binding of multivalent glycoprotein asialofetuin (ASF) (**4** in Fig. 1) to galectins was studies by ITC. Subsequently, ITC raw data were subjected to Hill plot analysis. Binding of ConA to multivalent glycoproteins [soybean agglutinin (SBA) (**5** in Fig. 1), avidin (**6** in Fig. 1), and invertase (**7** in Fig. 1)] was also studied by ITC. Details of these techniques are already described above.

5.2 Data Analysis and Interpretation

5.2.1 Glycan Epitopes of the Multivalent Glycoprotein ASF Possess a Gradient of Microscopic-Binding Affinities for Galectins

The Hill plots of ITC data for LacNAc binding to galectin-1, -2, -3, -4, -5, and -7 as well as truncated galectin-3 and -5 are essentially straight lines with slopes close to 0.93. These slope values are close to a value of 1.0 for noncooperative-binding interactions (Di Cera, 1995; Stryer, 1998), indicating that these galectins do not undergo cooperative-binding interactions with this monovalent sugar. The Hill plot of the ITC data for binding of ASF to galectin-3 is curvilinear rather than linear and disposed around the zero point on the ordinate, as observed for monovalent LacNAc (Dam et al., 2005), after correction for the functional valence of ASF. This provides confirmation of the ITC-derived functional valence of ASF for galectin-3. The tangent slopes values of progressive three-point intervals of the x-axis of the Hill plot for ASF decrease along the binding curve. The initial tangent slope value is close to 1.0; the final tangent slope value is approximately 0.15. These results show increasing negative cooperativity with increasing binding of ASF to galectin-3 (Dam et al., 2005). Similar results were obtained with other galectins tested (Dam et al., 2005).

5.2.2 Physical Basis for the Negative Cooperativity in Binding of Galectins to ASF

The physical basis of the negative-binding cooperativity of ASF to the galectins appears to be similar to that observed for the binding of ConA and DGL to di-,tri-, and tetraantennary carbohydrates as described above. Namely, the increasing negative cooperativity observed in the present study is due to the reduction in the functional valence of ASF as it binds an increasing number of galectin molecules.

5.2.3 Range of Microscopic K_a Values for Binding of ASF to the Galectins

Equation 6 can be written for binding of ASF to the galectins, in which there are nine microscopic ΔG values representing the nine microequilibria (for galectin binding to nine LacNAc epitopes).

$$\Delta G(\text{obs}) = (\Delta G1 + \Delta G2 + \cdots + \Delta G_9)/9 \tag{6}$$

The nine microequilibrium constants of ASF are represented by K_{a1}, K_{a2}, ..., and for binding of the first LacNAc epitope of ASF, binding of the second epitope of ASF, and binding of the last unbound LacNAc epitope of ASF to a galectin molecule, respectively. Hence, the observed ΔG values [ΔG(obs)] of ASF for the galectins are the average of the nine microscopic ΔG terms as shown in Eq. 6 for each galectin (Dam et al., 2005). The relative values of $\Delta G1$, $\Delta G2$, ..., and $\Delta G9$ decrease on the basis of the decreasing microscopic ΔG values of the epitopes of multivalent analogs with decreasing functional valence, as observed for multivalent carbohydrates binding to ConA and DGL (described above in sections 3 and 4). In kinetic terms, the microscopic off-rate (k^{-1}) for $K_{a1}(K_{a1}=k^1/k^{-1})$ would be expected to be slower than the microscopic off-rate for K_{a2}, etc., due to binding and recapture of the first bound galectin molecule by the remaining unbound LacNAc residues of ASF before full dissociation of the complex. An estimation showed that the microscopic affinity of the first of the nine LacNAc epitope of ASF was ~10 n*M*, which is ~6000-fold greater than the microscopic affinity of the last (ninth) LacNAc epitope of ASF (Dam et al., 2005).

In a recent study (Talaga et al., 2014), we have shown further proof of inequality between structural and functional valence as well as the evidence that when a lectin binds to a multivalent glycoprotein, the microscopic affinities of its glycan epitopes decrease (negative cooperativity). Invertase (**7** in Fig. 1) is a high-mannose glycan-containing glycoprotein with 14 accessible high-mannose glycans, whereas SBA (**5** in Fig. 1) possesses four high-mannose glycans. Thermodynamic-binding data and precipitation studies show that, compared to SBA, invertase has a higher valence for ConA. The functional valence of SBA for ConA is 4, whereas the functional valence of invertase for ConA is 12.5 (which is different from its structural valence 14). However, the affinities of invertase and SBA for ConA are comparable (similar ΔG values) (Talaga et al., 2014). The glycan epitope number on invertase is higher than that of SBA, but that does not translate into a higher macroscopic affinity primarily due to the decreasing microscopic affinities possessed by the glycan epitopes of invertase. Such negative cooperativity effect limits the enhancement of macroscopic-binding affinity.

5.2.4 ITC Can Determine the Thermodynamic Contributions of Scaffolds to Lectin–Glycan Interactions

In the same study (Talaga et al., 2014), we have shown that ITC can be used to investigate the role of glycoconjugate scaffolds in multivalent lectin–glycan interactions. For example, analogue **3** and SBA have similar valence and ΔG values for ConA, but their $T\Delta S$ values are significantly different (−185.8 kJ/mol for analogue **3** and −118.8 kJ/mol for SBA). While analogue **3** contains a synthetic scaffold, SBA possesses a protein scaffold. Therefore, the difference in the entropic values could be attributed to the structural differences of the scaffolds. These data suggest that the structures of the ligand scaffolds influence their lectin-binding thermodynamics. Data obtained with ConA and another tetravalent glycoprotein avidin (**6** in Fig. 1) further support this view (Talaga et al., 2014).

6. ITC STUDIES OF LECTIN BINDING TO POLYVALENT MUCINS

Mucins are heavily *O*-glycosylated glycoproteins that are secreted by higher organisms to protect and lubricate epithelial cell surfaces from biological, chemical, and mechanical insults. Mucin and mucin-type molecules exist as soluble and membrane-attached molecules that are involved in modulating immune response, inflammation, adhesion, and tumorigenesis (Byrd & Bresalier, 2004; Fukuda, 2002; Varki, 1993). The receptor glycoproteins for a number of endogenous lectins including selectins (Varki, 1997), siglecs (Varki & Angata, 2006), and galectins (Pace, Lee, Stewart, & Baum, 1999) possess mucin domains with a large number of glycan epitopes (Shogren, Gerken, & Jentoft, 1989). Despite their functional importance, the mechanism of binding of multivalent linear mucin glycoproteins to lectins is poorly understood. ITC has been applied to narrow that knowledge gap.

6.1 Methods

6.1.1 Isothermal Titration Calorimetry

Binding of lectins [soybean agglutinin (SBA) and *Vatairea macrocarpa* lectin (VML)] to porcine submaxillary mucins (PSM) (**8–10** in Fig. 1) was studied by ITC as described above. Hemagglutination inhibition assay was used to determine binding affinities of mucins before ITC measurements. Intact mucins (such as Tn-PSM, **8** in Fig. 1) with subnanomolar affinity and higher valence were used in very low concentration to avoid precipitation during ITC titration (Dam et al., 2007).

6.2 Data Analysis and Interpretation

6.2.1 Structural Valence of Mucins Is Different from Its Functional Valence

The $1/n$ value (functional valence) for Tn-PSM (**8**) binding of SBA is 540, which indicates that 540 α-GalNAc residues out of the ~2300 (structural valence) α-GalNAc residues of Tn-PSM bind to 540 monomers of SBA because each monomer of SBA binds one GalNAcα1-*O*-Ser molecule (Table 2) (for VML, the functional valence ($1/n$) of **8** is 833).

Table 2 Thermodynamic-Binding Data of SBA and VML Obtained with Porcine Submaxillary Mucins (PSM) of Different Length

	K_d (μM)	$-\Delta G$ (kJ/mol)	$-\Delta H$ (kJ/mol)	$-T\Delta S$ (kJ/mol)	n (sites/ monomer)
SBA					
GalNAcα1-*O*-Ser	170	21.8	33.1	11.3	1.0
38/40-mer Tn-PSM	1.4	33.5	134.7	101.3	0.2
81-mer Tn-PSM	0.06	41.0	234.7	193.7	0.12
Tn-PSM	0.0002	54.8	18033.0	17978.7	0.0018
VML					
GalNAcα1-*O*-Ser	130	22.2	26.8	4.6	1.0
38/40-mer Tn-PSM	0.20	38.1	154.0	115.9	0.2
81-mer Tn-PSM	0.012	45.2	220.5	175.3	0.1
Tn-PSM	0.0001	56.9	22066.4	22007.8	0.0012

ITC experiments were performed at 27 °C
Errors in K_d range from 1% to 7%. Errors in ΔG are less than 2%. Errors in ΔH are 1–4%. Errors in $T\Delta S$ are 1–7%. Errors in n are less than 4%.

6.2.2 Mechanisms of Binding of SBA and VML to Mucins of Different Lengths: The Internal Diffusion Jump ("Bind and Jump" Model)

The increasing affinities of SBA and VML for 38/40-mer Tn-PSM (**10**), 81-mer Tn-PSM (**9**), and Tn-PSM (**8**), respectively (Table 2), indicate that the lengths of the polypeptide chains and thus the total carbohydrate valence of these mucin analogs regulate their affinities for the lectins. The similarities in the ΔH per α-GalNAc-binding residue of Tn-PSM with that of GalNAcα1-*O*-Ser for the two lectins, respectively, and similar correlations with 81-mer Tn-PSM and 38/40-mer Tn-PSM, suggest that the binding mechanisms of these PSM analogs are common (Dam et al., 2007).

These results suggest that binding of a lectin to a multivalent carbohydrate or glycoprotein involves internal diffusion "jumps" of the lectin from carbohydrate epitope to epitope before complete dissociation. Kinetically, this has the effect of reducing the macroscopic off-rate of the lectin and hence increasing its affinity since the affinity constant is the ratio of the forward and reverse rate constants. The forward rate constant for lectin binding may also be enhanced as a result of the larger number of binding epitopes on the glycoprotein.

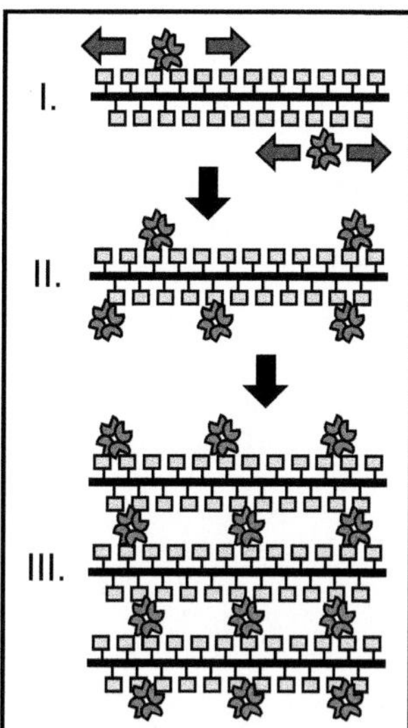

Figure 5 Schematic depiction of the "Bind and Jump" model. (I) SBA or VML binding (and rebinding through an internal diffusion mechanism) at low fractional occupancy to **8** (Tn-PSM), (II) SBA or VML binding at higher fractional occupancy to **8** (Tn-PSM), and (III) SBA or VML binding at saturation that involves cross-linking of **8** (Tn-PSM).

The "bind and jump" model suggests that as more lectin molecules bind to a mucin chain, the microscopic affinity of the lectin will decrease because of steric crowding and shorter diffusion distances on the polypeptide chain of the mucin (Fig. 5). In these cases, binding of the lectins to the multivalent and glycoproteins was associated with gradients of decreasing microaffinity constants (microscopic affinity) of the lectins for the multiple epitopes of the ligands. This suggests that the observed (macroscopic) dissociation constants (K_d) for SBA and VML binding to the three PSM analogs in Table 2 represent an average of the gradient of binding constants present in each case (Dam et al., 2007).

Importantly, the "Bind and Jump" model for lectins binding to mucins is distinct from the classical "lock and key" model of ligand binding to a receptor. This model is similar to the binding mechanism of a variety of ligands to DNA in which binding and sliding of ligands occur along the DNA backbone (von Hippel, 2007).

ACKNOWLEDGMENTS

This work was supported by a start-up fund and by Research Excellence Fund (T.K.D.) provided by Michigan Technological University and by Grant CA-16054 from the National Cancer Institute, Department of Health, Education and Welfare, and Core Grant P30 CA-13330 from the same agency (C.F.B.)

REFERENCES

Byrd, J. C., & Bresalier, R. S. (2004). Mucins and mucin binding proteins in colorectal cancer. *Cancer Metastasis Reviews*, *23*(1–2), 77–99.

Chervenak, M. C., & Toone, E. J. (1995). Calorimetric analysis of the binding of lectins with overlapping carbohydrate-binding ligand specificities. *Biochemistry*, *34*(16), 5685–5695.

Chervenak, M. C., & Toone, E. J. (1996). Analysis of the binding specificities of oligomannoside-binding proteins using methylated monosaccharides. *Bioorganic & Medicinal Chemistry*, *4*(11), 1963–1977.

Dam, T. K., Gabius, H. J., Andre, S., Kaltner, H., Lensch, M., & Brewer, C. F. (2005). Galectins bind to the multivalent glycoprotein asialofetuin with enhanced affinities and a gradient of decreasing binding constants. *Biochemistry*, *44*(37), 12564–12571.

Dam, T. K., Gerken, T. A., Cavada, B. S., Nascimento, K. S., Moura, T. R., & Brewer, C. F. (2007). Binding studies of alpha-GalNAc-specific lectins to the alpha-GalNAc (Tn-antigen) form of porcine submaxillary mucin and its smaller fragments. *The Journal of Biological Chemistry*, *282*(38), 28256–28263.

Dam, T. K., Oscarson, S., & Brewer, C. F. (1998). Thermodynamics of binding of the core trimannoside of asparagine-linked carbohydrates and deoxy analogs to Dioclea grandiflora lectin. *The Journal of Biological Chemistry*, *273*(49), 32812–32817.

Dam, T. K., Roy, R., Das, S. K., Oscarson, S., & Brewer, C. F. (2000). Binding of multivalent carbohydrates to concanavalin A and *Dioclea grandiflora* lectin. Thermodynamic analysis of "multivalency effect" *The Journal of Biological Chemistry*, *275*(19), 14223–14230.

Dam, T. K., Roy, R., Page, D., & Brewer, C. F. (2002a). Negative cooperativity associated with binding of multivalent carbohydrates to lectins. Thermodynamic analysis of the "multivalency effect" *Biochemistry*, *41*(4), 1351–1358.

Dam, T. K., Roy, R., Page, D., & Brewer, C. F. (2002b). Thermodynamic binding parameters of individual epitopes of multivalent carbohydrates to concanavalin a as determined by "reverse" isothermal titration microcalorimetry. *Biochemistry*, *41*(4), 1359–1363.

Di Cera, E. (1995). *Thermodynamic theory of site-specific binding processes in biological macromolecules*. New York: Cambridge University Press.

Fukuda, M. (2002). Roles of mucin-type O-glycans in cell adhesion. *Biochimica et Biophysica Acta*, *1573*(3), 394–405.

Indyk, L., & Fisher, H. F. (1998). Theoretical aspects of isothermal titration calorimetry. *Methods in Enzymology*, *295*, 350–364.

Lis, H., & Sharon, N. (1998). Lectins: Carbohydrate-specific proteins that mediate cellular recognition. *Chemical Reviews*, *98*(2), 637–674.

Mandal, D. K., Kishore, N., & Brewer, C. F. (1994). Thermodynamics of lectin-carbohydrate interactions. Titration microcalorimetry measurements of the binding of N-linked carbohydrates and ovalbumin to concanavalin A. *Biochemistry*, *33*(5), 1149–1156.

Pace, K. E., Lee, C., Stewart, P. L., & Baum, L. G. (1999). Restricted receptor segregation into membrane microdomains occurs on human T cells during apoptosis induced by galectin-1. *Journal of Immunology*, *163*(7), 3801–3811.

Rao, J., Lahiri, J., Isaacs, L., Weis, R. M., & Whitesides, G. M. (1998). A trivalent system from vancomycin.D-ala-D-Ala with higher affinity than avidin.biotin. *Science*, *280*(5364), 708–711.

Rozwarski, D. A., Swami, B. M., Brewer, C. F., & Sacchettini, J. C. (1998). Crystal structure of the lectin from Dioclea grandiflora complexed with core trimannoside of asparagine-linked carbohydrates. *The Journal of Biological Chemistry*, *273*(49), 32818–32825.

Shogren, R., Gerken, T. A., & Jentoft, N. (1989). Role of glycosylation on the conformation and chain dimensions of O-linked glycoproteins: Light-scattering studies of ovine submaxillary mucin. *Biochemistry*, *28*(13), 5525–5536.

Stryer, L. (1998). *Biochemistry* (3rd ed.). New York: W.H. Freeman.

Talaga, M. L., Fan, N., Fueri, A. L., Brown, R. K., Chabre, Y. M., Bandyopadhyay, P., et al. (2014). Significant other half of a glycoconjugate: Contributions of scaffolds to lectin-glycoconjugate interactions. *Biochemistry, 53*(27), 4445–4454.

Varki, A. (1993). Biological roles of oligosaccharides: All of the theories are correct. *Glycobiology, 3*(2), 97–130.

Varki, A. (1997). Selectin ligands: Will the real ones please stand up? *The Journal of Clinical Investigation, 100*(11 Suppl.), S31–S35.

Varki, A., & Angata, T. (2006). Siglecs—The major subfamily of I-type lectins. *Glycobiology, 16*(1), 1R–27R. http://dx.doi.org/10.1093/glycob/cwj008.

Varki, A., Cummings, R. D., Esko, J. D., Freeze, H. H., Stanley, P., & Bertozzi, C. R.et al. (2009). *Essentials of glycobiology*. (2nd ed.). New York: Cold Spring Harbor Laboratory Press.

von Hippel, P. H. (2007). From "simple" DNA-protein interactions to the macromolecular machines of gene expression. *Annual Review of Biophysics and Biomolecular Structure, 36,* 79–105.

Wiseman, T., Williston, S., Brandts, J. F., & Lin, L. N. (1989). Rapid measurement of binding constants and heats of binding using a new titration calorimeter. *Analytical Biochemistry, 179*(1), 131–137.

CHAPTER FIVE

Calorimetric and Spectroscopic Analysis of the Thermal Stability of Short Duplex DNA-Containing Sugar and Base-Modified Nucleotides

Kareem Fakhfakh[*,†], **Curtis B. Hughesman**[*], **A. Louise Creagh**[*,†], **Vincent Kao**[†], **Charles Haynes**[*,†,‡,1]

[*]Michael Smith Laboratories, University of British Columbia, Vancouver, British Columbia, Canada
[†]Department of Chemical and Biological Engineering, University of British Columbia, Vancouver, British Columbia, Canada
[‡]RES'EAU Water Research Network, Department of Chemical and Biological Engineering, University of British Columbia, Vancouver, British Columbia, Canada
[1]Corresponding author: e-mail address: israels@chbe.ubc.ca

Contents

Abstract

Base- and sugar-modified analogs of DNA and RNA are finding ever expanding use in medicine and biotechnology as tools to better tailor structured oligonucleotides by altering their thermal stability, nuclease resistance, base-pairing specificity, antisense activity, or cellular uptake. Proper deployment of these chemical modifications generally

Methods in Enzymology, Volume 567
ISSN 0076-6879
http://dx.doi.org/10.1016/bs.mie.2015.08.001

requires knowledge of how each affects base-pairing properties and thermal stabilities. Here, we describe in detail how differential scanning calorimetry and UV spectroscopy may be used to quantify the melting thermodynamics of short dsDNA containing chemically modified nucleosides in one or both strands. Insights are provided into why and how the presence of highly stable base pairs containing modified nucleosides can alter the nature of calorimetry or melting spectroscopy data, and how each experiment must therefore be conducted to ensure high-quality melting thermodynamics data are obtained. Strengths and weaknesses of the two methods when applied to chemically modified duplexes are also addressed.

1. INTRODUCTION

Applications of synthetic analogs of deoxyribonucleotides and ribonucleotides have experienced tremendous growth in the last two decades (Babu et al., 2005; Pinheiro & Hollinger, 2012), due in part to their ability to enhance diagnostic and therapeutic applications of structured oligonucleotides by altering their thermodynamic and/or chemical (e.g., resistance to nucleases) stability, base-pairing specificity, antisense activity, or cellular uptake (Krueger & Kool, 2009). The phosphodiester bonds of natural DNA and RNA are, by design, susceptible to hydrolysis by nucleases. Short natural oligonucleotides also tend to exhibit relatively poor plasma protein binding and transport within the cardiovasculature, while longer oligos complex their target slowly. For oligonucleotides used as antisense agents, slow-binding kinetics can be exacerbated by the tendency for their mRNA targets to be highly structured. As a result, unmodified DNA or RNA can exhibit poor pharmacokinetics and relatively rapid excretion when delivered as a parenteral therapeutic.

These limitations can be overcome through appropriate chemical modifications that serve to increase the specificity of the oligonucleotide for its target sequence, the stability of the complex/duplex formed, and/or the resistance to nuclease-catalyzed degradation. Reported chemical modifications to DNA and RNA now number in the 100s (Khakshoor & Kool, 2011; Sanghvi, 2011) and include chemistries that serve to alter helix structure (Kang, Heuberger, Chaput, Switzer, & Feigon, 2012), expand the genetic alphabet beyond the natural nucleosides *a*, *t*, *g*, and *c* (Henry & Romesberg, 2003; Kool, 2002), or alter cellular phenotypes (Krueger, Peterson, Chelliserry, Kleinbaum, & Kool, 2011). Unique biological functions, and improved diagnostic specificity or therapeutic potency may thereby be realized (Abramov, Schepers, Van Aerschot, Van Hummelen, & Herdewijn, 2008; Latorra, Campbell, Wolter, & Hurley, 2003; Leumann, 2001; Wilson & Keefe, 2006).

Many of the chemical modifications reported to date describe alteration of the backbone (Micklefield, 2001), furanose sugar (Astakhova & Wengel, 2014; Schoning et al., 2000), or base (Yang, Hutter, Sheng, Sismour, & Benner, 2006), or of multiple components of a nucleoside (e.g., Andersen, Anderson, Wengel, & Hrdlicka, 2013). Among the most effective sugar modifications is the locked nucleic acid (LNA) (Koshkin et al., 1998), where a methylene bridge has been introduced between the 4′-carbon and 2′-oxygen of the furanose sugar to "lock" it into a C3′-*endo* (RNA-like) conformation (Veedu & Wengel, 2010). By substituting DNA or RNA nucleotides with the corresponding LNA nucleotides, both the thermal stability and the melting temperature, T_m, of a duplex can be increased. A single LNA substitution in one strand generally increases T_m of a short complementary DNA duplex (dsDNA) by 1–8 °C (Hughesman, Fakhfakh, Bidshahri, Lund, & Haynes, 2015), while somewhat larger increases in T_m are seen when a single LNA substitution is made in short single-stranded DNA (ssDNA) paired with its complementary RNA strand (Kaur, Wengel, & Maiti, 2008).

Novel modifications to nucleoside bases include the so-called superbases 2-amino-adenine, 2′-*O*-(2-(methoxy)ethyl)-2-thiothymidine, and 2′-deoxy-2′-fluoro-2-thiothymidine (Law, Eritja, Goodman, & Breslauer, 1996; Rajeev, Prakash, & Manoharan, 2003). These modified bases pair more stably with their complementary DNA base (Hughesman, Turner, & Haynes, 2008), and exhibit remarkably high thermal stabilities when paired with their complementary RNA base (Rajeev et al., 2003). Further increases in thermal stability may be realized by introducing these base modifications into LNA nucleotides (Andersen et al., 2013). Moreover, the 2′-*O*-(2-(methoxy)ethyl)-2-thiothymidine modification has been shown to improve resistance to nucleases when substituted into DNA oligonucleotides (Diop-Frimpong, Prakash, Rajeev, Manoharan, & Egli, 2005). The general chemistries of LNAs and LNA superbases are provided in Fig. 1.

The capacity of nucleoside modifications to alter base-pairing properties and, as a result, duplex stability has motivated detailed studies of their melting thermodynamics using a variety of spectroscopic and calorimetric techniques. Our laboratory has experience in using two of these methods (e.g., Hughesman, Turner, & Haynes, 2011a), namely differential scanning

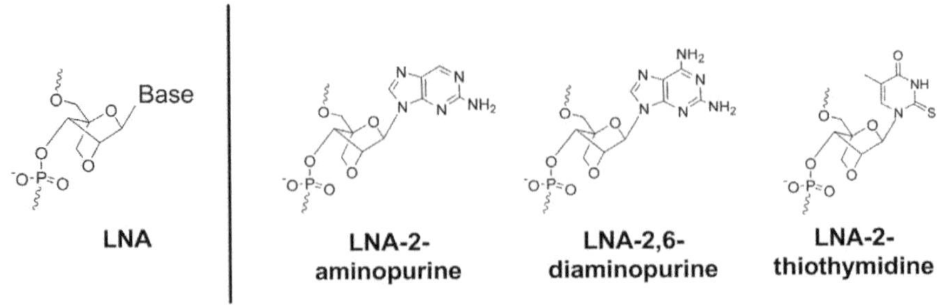

Figure 1 A standard locked nucleic acid (left) can be modified by replacing its natural nucleosidic base with one of several modified base chemistries (right) to further enhance duplex stability.

calorimetry (DSC) and UV-spectroscopy monitored melting (UVM), to study the melting thermodynamics of short dsDNA containing chemically modified nucleosides in one or both strands (e.g., Fakhfakh et al., 2015). Both methods have been widely employed to study melting transitions in natural dsDNA, and descriptions of how each technique is best applied to that application are available (Hughesman, Turner, & Haynes, 2013; Mergny & Lacroix, 2003; Wartell & Benight, 1985). Though some overlap with that prior art is unavoidable, our intent here is to detail how each method may best be applied to the study and interpretation of melting thermodynamics of short dsDNA containing chemically modified nucleotides in one or both strands. We address the central questions of—How is each experiment conducted? What are relative the strengths and weaknesses of each approach? What thermodynamic parameters can and cannot be measured by each method? How are raw data presented and analyzed, and what assumptions must be invoked to obtain high-quality values for each measurable thermodynamic parameter? And finally, how does one evaluate melting thermodynamics to properly understand the contribution to duplex stability provided by the specific nucleotide modification(s)?

2. MELTING THERMODYNAMICS FROM DSC

When applied to the determination of melting thermodynamics for modified (or natural) dsDNA, DSC experiments detect and characterize the helix-to-coil transition by measuring the difference in applied power dP (J s^{-1}) needed to maintain the DNA sample solution and a DNA-free reference solution at the same temperature, as that temperature is linearly increased at a chosen scan rate (C min^{-1}) through the melting transition (typically from ca. 1 to 100 °C, though larger ranges are possible). The measured differential power (d*P*) is divided by the scan rate (now expressed in K s^{-1}) to give the excess heat capacity C_p^{ex} (J K^{-1}), which is then plotted versus temperature (*T*) as illustrated in Fig. 2A. The quality of that raw data, as well as corresponding C_p^{ex}-vs.-*T* scans for the DNA-free reference solution, determines what melting thermodynamics data may be extracted to good accuracy. For each sample, multiple upscans at the chosen scan rate are conducted in secession to confirm reversibility in the melting transition. For a completely reversible transition, scans 2 and higher will substantially superimpose; the first scan may or may not.

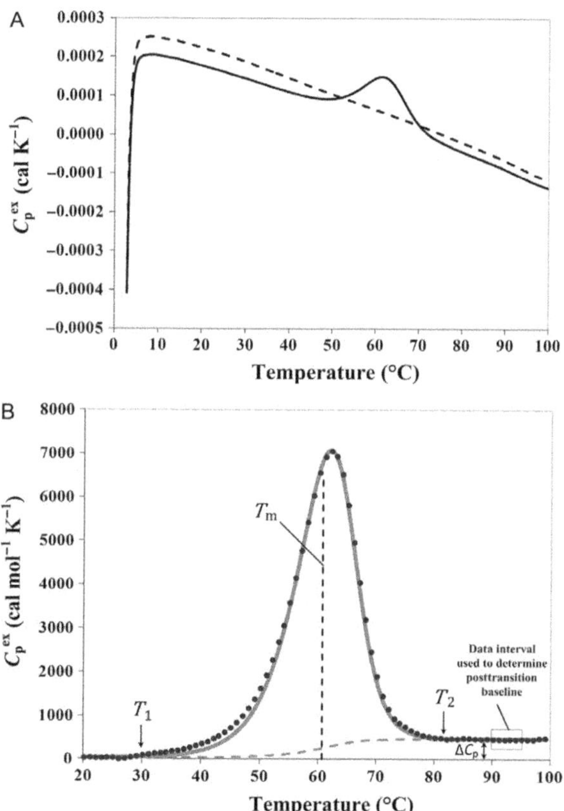

Figure 2 Raw C_p^{ex}-vs.-*T* scans for the sample and reference buffer (A), and reference-normalized thermogram (B) for the 5′-*gaDacagttDaag*-3′/5′-*ctttaactgtttc*-3′ duplex, where *D* is the 2-amino-adenine LNA super-base.

2.1 Strengths, Weaknesses, and What Can be Determined Reliably

The principle strength in determining dsDNA melting thermodynamics using DSC is that the method can provide a direct and highly accurate measure of T_m, as well as the melting enthalpy (ΔH_{cal}) and entropy (ΔS_{cal}) at T_m; here, the subscript "cal" indicates that the ΔH and ΔS values are determined directly from melting transition peak integration without invoking any model assumptions. If stable baselines are recorded, the heat capacity change (ΔC_p) for the melting transition may be directly estimated as well. Modern DSC instruments have served to make this latter measurement much more reliable and accurate through their increased sensitivity and greatly improved baseline stability. Those instruments include the Nano DSC now sold by TA Instruments Inc., and MicroCal VP-DSC and VP-Capillary DSC now sold by Malvern Inc.—our laboratory and associated Centre for Bio-Thermodynamics utilize the latter two instruments. Those instruments report energies in units of calories, but energies are expressed in Joules in this work in accordance with International System of Units (SI) convention.

As detailed below, a reference and concentration normalized C_p^{ex}-vs.-T plot, known as the "thermogram," is generated from the raw C_p^{ex}-vs.-T scans for the sample and reference. It serves as the basis for determining all relevant melting thermodynamics, either through direct integration of the melting transition or through regression to the two-state reaction model.[1] In the latter analysis, those melting thermodynamics include values of T_m, ΔH_{2-st}, and ΔS_{2-st} for the helix-to-coil transition, as well as the equilibrium constant $K(T)$ for that transition, which as noted is a temperature-dependent parameter. Here, the subscript "2−st" is used to signify parameters regressed assuming a two state transition. The ability to extract both ΔH_{cal} and ΔH_{2-st} from DSC data has led to the so-called calorimetric criterion, which assumes that the helix-to-coil transition obeys (and can therefore be described by) two-state melting thermodynamics if (i) the transition is reversible and (ii) ΔH_{cal} and ΔH_{2-st} agree to within their standard deviations (some researchers instead say to within $\pm 10\%$) (e.g., Movileanu, Benevides, & Thomas, 2002). Zhou, Hall, and Karplus (1999) have shown that satisfying the calorimetric criterion is necessary, but not wholly sufficient, to unequivocally verify two-state melting.

[1] The two-state melting transition model assumes that dsDNA denaturation occurs as an "all-or-none" process such that at any given T the fully duplexed state and the single-strand (random-coil) state are the only states present at equilibrium.

Nevertheless, a $\Delta H_{2-st}/\Delta H_{cal}$ value near unity appears to provide good evidence that the stability of a duplex (in the same reference) at temperatures away from T_m can be reliably estimated using two-state thermodynamics, the total molar strand concentration C_T, and the experimental values of ΔH, ΔS, and ΔC_p collected at T_m. This represents the second main advantage of the DSC method.

Relative to the UVM method, however, DSC determination of melting thermodynamics of chemically modified dsDNA is challenged by a number of issues. First, the DSC experiment generally requires more material. Moreover, high purities and sequence homogeneities are required of each equimolar strand of the duplex. This is because determination of ΔH and ΔS requires the raw C_p^{ex} data, which are provided in units of $J\ K^{-1}$, to be normalized by the total pretransition moles of duplex ($C_T/2$) in the sample cell volume to thereby express C_p^{ex} in $J\ mol^{-1}\ K^{-1}$ units. One must also assume that each strand is capable of participating in a fully complementary duplex at temperatures below and within the melting transition. If accurate melting thermodynamics are to be obtained, each synthesized strand must therefore be HPLC purified to homogeneity. While this does not represent a significant problem when studying short duplexes comprised of natural DNA, chemically modified nucleotides and oligonucleotides containing them are generally more expensive, often much more expensive, and the added cost of purifying them by HPLC can be prohibitive, particularly for duplexes containing a large number of nucleotide modifications. In our view, though the DSC method is superior to the UVM method in principle, its application to such sequences is therefore justified only when complete and highly accurate knowledge of melting thermodynamics, including second derivative thermodynamic properties such as ΔC_p, is required, or when one needs to be certain the duplex satisfies the calorimetric criterion.

Other more technical issues related to DSC analysis of chemically modified dsDNA are detailed below.

2.2 Sample Preparation

Modified dsDNA samples analyzed by DSC are generally prepared from stock solutions, each containing one of the two required HPLC-purified single strands. Given the proprietary nature of the chemical modifications, the required strands are most often purchased from a company licensed to produce and sell them. For strands containing LNAs, those include Exiqon (Vadbæk, Denmark) and Integrated DNA Technologies (IDT, Coralville,

IA). Exiqon can also reliably provide strands containing combinations of DNA nucleotides, LNAs, superbases and/or LNA superbases. For any strand comprised only of natural DNA, many reliable vendors are available.

Each HPLC-purified modified or natural DNA oligonucleotide is typically resuspended in aqueous buffer containing 1 *M* NaCl, 10 m*M* Na_2HPO_4 (pH 7), and 1 m*M* Na_2EDTA; this reference buffer is often referred to as "standard conditions" for melting analysis since much of the DNA melting thermodynamics data reported to date were collected in it (Hughesman et al., 2015; Sugimoto et al., 1995). However, large and important data sets (SantaLucia, 1998; SantaLucia, Allawi, & Seneviratne, 1996), including most of the data used to regress the most widely used nearest-neighbor stability parameters for Watson–Crick base pair doublets, have instead employed an aqueous reference buffer comprised of 1 *M* NaCl, 10 m*M* sodium cacodylate ($(CH_3)_2AsO_2Na{\cdot}3H_2O$) (pH 7), and 0.5 m*M* Na_2EDTA. Both natural and modified dsDNA is in general more stable in the phosphate-buffered reference. Amalgamation of melting thermodynamics collected in the two different reference buffers therefore requires conversion of data collected in one buffer to the corresponding values in the other. Alemayehu et al. (2009) have provided the equations needed to do this for T_m, ΔG, ΔH, and ΔS. Finally, for studies related to primer or probe stability, the NaCl (or possibly KCl) concentration in either reference is generally reduced to between 15 and 50 m*M* to mimic PCR conditions.

When melting dsDNA between ca. 6 and 25 bp in length (which is the focus here), the two strands, whether complementary or mismatched, are mixed at equimolar concentration to a C_T between 50 and 100 μ*M*, with C_T values at the higher end of this range employed for shorter duplexes (lower ΔH). That sample is allowed to equilibrate at room temperature for 5–10 min to ensure dsDNA formation. A stock of the DNA-free reference buffer is also prepared and filtered using a 0.2 μm syringe filter to avoid introducing any particulate matter into the DSC (capillary) cells (we have used the syringe filters from Pall Inc. (Port Washington, NY) with good results). The DNA sample (for which the buffer used to prepare it has also been filtered) and reference buffer are then degassed for 7–10 min under gentle stirring. In our facility, this is done by holding the sample at 80 °C for 5–10 min using a MicroCal Thermovac unit (Malvern Inc.).

2.3 Experimental Procedure

Generally, a DSC experiment on a given dsDNA sample is comprised of a set of two or more quality thermal scans for each of three different fluid pairs described below. Each scan consists of a prescan system equilibration (typically at 1 °C for 10 or 15 min), a thermal upscan in which dP data are collected from 1 to 100 °C (or possibly a higher final temperature if larger sample pressures are employed) as the temperature is linearly increased at a fixed rate (often at 1 °C min^{-1} or a lower rate), a downscan at a rate defined by the instrument (often ca. 3 °C min^{-1} for the instruments we employ), and a sample reequilibration. When a VP-DSC or VP-capillary DSC is employed, both cells are first syringe loaded (with the instrument on and idling at 25 °C) to overflow levels with degassed nanopure water (first fluid pair). Excess liquid in each cell is then removed using a syringe with a collared needle to ensure precise and equivalent liquid volumes in the sample and reference cells. The two cells are sealed with a Teflon top, the piston in which is then actuated to set the pressure in each cell to between 26 and 30 psig to allow scans up to 100 °C while suppressing bubble formation. The programmed scanning cycle of the instrument is then activated, starting with a downscan to the equilibration temperature, to continuously collect raw "water-vs.-water" $C_{\mathrm{p}}^{\mathrm{ex}}$-vs.-$T$ scans, typically overnight. Except for the first scan, which is influenced by the altered thermal history of the instrument, stable signals, and substantial superimposition of the raw water-vs.-water scans are expected. If they do not superimpose, it likely indicates that one or both of the cells of the calorimeter are contaminated or fouled. In that case, and at routine intervals, it is not just advisable but essential to carefully clean the cells according to guidelines specified by the manufacturer.

When satisfactory performance is observed and the instrument is either idling at 25 °C or in a downscan at a temperature between 35 and 25 °C, the water is removed from both cells, each of which is then twice rinsed, loaded with the reference buffer, and finally pressure sealed using the same procedure as described above. If dynamic loading is used, the process should be completed before the downscan temperature falls below 10 °C. Two substantially superimposed raw "buffer-vs.-buffer" (second fluid pair) $C_{\mathrm{p}}^{\mathrm{ex}}$-vs.-$T$ reference scans are then measured, which is usually achieved in the first two scans.

With the instrument idling at 25 °C, the degassed (modified) dsDNA sample is then loaded into the sample cell. In this case, first rinsing the cell volume with the sample ensures that C_{T} is known precisely in the subsequently loaded sample. However, the cost of the dsDNA sacrificed in the rinse step may be prohibitive. If buffer removal is followed directly by sample loading, without conducting sample washes, a 2% sample dilution may be assumed. As this is a crude correction/assumption, it will affect the accuracy of the melting thermodynamics data collected. A set of 2–4 raw "sample-vs.-buffer" (third fluid pair) C_p^{ex}-vs.-T scans are then recorded. These will usually be superimposable provided the maximum temperature of the scan is not too high.

Technical specifications and settings for the instrument used can influence data quality as well. For example, for the VP-DSC, the minimum thermal response time, baseline noise, and baseline repeatability are 5 s, 2 μJ °C^{-1}, and ±5 μJ °C^{-1}, respectively, for a 1–1.5 °C min^{-1} scan rate. The user may select both the filter period (data-acquisition interval) and gain of the instrument, and for dsDNA melting studies, we generally set these at 10 s and a null gain (passive feedback) setting, respectively, when using either MicroCal (Malvern) instrument.

2.4 Analyzing a DSC Thermogram

Analysis of dsDNA melting data acquired by DSC begins with subtraction of raw C_p^{ex}-vs.-T data for the buffer-vs.-buffer reference from that for each sample-vs.-buffer scan. The average of the two buffer-vs.-buffer scans may be used. However, we find that pre- and posttransition baselines generated in the resulting reference-normalized scan are often more linear and stable if a particular scan (see below), often the one collected immediately preceding sample loading, is used for this purpose. Each reference-normalized C_p^{ex}-vs.-T scan is then divided by the pretransition molar concentration of dsDNA ($C_T/2$) and the sample cell volume (unique to each instrument and calibrated by the manufacturer—V_{Cell} = 511.8 μL in our current VP-DSC) to create a set of reference and concentration normalized C_p^{ex}-vs.-T "melting thermograms." A representative example of a thermogram is provided in Fig. 2B.

As DNA duplexation is a bimolecular hybridization reaction, T_m depends on C_T. Thermograms displaying a melting transition with a maximum C_p^{ex} between ca. 55 and 70 °C tend to provide melting thermodynamics data of the highest quality, due in part to better defined pre- and posttransition baselines. If heats of melting are sufficiently large, the value of C_T should be set so that T_m lies within that range.

For a few representative duplexes in the full set of sequences to be studied, thermograms should be collected at several scan rates, typically between ca. 0.3 and 2 °C min^{-1}. Equilibrium thermodynamic data are being determined, and their values should therefore be scan rate independent.

For short natural or modified dsDNA, this is usually the case at scan rates ≤ 1 °C min^{-1}, but verification is prudent.

Both ΔH_{cal} and ΔS_{cal} for the helix-to-coil transition at T_m may be determined through direct integration of the thermogram:

$$\Delta H_{cal} = \int_{T_1}^{T_2} C_p^{ex}(T) dT \tag{1}$$

and

$$\Delta S_{cal} = \int_{T_1}^{T_2} \left(\frac{C_p^{ex}(T)}{T} \right) dT, \tag{2}$$

where T_1 and T_2 are the temperatures (integration limits) defined by the intersections of the $C_p^{ex}(T)$ melting transition with the pre- and posttransition baselines, respectively, as shown in Fig. 2B. If those two baselines are sufficiently stable, the value of ΔC_p may also be estimated as the difference in ordinate positions of the baselines (Fig. 2B). The duplex melting temperature, T_m, is computed as the temperature (Fig. 2B) bisecting the area midpoint of the melting transition bounded by T_1 and T_2.

DSC Insight #1: Although well-behaved reference-normalized baselines permitting observation and estimation of ΔC_p (such as those in Fig. 2B) and in virtually every example published by leading calorimetry experts (see, for example, Chalikian, Völker, Plum, & Breslauer, 1999; Privalov & Dragan, 2007) can be generated for some systems, they cannot always be generated, as baseline stability and shape are sensitive to many variables. One important, but poorly recognized variable is the dynamic range of the raw C_p^{ex}-vs.-T data for the buffer-vs.-buffer reference relative to that for the sample-vs.-buffer run. As shown in Fig. 3, stable horizontal pre- and posttransition baselines are often observed in the reference-normalized thermogram if the range of C_p^{ex} recorded in the two scans are similar in value. When the C_p^{ex} values for the sample-vs.-buffer scan are appreciably higher or lower than those recorded in the buffer-vs.-buffer scan, the baselines in the reference-normalized thermogram tend to be more poorly defined.

DSC Insight #2: Proper assignment of pre- and posttransition baselines can be difficult, and the uncertainties associated with those assignments are often the dominant source of error in the melting thermodynamics measured. The uncertainty in the slope of either baseline (most notably

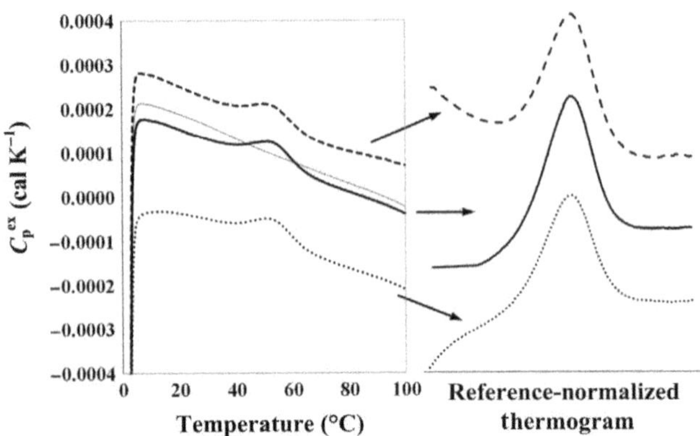

Figure 3 Inability to match the dynamic range of C_p^{ex} values recorded in the sample and reference buffer scans leads to nonlinearity in the pretransition baseline of the reference buffer-normalized thermogram. Generally, if C_p^{ex} values for the sample and reference buffer are close at each temperature, stable linear baselines (solid line) are observed.

the pretransition baseline) can typically contribute up to ±2% error in the value of ΔH_{cal} and up to ±100% error in ΔC_p.

DSC Insight #3: Use of Eqs. (1) and (2) requires specification of the shape of the baseline, $C_p^{ex}(T)$, through the melting transition (i.e., between T_1 and T_2). Several different methods have been proposed and used (Filimonov, Potekhin, Matveev, & Privolov, 1982). We generally compute $C_p^{ex}(T)$ as $\alpha(T)C_{p,\,dsDNA}^{ex*}(T) + (1-\alpha(T))C_{p,\,ssDNA}^{ex*}(T)$, a result that assumes the baseline moves from the $C_p^{ex*}(T)$ of the pure dsDNA state to that of the pure ssDNA state in proportion to the fraction $\alpha(T)$ of total strands in the dsDNA state at each T.

DSC Insight #4: Obtaining a stable pretransition baseline can be particularly challenging for highly modified dsDNA, and peak integration may not yield reliable melting thermodynamics when the presence or quality of a pretransition is lacking. LNA–LNA base pairs, for example, are unusually stable, increasing the potential for formation of hairpins and other stable structures in one or both single strands. Those structures can alter the pretransition baseline and thereby complicate or preclude data analysis using two-state theory. If possible, dsDNA sequences to be studied by DSC should be designed to avoid such structures. Both Exiqon Inc. (https://www.exiqon.com/oligo-tools) and Integrated DNA Technologies (https://www.idtdna.com/calc/analyzer) provide tools that predict/score the likelihood of oligonucleotides bearing modified nucleotides forming stable single-stranded structures.

DSC Insight #5: The width of the melting transition decreases with increasing melting enthalpy. Base pairs containing modified nucleotides (e.g., LNA–DNA base pairs) are generally more stable than their DNA–DNA counterparts. Thus, when compared to natural dsDNA, modified dsDNA of a shorter length can have the same T_m, but a smaller melting enthalpy. This leads to a broader melting transition, which further challenges assignment of baselines.

As noted, the thermogram can also be analyzed by fitting to the two-state reaction model. Historically, this has been done using a simplified form of the two-state model in which ΔC_p is set equal to zero (Breslauer, Frank, Blocker, & Marky, 1986). For much of the DSC data collected previously this assumption is wholly justified, as early generation calorimeters generally did not offer the sensitivity and baseline stability needed to quantify ΔC_p (with the custom-built instruments of Privalov (1980) being an exception). As a result, both ΔH and ΔS were taken to be temperature-independent quantities in spite of the theoretical inconsistencies presented by that assumption (Petruska et al., 1988).

In a two-state reaction, the melting enthalpy is given by the van't Hoff enthalpy $\Delta H_{\text{v.H.}}$, which may be computed from a thermogram as

$$\Delta H_{2-\text{st}} \equiv \Delta H_{\text{v.H.}} = -R\left[\frac{d(\ln K(T))}{d(1/T)}\right] \tag{3}$$

where R is the ideal gas constant 8.314 J mol^{-1} K^{-1}. For dsDNA that melts into two nonself-complementary sequences (i.e., $AB \overset{K}{\leftrightarrow} A + B$, where, for example, strand A might be the sequence 5′-*aaaaaaaaaaaa*-3′, which is nonself-complementary because it cannot base pair with itself), the temperature dependence of K is given by

$$K(T) = \frac{C_T}{2}\frac{(1-\alpha(T))^2}{\alpha(T)} \tag{4A}$$

and for dsDNA that melts into two self-complementary strands (i.e., $A_2 \overset{K}{\leftrightarrow} 2A$) by

$$K(T) = 2C_T\frac{(1-\alpha(T))^2}{\alpha(T)} \tag{4B}$$

Here, $\alpha(T)$, the fraction of total strands in the dsDNA state at each T, is given by

$$\alpha(T) = \frac{\int_{T_1}^{T} C_p^{ex}(T)\mathrm{d}T}{\Delta H_{cal}} \tag{5}$$

Use of Eq. (3) to determine $\Delta H_{v.H.}$ allows estimation of the ratio $\Delta H_{v.H.}/\Delta H_{cal}$. As noted, a ratio near unity provides evidence a dsDNA melting transition follows two-state thermodynamics, while systems exhibiting a ratio <1 are generally characterized by significant populations of intermediate states within the melting transition. Ratios >1 occur infrequently and are less understood.

However, determination of ΔC_p is entirely possible using modern DSC instruments, and such measurements show that ΔC_p is not zero, but instead positive in value and to a first approximation temperature independent at temperatures above ca. 50 °C (Chalikian et al., 1999; Hughesman, Turner, & Haynes, 2011b; Mikulecky & Feig, 2006a, 2006b; Rouzina & Bloomfield, 1999). As we have shown previously (Hughesman et al., 2015), thermograms may then be fit to the full two-state transition model

$$C_p^{ex}(T) = b_{post}^{cal} + m_{post}^{cal}T - \Delta C_p \alpha(T) + \frac{\left[\Delta H_{2-st} + \Delta C_p(T - T_m)\right]^2}{RT^2}\left(\frac{\alpha(T) - \alpha(T)^2}{1 + \alpha(T)}\right) \tag{6}$$

$$\alpha(T) = 1 + \frac{K(T) - \sqrt{[K(T)]^2 + 2K(T)C_T}}{C_T} \tag{7}$$

$$K(T) = K(\alpha = 0.5) * \exp\left[-\frac{\Delta H_{2-st}}{RT}\left(1 - \frac{T}{T_m}\right) - \frac{\Delta C_p}{RT}\left(T - T_m - T\ln\left(\frac{T}{T_m}\right)\right)\right] \tag{8}$$

to obtain a complete description of melting thermodynamics, including an estimate of ΔC_p; m_{post} and b_{post} are the slope and intercept of the post-transition baseline, respectively. Here, we have expressed the full two-state melting model using the posttransition baseline as a reference, as it is usually stable. More reliable melting thermodynamic data (T_m, $\Delta H_{2-st}(T_m)$, $K(T)$) are thereby obtained and can be used to compute additional melting thermodynamics, including—$\Delta G_{2-st}(T_m) = -RT_m \ln K(T_m)$ and $\Delta S_{2-st}(T_m) = (\Delta H_{2-st}(T_m) - \Delta G_{2-st}(T_m))/T_m$. In our lab, a nonlinear least squares fit of Eqs. (6)–(8) to the melting thermogram is achieved using the Levenberg–Marquardt method (Press, Teukolsky, Vetterling, & Flannery,

2007) with local (linear least squares) estimates used as initial guesses. A thermogram fit using this method is provided in Fig. 2B. The fit is very good (which is required and typical), *but not perfect*, despite the selective (and misleading) display of only perfect data generally reported in past publications.

Finally, if the pretransition baseline is not sufficiently defined to permit direct estimation of ΔC_p, the thermogram may be fit to Eqs. (6)–(8) using a fixed value for ΔC_p. DSC-based measurements of the heat capacity change per base pair ΔC_p^{bp} typically lie between ca. 125 and 300 J (mol bp)$^{-1}$ K^{-1} (Chalikian et al., 1999; Tikhomirova, Beletskaya, & Chalikian, 2006), with our group reporting and typically using a value of 175 ± 70 J (mol bp)$^{-1}$ K^{-1} (Hughesman et al., 2011b), from which ΔC_p can be estimated as

$$\Delta C_p = n_{bp} \Delta C_p^{bp}, \tag{9}$$

where n_{bp} is the number of base pairs (natural or modified) in the duplex. Table 1 shows that melting thermodynamics obtained from a two-state model fit of the thermogram shown in Fig. 2B using these various approaches are quite comparable for good-quality DSC data—the values are generally equivalent within reported experimental errors. For DSC data of poorer quality, including those showing unstable baselines, use of Eqs. (6)–(8) with a fixed value of ΔC_p computed from Eq. (9) is recommended.

DSC Insight #6: Accurate DSC-based determination of melting thermodynamics for dsDNA samples is only possible on a properly functioning and calibrated instrument. Collection of "calibration" data for a duplex for which melting thermodynamics at fully specified conditions have been reported by one or more reputable laboratories (e.g., see Chiu et al., 2003) should precede all studies on new sequences. Parallel studies on a sequence of interest using your instrument and within a reputable

Table 1 Calorimetric and Two-State Thermodynamic Parameters

Analysis Method	T_m (°C)	$\Delta H(T_m)$ (kJ mol^{-1})	$\Delta S(T_m)$ (J mol^{-1} K^{-1})	ΔC_p (J mol^{-1} K^{-1})
Calorimetric	61.1 (±0.5)	412 (±13)	1140 (±34)	1790 (±760)
Two-state	60.8 (±0.5)	390 (±12)	1077 (±33)	1340 (±335)

Determined from DSC data for the helix-to-coil transition for the duplex 5′-*gaDacagttDaag*-3′/5′-*ctttaactgtttc*-3′, where *D* is the 2-amino-adenine LNA superbase. DSC data obtained at a $C_T = 69.3$ μ*M* in buffer containing 1 *M* NaCl, 10 m*M* Na_2HPO_4 (pH 7.0), and 1 m*M* Na_2EDTA.

core facility providing DSC expertise (e.g., The UBC Centre for Bio-Thermodynamics) may also serve this purpose well.

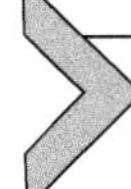

3. MELTING THERMODYNAMICS FROM UV SPECTROSCOPY

Though modern DSC instruments are motivating measurements of DNA melting thermodynamics by calorimetry, spectroscopy, most notably UVM, remains the far more commonly used technique. In a standard dsDNA melting study by UVM, absorbance (A) at a single wavelength, typically 260 nm (A_{260nm}) but sometimes 268 nm (where the same extinction coefficient applies to melting of either *g*/*c* or *a*/*t* base pairs), is measured as a function of temperature. The $A_{260nm}(T)$ data so collected, when interpreted based on the fact that the A_{260nm} of dsDNA is lower than that of the constituent single-stranded oligonucleotides, allow one to monitor the melting transition and from it the value of α at each temperature. The observed change in absorbance accompanying the melting transition (Fig. 4A)—a phenomenon known as the hyperchromic effect—arises in part through the loss of hydrogen bonding between paired bases and disruption of stacking interactions between neighboring base pairs (Wartell & Benight, 1985). The resulting change in A_{260nm} across a melting transition is typically little more than 0.1 absorbance units, but this is sufficient resolution to regress accurate melting thermodynamics at T_m.

3.1 Strengths, Weaknesses, and What Can be Determined Reliably

As detailed below, the standard procedure for computing melting thermodynamics from raw $A_{260nm}(T)$ data requires one to assume that the A_{260nm} reading at any temperature represents a linear combination of two spectral components—one specific to the DNA duplex and the other to the two single strands. Thus, the analysis assumes that the melting transition obeys the two-state reaction model. This assumption is appropriate for many natural and modified dsDNA sequences, but Haq, Chowdhry, and Chaires (1997) have shown that it is not necessarily true, as spectral analyses of some duplexes indicate the presence of multiple intermediate species (whose concentrations vary) within the melting envelop. Much more complex spectroscopic and data analysis methods (e.g., singular value decomposition methods) are then required to properly interpret the melting transition and extract useful thermodynamic data from it (Haq et al., 1997). In these

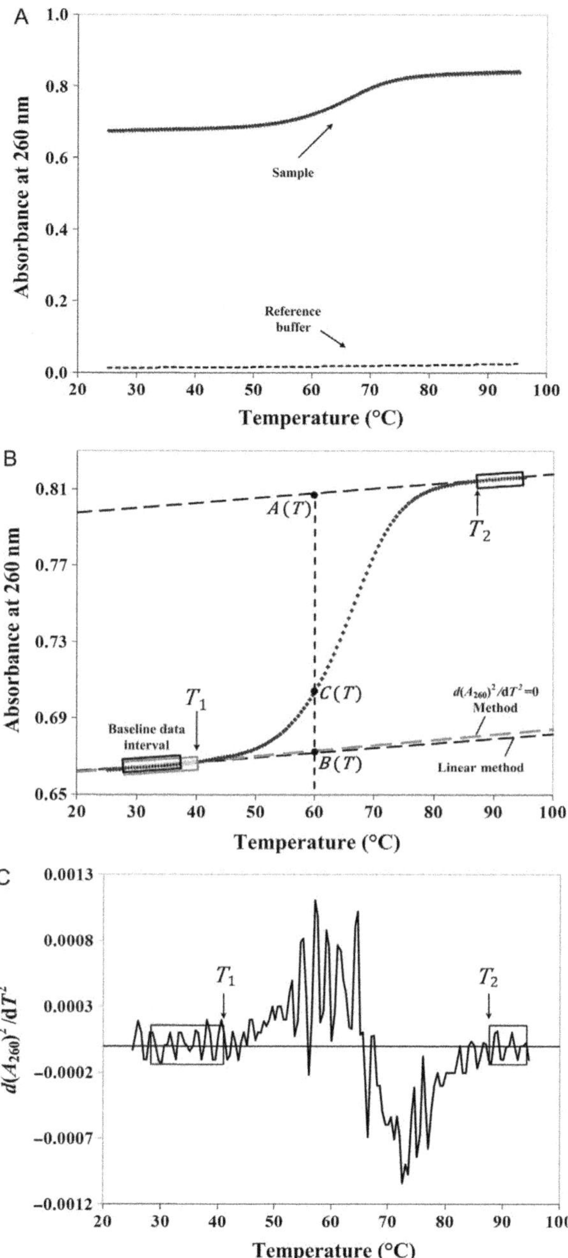

Figure 4—Cont'd

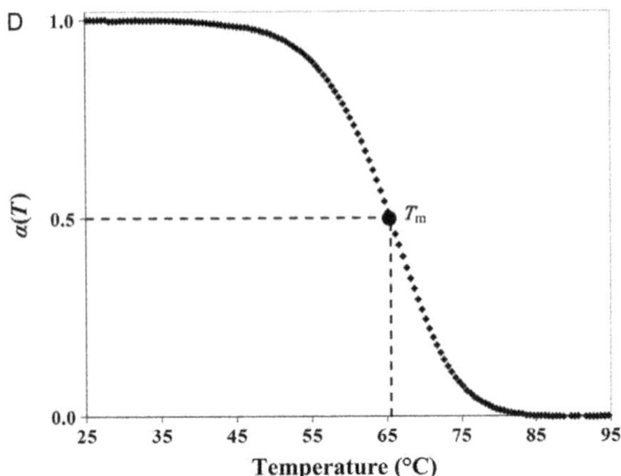

Figure 4 UVM data for LNA-substituted duplex 5′-*gaDacagttDaag*-3′/5′-*ctttaactgtttc*-3′ obtained at a $C_T = 75$ μ*M* in buffer containing 1 *M* NaCl, 10 m*M* Na_2HPO_4 (pH 7.0), and 1 m*M* Na_2EDTA. (A) Raw $A_{260nm}(T)$ data for sample and reference buffer, (B) reference buffer-normalized $A_{260nm}(T)$ data, (C) $d^2(A_{260nm})/dT^2$ values with regions where $d^2(A_{260nm})/dT^2 = 0$ identified, and (D) $\alpha(T)$ data computed using standard baseline fit.

cases, the main advantages in analyzing dsDNA melting thermodynamics by UVM—namely (i) the availability and generally low cost of the required equipment and (ii) the relative simplicity of the data analysis algorithms—are lost or at least substantially reduced.

For dsDNA sequences whose melting transition can be described by two-state theory, a UVM-measured melting transition may be analyzed to obtained accurate melting thermodynamics for a duplex at a given C_T. Ideally, one would like to regress that UVM data to the fundamental thermodynamic potential for the melting reaction, ΔG:

$$\begin{aligned}\Delta G(T) &= -RT\ln(K(T)) \\ &= \Delta H^{\circ}_{2-st} + \Delta C_p(T - T_o) - T\left[\Delta S^{\circ}_{2-st} + \Delta C_p \ln\left(\frac{T}{T_o}\right)\right]\end{aligned} \tag{10}$$

where $K(T)$ is given in Eq. (4A) or (4B) depending on whether the dsDNA is formed from nonself-complementary or self-complementary strands, respectively. This would permit one to not only determine T_m, ΔH°_{2-st}, and ΔS°_{2-st}, but ΔC_p as well. However, while the UVM method accurately monitors the melting transition, it does not measure melting thermodynamics directly. As a result, only T_m, ΔG and first-derivative properties of ΔG (i.e., ΔH_{2-st} and ΔS_{2-st}) can be determined from direct fitting of the two-state reaction model to a given UVM-derived melting transition; ΔC_p cannot (Chaires, 1997).

3.2 Sample Preparation

Chemically modified oligonucleotides purchased from a licensed vendor (e.g., IDT Inc.; Coralville, IA or Exiqon Inc.; Vadbæk, Denmark) are resuspended in phosphate (1 M NaCl, 10 mM $NaHPO_4$ (pH 7), and 1 mM Na_2EDTA) or cacodylate (1 M NaCl, 10 mM sodium cacodylate ($(CH_3)_2AsO_2Na{\cdot}3H_2O$) (pH 7), and 0.5 m$M$ Na_2EDTA) reference buffer. A_{260nm} readings for each such stock solution are then recorded at 25 and 80 °C to define the ssDNA concentration and, if necessary, adjust it (by buffer addition) to a desired value. The required extinction coefficient is sequence specific and may be computed using any of a number of accurate methods (see, for example, Tataurov, You, & Owczarzy, 2008). We generally use the method (http://www.owczarzy.net/extinctionDNA.htm) provided by IDT Inc. Significant difference in the A_{260nm} readings (or C_T estimates) at 25 and 80 °C likely indicates formation of stable secondary structure(s) (e.g., a hairpin) in the oligonucleotide at lower temperatures, which should be avoided if at all possible for reasons defined in Section 2. The two stocks of ssDNA are then prepared and mixed at equimolar concentration as before using filtered buffer to remove insoluble materials that might distort the spectral readings. In preparing the dsDNA sample, the C_T should be set so as to record A_{260nm} readings for the complete melting transition (baselines included) within the linear range (ideally at the higher end of that range) of the spectrophotometer when employing either a 1 cm or a lower volume optical path-length quartz cuvette. For melting of dsDNA 8–20 bp in length in a standard 1-cm path-length cuvette, a C_T between 2 and 10 μM is often appropriate.

3.3 Experimental Procedure

UVM-based monitoring of a dsDNA melting transition requires a UV/VIS spectrophotometer having high-quality optics, excellent signal stability in the UV/far-UV region, and a sample compartment equipped with a peltier-type thermopile (heating block) that simultaneously heats two or more cuvette-held samples at a user-programmed rate with precise temperature monitoring. Our laboratory and associated UBC Centre for Bio-Thermodynamics employ a custom 12-cell Cary 100E instrument (Agilent).

A solution of the dsDNA-free buffer is prepared and then degassed at 80 °C (Thermovac heating block) for 5–10 min so as to avoid formation of bubbles at higher temperatures that might interfere with $A_{260\text{nm}}$ readings. Each cuvette to be used is cleaned, loaded with degassed reference buffer and then sealed, in our lab by tightly securing each cuvette cap using high-T resistant Teflon tape (other methods to avoid evaporation of solution during heating are also used; Mergny & Lacroix, 2003). One of the caps bears a thermocouple that you have precalibrated to record the temperature of the sample it seals. Each sealed buffer solution is loaded into the spectrophotometer idling at 25 °C, taking careful precaution to properly align the cuvette, and the cuvette-held solutions are equilibrated until the thermocouples in the heating block and sample read 25.0 ± 0.3 °C. The user-programmed scanning cycle is then initiated, starting with a temperature downscan, usually from 25 to 20 °C. In UVM, we typically avoid scanning below 20 °C, as condensation events at temperatures below ambient can mar melting transition data. Methods for solving this problem have been proposed (e.g., continuously flooding the sample block chamber with dry nitrogen gas), but have proven unreliable in our experience. In a typical cycle, the samples are reequilibrated at 20 °C for 10 min, then subjected to an upscan from 20 to 95 °C at a fixed linear scan rate of no more than 0.8 °C min^{-1}, during which $A_{260\text{nm}}(T)$ data are recorded at the chosen filter period for the instrument (we collect data every 0.5 °C). Finally, the sample is subjected to a downscan and the full cycle repeated two or more times.

UVM Insight #1: Compared to DSC, accurate determination of melting thermodynamics by UVM generally requires the use of a slower scan rate. This is due in part to the thermal mass and insular properties of the cuvettes, which together serve to slow the rate of sample heating. Scan rates ≤ 0.5 °C are therefore often required.

Ideally, the buffer used in UVM studies shows no absorbance at the chosen wavelength across the entire temperature range, and this is the case for the standard phosphate and cacodylate buffers most often employed. If changes in absorbance are recorded in the pure buffer scans, it suggests that the cuvette(s) is not clean or the buffer is contaminated. Data for the pure buffer scans are also used to verify that the thermocouples in the sample heating block and the selected sample are functioning properly, and that the rate of heating is sufficiently slow to keep the block and the sample(s) in thermal equilibrium.

Cuvettes other than that for the thermocoupled reference may then be cleaned and loaded with a dsDNA sample that has been degassed at 80 °C for

5–10 min. Loading and sealing of the cuvettes is as described for the reference buffer. Prior to data acquisition, the sample is first heated to 95 °C for 3 min, then cooled to 20 °C, typically at 1 or 3 °C min^{-1}, and held at 20 °C until the thermocouples in the heating block and reference equilibrate. The user-programmed scanning cycle is then initiated. Melting transition data for three upscans are recorded and overlaid to verify that they are substantially superimposable—this confirms both proper sample handling and thermodynamic reversibility. If they do not, the scan rate is reduced and the study repeated.

Following the last scanning cycle, the sample temperature is raised to 80 °C and the $A_{260\mathrm{nm}}$ reading at that temperature used to check C_{T}.

3.4 Analysis of UVM Data

Buffer-normalized $A_{260\mathrm{nm}}$ data for each scan are plotted against T to visualize the melting transition and regress melting thermodynamics. If the reference buffer does not absorb, the raw (Fig. 4A) and buffer-normalized (Fig. 4B) melting transition data are essentially indistinguishable. Here, we show that one may analyze buffer-normalized $A_{260\mathrm{nm}}(T)$ data using either a generally more accurate modern nonlinear approach or any one of the classic methods.

Modern analysis method: Provided both the pre- and posttransition baselines are stable and linear, buffer-normalized $A_{260\mathrm{nm}}$-vs.-T may be used to compute $\alpha(T)$ through nonlinear least squares fit to the relation

$$A_{260\mathrm{nm}}(T) = \left(m_{\mathrm{pre}}^{\mathrm{UVM}} T + b_{\mathrm{pre}}^{\mathrm{UVM}}\right)\alpha(T) + \left(m_{\mathrm{post}}^{\mathrm{UVM}} T + b_{\mathrm{post}}^{\mathrm{UVM}}\right)(1-\alpha(T)) \tag{11}$$

where $m_{\mathrm{pre}}^{\mathrm{UVM}}$ and $b_{\mathrm{pre}}^{\mathrm{UVM}}$, and $m_{\mathrm{post}}^{\mathrm{UVM}}$ and $b_{\mathrm{post}}^{\mathrm{UVM}}$ are the slopes and intercepts of the pre- and posttransition baselines, respectively. Fitting of Eqs. (7) and (8) to the resulting $\alpha(T)$ generally yields high-quality values of $K(T)$, $\Delta H_{2-\mathrm{st}}^{\mathrm{o}}$ and T_{m}. In Eq. (8), $K(\alpha=0.5)$ is given by either Eq. (4A) or (4B), and the Levenberg–Marquardt algorithm is used to fit the model to the data. As a UVM-measured melting transition cannot be used to determine ΔC_{p}, its value in Eq. (8) must either be computed using Eq. (9) or set equal to zero for this regression, which we find converges well if $T_{\max}$ (given by the temperature at which $d(A_{260\mathrm{nm}})/\mathrm{d}T$ is a maximum) is used as the value of T_{m} in the first iteration. $\Delta S_{2-\mathrm{st}}^{\mathrm{o}}$ is then computed as $\left(\Delta H_{2-\mathrm{st}}^{\mathrm{o}} + RT_{\mathrm{m}} \ln[K(\alpha=0.5)]\right)/T_{\mathrm{m}}$.

Different methods have been proposed to select and fit the two required baselines. The standard approach, reflected in Eq. (11), is to compare linear least squares fits of 8–10 °C spans of the putative baseline region to identify that span having the highest R^2 value—this sets the slope and intercept values required in Eq. (11). If applied to the pretransition baseline, it also sets the temperature T_1 defining the start of the melting transition, which is given by highest temperature within that 8–10 °C span. Setting of T_2, the transition end point, and $m_{\text{post}}^{\text{UVM}}$ and $b_{\text{post}}^{\text{UVM}}$ is achieved in an analogous manner.

UVM Insight #2: Other more advanced methods for fitting pre- and posttransition baselines may be used. In our view, among the most useful is that developed by Owczarzy (2005), who found that high-quality melting thermodynamics can often be obtained using linear baselines fit to regions where $d^2(A_{260\text{nm}})/\text{d}T^2=0$. An example of this approach is shown in Fig. 4C for the data reported in Fig. 4B. One challenge to this approach is the noise created by numerical determination of second derivatives, which can make identification of the baseline regions and the values of T_1 and T_2 challenging. A data smoothing filter is therefore sometimes applied, which can ease assignments but also introduce error.

Melting thermodynamic parameters determined for the example sequence (Fig. 4B) using these two different baseline fit strategies are shown in Table 2.

Classic analysis methods: In most published analyses of dsDNA melting transition data acquired by UVM, ΔC_{p} is usually ignored for the reasons given in Section 3.1. The fundamental potential used for regressing melting thermodynamics is then given by

$$\Delta G(T) = -RT\ln(K(T)) = \Delta H^{\text{o}}_{2-\text{st}} - T\Delta S^{\text{o}}_{2-\text{st}}, \tag{12}$$

where, because we have assumed $\Delta C_{\text{p}}=0$, $\Delta H_{2-\text{st}}$ and $\Delta S_{2-\text{st}}$ are now temperature independent. As in Eq. (10), the superscript o denotes that $\Delta H_{2-\text{st}}$ and $\Delta S_{2-\text{st}}$ are defined at the standard state, which for UVM experiments is typically DNA suspended in either the standard phosphate or cacodylate reference buffer (pH 7) at atmospheric pressure. $K(T)$ is given by Eq. (4A) for dsDNA melting into two nonself-complementary strands, and by Eq. (4B) for two self-complementary strands. K can thereby be expressed as a function of $\alpha(T)$, and Eq. (12) rearranged to give

$$T = \frac{\Delta H^{\text{o}}_{2-\text{st}}}{\Delta S^{\text{o}}_{2-\text{st}} - R\ln[K(\alpha(T))]} \tag{13}$$

Table 2 Melting Thermodynamics for the LNA-Substituted Duplex 5′-*ttcatagCcgt*-3′/5′-*acggctatgaa*-3′

Baseline Fit Method	T_m (°C)	$\Delta H(T_m)$ (kJ mol^{-1})	$\Delta S(T_m)$ (J mol^{-1} K^{-1})	$\Delta G(T_m)$ (kJ mol^{-1})
Standard	65.5 (±0.2)	328 (±7)	878 (±18)	30.6 (±0.1)
$\frac{d(A_{260nm})^2}{dT^2}=0$	65.6 (±0.3)	334 (±7)	895 (±19)	30.7 (±0.1)

Determined from UVM data by regression to the two-state reaction model (Eqs. 4A and 4B, 7, 8, and 11) using the standard and Owczarzy (2005) baseline fitting methods. UVM data obtained at a $C_T=75$ μ*M* in buffer containing 1 *M* NaCl, 10 m*M* Na_2HPO_4 (pH 7.0), and 1 m*M* Na_2EDTA.

In classic analyses of UVM data, the regression process requires one to compute $\alpha(T)$ from the buffer-normalized UVM data. Provided the melting transition is centered at a temperature between ca. 50 and 70 °C, the UVM experiment will typically yield a stable posttransition baseline, making its fitting with good certainty fairly straightforward. But for chemically modified duplexes, particularly ones in which secondary structures are possible in one or both strands, fitting of the pretransition baseline can be problematic.

Once selected, each baseline fit is extrapolated through the melting transition, permitting the lever rule to be used to determine α at each temperature within the span bounded by T_1 (where $\alpha=1$) and T_2 (where $\alpha=0$):

$$a(T)=\frac{A(T)-C(T)}{A(T)-B(T)} \tag{14}$$

As shown in Fig. 4B, $A(T)$, $B(T)$, and $C(T)$ are the y-axis values at T of the extrapolated posttransition baseline, the extrapolated pretransition baseline, and the melting transition, respectively. One may thereby plot $\alpha(T)$ versus T for the transition (Fig. 4D). T_m is given by the temperature within the melting transition at which $\alpha=0.5$, and

$$\Delta G(T_m)=-RT\ln[K(\alpha=0.5)] \tag{15}$$

The inverse tangent ($dT/d\alpha$) of the melting transition shown in Fig. 4D is then evaluated at each T to compute ($dT/d\alpha$) values. Those data may be fit using the nonlinear least squares (Levenberg–Marquardt) method to the two-state model relation derived by differentiation of Eq. (13)

$$\frac{\mathrm{d}T}{\mathrm{d}\alpha} = -R\frac{\Delta H^{\circ}_{2-\mathrm{st}}}{\left(\Delta S^{\circ}_{2-\mathrm{st}} - R\ln[K(\alpha)]\right)^2}\frac{\left(\frac{\mathrm{d}[K(\alpha)]}{\mathrm{d}\alpha}\right)}{K(\alpha)}$$
$$= -\frac{RT^2}{\Delta H^{\circ}_{2-\mathrm{st}}}\frac{\left(\frac{\mathrm{d}[K(\alpha)]}{\mathrm{d}\alpha}\right)}{K(\alpha)} \quad (16)$$

A value of $\Delta H^{\circ}_{2-\mathrm{st}}$ may thereby be regressed, and a value of $\Delta S^{\circ}_{2-\mathrm{st}}$ then be determined as $\Delta S^{\circ}_{2-\mathrm{st}} = \left(\Delta H^{\circ}_{2-\mathrm{st}} - \Delta G(T_{\mathrm{m}})\right)/T_{\mathrm{m}}$. For melting of dsDNA into either nonself-complimentary strands, where $K(\alpha)$ is given by Eq. (4A), or self-complementary strands ($K(\alpha)$) given by Eq. (4B), Eq. (16) can be expressed as

$$\frac{\mathrm{d}T}{\mathrm{d}\alpha} = -\frac{RT^2}{\Delta H^{\circ}_{2-\mathrm{st}}}\frac{1+\alpha}{\alpha(1-\alpha)} \quad (17)$$

Provided the pre- and posttransition baselines are well defined and stable, the above regression approach generally represents a reliable method of determining melting thermodynamics from UVM data. In practice, the creation and use of plots such as those shown in Fig. 4C and D is not done, as the numerical methods needed to generate $\alpha(T)$ and $\mathrm{d}\alpha/\mathrm{d}T$ values from the buffer-normalized melting transition are readily available in many commercial programs, including Matlab and Excel.

UVM Insight #3: Equation (17) does not depend on C_{T}. As a result, relative to DSC studies, the accuracy of $\Delta H^{\circ}_{2-\mathrm{st}}$ values determined from UVM data is less sensitive to precise knowledge of strand concentration. HPLC-purified strands therefore may not be required, which is a significant advantage when studying melting of expensive duplexes comprised of chemically modified oligonucleotides.

A simplified version of this data regression algorithm may also be used, often to good effect. It is based on the fact that $T = T_{\mathrm{m}}$, when $\alpha = 0.5$, and at that temperature Eq. (17) reduces to

$$\Delta H^{\circ}_{2-\mathrm{st}} = -6RT^2_{\mathrm{m}}\left.\frac{\partial\alpha}{\mathrm{d}T}\right|_{T_{\mathrm{m}}} \quad (18)$$

Here, simple linear least squares fitting of $\alpha(T)$ data within a narrow temperature range centered about $\alpha = 0.5$ may be used to estimate $\left.\frac{\mathrm{d}\alpha}{\mathrm{d}T}\right|_{T_{\mathrm{m}}}$, which when introduced into Eq. (18) allows one to determine $\Delta H^{\circ}_{2-\mathrm{st}}$.

Finally, and particularly for dsDNA sequences displaying ill-defined pretransition (or posttransition) baselines, we note that accurate determination of T_m from UVM data may not be possible. In this case, T_m is usually assumed to be equal to T_{max} for the buffer-normalized melting transition. Measurement of T_{max} (or corresponding T_m values) as a function of C_T then allows determination of ΔH^o_{2-st} and ΔS^o_{2-st} through least squares fitting of a linear form of Eq. (13)

$$\frac{1}{T_m} = -\frac{R}{\Delta H^o_{2-st}} \ln[K(\alpha=0.5)] + \frac{\Delta S^o_{2-st}}{\Delta H^o_{2-st}} \tag{19}$$

where $K(\alpha=0.5)$ equals C_T or $C_T/4$ for melting of self-complementary or nonself-complementary strands, respectively. The display of $1/T_m$ data as a function of C_T is known as a van't Hoff plot, from which a fairly accurate estimate of ΔH^o_{2-st} can be obtained from the slope.

UVM Insight #4: In theory, ΔS^o_{2-st} can be estimated from the intercept of a van't Hoff plot. However, data extrapolation is required, and this generally leads to unacceptably large errors. One is therefore better served to compute ΔS^o_{2-st} as $\left(\Delta H^o_{2-st} + RT_m \ln[K(\alpha=0.5)]\right)/T_m$.

Curvature observed in a van't Hoff plot can be theoretically related to a non-zero ΔC_p. We find, however, that UVM data are rarely of sufficient quality to yield a reliable estimate of ΔC_p by van't Hoff plot analysis due to a combination of instrument noise, uncertainties in baseline assignments, and the compensating nature of the T dependencies of ΔH and ΔS that result in a weak T dependence of ΔG.[2]

Finally, though arguably the most popular method, analysis of UVM data through a van't Hoff plot is error prone. T_m and T_{max} are not equal, though they generally are close in value. Additional sources of error are known (Breslauer, 1995), and together they serve to reduce the precision of melting thermodynamics obtained using this classic method.

4. ERRORS ANALYSIS

Determination of total errors in melting thermodynamic values determined from DSC and UVM experiments is done in a similar manner. For

[2] Though it cannot be done for melting of either short duplex DNA or hairpin structures within ssDNA (the T_m for unimolecular melting reactions has no dependence on C_T), UVM data can be used to estimate ΔC_p, in some cases with good accuracy, when applied to melting of triplex complexes and higher molecularity complexes (Breslauer, 1995).

brevity, the following methodology specifically applies to UVM data, but the changes needed to apply it to DSC data are minor and obvious.

Sources of error in any thermodynamic parameter (e.g., T_m, ΔH, ΔS, etc.) determined from analysis of a single buffer-normalized UVM scan include uncertainties (standard deviations σ) in the subtracted reference buffer scan (σ_{buffer}), the model fit (σ_{model}), and the baselines selected ($\sigma_{baseline}$). For ΔH, for example, these uncertainties are typically of order ± 0.0003 AU, ± 1.7 kJ mol^{-1}, and ± 6.7 kJ mol^{-1}, respectively. It is, however, easily shown that uncertainty in the subtracted reference buffer scan makes a negligible contribution to the total error (σ_i) in any parameter derived from analysis of an individual UVM scan i. Thus,

$$\sigma_i = \sqrt{\sigma^2_{model} + \sigma^2_{baseline}} \tag{20}$$

The propagated error (σ_j) in the parameter (in this example ΔH) determined from independent analyses of N replicate scans of the same sample is then given by

$$\sigma_j = \sqrt{\sigma^2_{instrument} + \frac{1}{N-1}\sum_{i=1}^{N}\left(\Delta H_i - \Delta H_j\right)^2}, \tag{21}$$

where $\sigma_{instrument}$ gives the variance in instrument performance (temperature control and scan rate uncertainties, etc.), ΔH_i is the value of the parameter determine from scan i, and ΔH_j ($=\overline{\Delta H_i}$) is the mean value computed from the ΔH_i for all scans analyzed for that sample.

If M completely independent preparations of the same sample are made, the total propagated error (σ_k) in the parameter is computed as

$$\sigma_k = \sqrt{\sigma^2_{sample} + \frac{1}{M-1}\sum_{i=1}^{M}\left(\Delta H_j - \Delta H_k\right)^2}, \tag{22}$$

where σ_{sample} gives the uncertainty in sample preparation (pipetting error, C_T value, etc.) and ΔH_k ($=\overline{\Delta H_j}$).

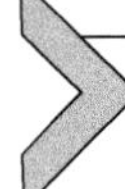

5. UNDERSTANDING THE CONTRIBUTION TO DUPLEX STABILITY OF SPECIFIC NUCLEOTIDE MODIFICATIONS

As described here, DSC and UVM experiments provide melting thermodynamics for natural and modified dsDNA at the measured T_m for the specified C_T. In addition, DSC studies have and can provide knowledge of ΔC_p. However, in the vast majority of studies reporting melting thermodynamics for these systems, the analyses ignore ΔC_p entirely. This is both regrettable and problematic. To understand why, let us consider the question of the precise contribution that replacing a single nucleotide in a natural DNA duplex with its corresponding LNA makes to the thermal stability of the duplex. As noted, the primary function of the LNA modification is to lock the furanose sugar into a C3′-*endo* conformation. This lowers the entropy of the nucleotide in its ssDNA state. To a first approximation then, one would expect that an LNA substitution enhances the thermal stability of dsDNA by reducing the entropy loss accompanying base pair formation. Other factors may contribute, but one would therefore expect to observe an entropic contribution to the stability enhancement.

To rigorously determine whether this is indeed the case, one must compare measured melting thermodynamics for the un-substituted and LNA-substituted duplexes at the same temperature—the T_m of the unsubstituted duplex, say. As the T_m of the LNA-substituted duplex ($T_{m,LNA}$) will differ from that of the unsubstituted duplex ($T_{m,DNA}$), thermodynamics stipulates that knowledge of ΔC_p is required. For the melting enthalpy and entropy, for example,

$$\Delta H_{LNA}(T_{m,DNA}) = \Delta H_{LNA}(T_{m,LNA}) + \Delta C_p(T_{m,DNA} - T_{m,LNA}) \quad (23)$$

and

$$\Delta S_{LNA}(T_{m,DNA}) = \Delta S_{LNA}(T_{m,LNA}) + \Delta C_p \ln\left(\frac{T_{m,DNA}}{T_{m,LNA}}\right) \quad (24)$$

One may then determine the net enthalpic and entropic contributions to thermal stability provided by the LNA substitution through the difference relations

$$\Delta\Delta H_{LNA}(T_{m,DNA}) = \Delta H_{LNA}(T_{m,DNA}) - \Delta H_{DNA}(T_{m,DNA}) \quad (25)$$

and

$$\Delta\Delta S_{LNA}(T_{m,DNA}) = \Delta S_{LNA}(T_{m,DNA}) - \Delta S_{DNA}(T_{m,DNA}) \quad (26)$$

Table 3 reports the net contribution of a single LNA substitution to the stability of a representative duplex using this analysis and melting thermodynamics data we measured and reported previously (Hughesman et al., 2011b). As expected, the LNA substitution results in a $\Delta\Delta S_{LNA} < 0$,

Table 3 Net (Incremental) Enthalpic and Entropic Contributions of an LNA Substitution

Parameters	Units	*5′-ggcacgcttcg-3′*[a]	*5′ggcaCgcttcg-3′*[a]
T_m	(°C)	68.2 (±0.5)	73.2 (±0.5)
ΔG (37 °C)	(kJ mol^{-1})	61.9 (±1.3)	67.4 (±1.3)
$\Delta H(T_m)$	(kJ mol^{-1})	370 (±11)	377 (±11)
$\Delta S(T_m)$	(J mol^{-1} K^{-1})	993 (±29)	999 (±30)
ΔC_p[b]	(J mol^{-1} K^{-1})	1960 (±490)	1440 (±360)
Incremental changes		$\Delta C_p = 0$ Assumption	$\Delta C_p > 0$[c]
$\Delta\Delta G$ (37 °C)	(kJ mol^{-1})	5.4	5.0
$\Delta\Delta H(T_{m,DNA})$	(kJ mol^{-1})	7.5	0.4
$\Delta\Delta S(T_{m,DNA})$	(J mol^{-1} K^{-1})	6.3	−14.6
ΔT_m	(°C)	5.0	

[a]Duplex melting was conducted at a $C_T = 75$ μ*M* in buffer containing 1 *M* NaCl 10 m*M* Na_2HPO_4 (pH 7.0), and 1 m*M* Na_2EDTA.
[b]Reported values are those determined from model regression to the DSC thermogram for the specified duplex.
[c]Values for $\Delta\Delta G$(37 °C), $\Delta\Delta H(T_{m,DNA})$, and $\Delta\Delta S(T_{m,DNA})$ were computed using the measured ΔC_p for the LNA-substituted duplex.
To the helix-to-coil melting thermodynamics of a complementary duplex when computed using a measured ΔC_p or assuming $\Delta C_p = 0$.

indicating that the substitution does indeed thermally stabilize the duplex by reducing the entropy gain accompanying the helix-to-coil transition. This conclusion is supported by the value of $\Delta\Delta H_{LNA}$, which is statistically negligible, indicating the substitution provides no net enthalpic contribution to the observed enhancement in duplex stability.

One may then ask how this analysis would change if ΔC_p were ignored. The corresponding values of $\Delta\Delta H_{LNA}$ and $\Delta\Delta S_{LNA}$ computed based on this assumption are also reported in Table 3. Because ΔH increases more rapidly with temperature than does ΔS, this traditional (but thermodynamically imprecise) approach will often predict, as observed here, that the net contribution of the LNA substitution to the increase in thermal stability is largely enthalpically driven. Indeed, for this particular duplex, $\Delta\Delta H_{LNA}$ is so favorable that locking the furanose sugar is predicted *to destabilize the duplex entropically*.

This single example therefore highlights the fact that the ability to accurately measure melting thermodynamics, including ΔC_p, can serve to greatly improve our understanding of exactly how nucleotide modifications affect duplex stabilities. And this in turn should help to improve and extend the uses of these novel reagents.

REFERENCES

Abramov, M., Schepers, G., Van Aerschot, A., Van Hummelen, P., & Herdewijn, P. (2008). HNA and ANA high-affinity arrays for detection of DNA and RNA single-base mismatches. *Biosensors and Bioelectronics*, *23*, 1728–1732.

Alemayehu, S., Fish, D. J., Brewood, G. P., Horne, M. T., Manyanga, F., Dickman, R., et al. (2009). Influence of buffer species on the thermodynamics of short DNA duplex melting: Sodium phosphate versus sodium caodylate. *The Journal of Physical Chemistry B*, *113*, 2578–2586.

Andersen, N. K., Anderson, B. A., Wengel, J., & Hrdlicka, P. J. (2013). Synthesis and characterization of oligodeoxyribonucleotides modified with 2'-amino-alpha-L-LNA adenine monomers: High-affinity targeting of single-stranded DNA. *The Journal of Organic Chemistry*, *78*(24), 12690–12702.

Astakhova, I., & Wengel, J. (2014). Scaffolding along nucleic acid duplexes using 2'-amino locked nucleic acids. *Accounts of Chemical Research*, *47*(6), 1768–1777.

Babu, B. R., Raunek, A., Poopelko, N. E., Juhl, M., Bond, A. D., Parmar, V. S., et al. (2005). XNA (xylo nucleic acid): A summary and new derivatives. *European Journal of Organic Chemistry*, *11*, 2297–2321.

Breslauer, K. J. (1995). Extracting thermodynamic data from equilibrium melting curves for oligonucleotide order-disorder transitions. *Methods in Enzymology*, *259*, 221–242.

Breslauer, K. J., Frank, R., Blocker, H., & Marky, L. A. (1986). Predicting DNA duplex stability from the base sequence. *Proceedings of the National Academy of Sciences of the United States of America*, *83*(11), 3746–3750.

Chaires, J. B. (1997). Possible origin of differences between van't Hoff and calorimetric enthalpy estimates. *Biophysical Chemistry*, *64*, 15–23.

Chalikian, T. V., Völker, J., Plum, E., & Breslauer, K. J. (1999). A more unified picture for the thermodynamics of nucleic acid duplex melting: A characterization by calorimetric and volumetric techniques. *Proceedings of the National academy of Sciences of the United States of America*, *96*(14), 7853–7858.

Chiu, W. L. A. K., Sze, C. N., Ma, N. T., Chiu, L. F., Leung, C. W., & Au-Yeung, S. C. F. (2003). NTDB: Thermodynamic data-base for nucleic acids, version 2.0. *Nucleic Acids Research*, *31*, 2461–2474.

Diop-Frimpong, B., Prakash, T. P., Rajeev, K. G., Manoharan, M., & Egli, M. (2005). Stabilizing contributions of sulfur-modified nucleotides: crystal structure of a DNA duplex with 2'-O-(2-(methoxy)ethyl)-2-thiothymidines. *Nucleic Acids Research*, *33*(16), 5297–5307.

Fakhfakh, K., Marais, O., Cheng, X. B. J., Casteñeda, J. R., Hughesman, C. B., & Haynes, C. (2015). Molecular thermodynamics of LNA:LNA base pairs and the hyperstabilizing effect of 5'-proximal LN:DNA base pairs. *AIChE Journal*, *61*(9), 2711–2731.

Filimonov, V. V., Potekhin, S. A., Matveev, S. V., & Privolov, P. L. (1982). Thermodynamic analysis of scanning micro-calorimetric data. 1. Algorithms for deconvolution of complex heat absorption curves. *Molecular Biology*, *16*(3), 435–444.

Haq, I., Chowdhry, B. Z., & Chaires, J. B. (1997). Singular value decomposition of 3D DNA melting curves reveals complexity in the melting process. *European Biophysics Journal*, *26*, 419–426.

Henry, A. A., & Romesberg, F. E. (2003). Beyond A, C, G and T: Augmenting nature's alphabet. *Current Opinion in Chemical Biology*, *7*, 727–733.

Hughesman, C. B., Fakhfakh, K., Bidshahri, R., Lund, H. L., & Haynes, C. (2015). A new general model for predicting melting thermodynamics of complementary and mismatched B-form duplexes containing locked nucleic acids: Application to probe design for digital PCR detection of somatic mutations. *Biochemistry, 54,* 1338–1352.

Hughesman, C. B., Turner, R. F. B., & Haynes, C. (2008). Stability and mismatch discrimination of DNA duplexes containing 2,6-diaminopurine and 2-thiothymidine locked nucleic acid bases. *Nucleic Acids Symposium Series (Oxford), 52*(1), 245–246.

Hughesman, C. B., Turner, R. F. B., & Haynes, C. (2011a). Correcting for heat capacity and 5'-TA type terminal nearest neighbors improves prediction of DNA melting temperatures using nearest neighbor thermodynamic models. *Biochemistry, 50,* 2642–2649.

Hughesman, C. B., Turner, R. F. B., & Haynes, C. (2011b). Role of the heat capacity change in understanding and modeling melting thermodynamics of complementary duplexes containing standard and nucleobase-modified LNA. *Biochemistry, 50*(23), 5354–5368.

Hughesman, C. B., Turner, R. F. B., & Haynes, C. (2013). Measuring, interpreting and modeling the stabilities and melting temperatures of B-form DNA that exhibits a two-state helix-to-coil transition. In U. von Stocker & L. A. M. van der Wielen (Eds.), *Biothermodynamics: The role of thermodynamics in biochemical engineering* (pp. 355–389). Lausanne, Switzerland: EPFL Press.

Kang, M., Heuberger, B., Chaput, J. C., Switzer, C., & Feigon, J. (2012). Solution structure of a parallel-stranded oligoisoguanine DNA pentaplex formed by d(T(iG)$_4$ T) in the presence of Cs$^+$ ions. *Angewandte Chemie International Edition in English, 51,* 7952–7955.

Kaur, H., Wengel, J., & Maiti, S. (2008). Thermodynamics of DNA-RNA hetero-duplex formation: Effects of locked nucleic acid nucleotides incorporated into the DNA strand. *Biochemistry, 47*(4), 1218–1227.

Khakshoor, O., & Kool, E. T. (2011). Chemistry of nucleic acids: Impacts in multiple fields. *Chemical communications (Cambridge), 47*(25), 7018–7024.

Kool, E. T. (2002). Replacing the nucleobases in DNA with designer molecules. *Accounts of Chemical Research, 35,* 936–943.

Koshkin, A. A., Nielsen, P., Meldgaard, M., Rajwanshi, V. K., Singh, S. K., & Wengel, J. (1998). LNA (locked nucleic acid): An RNA mimic forming exceedingly stable LNA: LNA duplexes. *Journal of the American Chemical Society, 120,* 13252–13253.

Krueger, A. T., & Kool, E. T. (2009). Redesigning the architecture of the base pair: Toward biochemical and biological function of new genetic sets. *Chemistry and Biology, 16,* 242–248.

Krueger, A. T., Peterson, L. W., Chelliserry, J., Kleinbaum, D. J., & Kool, E. T. (2011). Encoding phenotype in bacteria with an alternative genetic set. *Journal of the American Chemical Society, 133,* 18447–18451.

Latorra, D., Campbell, K., Wolter, A., & Hurley, J. M. (2003). Enhanced allele-specific PCR discrimination in SNP genotyping using 3' locked nucleic acid (LNA) primers. *Human Mutation, 22*(1), 79–85.

Law, S. M., Eritja, R., Goodman, M. F., & Breslauer, K. J. (1996). Spectroscopic and calorimetric characterization of DNA duplexes containing 2-aminopurine. *Biochemistry, 35*(38), 12329–12337.

Leumann, C. J. (2001). DNA analogues: From supermolecular principles to biological properties. *Bioorganic and Medicinal Chemistry, 10*(4), 841–854.

Mergny, J. L., & Lacroix, L. (2003). Analysis of thermal melting curves. *Oligonucleotides, 13,* 515–537.

Micklefield, J. (2001). Backbone modification of nucleic acids: Synthesis, structure and therapeutic applications. *Current Medicinal Chemistry, 8,* 1157–1179.

Mikulecky, P. J., & Feig, A. L. (2006a). Heat capacity changes associated with nucleic acid folding. *Biopolymers, 82,* 38–58.

Mikulecky, P. J., & Feig, A. L. (2006b). Heat capacity changes associated with DNA duplex formation: Salt and sequence-dependent effects. *Biochemistry, 45,* 604–616.

Movileanu, L., Benevides, J. M., & Thomas, G. J. (2002). Determination of base and backbone contributions to the thermodynamics of pre-melting and melting transitions in B-DNA. *Nucleic Acids Research, 30*(17), 3767–3777.

Owczarzy, R. (2005). Melting temperatures of nucleic acids: Discrepancies in analysis. *Biophysical Chemistry, 117*, 207–215.

Petruska, J., Goodman, M. F., Boosalis, M. S., Sowers, L. C., Cheong, C., & Tinoco, I., Jr. (1988). Comparison between DNA melting thermodynamics and DNA polymerase fidelity. *Proceedings of the National Academy of Sciences of the United States of America, 85*, 6252–6256.

Pinheiro, V. B., & Hollinger, P. (2012). The XNA world: Progress towards replication and evolution of synthetic genetic polymers. *Current Opinion in Chemical Biology, 16*(3-4), 245–252.

Press, W. H., Teukolsky, S. A., Vetterling, W. T., & Flannery, B. P. (2007). *Numerical recipes: The art of scientific computing* (3rd ed.). Cambridge, UK: Cambridge University Press.

Privalov, P. L. (1980). Scanning microcalorimeters for studying macromolecules. *Pure and Applied Chemistry, 52*(2), 479–497.

Privalov, P. L., & Dragan, A. I. (2007). Microcalorimetry of biological macromolecules. *Biophysical Chemistry, 126*, 16–24.

Rajeev, K. G., Prakash, T. P., & Manoharan, M. (2003). 2'-Modified-2-thiothymidine oligonucleotides. *Organic Letters, 5*(17), 3005–3008.

Rouzina, I., & Bloomfield, V. A. (1999). Heat capacity effects on the melting of DNA.1. general aspects. *Biophysical Journal*, 77(6), 3242–3251.

Sanghvi, Y. S. (2011). A status update of modified oligonucleotides for chemo-therapeutics applications. In S. L. Beaucage, D. E. Bergstrom, P. Herdewijn, & A. Matsuda (Eds.), *Current protocols in nucleic acid chemistry* (pp. 4.1.1–4.1.2). Hoboken, NJ: John Wiley & Sons.

SantaLucia, J., Jr. (1998). A unified view of polymer, dumbbell, and oligonucleotide DNA nearest-neighbor thermodynamics. *Proceedings of the National Academy of Sciences of the United States of America, 95*, 1460–1465.

SantaLucia, J., Jr., Allawi, H. T., & Seneviratne, A. (1996). Improved nearest-neighbor parameters for predicting DNA duplex stability. *Biochemistry, 35*(11), 3555–3562.

Schoning, K., Scholz, P., Guntha, S., Wu, X., Krishnamurthy, R., & Eschenmoser, A. (2000). Chemical etiology of nucleic acid structure: The alpha-threofuranosyl-(3'->2') oligonucleotide system. *Science, 290*, 1347–1351.

Sugimoto, N., Nakano, S., Katoh, M., Matsumura, A., Nakamuta, H., Ohmichi, T., et al. (1995). Thermodynamic parameters to predict stability of RNA/DNA hybrid duplexes. *Biochemistry, 34*(35), 11211–11216.

Tataurov, A. V., You, Y., & Owczarzy, R. (2008). Predicting ultraviolet spectrum of single stranded and double stranded deoxyribonucleic acids. *Biophysical Chemistry, 133*, 66–70.

Tikhomirova, A., Beletskaya, I. V., & Chalikian, T. V. (2006). Energetics of nucleic acid stability: The effect of ΔC_p. *Journal of the American Chemical Society, 126*, 16387–16394.

Veedu, R. N., & Wengel, J. (2010). Locked nucleic acids: Promising nucleic acid analogues for therapeutic applications. *Chemistry and Biodiversity*, 7(3), 536–542.

Wartell, R. M., & Benight, A. S. (1985). Thermal denaturation of DNA molecules—A comparison of theory with experiment. *Physics Reports, 126*(2), 67–107.

Wilson, C., & Keefe, A. D. (2006). Building oligonucleotide therapeutics using non-natural chemistries. *Current Opinion in Chemical Biology, 10*, 607–614.

Yang, Z., Hutter, D., Sheng, P., Sismour, A. M., & Benner, S. A. (2006). Artificially expanded genetic information system: A new base pair with an alternative hydrogen bonding pattern. *Nucleic Acids Research, 34*, 6095–6101.

Zhou, Y. Q., Hall, C. K., & Karplus, M. (1999). The calorimetric criterion for a two-state process revisited. *Protein Sciences, 8*(5), 1064–1074.

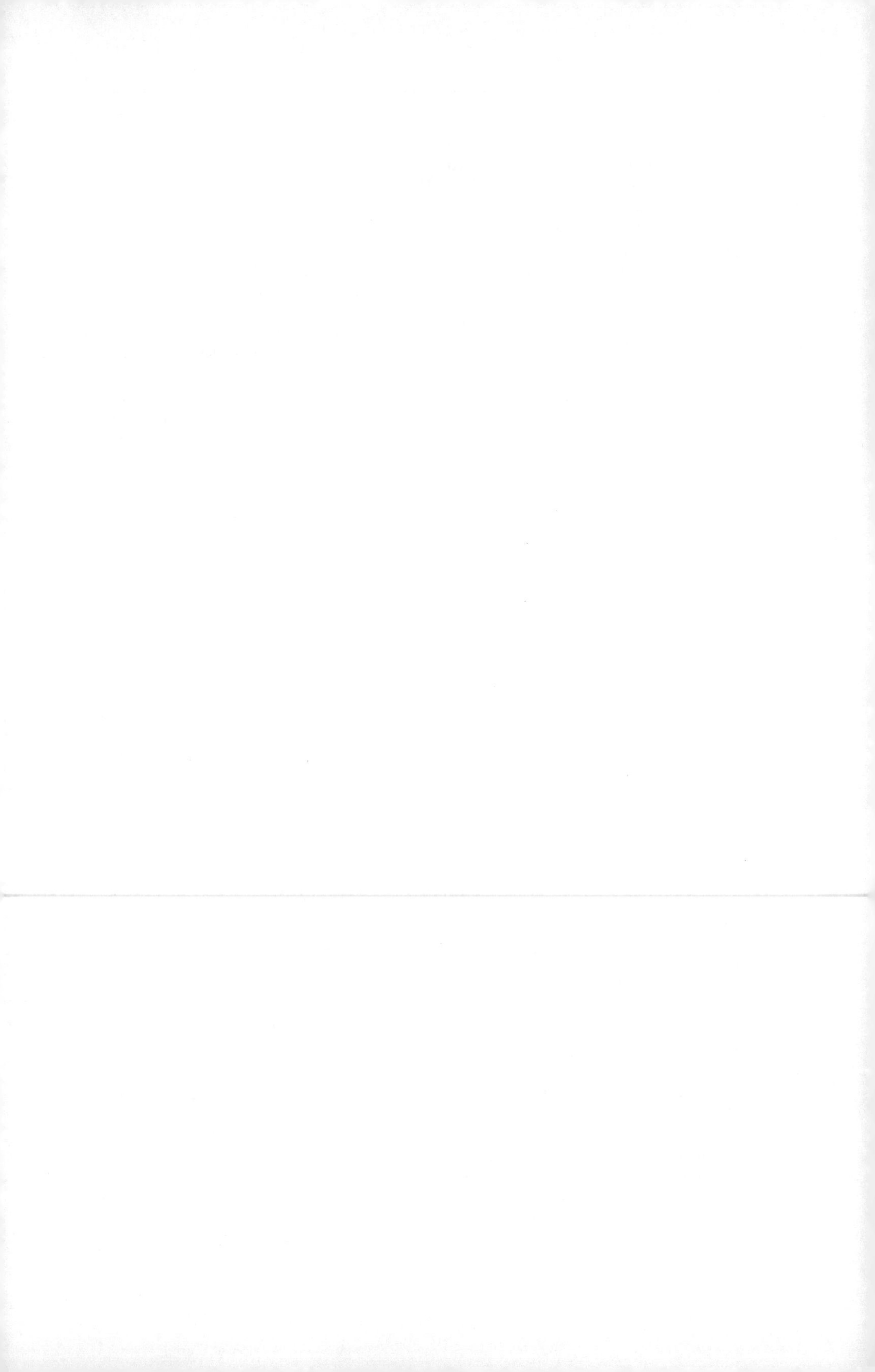

CHAPTER SIX

Calorimetric Quantification of Cyclodextrin-Mediated Detergent Extraction for Membrane-Protein Reconstitution

Martin Textor, Sandro Keller[1]
Molecular Biophysics, University of Kaiserslautern, Kaiserslautern, Germany
[1]Corresponding author: e-mail address: mail@sandrokeller.com

Contents

Abstract

For many *in vitro* studies, purified membrane proteins need to be reconstituted from detergent micelles into lipid bilayers to regain their native structures and functions. Stoichiometric complexation of detergent by cyclodextrin provides a tightly controllable strategy for detergent extraction. Here, we describe a practical approach making use of isothermal titration calorimetry to obtain a complete set of thermodynamic parameters that allows for quantitative prediction of the transition from micelles to bilayer membranes during reconstitution. These parameters include the dissociation constant of the cyclodextrin/detergent inclusion complex, the critical micellar concentration of the detergent, and the phase boundaries of the lipid/detergent phase diagram. The underlying theoretical framework involves linked equilibria among all pseudophases, as described previously (Textor, Vargas, & Keller, 2015). This chapter focuses on practical aspects of the approach and discusses caveats and calorimetry-specific details of data

Methods in Enzymology, Volume 567
ISSN 0076-6879
http://dx.doi.org/10.1016/bs.mie.2015.07.033

analysis. With the entire parameter set at hand, exploration of different reconstitution trajectories within the lipid/detergent phase diagram is possible. Together with the straightforward control over the rate of detergent extraction offered by cyclodextrin complexation, this opens the possibility of systematically tuning and optimizing the reconstitution process of membrane proteins. Provided some particular precautions are taken, the approach can be adapted to many other combinations of proteins, lipids, detergents, and cyclodextrins.

1. INTRODUCTION

Membrane proteins comprise more than half of all drug targets and, thus, are of outstanding scientific and medical importance (Overington, Al Lazikani, & Hopkins, 2006). However, membrane proteins are usually not readily amenable to *in vitro* studies because they require membrane-mimetic environments such as detergent micelles or lipid bilayers to retain their native structures and functions. In particular, membrane proteins with vectorial functions such as channels and transporters need to be reconstituted into lipid bilayer membranes after detergent-mediated solubilization and purification (Racker, 1979). To this end, proteins are transferred from micelles into bilayers by decreasing the concentration of detergent in a ternary protein/lipid/detergent mixture to a level that allows for the formation of bilayer vesicles (Seddon, Curnow, & Booth, 2004), which oftentimes is followed by complete removal of detergent. The stage between these two steps is crucial for the success of a reconstitution trial because destabilization of vesicles by detergent saturation leads to increased permeability of the lipid bilayer, thereby promoting membrane insertion of proteins (Herrmann, Danielczak, Textor, Klement, & Keller, 2015; Rigaud & Lévy, 2003). As the detergent content in the lipid bilayer depends on the detergent's water/membrane partition equilibrium, detailed knowledge of the interactions between the lipid(s) and detergent(s) used is required to steer the reconstitution process in a quantitative manner.

2. MEMBRANE-PROTEIN RECONSTITUTION

2.1 Reconstitution as a Trajectory in a Lipid/Detergent Phase Diagram

To a first approximation, reconstitution can be described thermodynamically by considering the aqueous bulk solution as well as the supramolecular aggregates that occur during the process, namely, micelles and bilayer vesicles, as pseudophases (Heerklotz, Lantzsch, Binder, Klose, & Blume, 1995). Then, pairs of lipid and detergent concentrations at the transitions between these pseudophases can be plotted as phase-boundary coordinates in a lipid/ detergent phase diagram. In the simplest case, which applies to many pseudobinary lipid/detergent systems in aqueous solvent, the phase boundaries follow two straight lines (Heerklotz et al., 1995; Heerklotz, Tsamaloukas, & Keller, 2009; Lichtenberg, 1985). Thus, the phase boundaries divide the phase diagram into three distinct ranges, each of which is characterized by the presence of a particular set of pseudophases: (i) at a given lipid concentration, the vesicular range features bilayer vesicles at low detergent concentrations; (ii) in the coexistence range between the two phase boundaries, both mixed vesicles and mixed micelles are present; and (iii) at high detergent concentrations, mixed micelles exist. In addition to the above supramolecular aggregates, detergent monomers are present in the aqueous phase throughout all three ranges. The vesicular and coexistence ranges are separated by the saturation boundary and the coexistence and micellar ranges by the solubilization boundary. The two phase boundaries are defined by their slopes, $R_D^{b,\,SAT}$ and $R_D^{m,\,SOL}$, which correspond to the detergent/lipid ratios in detergent-saturated bilayers and lipid-saturated micelles, respectively, and a common ordinate intercept, $c_D^{aq^\circ}$. The latter denotes the concentration of detergent monomers in the aqueous phase, which is constant within the coexistence range. Also, the bilayer and micelle compositions remain constant within the coexistence range with detergent/ lipid ratios of $R_D^{b,\,SAT}$ and $R_D^{m,\,SOL}$, respectively. For complete reconstitution, both phase boundaries have to be crossed, and the method of detergent extraction decides on the trajectory that is followed within the phase diagram.

2.2 Methods of Detergent Extraction

Different strategies have been established to extract detergents from reconstitution mixtures (Ollivon, Lesieur, Grabielle-Madelmont, & Paternostre, 2000; Rigaud, Chami, Lambert, Lévy, & Ranck, 2000; Rigaud & Lévy, 2003), including dilution (Jahnke et al., 2014; Schurtenberger, Mazer, Waldvogel, & Känzig, 1984), dialysis (Milsmann, Schwendener, & Weder, 1978; Rhoden & Goldin, 1979), gel filtration (Brunner, Skrabal, & Hauser, 1976), adsorption to bio-beads (Rigaud, Lévy, Mosser, & Lambert, 1998; Rigaud et al., 1997), and complexation by cyclic oligosaccharides called cyclodextrins (CDs) (DeGrip, VanOostrum, & Bovee-Geurts, 1998; Signorell, Kaufmann, Kukulski, Engel, & Rémigy, 2007; Textor, Vargas, & Keller, 2015). These methods result in different reconstitution trajectories, that is, simultaneous changes in lipid and detergent concentrations in the lipid/detergent phase diagram.

(i) *Dilution*: When the detergent concentration is decreased by dilution (Jiskoot, Teerlink, Beuvery, & Crommelin, 1986; Racker, Violand, O'Neal, Alfonzo, & Telford, 1979), the trajectory traced during reconstitution follows a straight line going through the origin of the phase diagram. Upon dilution, the concentrations of all components are concomitantly decreased, and the protein concentration, in particular, may be reduced to levels that are insufficient for subsequent studies such as spectroscopic measurements. Additionally, dilution needs to be followed by other methods such as gel filtration or dialysis to achieve complete detergent removal. For better control over vesicle size or to cope with low critical micellar concentrations (CMCs), which would require excessive dilution to reach subsolubilizing detergent concentrations, dilution can be accompanied by addition of preformed liposomes, resulting in a trajectory almost parallel to the ordinate of the phase diagram (Jahnke et al., 2014). In the latter case, however, the protein is still diluted, and the protein/lipid ratio does not remain constant.

(ii) *Gel filtration*: Detergent removal by size exclusion chromatography is not widely used anymore because it has been shown (Rigaud & Lévy, 2003) to result in incomplete or inhomogeneous incorporation of membrane proteins as well as heterogeneous size distributions of proteoliposomes. Moreover, gel filtration suffers from substantial lipid retention on the column (Ruysschaert et al., 2005).

(iii) *Dialysis*: In contrast with dilution and gel filtration, only the detergent concentration is changed when detergent removal is performed by dialysis. However, application of this method is limited to detergents having relatively high CMC values, because low-CMC detergents would require excessively long times for equilibration across dialysis membranes.

(iv) *Bio-beads*: A popular method of detergent extraction involves hydrophobic adsorption to polystyrene beads known as bio-beads (Lévy, Bluzat, Seigneuret, & Rigaud, 1990; Rigaud et al., 1997, 1998). Analogously to dialysis, bio-beads are generally expected to reduce only the detergent concentration in a reconstitution mixture. However, a known issue is that bio-beads may also extract phospholipids to a significant amount. Although Rigaud and colleagues (Rigaud et al., 1997) have shown that the adsorptive capacity of bio-beads for egg yolk phosphatidylcholine is two orders of magnitude lower than that for most detergents, phospholipid losses of 40% and 55% have been reported upon hydrophobic adsorption of Triton X-100 or octyl glucoside, respectively (Philippot, Mutaftschiev, & Liautard, 1983). As long as unspecific adsorption can be neglected, bio-beads allow for detergent removal at constant protein and lipid concentrations, which results in a reconstitution trajectory parallel to the ordinate of the phase diagram. The amount of bio-beads required for quantitative detergent removal can be inferred from their binding capacity for the detergent of interest (Rigaud et al., 1997). Nevertheless, it is common practice to ultimately add bio-beads in excess for complete detergent extraction; hence, exact knowledge of the total amount of bead slurry added is unnecessary. However, slow rates of detergent removal by stepwise addition of bio-beads require rather small amounts of beads that are difficult to deploy in a reproducible manner because of their finite and heterogeneous size. For example, for slow reconstitution of Ca^{2+}-ATPase (Young, Rigaud, Lacapère, Reddy, & Stokes, 1997), amounts as low as 1 mg are required per addition, which corresponds to approximately four medium-sized beads. More control over the reconstitution process can be achieved by combining multiple methods of detergent removal, such as "indirect" use of bio-beads outside a dialysis bag (Philippot et al., 1983).

(v) *Complexation by CDs*: In this chapter, we provide a practical guide detailing complexation by CDs as an alternative method of detergent extraction successfully used for reconstitution (DeGrip et al., 1998) and 2D crystallization of membrane proteins (Signorell et al., 2007) as well as for probing protein stability in the absence of a membrane-mimetic system (Broecker, Fiedler, Gimpl, & Keller, 2014). CDs are relatively small, ring-shaped oligosaccharides consisting of 6, 7, or 8 α-D-glucopyranoside units denoted α-, β-, and γ-CD, respectively. The nonpolar cavities of CDs bind a plethora of small hydrophobic molecules and moieties, including detergent alkyl chains (Rekharsky & Inoue, 1998). The aqueous solubility of CDs and CD inclusion complexes is due to the glucopyranose hydroxyl groups on the exterior of the ring and, together with ligand specificity, can be tuned by substitution of these hydroxyl groups by a broad range of other moieties. For a highly concentrated CD titrant, the dilution effect accompanying CD-mediated detergent extraction is negligible, and the reconstitution trajectory is comparable to that resulting from addition of bio-beads. Although the possibility of unspecific binding of proteins and lipids, as sometimes observed for bio-beads (Philippot et al., 1983), also needs to be taken into consideration for CDs, such issues usually can be circumvented by using an appropriate CD derivative. Moreover, application of CDs for detergent extraction allows the user to monitor the reconstitution process in real time with the aid of spectroscopic methods or isothermal titration calorimetry (ITC), as detailed below (Jahnke et al., 2014; Textor et al., 2015).

The methods for detergent extraction described above differ substantially in their capabilities of controlling the rate of detergent removal, which is of particular importance because it substantially affects the size and homogeneity of vesicles formed during reconstitution and can also influence protein incorporation with regard to topology (Helenius, Sarvas, & Simons, 1981). For instance, reconstitution by rapid dilution has been reported to result in heterogeneously sized and unstable proteoliposome dispersions (Jiskoot et al., 1986). By contrast, two vesicle populations of different protein/lipid ratios are obtained at slow rates of bio-bead-mediated reconstitution of Ca^{2+}-ATPase, whereas rapid detergent removal gives rise to homogeneous proteoliposomes with a uniform protein/lipid ratio (Young et al., 1997). This has been attributed to a mechanism proposed by Rigaud and colleagues (Lévy, Gulik, Bluzat, & Rigaud, 1992) that assumes more pronounced coalescence of binary lipid/detergent micelles with ternary protein/lipid/detergent micelles at fast rates, which is expected to produce a more homogeneous protein distribution. However, fast rates have been demonstrated to lead to asymmetric protein incorporation, with 20% of the cytoplasmic domains facing the vesicle interior (Young et al., 1997). Taken together, it appears that proteoliposome homogeneity and membrane-protein orientation depend not only on the rate of detergent removal but also on the method of detergent extraction. In general, however, according to the mechanism of vesicle formation as proposed by Lassic (Lasic, 1988), smaller vesicles are expected to form at faster rates of detergent depletion because of the relatively slow kinetics of micelle fusion. Reasonable rate control can be achieved by applying dilution in a stepwise manner, by adjusting the volume against which the reconstitution mixture is dialyzed, or by using a continuous flow-through dialysis apparatus (Engel et al., 1992; Milsmann et al., 1978). The rate of detergent extraction with bio-beads can be controlled to some extent by stepwise addition of increasing amounts of bio-beads to the reconstitution mixture (Rigaud et al., 1997). Moreover, bio-bead-mediated reconstitution involving Triton X-100 can be slowed down by lowering the temperature (Lévy et al., 1990). In general, however, CDs allow for a much more accurate and precise control of the rate of detergent removal because they can be dispensed in small volumes and, importantly, sequester detergent molecules in a stoichiometric manner with defined and adjustable affinities (Textor et al., 2015).

2.3 Quantitative Model of CD-Mediated Reconstitution

To fully exploit the advantages offered by CD-mediated reconstitution, a quantitative foundation is required that models the transitions among all monomeric and supramolecular states in a complex protein/CD/lipid/detergent mixture (Fig. 1). The theory underlying this model is based on a set of linked equilibria, as previously explained in detail and exemplified for the four-helix membrane protein Mistic (Textor et al., 2015; Roosild et al., 2005). In the present contribution, the focus lies on practical aspects of the approach, with a particular emphasis on those pertaining to the use of ITC as a highly sensitive and informative analytical method ideally suited for monitoring the reconstitution process in real time. In brief, the model predicts the supramolecular state(s) present in a complex reconstitution mixture for given concentrations of CD, lipid, and detergent on the basis of the corresponding lipid/detergent phase diagram. To this end, the concentration of free

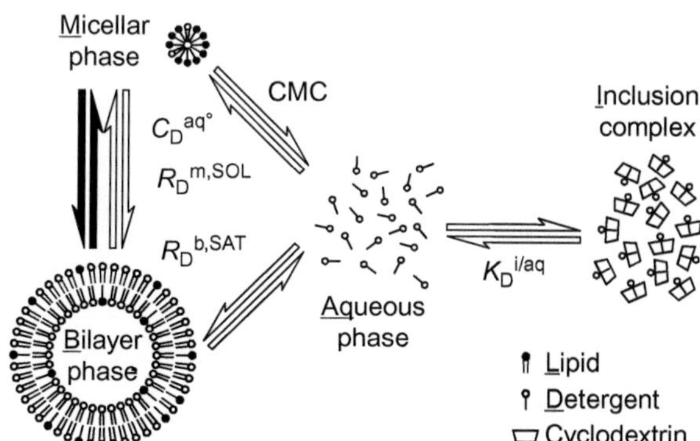

Figure 1 Overview of chemical species, pseudophases, and linked equilibria involved in CD-mediated reconstitution and thermodynamic parameters required for its quantitative description. (See the color plate.)

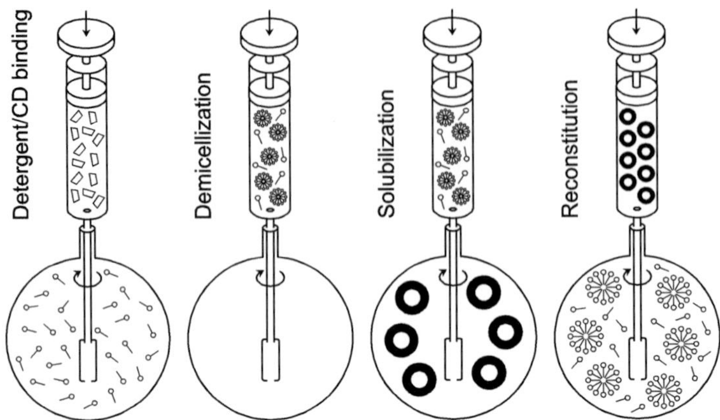

Figure 2 Calorimetric experiments for obtaining thermodynamic parameters describing CD-driven reconstitution. Molecular species depicted include CD (trapezoids), lipid vesicles (circles), and detergent (sticks). (See the color plate.)

detergent, that is, detergent not sequestered by CD, is calculated from a set of linked equilibria among bilayer, micellar, and aqueous phases. These equilibria are characterized in terms of the dissociation constant of the CD/detergent inclusion complex, the detergent's CMC, and partition coefficients reflecting the transfer of detergent and lipid between bilayer and micellar phases. All of these parameters can conveniently be obtained by ITC, which provides unrivaled sensitivity, resolution, and reproducibility and is attracting increasing attention in membrane-protein research in particular (Draczkowski, Matosiuk, & Jozwiak, 2014). The calorimetric titrations required to derive the parameters for the quantitative model include a CD/detergent binding experiment, detergent demicellization, and multiple solubilization and reconstitution experiments, which furnish the lipid/detergent phase diagram in the absence of CD (Fig. 2). Here, we elaborate on the requirements and caveats of these experiments and demonstrate how the thermodynamic parameters thus obtained can be used to predict the trajectory of CD-driven reconstitution within the phase diagram, including its crossing points with the phase boundaries.

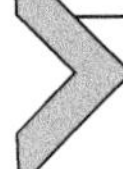

3. PRACTICAL APPROACH TO CD-DRIVEN RECONSTITUTION

3.1 Instrumentation, Materials, and Sample Preparation

For preparing and performing the calorimetric titrations furnishing the model parameters, the following instruments are required: pipettes, an analytical high-precision microbalance (1 μg readability) such as the XP205

(Mettler Toledo, Greifensee, Switzerland), an extruder such as the LiposoFast extruder (Avestin, Ottawa, Canada), and an ITC instrument such as the VP-ITC or iTC$_{200}$ (Malvern Instruments, Worcestershire, UK).

To quantify the above-mentioned set of linked equilibria, all calorimetric experiments have to be performed at the same temperature and in the same buffered aqueous solution. Since the experiments are aimed at providing a quantitative foundation for membrane-protein reconstitution, the temperature, buffer, pH, and other experimental settings should be chosen such as to provide conditions optimal for the protein of interest. Likewise, the lipid(s) and detergent(s) most suitable for preserving protein stability and activity should be established beforehand. In doing so, one should be aware that a detergent optimal for solubilization of a given membrane protein need not necessarily represent the best choice for reconstitution. Many CD derivatives lend themselves for detergent extraction; the choice should be governed by two major considerations, namely, ligand selectivity and aqueous solubility. On the one hand, the CD derivative must be chosen carefully to avoid complexation of lipid. The selectivity of a CD derivative for a particular lipid can be investigated, for instance, by ITC or in a solubilization assay by monitoring the intensity of scattered light upon titration of lipid vesicles into a CD solution (Fig. 3; Anderson, Tan, Ganz, & Seelig, 2004). A shift in the increase in scattering intensity to higher lipid concentrations is indicative of vesicle solubilization by CD and, hence, formation of

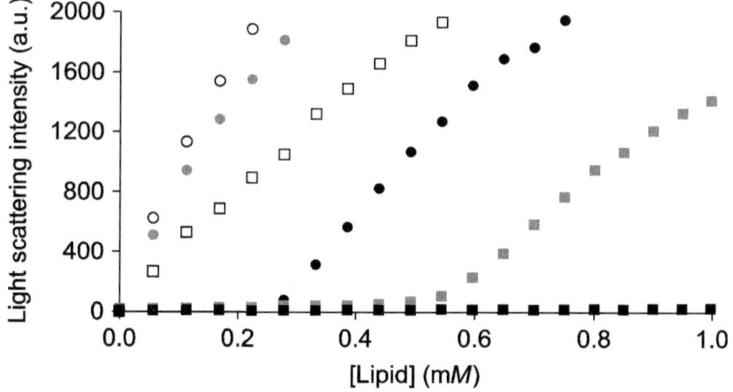

Figure 3 Lipid complexation by CDs. Formation of CD/lipid inclusion complexes was assessed by monitoring the intensity of light scattered at 90° during titration of various CD solutions with lipid vesicles. A steep increase in scattering intensity indicates presence of vesicles, (open circles) POPC to buffer, (open squares) DLPC to buffer, (gray circles) POPC to 50 m*M* HPβCD, (gray squares) DLPC to 50 m*M* HPβCD, (black circles) POPC to 50 m*M* MβCD, (black squares) DLPC to 50 m*M* MβCD. In all experiments, the titrant contained the same CD concentration as the titrand to maintain a constant CD concentration. 25 °C, 50 m*M* Tris, 50 m*M* NaCl, pH 7.4.

a CD/lipid complex. For example, randomly methylated β-cyclodextrin (MβCD), a CD derivative frequently used for transferring cholesterol from or into lipid bilayers (Mahammad & Parmryd, 2015), is suitable only at low concentrations in combination with 1-palmitoyl-2-oleoyl-*sn*-glycero-3-phosphocholine (POPC) because it complexes this lipid to a significant extent (Huang & London, 2013; Tsamaloukas, Szadkowska, Slotte, & Heerklotz, 2005). By contrast, the more hydrophilic CD derivative 2-hydroxypropyl-β-cyclodextrin (HPβCD) does not bind POPC but forms an inclusion complex with the shorter-chain lipid 1,2-dilauroyl-*sn*-glycero-3-phosphocholine (DLPC), albeit with lower affinity than MβCD (Fig. 3). The lipid-solubilizing capacity of CDs increases with decreasing acyl chain length, as becomes apparent from a comparison of POPC and DLPC complexation. On the other hand, the solubility of the CD derivative has to be taken into consideration because it imposes an upper bound on the titrant concentration for reconstitution, which, in turn, limits the amount of detergent that can be complexed. Thus, the phase diagram can be explored up to a certain lipid concentration only, so that poor CD solubility ultimately constrains the accessible range of protein/lipid ratios. Again, HPβCD is favored over MβCD because the solubility is >600 mg/L for the former as compared with >500 mg/L for the latter in pure water at 25 °C (Loftsson, Jarho, Másson, & Järvinen, 2005).

Besides choosing suitable compounds, several measures have to be taken to ensure that sample preparation results in accurate and precise stock concentrations. For preparing CD stocks, be aware that dissolving CDs significantly increases the volume of the solution (e.g., by ~0.7 μL/mg for both MβCD and HPβCD). Also, just like detergents, most CDs are surface-active and require particular care to avoid excessive foaming. *Thus, refrain from degassing solutions containing detergent or CD and, instead, consider short low-speed centrifugation to remove bubbles.*

For preparing lipid stocks, allow the lipid powder to thermally equilibrate to room temperature, weigh, suspend in buffer, and vortex vigorously for at least 10 min to generate multilamellar vesicles (MLVs). Alternatively, dry down lipid supplied in chloroform under a stream of nitrogen and agitation to produce a lipid film, desiccate overnight, and resuspend in buffer. From this MLV suspension, prepare large unilamellar vesicles (LUVs) by extrusion through two stacked polycarbonate filters with a defined pore diameter in at least 35 extrusion steps. Make sure to perform an odd number of passes to avoid contamination of the LUV suspension by MLVs that might not have passed through the filters. Moreover, make sure to perform the extrusion above the lipid's main phase transition temperature; for lipids with high transition temperatures such as 1,2-dimyristoyl-*sn*-glycero-3-phosphocholine, this might require use of a thermostattable extruder. Check for a unimodal and reasonably narrow size distribution of extruded LUVs with the aid of dynamic light scattering.

3.2 Calorimetric Experiments

ITC is routinely used to characterize the energetics of biomolecular interactions, and the reader is referred to the many reviews summarizing general applications (Draczkowski et al., 2014; Ghai, Falconer, & Collins, 2012; Kabiri & Unsworth, 2014) and data analysis in particular (Broecker, Vargas, & Keller, 2011; Keller et al., 2012; Tellinghuisen, 2004). The reader is expected to be familiar with standard calorimetric experiments and basic data analysis such as baseline determination, peak integration, and normalization of raw heats. Thus, in the following, the focus lies on specific aspects pertinent to the approach presented in this chapter.

3.2.1 Membrane Partitioning and Translocation by Uptake and Release

In both solubilization and reconstitution titrations required for establishing the lipid/detergent phase diagram, complete equilibration after each injection is reached within an acceptable experimental timescale only if the kinetics of detergent translocation between the two bilayer leaflets is sufficiently fast. The ability of a detergent to flip-flop between the bilayer leaflets can be confirmed with the aid of ITC by uptake and release experiments (Heerklotz, Binder, & Epand, 1999; Heerklotz & Seelig, 2000; Tsamaloukas, Keller, & Heerklotz, 2007). These experiments probe the accessibility of the inner vesicle leaflet to detergent from the exterior aqueous solution. Whereas fast kinetics are expected for nonionic detergents bearing small headgroups and some zwitterionic detergents such as the dipolar compound lauryldimethylamine *N*-oxide (LDAO) (Fig. 4), transbilayer equilibration of nonionic detergents with large headgroups, other zwitterionic detergents, or ionic detergents such as sodium dodecyl sulfate is slow at ambient temperature but can be accelerated at elevated temperature (Frotscher et al., 2015; Keller, Heerklotz, & Blume, 2006; Keller, Heerklotz, Jahnke, & Blume, 2006). If the flip-flop kinetics of the detergent is unknown, refer to Tsamaloukas et al. (2007) for a detailed uptake and

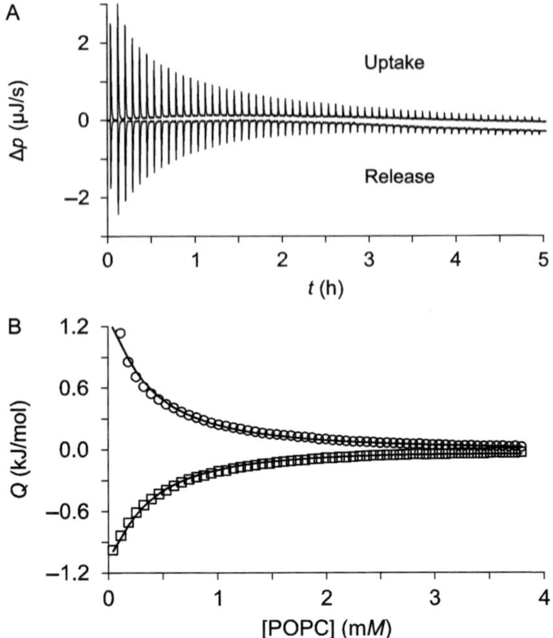

Figure 4 ITC uptake and release experiments with POPC. For uptake, 20 m*M* POPC in the form of LUVs was titrated to 100 μ*M* LDAO in a series of 60 injections. For release, 20 m*M* POPC vesicles preloaded with 2 m*M* LDAO were titrated into buffer in a series of 60 injections. (A) Raw thermograms depicting differential heating power, Δp, versus time, t. (B) Isotherms showing integrated and normalized heats of reaction, Q, versus POPC concentration. Experimental data of uptake (open circles) and release (open squares) and fits (solid lines) as described in Tsamaloukas et al. (2007), yielding an accessibility factor of $\gamma = 1$. For both uptake and release, the injection volume was set to 5 μL, the time spacing to 300 s, the stirrer speed to 310 rpm, the filter period to 2 s, and the reference power to 75 μJ/s. 25 °C, 50 m*M* Tris, 50 m*M* NaCl, pH 7.4.

release protocol before performing the following calorimetric experiments that are required for obtaining the parameter values of the CD-mediated reconstitution approach.

3.2.2 *CD/Detergent Binding*

Formation of a CD/detergent inclusion complex can be determined in a simple binding experiment using a one-site binding model for most detergents (Fig. 5; Bernat et al., 2008; Du, Chen, Lu, & Hou, 2004). However, besides equimolar complex formation, some CD ligands bind also with other stoichiometries. For example, four MβCD molecules bind to one

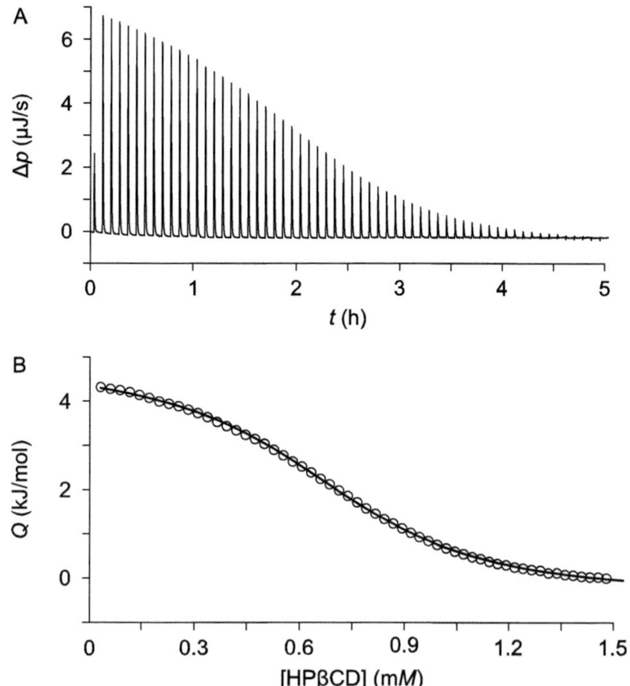

Figure 5 Determination of $K_D^{i/aq}$ for the binding of LDAO to HPβCD by ITC. 10 m*M* HPβCD was titrated to 1 m*M* LDAO in a series of 60 injections. (A) Raw thermogram depicting differential heating power, Δp, versus time, t. (B) Isotherm showing integrated and normalized heats of reaction, Q, versus HPβCD concentration. Experimental data (circles) and fit (solid line) according to a one-site binding model. The injection volume was set to 5 μL, the time spacing to 900 s, the stirrer speed to 310 rpm, the filter period to 2 s, and the reference power to 75 μJ/s. 25 °C, 50 m*M* Tris, 50 m*M* NaCl, pH 7.4.

POPC molecule (Anderson et al., 2004). Other stoichiometries have also been observed for some detergents, such as 2:1 or 1:2 stoichiometries occurring in addition to the predominant 1:1 complex for *n*-octyl-β-D-glucopyranoside (Haller & Kaatze, 2009), a detergent with a short alkyl chain and a high CMC. Dissociation constants of CD/detergent inclusion complexes are expected to lie in the micromolar to lower millimolar range (Mehta, Bhasin, Dham, & Singla, 2008; Rekharsky & Inoue, 1998), depending on the CD derivative used. Note that there are CDs bearing different molar substitutions, which often differ in the fraction of substituted hydroxyl groups; if possible, homogeneous samples with complete or, at least, defined substitution are preferred over less well-defined mixtures.

3.2.3 Detergent Demicellization

To obtain the CMC of a detergent by ITC, a micellar detergent suspension is titrated into solvent (i.e., aqueous buffer), resulting in a sigmoidal demicellization isotherm (Fig. 6; Bouchemal, Agnely, Koffi, Djabourov, & Ponchel, 2010; Paula, Süs, Tuchtenhagen, & Blume, 1995; Olofsson & Loh, 2009). Calorimetric determination of CMCs is often used for detergents and other surfactants, and data analysis is straightforward (Textor & Keller, 2015). The CMC is deduced from either the first derivative of a demicellization isotherm or a generic sigmoidal fit to the isotherm; ideally, both of these approaches are used in combination to extract the CMC from calorimetric demicellization data (Textor & Keller, 2015). CMC values for many detergents are available in the literature; thus, the demicellization

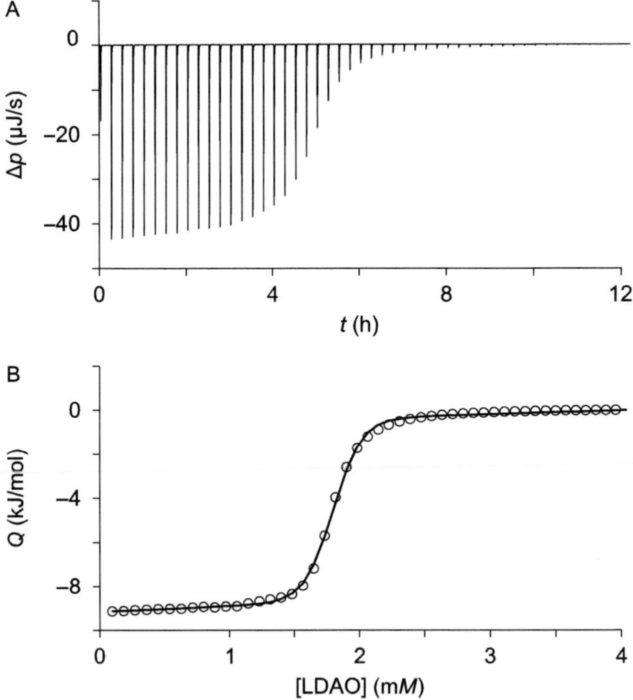

Figure 6 Determination of the CMC of LDAO by ITC demicellization. 25 m*M* LDAO was titrated into buffer in a series of 60 injections (not all injections shown). (A) Raw thermogram depicting differential heating power, Δp, versus time, t. (B) Isotherm showing integrated and normalized heats of reaction, Q, versus LDAO concentration. Experimental data (circles) and generic sigmoidal fit (solid line) as explained in Textor and Keller (2015). The injection volume was set to 5 μL, the time spacing to 900 s, the stirrer speed to 310 rpm, the filter period to 2 s, and the reference power to 75 μJ/s. 25 °C, 50 m*M* Tris, 50 m*M* NaCl, pH 7.4.

experiment may not be necessary at all. However, it is imperative that the CMC be derived from a demicellization measurement performed at the same temperature and using the same aqueous solvent as the other ITC experiments required for the quantitative model underlying CD-driven reconstitution because self-aggregation phenomena can be substantially influenced by temperature and ionic strength (Garidel, Hildebrand, Neubert, & Blume, 2000; Chattopadhyay & Harikumar, 1996). In principle, the detergent partition equilibrium between the aqueous and micellar phases is also reflected in and, thus, could be derived from the phase diagram (Textor et al., 2015). Nonetheless, obtaining the CMC independently in a separate demicellization experiment allows for more reliable determination of the slopes of the phase boundaries derived from solubilization and reconstitution experiments.

3.2.4 Lipid/Detergent Phase Diagram from Solubilization and Reconstitution

The lipid/detergent phase diagram is obtained from multiple solubilization and reconstitution experiments, in the course of which detergent is titrated to lipid vesicles and vice versa, respectively (Heerklotz & Seelig, 2000; Heerklotz et al., 2009; Jahnke et al., 2014). In both kinds of experiments, the phase boundaries manifest as abrupt changes in the thermograms and corresponding isotherms and can be identified as extrema in the first derivatives of the latter (Fig. 7). For establishing a new lipid/detergent phase diagram, the locations of the phase boundaries have to be estimated for choosing suitable concentration ranges in the initial solubilization and reconstitution experiments. In general, one should try to cover a rather broad range of titrant concentrations in the first titration in order to increase the probability of crossing both phase boundaries. The ordinate intercept of the phase boundaries, $c_{\mathrm{D}}^{\mathrm{aq}^{\circ}}$, is expected slightly below the CMC of the detergent because the phase diagram applies to the partitioning of detergent between the aqueous phase and mixed lipid/detergent micelles rather than pure detergent micelles, which are in equilibrium with detergent monomers at the CMC. Thus, choose a final detergent concentration that is only slightly higher than the CMC for a first reconstitution experiment or, alternatively, perform a first solubilization experiment at low lipid concentration and then increase the detergent concentration to about two times the CMC. To estimate the slopes of the phase boundaries, it is advisable to refer to known phase diagrams of similar binary lipid/detergent mixtures (Heerklotz & Seelig, 2000).

In total, a minimum of three solubilization and three reconstitution experiments should be performed, and these experiments should cover a concentration range of at least some millimolar for sufficient precision in

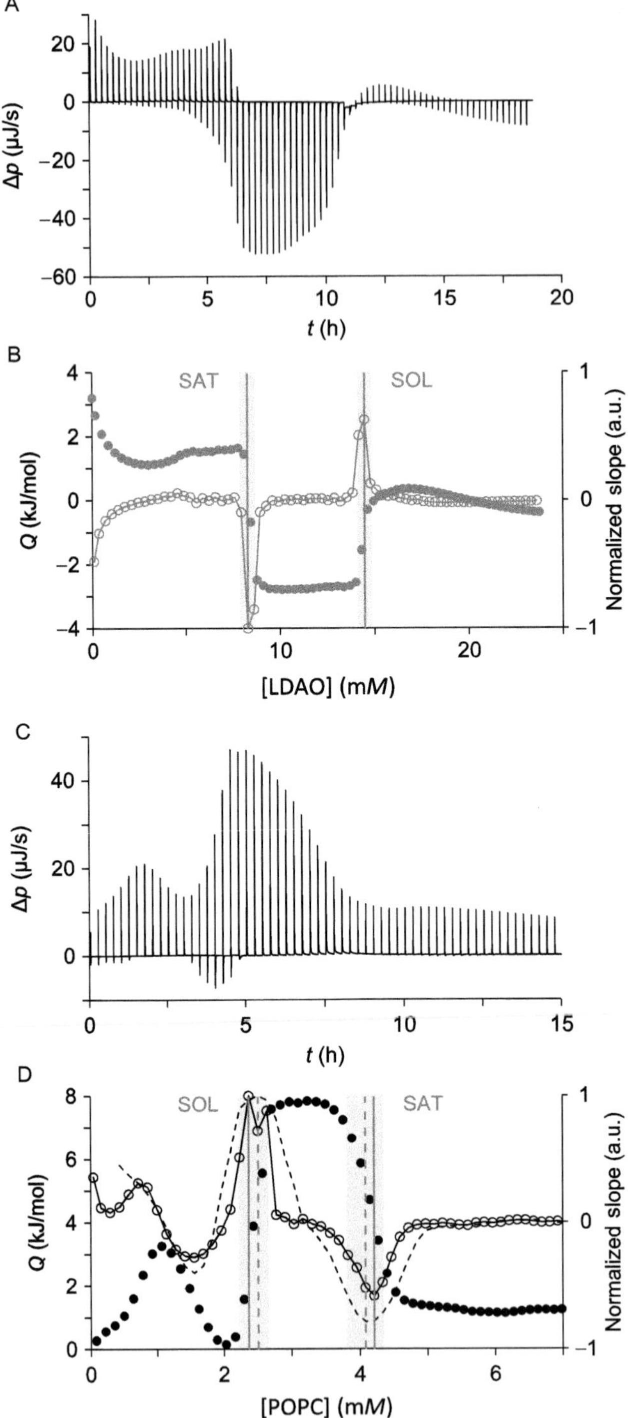

Figure 7 See legend on opposite page.

the subsequent fitting procedure. Be aware that the membrane protein of interest might affect the partitioning of lipid and detergent, as has been observed, for instance, for the K^+ channel KcsA (Jahnke et al., 2014). Significant deviations of the phase boundaries from those determined for simple lipid/detergent mixtures due to incomplete reconstitution or precipitation of the membrane protein can be revealed by solubilization and reconstitution experiments in the presence of the protein (Jahnke et al., 2014).

3.3 Removal of CD/Detergent Inclusion Complexes

Different approaches have been investigated that serve to physically remove both CD and detergent in the form of inclusion complexes from reconstituted samples. High-speed centrifugation in a sucrose step gradient has been found to be most suitable for separating CD/detergent inclusion complexes from proteoliposomes (DeGrip et al., 1998). However, we did not succeed in using this approach for separating HPβCD/LDAO inclusion complexes from reconstituted Mistic, presumably because of the relatively low molar mass of this protein. Similarly, physical removal of inclusion complexes from the reconstitution mixture also constitutes a considerable challenge to other methods. In our experience, size exclusion chromatography is suboptimal for separating inclusion complexes from proteoliposomes because of pronounced lipid retention on the column matrix, as mentioned in Section 2.2 (Ruysschaert et al., 2005). We typically encounter losses of up to 40% of proteoliposomes with this method, and similar issues are observed for ultrafiltration. Dialysis, in particular, has proven unsuccessful for

Figure 7—Cont'd Determination of phase boundaries from ITC solubilization and reconstitution experiments. (A, B) For solubilization, 125 m*M* LDAO was titrated to 6 m*M* of POPC in the form of LUVs in a series of 75 injections. (C, D) For reconstitution, 40 m*M* POPC was titrated to 10 m*M* LDAO in a series of 60 injections. (A, C) Raw thermograms showing differential heating power, Δp, versus time, t. (B, D) Shown are integrated and normalized heats of reaction, Q (filled circles) and corresponding numeric first derivatives (open circles) as functions of titrant concentration together with solubilization (SOL) and saturation boundaries (SAT) derived therefrom. Uncertainties of phase boundaries (shaded bands) correspond to the titrant concentration range over which the first derivative of the isotherm deviates by less than 10% from its extremum. In (D), the phase boundaries (vertical dashed lines) derived from a smoothed derivative (black dashed line, moving average of five injections) are shown. The injection volume was set to 4 μL for solubilization and to 5 μL for reconstitution. For both experiments, the time spacing was set to 900 s, the stirrer speed to 310 rpm, the filter period to 2 s, and the reference power to 75 μJ/s. 25 °C, 50 m*M* Tris, 50 m*M* NaCl, pH 7.4.

removing CD from proteoliposomes, as the permeation behavior of dialysis membranes changes in the presence of CD, resulting in the escape of protein through the dialysis membrane (DeGrip et al., 1998). Irrespective of the method used for physical removal of inclusion complexes after CD-mediated reconstitution, quantitation of submillimolar residual concentrations of CD with the aid of thin-layer chromatography is challenging; hence, more elaborate methods such as high-performance liquid chromatography or mass spectrometry are required to demonstrate complete removal of CD and detergent. Nevertheless, as the vast majority of CD derivatives are small and fairly inert and do not absorb considerably in the UV/VIS range, downstream biophysical or biochemical investigations involving spectroscopic methods such as circular dichroism spectroscopy or scattering techniques such as dynamic light scattering are not impeded by residual CD/detergent complexes (Broecker et al., 2014; Textor et al., 2015).

3.4 Data Analysis

Although all parameters required for describing CD-driven reconstitution for a given set of lipid, detergent, and CD are derived with the aid of ITC, data analysis differs considerably among the various experimental protocols sketched out in Section 3.2.

The dissociation constant of the CD/detergent complex is obtained on the basis of a binding model for stoichiometric formation of a 1:1 inclusion complex, which allows—and requires—a detailed, physically meaningful fit to the entire binding isotherm (Houtman et al., 2007). By contrast, for determination of the CMC of the detergent, a generic sigmoidal fit is applied that does not take into account a particular model for micelle formation in terms of fitting parameters (Broecker & Keller, 2013; Textor & Keller, 2015). Alternatively, the extremum of the first derivative of the isotherm is taken as the CMC (Bouchemal et al., 2010; Heerklotz & Seelig, 2000; Textor & Keller, 2015). Determination of phase boundaries from solubilization and reconstitution titrations is even less amenable to a detailed fit because such experiments involve series of rather complex processes. Therefore, unless the system is defined exceptionally well (Heerklotz, Lantzsch, Binder, Klose, & Blume, 1996; Keller, Heerklotz, et al., 2006), one has to resort to semiquantitative analysis (Heerklotz et al., 2009). Since the analysis of 1:1 binding isotherms (Fig. 5) and demicellization experiments (Fig. 6) is straightforward, the reader is referred to the literature, and only the derivation of the phase diagram shall be described in more detail in the following.

3.4.1 Lipid/Detergent Phase Diagram

To establish a lipid/detergent phase diagram from solubilization and reconstitution experiments, both the titrand and the titrant concentrations at which the two phase boundaries are crossed in the course of a titration are identified, and the detergent concentrations thus obtained are plotted versus the corresponding lipid concentrations (Heerklotz & Seelig, 2000). The titrant concentrations at the phase boundaries are given by the inflection points of the isotherm. Therefore, they can be determined by calculating the first derivative of the isotherm and reading the titrant concentrations at the extrema of the derivative. The saturation boundary corresponds to a minimum in the first derivative, whereas the solubilization boundary is represented by a maximum (Fig. 7) in both solubilization and reconstitution experiments. The first derivative is most easily obtained numerically by calculating the slope between each pair of consecutive injections. In the first part of reconstitution titrations, where only micellar aggregates are present, an additional peak in the experimental heat is observed in some cases (Fig. 7D), which usually stems from sphere-to-rod transitions of mixed lipid/detergent micelles (Heerklotz, 2008). This does not impede analysis if care is taken not to mistake the corresponding local extremum of the first derivative for one of the two phase boundaries delimiting the coexistence range. Moreover, if only relatively few injections are performed at mediocre signal/noise ratio, more than one local extremum of the first derivative may be encountered (Fig. 7D). In such cases, smoothing of the first derivative may be an option, although repeating the measurement with a larger number of injections would be preferable. Once the lipid and detergent concentrations at the phase boundaries are extracted from the extrema of the first derivative and the pairwise lipid and detergent concentrations are plotted, the parameters defining the phase boundaries can be determined in terms of their slopes, $R_D^{b,\,SAT}$ and $R_D^{m,\,SOL}$, and their common ordinate intercept, $c_D^{aq^\circ}$ (Lichtenberg, Robson, & Dennis, 1983). Hence, the phase boundaries are obtained by applying a global linear least-squares fit comprising the phase-boundary slopes as local fitting parameters and the ordinate intercept as a global fitting parameter. If the CMC is determined from a demicellization titration, the number of fitting parameters in the analysis of the phase diagram is further reduced by linking $c_D^{aq^\circ}$ to the CMC through $R_D^{m,\,SOL}$, as expressed in Eq. (5) in Section 3.4.2 (Textor et al., 2015).

For the POPC/LDAO phase diagram (Textor & Keller, 2015), slightly higher slopes of the phase boundaries are suggested by reconstitution experiments than by solubilization experiments (Fig. 8). This observation cannot be ascribed to hysteresis (Ollivon et al., 2000) because this would result in a decrease of both phase-boundary slopes in reconstitution as compared with solubilization titrations. It should also be noted that the transitions in the isotherms occur over narrow but finite titrant concentration ranges rather than at a single titrant concentration. This is similar to the demicellization process, which also extends over a finite concentration range, thus complicating determination (or even definition) of the CMC (Textor & Keller, 2015). In our approach, this uncertainty is accounted for by considering the titrant concentration range in which the first derivative of the isotherm deviates by less than 10% from its extremum (Fig. 8). Although all of the uncertainties determined in this way are relatively small, they need to be considered for membrane-protein reconstitution experiments. In particular, the start and

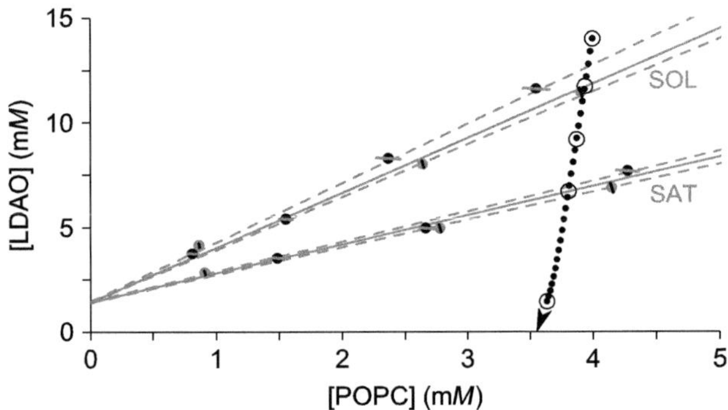

Figure 8 POPC/LDAO phase diagram with exemplary trajectory predicted for CD-driven reconstitution. Depicted are the solubilization boundary (SOL), the saturation boundary (SAT), and a typical reconstitution trajectory (dotted line) with individual titration steps according to Table 1 (open circles). Also shown are experimental data of solubilization (gray dots) and reconstitution (black dots) titrations. Uncertainties of individual data points (local error margins) were derived from the widths of the transition regions in the corresponding isotherms, while uncertainties of phase boundaries (dashed lines) were estimated by fitting data from either solubilization or reconstitution experiments alone. For the trajectory, the concentration of free detergent, $c_D^{free} = c_D - c_D^i$, is plotted against the total lipid concentration, c_L, as calculated for a 500-μL reconstitution mixture containing $c_L^0 = 4$ m*M* POPC and $c_D^0 = 14$m*M* LDAO titrated with $c_{CD}^0 = 150$m*M*. The trajectory corresponds to a total titrant volume of 50 μL, and each point in the trajectory reflects a 2-μL injection. At the end of the titration, the mole fraction of detergent in the mixed lipid/detergent bilayer amounts to $X_D^b = 0.205$. 25 °C, 50 m*M* Tris, 50 m*M* NaCl, pH 7.4.

end points of a reconstitution trajectory should be located clearly within the ranges of interest and outside the ranges of uncertainty.

Finally, one should also be aware of more fundamental limitations of the quantitative model, which does not consider nonideal mixing of lipid and detergent (Heerklotz, 2008) or the possibility of CD partitioning into the micellar core (Tsianou & Fajalia, 2014). Any additional partition equilibria not captured by the model can complicate or even prevent its application, although the model might be expanded if additional quantitative data are available.

3.4.2 Prediction of Phase-Boundary Crossing Points

Once the parameter values that describe the linked equilibria in a given CD/lipid/detergent mixture are available from the range of ITC experiments outlined in Section 3.2, they can be used to predict how much CD needs to be added to a reconstitution mixture to reach any desired free detergent concentration at a given lipid concentration, for example, a concentration corresponding to one of the two phase boundaries. At the phase boundaries, the lipid is found in only one pseudophase, that is, in micelles at the solubilization boundary or in bilayers at the saturation boundary. This is reflected in the following equations:

$$c_{\mathrm{L}}^{\mathrm{b}} = c_{\mathrm{L}} - c_{\mathrm{L}}^{\mathrm{m}} = 0 \tag{1}$$

$$c_{\mathrm{L}}^{\mathrm{m}} = 0, \tag{2}$$

with Eq. (1) applying to the solubilization boundary and Eq. (2) to the saturation boundary. As derived in Textor et al. (2015), the micellar lipid concentration in the coexistence range, $c_{\mathrm{L}}^{\mathrm{m}}$, and the concentration of detergent sequestered in inclusion complexes, $c_{\mathrm{D}}^{\mathrm{i}}$, are given by

$$c_{\mathrm{L}}^{\mathrm{m}} = \frac{c_{\mathrm{D}} - c_{\mathrm{D}}^{\mathrm{aq}^{\circ}} - c_{\mathrm{D}}^{\mathrm{i}} - R_{\mathrm{D}}^{\mathrm{b,SAT}} c_{\mathrm{L}}}{R_{\mathrm{D}}^{\mathrm{m,SOL}} - R_{\mathrm{D}}^{\mathrm{b,SAT}}} \tag{3}$$

$$c_{\mathrm{D}}^{\mathrm{i}} = \frac{c_{\mathrm{D}}^{\mathrm{aq}}}{c_{\mathrm{D}}^{\mathrm{aq}} + K_{\mathrm{D}}^{\mathrm{i/aq}}} c_{\mathrm{CD}}. \tag{4}$$

In the coexistence range, the detergent concentration in the aqueous phase is constant, $c_{\mathrm{D}}^{\mathrm{aq}} = c_{\mathrm{D}}^{\mathrm{aq}^{\circ}}$, and is related to the CMC through:

$$c_{\mathrm{D}}^{\mathrm{aq}^{\circ}} = \mathrm{CMC} \cdot \frac{R_{\mathrm{D}}^{\mathrm{m,SOL}}}{1 + R_{\mathrm{D}}^{\mathrm{m,SOL}}} \tag{5}$$

By combining Eqs. (3) and (4), inserting into (1) and (2), and solving for c_{CD}, the target concentration of CD required to reach the phase boundaries for a given pair of c_{D} and c_{L} can be calculated as:

$$c_{\mathrm{CD}} = \left(c_{\mathrm{D}} - R_{\mathrm{D}} c_{\mathrm{L}} - c_{\mathrm{D}}^{\mathrm{aq}^{\circ}}\right) \frac{c_{\mathrm{D}}^{\mathrm{aq}^{\circ}} + K_{\mathrm{D}}^{\mathrm{i/aq}}}{c_{\mathrm{D}}^{\mathrm{aq}^{\circ}}}, \tag{6}$$

with $R_{\mathrm{D}} = R_{\mathrm{D}}^{\mathrm{m,SOL}}$ for the solubilization boundary and $R_{\mathrm{D}} = R_{\mathrm{D}}^{\mathrm{b,SAT}}$ for the saturation boundary.

Equation (6) holds true for a given set of CD, lipid, and detergent concentrations. For an actual reconstitution experiment, dilution resulting from titrant addition needs to be taken into account. For a reconstitution titration monitored by ITC in a completely filled cell, the effective calorimetric volume of the sample cell remains constant, but a small volume is displaced from the sample cell into the calorimetrically inert neck upon each injection. When the sample is finally recovered from the calorimeter after ITC-monitored reconstitution, the volume in the sample cell and that in the neck are recombined, such that only the total volume of the reconstitution mixture shall be considered in the following. Dilution of CD, lipid, and detergent is accounted for by substituting the corresponding total concentrations by the following expressions:

$$c_{\mathrm{CD}} = c_{\mathrm{CD}}^{0} \frac{\Delta V}{\Delta V + V^{0}} \tag{7}$$

$$c_{\mathrm{D}} = c_{\mathrm{D}}^{0} \frac{V^{0}}{\Delta V + V^{0}} \tag{8}$$

$$c_{\mathrm{L}} = c_{\mathrm{L}}^{0} \frac{V^{0}}{\Delta V + V^{0}}, \tag{9}$$

where c_{CD}^{0} is the concentration of the CD titrant in the syringe, while c_{L}^{0} and c_{D}^{0} denote the initial lipid and detergent concentrations, respectively, in the sample cell. V^{0} is the volume of the sample cell, and ΔV is the total volume of CD titrant injected throughout the reconstitution titration. By inserting Eqs. (7)–(9) into Eq. (6) and solving for ΔV, the CD titrant volume required to reach either of the two phase boundaries, R_{D}, is obtained as:

$$\Delta V = V^{0} \frac{\left(c_{\mathrm{D}}^{0} - c_{\mathrm{L}}^{0} R_{\mathrm{D}} - c_{\mathrm{D}}^{\mathrm{aq}^{\circ}}\right)\left(c_{\mathrm{D}}^{\mathrm{aq}^{\circ}} + K_{\mathrm{D}}^{\mathrm{i/aq}}\right)}{c_{\mathrm{D}}^{\mathrm{aq}^{\circ}}\left(c_{\mathrm{CD}}^{0} + c_{\mathrm{D}}^{\mathrm{aq}^{\circ}} + K_{\mathrm{D}}^{\mathrm{i/aq}}\right)} \tag{10}$$

To exemplify this calculation, let us assume a reconstitution mixture comprising POPC and LDAO that is titrated with HPβCD to form an HPβCD/LDAO inclusion complex (Fig. 8 and Table 1). The model parameters obtained by ITC are $R_D^{b,\,SAT} = 1.39$, $R_D^{m,\,SOL} = 2.62$, $c_D^{aq^\circ} = 1.40$ mM, and $K_D^{i/aq} = 0.10$mM (Textor et al., 2015). If the reconstitution mixture has a total volume of 500 μL with $c_L^0 = 4$ mM and $c_D^0 = 14$ mM, and the CD titrant has a concentration of $c_{CD}^0 = 150$ mM, then inserting all of these values into Eq. (10) results in a titrant volume of $\Delta V = 7.5$ μL required to reach the solubilization boundary and $\Delta V = 24.9$ μL to reach the saturation boundary. In this example, the sample volume increases by ~5% upon reaching the saturation boundary and by ~10% when decreasing the concentration of free detergent below 10% of the total detergent concentration (i.e., to about 0.5 mM below the CMC). These values demonstrate that, for highly concentrated CD titrants, the dilution due to titrant addition is small, as further reflected in the almost vertical reconstitution trajectory (Fig. 8). If larger titration volumes are preferred for technical reasons, a lower titrant concentration can be used at the expense of more pronounced dilution of the reconstitution mixture. Nevertheless, even with lower titrant concentrations, the dilution factor of CD-driven detergent extraction remains orders of magnitudes lower as compared with the dilution strategy. For example, reconstitution of the mitochondrial proton pump solubilized with cholate required 25-fold dilution (Racker, Chien, & Kandrach, 1975).

In addition to the prediction of the phase-boundary crossing points, the quantitative model allows one to map the complete reconstitution trajectory and to calculate the concentrations of all species involved in the different pseudophases to obtain a so-called speciation plot. To this end, the total concentrations of CD, lipid, and detergent are calculated for each injection, thereby accounting for sample dilution, and the quantitative model is applied to each postinjection composition of the sample cell (Textor et al., 2015). An implementation of the model in a Microsoft Excel worksheet is available from the authors upon request.

4. CONCLUDING REMARKS

The practical guide provided herein explains how a combination of various ITC experiments can be performed to yield all thermodynamic parameters required for a quantitative description of CD-mediated detergent extraction from a complex reconstitution mixture. These parameters

Table 1 Concentrations Expected for Exemplary CD-driven Reconstitution

Phase Diagram Location	ΔV_{CD} (μL)	c_L	c_D	c_{CD}	c_L^m	c_L^b	c_D^{aq}	c_D^m	c_D^b	c_D^{free}	c_D^i	c_{CD}^{free}
Micellar range	0	4.00	14.0	0	4.00	0	1.47	12.5	0	14.0	0	0
SOL boundary	7.54	3.94	13.8	2.23	3.94	0	1.40	10.3	0	11.7	2.07	0.15
Coexistence range	16.2	3.87	13.6	4.71	1.94	1.93	1.40	5.09	2.69	9.18	4.39	0.32
SAT boundary	24.9	3.81	13.3	7.12	0	3.81	1.40	0.00	5.30	6.70	6.63	0.49
Vesicle range	50	3.64	12.7	13.6	0	3.64	0.50	0	0.94	1.44	11.30	2.35

Concentrations (in units of millimolar) of lipid (L), detergent (D), and cyclodextrin (CD) in inclusion complexes (i) or in the aqueous (aq), bilayer (b), and micellar (m) phases at several points during CD-driven reconstitution. Concentrations were calculated on the basis of the quantitative model assuming titration of 150 m*M* HPβCD to 500 μL of a POPC/LDAO reconstitution mixture, as exemplified by the trajectory (dot with arrowhead) in Fig. 8.

include the dissociation constant of the CD/detergent inclusion complex, $K_D^{i/aq}$, the detergent's CMC, and the phase boundaries of the lipid/detergent phase diagram defined by $R_D^{b,\,SAT}$, $R_D^{m,\,SOL}$, and $c_D^{aq^\circ}$. This calorimetric approach aims at facilitating the use of CDs for extracting detergent for membrane-protein reconstitution and allows for systematic exploration of parameters that affect the outcome of the reconstitution process. The possibility of tailoring reconstitution trajectories in a quantitative manner and, if necessary, adjusting the rate of detergent extraction is expected to provide better control over membrane-protein reconstitution in terms of the final concentrations of all species involved. Improved control over the size and homogeneity of proteoliposomes and the orientation of membrane-embedded proteins can also be envisioned because these properties depend on the rate of detergent extraction.

ACKNOWLEDGMENTS

We thank Dr. Carolyn Vargas (University of Kaiserslautern) for helpful comments on the manuscript. This work was supported by a scholarship from the German National Academic Foundation (to M.T.) and by the International Research Training Group 1830 funded by the Deutsche Forschungsgemeinschaft (DFG).

Note: The authors declare no competing financial interest.

REFERENCES

Anderson, T., Tan, A., Ganz, P., & Seelig, J. (2004). Calorimetric measurement of phospholipid interaction with methyl-β-cyclodextrin. *Biochemistry*, *43*, 2251–2261.

Bernat, V., Ringard-Lefebvre, C., Bas, G., Perly, B., Djedaïni-Pilard, F., & Lesieur, S. (2008). Inclusion complex of n-octyl β-D-glucopyranoside and α-cyclodextrin in aqueous solutions: Thermodynamic and structural characterization. *Langmuir*, *24*, 3140–3149.

Bouchemal, K., Agnely, F., Koffi, A., Djabourov, M., & Ponchel, G. (2010). What can isothermal titration microcalorimetry experiments tell us about the self-organization of surfactants into micelles? *Journal of Molecular Recognition*, *23*, 335–342.

Broecker, J., Fiedler, S., Gimpl, K., & Keller, S. (2014). Polar interactions trump hydrophobicity in stabilizing the self-inserting membrane protein Mistic. *Journal of the American Chemical Society*, *136*, 13761–13768.

Broecker, J., & Keller, S. (2013). Impact of urea on detergent micelle properties. *Langmuir*, *29*, 8502–8510.

Broecker, J., Vargas, C., & Keller, S. (2011). Revisiting the optimal *c* value for isothermal titration calorimetry. *Analytical Biochemistry*, *18*, 307–309.

Brunner, J., Skrabal, P., & Hauser, H. (1976). Single bilayer vesicles prepared without sonication. Physic-chemical properties. *Biochimica et Biophysica Acta*, *455*, 322–331.

Chattopadhyay, A., & Harikumar, K. G. (1996). Dependence of critical micelle concentration of a zwitterionic detergent on ionic strength: Implications in receptor solubilization. *FEBS Letters*, *391*, 199–202.

DeGrip, W., VanOostrum, J., & Bovee-Geurts, P. (1998). Selective detergent-extraction from mixed detergent/lipid/protein micelles, using cyclodextrin inclusion compounds: A novel generic approach for the preparation of proteoliposomes. *Biochemical Journal*, *330*, 667–674.

Draczkowski, P., Matosiuk, D., & Jozwiak, K. (2014). Isothermal titration calorimetry in membrane protein research. *Journal of Pharmaceutical and Biomedical Analysis*, *87*, 313–325.

Du, X., Chen, X., Lu, W., & Hou, J. (2004). Spectroscopic study on binding behaviors of different structural nonionic surfactants to cyclodextrins. *Journal of Colloid and Interface Science*, *274*, 645–651.

Engel, A., Hoenger, A., Hefti, A., Henn, C., Ford, R., Kistler, J., et al. (1992). Assembly of 2-D membrane protein crystals: Dynamics, crystal order, and fidelity of structure analysis by electron microscopy. *Journal of Structural Biology*, *109*, 219–234.

Frotscher, E., Danielczak, B., Vargas, C., Meister, A., Durand, G., & Keller, S. (2015). A fluorinated detergent for membrane-protein applications. *Angewandte Chemie International Edition*, *54*, 5069–5073.

Garidel, P., Hildebrand, A., Neubert, R., & Blume, A. (2000). Thermodynamic characterization of bile salt aggregation as a function of temperature and ionic strength using isothermal titration calorimetry. *Langmuir*, *16*, 5267–5275.

Ghai, R., Falconer, R., & Collins, B. (2012). Applications of isothermal titration calorimetry in pure and applied research—Survey of the literature from 2010. *Journal of Molecular Recognition*, *25*, 32–52.

Haller, J., & Kaatze, U. (2009). Octylglucopyranoside and cyclodextrin in water. Self-aggregation and complex formation. *The Journal of Physical Chemistry B*, *113*, 1940–1947.

Heerklotz, H. (2008). Interactions of surfactants with lipid membranes. *Quarterly Reviews of Biophysics*, *41*, 205–264.

Heerklotz, H., Binder, H., & Epand, R. (1999). A "release" protocol for isothermal titration calorimetry. *Biophysical Journal*, *76*, 2606–2613.

Heerklotz, H., Lantzsch, G., Binder, H., Klose, G., & Blume, A. (1995). Application of isothermal titration calorimetry for detecting lipid membrane solubilization. *Chemical Physics Letters*, *235*, 517–520.

Heerklotz, H., Lantzsch, G., Binder, H., Klose, G., & Blume, A. (1996). Thermodynamic characterization of dilute aqueous lipid/detergent mixtures of POPC and $C_{12}EO_8$ by means of isothermal titration calorimetry. *The Journal of Physical Chemistry*, *100*, 6764–6774.

Heerklotz, H., & Seelig, J. (2000). Titration calorimetry of surfactant-membrane partitioning and membrane solubilization. *Biochimica et Biophysica Acta*, *1508*, 69–85.

Heerklotz, H., Tsamaloukas, A., & Keller, S. (2009). Monitoring detergent-mediated solubilization and reconstitution of lipid membranes by isothermal titration calorimetry. *Nature Protocols*, *4*, 686–697.

Helenius, A., Sarvas, M., & Simons, K. (1981). Asymmetric and symmetric membrane reconstitution by detergent elimination. Studies with Semliki-forest-virus spike glycoprotein and penicillinase from the membrane of *Bacillus licheniformis*. *European Journal of Biochemistry*, *116*, 27–35.

Herrmann, M., Danielczak, B., Textor, M., Klement, J., & Keller, S. (2015). Modulating bilayer mechanical properties to promote the coupled folding and insertion of an integral membrane protein. *European Biophysics Journal*. http://dx.doi.org/10.1007/s00249-015-1032-y.

Houtman, J., Brown, P., Bowden, B., Yamaguchi, H., Appella, E., Samelson, L., et al. (2007). Studying multisite binary and ternary protein interactions by global analysis of isothermal titration calorimetry data in SEDPHAT: Application to adaptor protein complexes in cell signaling. *Protein Science*, *16*, 30–42.

Huang, Z., & London, E. (2013). Effect of cyclodextrin and membrane lipid structure upon cyclodextrin-lipid interaction. *Langmuir*, *29*, 14631–14638.

Jahnke, N., Krylova, O., Hoomann, T., Vargas, C., Fiedler, S., Pohl, P., et al. (2014). Real-time monitoring of membrane-protein reconstitution by isothermal titration calorimetry. *Analytical Chemistry*, *86*, 920–927.

Jiskoot, W., Teerlink, T., Beuvery, E., & Crommelin, D. (1986). Preparation of liposomes via detergent removal from mixed micelles by dilution. The effect of bilayer composition and process parameters on liposome characteristics. *Pharmaceutisch Weekblad*.

Scientific Edition, 8, 259–265. http://link.springer.com/article/10.1007%2Fs00249-015-1032-y.

Kabiri, M., & Unsworth, L. (2014). Application of isothermal titration calorimetry for characterizing thermodynamic parameters of biomolecular interactions: Peptide self-assembly and protein adsorption case studies. *Biomacromolecules, 15*, 3463–3473.

Keller, S., Heerklotz, H., & Blume, A. (2006a). Monitoring lipid membrane translocation of sodium dodecyl sulfate by isothermal titration calorimetry. *Journal of the American Chemical Society, 128*, 1279–1286.

Keller, S., Heerklotz, H., Jahnke, N., & Blume, A. (2006b). Thermodynamics of lipid membrane solubilization by sodium dodecyl sulfate. *Biophysical Journal, 90*, 4509–4521.

Keller, S., Vargas, C., Zhao, H., Piszczek, G., Brautigam, C., & Schuck, P. (2012). High-precision isothermal titration calorimetry with automated peak-shape analysis. *Analytical Chemistry, 84*, 5066–5073.

Lasic, D. (1988). The mechanism of vesicle formation. *Biochemical Journal, 256*, 1–11.

Lévy, D., Bluzat, A., Seigneuret, M., & Rigaud, J. (1990). A systematic study of liposome and proteoliposome reconstitution involving Bio-Bead-mediated Triton X-100 removal. *Biochimica et Biophysica Acta, 1025*, 179–190.

Lévy, D., Gulik, A., Bluzat, A., & Rigaud, J. (1992). Reconstitution of the sarcoplasmic reticulum $Ca^{(2+)}$-ATPase: Mechanisms of membrane protein insertion into liposomes during reconstitution procedures involving the use of detergents. *Biochimica et Biophysica Acta, 1107*, 283–298.

Lichtenberg, D. (1985). Characterization of the solubilization of lipid bilayers by surfactants. *Biochimica et Biophysica Acta, 821*, 470–478.

Lichtenberg, D., Robson, R., & Dennis, E. (1983). Solubilization of phospholipids by detergents: Structural and kinetic aspects. *Biochimica et Biophysica Acta, 737*, 285–304.

Loftsson, T., Jarho, P., Másson, M., & Järvinen, T. (2005). Cyclodextrins in drug delivery. *Expert Opinion on Drug Delivery, 2*, 335–351.

Mahammad, S., & Parmryd, I. (2015). Cholesterol depletion using methyl-β-cyclodextrin. *Methods in Molecular Biology, 1232*, 91–102.

Mehta, S., Bhasin, K., Dham, S., & Singla, M. (2008). Micellar behavior of aqueous solutions of dodecyldimethylethylammonium bromide, dodecyltrimethylammonium chloride and tetradecyltrimethylammonium chloride in the presence of α-, β-, HPβ- and γ-cyclodextrins. *The Journal of Colloid and Interface Science, 321*, 442–451.

Milsmann, M., Schwendener, R., & Weder, H. (1978). The preparation of large single bilayer liposomes by a fast and controlled dialysis. *Biochimica et Biophysica Acta, 512*, 147–155.

Ollivon, M., Lesieur, S., Grabielle-Madelmont, C., & Paternostre, M. (2000). Vesicle reconstitution from lipid-detergent mixed micelles. *Biochimica et Biophysica Acta, 1508*, 34–50.

Olofsson, G., & Loh, W. (2009). On the use of titration calorimetry to study the association of surfactants in aqueous solutions. *Journal of the Brazilian Chemical Society, 20*, 577–593.

Overington, J., Al-Lazikani, B., & Hopkins, A. (2006). How many drug targets are there? *Nature Reviews Drug Discovery, 5*, 993–996.

Paula, S., Süs, W., Tuchtenhagen, J., & Blume, A. (1995). Thermodynamics of micelle formation as a function of temperature: A high sensitivity titration calorimetry study. *The Journal of Physical Chemistry, 99*, 11742–11751.

Philippot, J., Mutaftschiev, S., & Liautard, J. (1983). A very mild method allowing the encapsulation of very high amounts of macromolecules into very large (1000 nm) unilamellar liposomes. *Biochimica et Biophysica Acta (BBA) - Biomembranes, 734*, 137–143.

Racker, E. (1979). Reconstitution of membrane processes. *Methods in Enzymology, 55*, 699–711.

Racker, E., Chien, T., & Kandrach, A. (1975). A cholate-dilution procedure for the reconstitution of the Ca^{++} pump, $^{32}P_i$—ATP exchange, and oxidative phosphorylation. *FEBS Letters, 57*, 14–18.

Racker, E., Violand, B., O'Neal, S., Alfonzo, M., & Telford, J. (1979). Reconstitution, a way of biochemical research; some new approaches to membrane-bound enzymes. *Archives of Biochemistry and Biophysics, 198*, 470–477.

Rekharsky, M., & Inoue, Y. (1998). Complexation thermodynamics of cyclodextrins. *Chemical Reviews, 98*, 1875–1917.

Rhoden, V., & Goldin, S. (1979). Formation of unilamellar lipid vesicles of controllable dimensions by detergent dialysis. *Biochemistry, 18*, 4173–4176.

Rigaud, J., Chami, M., Lambert, O., Lévy, D., & Ranck, J. (2000). Use of detergents in two-dimensional crystallization of membrane proteins. *Biochimica et Biophysica Acta, 1508*, 112–128.

Rigaud, J., & Lévy, D. (2003). Reconstitution of membrane proteins into liposomes. *Methods in Enzymology, 372*, 65–86.

Rigaud, J., Lévy, D., Mosser, G., & Lambert, O. (1998). Detergent removal by non-polar polystyrene beads. Applications to membrane protein reconstitution and two-dimensional crystallization. *European Biophysics Journal, 27*, 305–319.

Rigaud, J., Mosser, G., Lacapere, J., Olofsson, A., Lévy, D., & Ranck, J. (1997). Bio-beads: An efficient strategy for two-dimensional crystallization of membrane proteins. *Journal of Structural Biology, 118*, 226–235.

Roosild, T., Greenwald, J., Vega, M., Castronovo, S., Riek, R., & Choe, S. (2005). NMR structure of Mistic, a membrane-integrating protein for membrane protein expression. *Science, 307*, 1317–1321.

Ruysschaert, T., Marque, A., Duteyrat, J., Lesieur, S., Winterhalter, M., & Fournier, D. (2005). Liposome retention in size exclusion chromatography. *BMC Biotechnology, 5*, 1–13.

Schurtenberger, P., Mazer, N., Waldvogel, S., & Känzig, W. (1984). Preparation of monodisperse vesicles with variable size by dilution of mixed micellar solutions of bile salt and phosphatidylcholine. *Biochimica et Biophysica Acta, 775*, 111–114.

Seddon, A., Curnow, P., & Booth, P. (2004). Membrane proteins, lipids and detergents: Not just a soap opera. *Biochimica et Biophysica Acta, 1666*, 105–117.

Signorell, G., Kaufmann, T., Kukulski, W., Engel, A., & Rémigy, H. (2007). Controlled 2D crystallization of membrane proteins using methyl-β-cyclodextrin. *Journal of Structural Biology, 157*, 321–328.

Tellinghuisen, J. (2004). Statistical error in isothermal titration calorimetry. *Methods in Enzymology, 383*, 245–282.

Textor, M., & Keller, S. (2015). Automated analysis of calorimetric demicellization titrations. *Analytical Biochemistry, 485*, 119–121.

Textor, M., Vargas, C., & Keller, S. (2015). Calorimetric quantification of linked equilibria in cyclodextrin/lipid/detergent mixtures for membrane-protein reconstitution. *Methods, 76*, 183–193.

Tsamaloukas, A., Keller, S., & Heerklotz, H. (2007). Uptake and release protocol for assessing membrane binding and permeation by way of isothermal titration calorimetry. *Nature Protocols, 2*, 695–704.

Tsamaloukas, A., Szadkowska, H., Slotte, P., & Heerklotz, H. (2005). Interactions of cholesterol with lipid membranes and cyclodextrin characterized by calorimetry. *Biophysical Journal, 89*, 1109–1119.

Tsianou, M., & Fajalia, A. (2014). Cyclodextrins and surfactants in aqueous solution above the critical micelle concentration: Where are the cyclodextrins located? *Langmuir, 30*, 13754–13764.

Young, H., Rigaud, J., Lacapère, J., Reddy, L., & Stokes, D. (1997). How to make tubular crystals by reconstitution of detergent-solubilized $Ca^{2(+)}$-ATPase. *Biophysical Journal, 72*, 2545–2558.

CHAPTER SEVEN

Extending ITC to Kinetics with kinITC

Philippe Dumas*,[1], Eric Ennifar*, Cyrielle Da Veiga*, Guillaume Bec*, William Palau[†,‡], Carmelo Di Primo[†,‡], Angel Piñeiro[§,¶], Juan Sabin[§], Eva Muñoz[§], Javier Rial[§]

*Biophysics & Structural Biology Team, IBMC, UPR9002-CNRS, University of Strasbourg, Strasbourg, France
[†]University of Bordeaux, Laboratoire ARNA, Pessac, France
[‡]INSERM U869, Laboratoire ARNA, Pessac, France
[§]AFFINImeter Scientific & Development Team, Software 4 Science Developments, S. L. Ed. Emprendia, Campus Vida, Santiago de Compostela, A Coruña, Spain
[¶]Department of Applied Physics, Faculty of Physics, University of Santiago de Compostela, Campus Vida, Santiago de Compostela, A Coruña, Spain
[1]Corresponding author: e-mail address: p.dumas@ibmc-cnrs.unistra.fr

Contents

Abstract

Isothermal titration calorimetry (ITC) has long been used for kinetic studies in chemistry, but this remained confined to enzymatic studies in the biological field. In fact, the biological community has long had the tendency of ignoring the kinetic possibilities of ITC considering it solely as a thermodynamic technique, whereas surface plasmon resonance is seen as the kinetic technique *par excellence*. However, the primary signal recorded by ITC is a heat power which is directly related to the kinetics of the reaction. Here, it is shown how this kinetic signal can be recovered by using *kinITC*, the kinetic

Methods in Enzymology, Volume 567
ISSN 0076-6879
http://dx.doi.org/10.1016/bs.mie.2015.08.026

extension of ITC. The theoretical basis of *kinITC* is detailed for the most common situation of a second-order reaction $A+B\ \Omega\ C$ characterized by kinetic parameters k_{on}, k_{off}. A simplified *kinITC–ETC* method based upon the determination of an "Equilibration Time Curve" (ETC) is presented. The ETC is obtained by automatic determination of the "effective end" of each injection. The method is illustrated with experimental results with a comparison to Surface Plasmon Resonance (SPR) data. k_{on} values were obtained in a wide range, from 10^3 to $0.5 \times 10^6\ M^{-1}\ s^{-1}$. All procedures were implemented in the program AFFINImeter (https://www.affinimeter.com/).

1. INTRODUCTION

Classical thermodynamics mentions time only when it comes to defining "reversible processes" that are necessary to evaluate state functions along the path of finite transformations. It is then imagined that the finite transformation takes place, in theory, infinitely slowly. This means that real kinetic considerations are in fact absent from classical thermodynamics. It is thus commonplace to view isothermal titration calorimetry (ITC) as the standard for thermodynamic studies by its ability to measure directly the enthalpy variation, but not as a technique from which can be derived kinetic information. Such a view is totally unjustified as the primary signal of ITC is a heat power in Joules per second (Watts), not the heat itself in Joules. It should be mentioned that this false view is more commonplace in biology than in chemistry where the link between the measured heat power and the kinetics of the reaction has long been investigated (Calvet & Prat, 1963; Garcia-Fuentes, Baron, & Mayorga, 1998; Lopez-Mayorga, Mateo, & Cortijo, 1987). Historically, this is somewhat strange as the first application of calorimetry by Lavoisier and Laplace was for measuring the heat power evolved by living animals. However, with studies on isolated biological processes (e.g., oxygen binding by hemoglobin) rather than on global physiological functions like respiration, an immediate limitation was that of the amount of biological material and the extreme smallness of the signal to be measured. This led to consider calorimetry as an almost inaccessible technique in biology, not even mentioning kinetic applications. These obvious facts explain why the usage of calorimetry remained for a long domain of pioneers in biology (Buzzell & Sturtevant, 1951; Langerman & Biltonen, 1979; Privalov, 1979; Tian & Cotie, 1924; Watt & Sturtevant, 1969; Wiseman, Williston, Brandts, & Lin, 1989) and why its introduction as a daily technique in biological labs lagged so much behind its usage in chemistry. Illustrative of this state of affair is the following statement by Edsall and

Gutfreund (1983), p. 210: "There are probably as many spectrophotometers [...] in a single major university as there are calorimeters in all the biochemical laboratories in the world; but the number of calorimeters is sure to increase greatly in the next decade or two". However, even in chemistry, the large sample volumes and the long response times of the instruments, made that measurements remained limited to slow, and sometimes very slow reactions (Willson, Beezer, Mitchell, & Loh, 1995). A remarkable exception is the work by Johnson & Biltonen (1975). More recently (Hansen et al., 2010), kinetic measurements were performed on a textbook example of a very slow reaction lasting for 10 h (acid-catalyzed hydrolysis of sucrose) to illustrate the different methods of retrieving the kinetics of the reaction from the raw signal. This means that, even in chemistry, kinetic measurements by ITC do not seem to be widely spread.

It now appears that a new impetus is coming from biology after ITC has been rediscovered as a kinetic technique (Burnouf et al., 2012; Egawa, Tsuneshige, Suematsu, & Yonetani, 2007; Vander Meulen & Butcher, 2012). Such an emergence was obviously facilitated by the availability of modern instruments with cell volumes as small as 200 μL and response times of order of 3.5 s. However, in Vander Meulen and Butcher (2012), slow RNA folding could be studied efficiently with a VP-ITC (Microcal-Malvern) having a larger response time (ca. 10–12 s) (see Chapter "Measuring the kinetics of molecular association by isothermal titration calorimetry" by Vander Meulen et al., in this volume). One can thus anticipate that *kinITC*, as it was coined in Burnouf et al. (2012), has the potentiality of becoming a standard kinetic technique in biology.

We showed in a previous work that, not only simple reactions like $A+B \leftrightharpoons C$ but also more complex reactions involving a two-step kinetic mechanism (e.g., "induced-fit mechanism"), could be addressed by *kinITC* (Burnouf et al., 2012). In this chapter, we will focus exclusively on simple bimolecular reactions and, particularly, on the analysis of a simplified *kinITC* method that was recently introduced in the program AFFINImeter (S4SD, Santiago de Compostela, Spain; https://www.affinimeter.com/). We called this simplified method *kinITC–ETC* as it is based upon the determination of the effective end of each injection which yields an "Equilibration Time Curve", in short *ETC*, allowing to retrieve the kinetic parameters k_{on}, k_{off}. We will first present the theoretical basis of *kinITC–ETC* and the limitations that may be anticipated. The method will be illustrated by three experimental examples for which SPR measurements are available: (1) interaction of HIV-1 Reverse Transcriptase (RT) with its inhibitor

Nevirapine, (2) interaction of carbonic anhydrase with its inhibitor 4-carboxybenzenesulfonamide (4-CBS), (3) interaction of two complementary DNA strands. Cases (1) and (3) not only provided reasonable results but also allowed illustrating the problems occurring for opposite reasons, respectively, very slow association ($k_{on} \sim 1500\ M^{-1}\ s^{-1}$) being the cause of poor affinity and too fast association ($k_{on} \sim 10^6\ M^{-1}\ s^{-1}$ above 32 °C). Case (2) is illustrative of a typical favorable situation at all temperatures.

2. THE THEORETICAL BASIS OF *kinITC–ETC*

The basis of *kinITC* only requires textbook considerations on chemical kinetics. In simple situations corresponding to a bimolecular reaction:

$$A + B \leftrightharpoons C \quad \text{with parameters } k_{on}, k_{off} \tag{1}$$

The kinetics of the reaction is represented by the differential equation:

$$\frac{d[C]}{dt} = k_{on}[A][B] - k_{off}[C] \tag{2a}$$

We will immediately transform it by using the reduced concentrations $A = [A]/[A]_{tot}$, $B = [B]/[A]_{tot}$, and $C = [C]/[A]_{tot}$, $[A]_{tot}$ being the total concentration of the titrand in the measurement cell:

$$\frac{dC}{dt} = k_{on}[A]_{tot}AB - k_{off}C \tag{2b}$$

This allows manipulating molar fractions that have a clear meaning, rather than absolute concentrations. The terms A, B, C, therefore, represent dimensionless "concentrations" when not used explicitly as "compound A, B, or C." In the following, it is assumed that compound B is in the syringe. Equation (2b) represents perfectly the kinetics of the reaction at any step of a titration, that is at any injection, if one considers the evolution of the concentrations after compound B has been injected (to be rigorous, the injection time should be increased slightly to take into account the mixing time). For this simplified *kinITC* method (but not in the full *kinITC* method), the small concentration variations resulting from the course of the reaction during the injection period are neglected, but the concentration variations due to dilution are not.

By using conservation equations $[A] + [C] = [A]_{tot}$ and $[B] + [C] = [B]_{tot}$, where $[A]_{tot}$ and $[B]_{tot}$ are, respectively, the total concentrations of compounds A and B after injection, one is led by classical methods (see Burnouf et al., 2012; Dumas, 2015) to:

$$\frac{dC}{dt} = k_{on}[A]_{tot}(C - C_1)(C - C_2) \tag{3a}$$

C_1 and C_2 being the roots of $k_{on}[A]_{tot}(1 - C)(s - C) - k_{off}C = 0$ (with $C_1 > C_2$):

$$C_1 = \frac{1}{2}\left(p + \sqrt{p^2 - 4s}\right), \quad C_2 = \frac{1}{2}\left(p - \sqrt{p^2 - 4s}\right); \quad p = 1 + s + \frac{K_d}{[A]_{tot}} \tag{3b}$$

with $s = [B]_{tot}/[A]_{tot}$ being the stoichiometric ratio for the current injection and $K_d = k_{off}/k_{on}$ the dissociation constant. Again, note that $[A]_{tot}$ varies from injection to injection due to the dilution resulting from injection of compound B.

Interestingly, the difference $(C_1 - C_2)$ is equal to $\widetilde{A} + \widetilde{B} + c^{-1}$ where $\widetilde{A}, \widetilde{B}$ are the reduced equilibrium concentrations (i.e., divided by $[A]_{tot}$) of, respectively, compounds A and B, and $c = [A]_{tot}/K_d$ is the Wiseman parameter (Wiseman et al., 1989). By integration of the differential Eq. (3a) following standard methods, it is obtained:

$$C(t) = C_1 - \frac{C_1 - C_2}{1 - K^{-1}e^{-t/\tau}}; \quad K^{-1} = \frac{C_2 - C_0}{C_1 - C_0} \tag{4a}$$

$$\tau^{-1} = k_{on}[A]_{tot}(C_1 - C_2) = k_{on}[A]_{tot}\left[\left(1 + c^{-1} + s\right)^2 - 4s\right] \tag{4b}$$

with $C_0 = C(t = 0)$ for the current injection and τ a characteristic time of the evolution of concentrations. From Eq. (4a), the following approximation $C(t) \approx C_2 - (C_1 - C_2)K^{-1}e^{-t/\tau}$ holds for sufficiently large values of t/τ, which means that τ is representative of a single exponential, but only for sufficiently large values of t/τ. In order to make use of a quick and simplified *kinITC* method based upon the determination of the time necessary to return to baseline, it is important to evaluate how the effective characteristic time evolves during the equilibration process, particularly close to the end of the equilibration process. This is performed in the following.

It can be shown (Dumas, 2015) that $C(t)$ can be expressed rigorously as:

$$C(t) = C_1 - (C_1 - C_2)\sum_{n=0}^{n=\infty} K^{-n}e^{-nt/\tau} \tag{5}$$

and since the heat power evolved during the reaction is $P_s(t) = V_{cell}\ \Delta H\ [A]_{tot} \mathrm{d}C/\mathrm{d}t$, by considering the expression for τ (Eq. 4b), it is obtained:

$$P_s(t) = V_{cell} \Delta H k_{on} [A]_{tot}^2 (C_1 - C_2)^2 \sum_{n=1}^{n=\infty} nK^{-n} \mathrm{e}^{-nt/\tau} \tag{6}$$

This corresponds to the instantaneous heat power evolved in the measurement cell, and not to the actually measured heat power $P_m(t)$ due to the finite response time τ_{ITC} of the instrument. Taking this into account leads to (Dumas, 2015):

$$P_m(t) = V_{cell} \Delta H\ k_{on} [A]_{tot}^2 (C_1 - C_2)^2 \sum_{n=1}^{n=\infty} nK^{-n} \frac{\mathrm{e}^{-nt/\tau} - \mathrm{e}^{-t/\tau_{ITC}}}{1 - n\tau_{ITC}/\tau} \tag{7}$$

Admittedly, invoking a single time to characterize the response of an instrument may be a simplification (Garcia-Fuentes et al., 1998; Lopez-Mayorga et al., 1987; Tachoire, Macqueron, & Torra, 1986). However, with our instrument (ITC200 from Malvern-Microcal), we have observed that this is quite a reasonable assumption (see Section 5). We do not claim that all instruments behave exactly in the same way.

It is of interest that the comparison of Eqs. (6) and (7) shows directly the influence of the instrument response time on each component of the infinite sum through the replacement of $\mathrm{e}^{-nt/\tau}$ with $\left(\mathrm{e}^{-nt/\tau} - \mathrm{e}^{-t/\tau_{ITC}}\right)/(1 - n\tau_{ITC}/\tau)$. It can be shown that $K^{-1} < 1$ and it follows that the successive terms in Eq. (7) become less and less important and, practically, a finite number of terms is sufficient (Fig. 1). Note that at the scale of the figure, at most five components among seven are visible.

After a detailed examination (Dumas, 2015), it was shown that, in most occasions, the first component alone governs the return to baseline of $P_m(t)$ after sufficient time following injection (ca. $t > 2\tau$, with τ being determined by Eq. 4b). This means that locating the effective end time of an injection, i.e., the time after which $P_m(t)$ may be considered as null, provides direct information on the time τ and thus on the kinetic parameters. The simplified *kinITC* method, therefore, is based upon the determination of the effective end time or, in other words, of the equilibration time of each injection, which yields a more or less bell-shaped ETC. Examples of such ETCs are in Figs. 4, 6, and 9. All links of the geometric features of an ETC to few relevant parameters are given in Appendix.

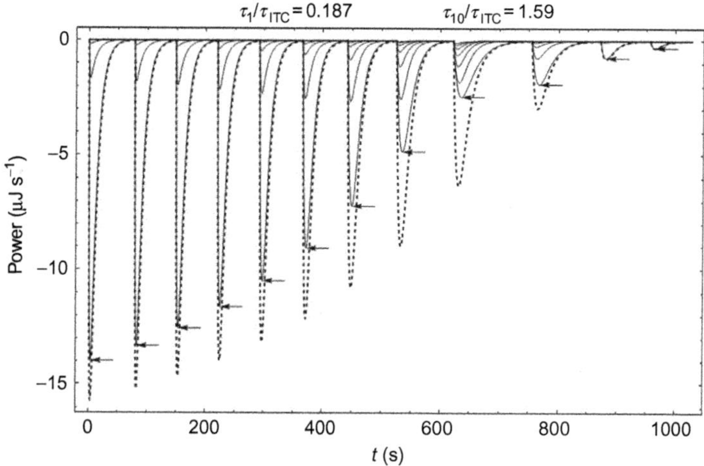

Figure 1 Illustration of the successive components in Eq. (7). The components from $n=1$ to $n=7$ in Eq. (7) are superimposed (thin curves). The max amplitude of each first component is indicated by an arrow. The sum of all components corresponds to the measured signal (dashed curve). The simulation was performed with the values: cell volume = 1.4 mL, injected volume at each injection = 12 μL, $[A]_0=20\ \mu M$ (cell), $[B]_0=240\ \mu M$ (syringe), $k_{on}=3\times10^4\ M^{-1}\ s^{-1}$, $k_{off}=10^{-3}\ s^{-1}$, $\Delta H=-80\ kJ\ mol^{-1}$, $\tau_{ITC}=10$ s. The term τ_1 and τ_{10} corresponds, respectively, to the characteristic times defined in Eq. (4b) for injections #1 and #10 (unit stoichiometry).

3. PRACTICAL ASPECTS OF THE METHOD

3.1 ETC Determination in Practice

The necessary methods (automatic baseline determination, integration of the injection power curves and ETC determination) allow determining automatically the "effective end" of each injection and the result matches closely what would have been obtained "by eye" (arrows in Fig. 2). It is obvious from inspection of Fig. 2 that the uncertainty on the equilibration time depends essentially on the signal/noise ratio of the base line.

3.2 Determination of k_{on} and k_{off} from the ETC

Once the ETC has been determined, a link has to be made between the "effective end time" $t_{end}(N_{inj})$ determined at injection $\#N_{inj}$ to the time $\tau(N_{inj})$ defined by Eq. (4b). The following approximation is used:

$$t_{end}\left(N_{inj}\right)=\alpha\left[\tau\left(N_{inj}\right)+\tau_{ITC}\right]+\tau_{inj}+\tau_{mix} \tag{8}$$

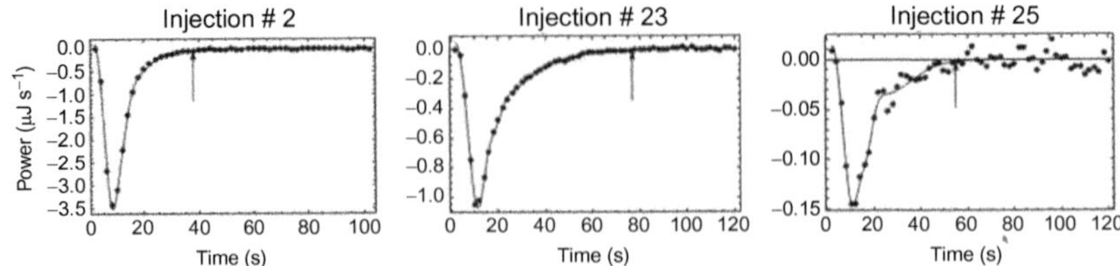

Figure 2 Determination of the "effective end" of each injection $P_m(t)$. Three particular injections of a titration of the oligoDNA 5′-AAG AAG AGG AG-3′ with its complementary strand are shown. The arrows mark the automatically determined "effective end" of each injection. Injection #23 corresponds to the injection with the longest equilibration time. Experimental conditions: $[A]_0 = 11.6\mu M$, $[B]_0 = 100\mu M$, injections of 1 μL, cell volume = 203.6 μL, sampling time = 2 s.

where α was tuned to 4.5 after trials and errors and τ_{inj} and τ_{mix} are, respectively, the injection time and the mixing time. Normally, there is a rule given by the manufacturer for the injection time (e.g., 2 s/μL injected with the ITC200 from Microcal-Malvern). The mixing time (of the order of 1 s) corresponds to the characteristic time of the exponential governing the approach to complete mixing of compound B after its injection in the cell. For the complete *kinITC* method with the aim of simulating exactly, the whole injections, τ_{inj} and τ_{mix}, have an important influence during, and immediately after, the injection (see realistic simulations on our web site at http://www-ibmc.u-strasbg.fr:8080/webMathematica/kinITCdemo/); for the simplified *kinITC-ETC* method, these two parameters are solely involved in the approximate correction to $t_{\mathrm{end}}(N_{\mathrm{inj}})$ by Eq. (8).

Importantly, if τ_{inj}, τ_{mix}, and τ_{ITC} are known, only k_{on} is unknown in the fit of an experimental ETC with Eqs. (4b) and (8), since $c = [A]_{\mathrm{tot}}/K_{\mathrm{d}}$ is known from the usual processing of the titration curve and since, obviously, the stoichiometric ratio s is known at each injection. In fact, even though τ_{ITC} can be determined experimentally (see below), it appeared that an efficient pragmatic solution was (most often) to consider τ_{ITC} as a free parameter too (Dumas, 2015). There are several reasons for this choice. First, τ_{ITC} may not be known accurately if it was measured too early before the experiment. Second, the conditions used for the determination of τ_{ITC} may not be exactly relevant for the conditions of the experiment (in particular, due to differences in the viscosity and the heat conductivity of the solution). Third, Eq. (8) is only approximate and, for this reason, it may be necessary to adjust τ_{ITC}.

4. RESULTS

4.1 Results with Carbonic Anhydrase and Its Inhibitor 4-CBS

Carbonic anhydrase is a zinc-containing enzyme that interconverts H^+ and bicarbonate HCO_3^- to CO_2 and water. As such, it has an essential role in pH regulation and respiration. This was an obvious choice as it was used in an extensive SPR benchmark involving 22 laboratories (Navratilova et al., 2007). In this way, the best possible kinetic parameters were available for comparison with the results from *kinITC–ETC*. The enzyme (bovine carbonic anhydrase II from Sigma Chemical), inhibitor 4-CBS (Acros Organics), and the buffer conditions used for ITC were identical to those used for the SPR study. Protein and ligand concentrations as well as the temperatures used are in the legend of Fig. 4. Baseline-corrected data are shown for two temperatures in Fig. 3 and the experimental ETCs with their fits with Eqs. (4b) and (8) are shown in Fig. 4. Since different temperatures were used in both types of experiments, the results are presented in the form of Arrhenius plots (Fig. 5).

Although the results are noisier for *kinITC-ETC*, the straight lines from the fit of $\ln(k_{on/off})$ versus T^{-1} are parallel within experimental errors, which points to identical activation energies and, thus, identical temperature dependence of k_{on} and k_{off} from the two methods. The straight lines from *kinITC-ETC* are systematically lower than those from SPR but at most by 0.5, which corresponds to $k_{on}(\mathrm{SPR})/k_{on}(\mathrm{kinITC-ETC}) \approx k_{off}(\mathrm{SPR})/k_{off}(\mathrm{kinITC-ETC}) \leq 1.65$ in the temperature range common to the two types of experiments (6.1–25 °C). Notably, this ratio was significantly higher (ca. 2.5) when τ_{ITC} was not considered as a free parameter and

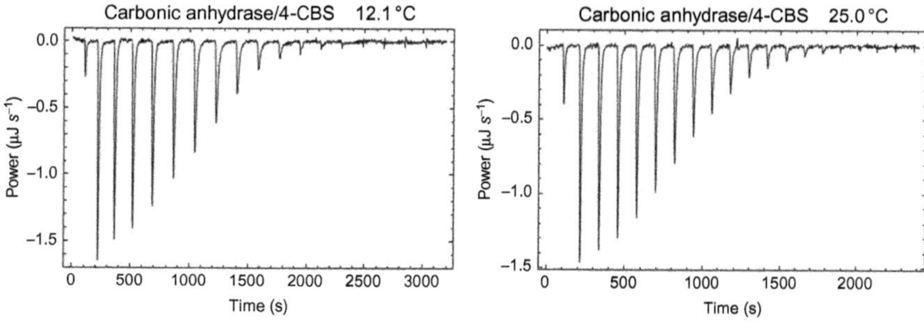

Figure 3 Example of baseline-corrected injection curves for carbonic anhydrase/4-CBS.

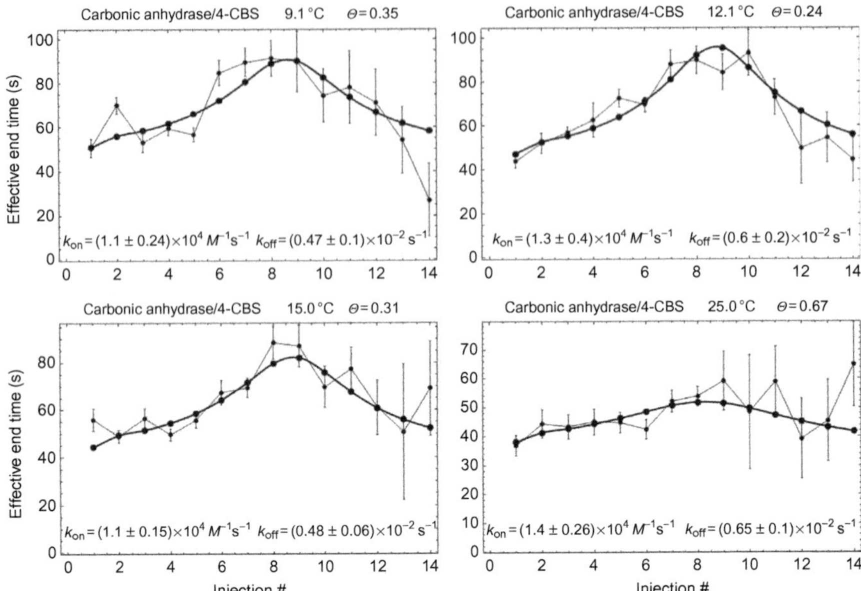

Figure 4 Fit of carbonic anhydrase ETCs at various temperatures. The initial concentrations of carbonic anhydrase (compound A) were $[A]_0 = 24.4$, 24.2, 24.5, 24.6, and 17.6μM at, respectively, 6.1 °C (not shown), 9.1, 12.1, 15, and 25 °C. The concentration of 4-CBS was $[B]_0 = 315$μM at all temperatures and the injected volumes were 1.9 μL up to 15 °C and 1.4 μL at 25 °C. The parameter Θ indicated on top of each figure is defined in Table 1 and discussed in Appendix.

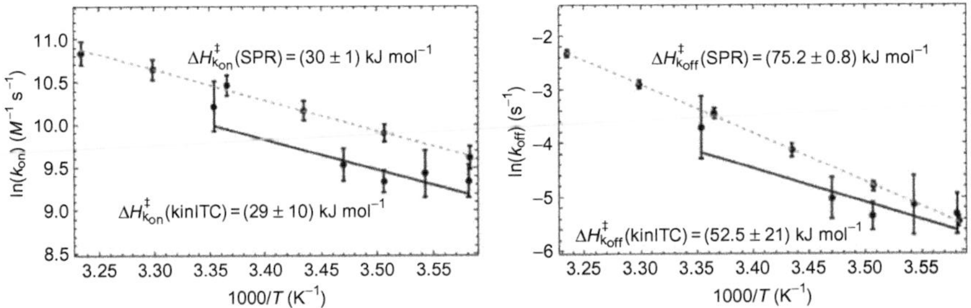

Figure 5 Arrhenius plots for the comparison of *kinITC-ETC* and SPR results. The *kinITC-ETC* results (continuous lines) are compared to the SPR results (dashed lines). The corresponding activation energies are indicated.

fixed to 3.5 s (not shown). This experimental case, therefore, appears to be representative of a favorable situation leading to kinetic parameters comparing well with those from SPR results of the best possible quality due to the unusually great number of independent replicate experiments.

4.2 Results with HIV-1 RT and Its Inhibitor Nevirapine

HIV-1 RT is a key enzyme in the life cycle of the virus and Nevirapine was the first discovered nonnucleoside inhibitor (De Clercq, 1998). This is a hydrophobic compound which is deeply buried within the protein. Interestingly, the binding site revealed by X-ray crystallography (PDB: 1VRT) is not preformed (Ren et al., 1995), which means that the ligand has to select transient favorable conformations to "find its way." We already used RT/nevirapine in the work for the original *kinITC* method for which we also had kinetic measurements by SPR (Burnouf et al., 2012). Three temperatures (25, 30, and 35 °C) were used for the ITC experiments, but SPR was performed only at 25 °C. Due to the hydrophobic character and limited solubility of Nevirapine, RT was used as compound B in the syringe to avoid using Nevirapine at too high concentrations. Here, we simply reprocessed the ITC data by *kinITC–ETC* (Fig. 6). Comparison of the results obtained here by *kinITC–ETC*, and by *kinITC* and SPR in Burnouf et al. (2012) is shown in Fig. 7. Recall that the results obtained by the full *kinITC* method are equations that can be plotted as continuous curves.

Overall, the different results are compatible with the largest discrepancy being a factor of three between *kinITC–ETC* and *kinITC*/SPR for k_{off} at 25 °C. Notably, the replicates of the SPR experiments also differed by a factor of two for k_{on}. Nevertheless, these results are all in agreement with the expected slow binding due to the necessary conformational change of the protein to accommodate the ligand. This interpretation is also in line with

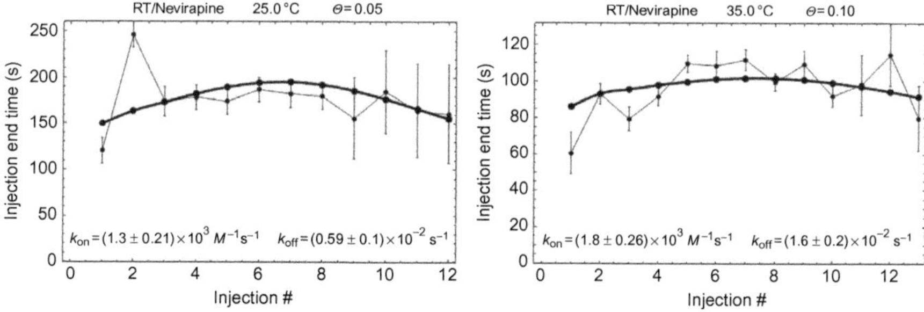

Figure 6 Fit of RT/Nevirapine ETCs. The initial concentrations were 20 μ*M* for nevirapine (compound *A*), 240 μ*M* for RT and the injected volumes were 2.33 μL. The different temperatures were 25, 30 (not shown), and 35 °C. Notably, the response time was kept fixed to 3.5 s as considering it as a free parameter in the fit led to aberrant negative values at 30 and 35 °C. The parameter Θ indicated on top of each figure is defined in Table 1 and discussed in Appendix.

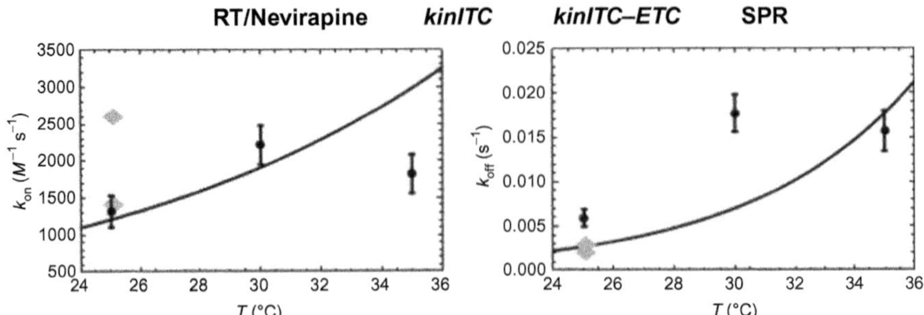

Figure 7 Comparison of *kinITC*, *kinITC–ETC*, and SPR results for RT/Nevirapine. The *kinITC* results from Burnouf et al. (2012) (solid line) and *kinITC-ETC* results (dots with error bar) are compared to the SPR results (diamonds).

the unfavorable large negative value of ΔS(binding) $= -145$ J mol^{-1} K^{-1} at 25 °C (Burnouf et al., 2012), which marks an important stabilization of the protein upon ligand binding.

These results also show that very slow binding is not synonymous with great accuracy in *kinITC-ETC*. In fact, this slow binding implies a rather low affinity ($K_d = 4.4$ μM at 25 °C and 8.6 μM at 35 °C) which leads to very low c values (4.9 at 25 °C and 3.2 at 35 °C). In agreement with the theoretical results (Lines 2 and 3 of Table 1 in Appendix), the relative variation of the equilibration time is at most 20% and the characteristic bell shape of the ETC appears little due to the errors on the determination of the equilibration times (Fig. 6). The greatest part of the information, therefore, lies in the large equilibration time (well above $4.5\,\tau_{ITC}$) right from the first injections.

4.3 Results with the Annealing of Two Complementary DNA Strands

In order to test *kinITC-ETC* with fast association rates, we studied duplex formation from the polypurine DNA oligomer $ODN_1 = 5'$-AAG AAG AGG AG-3′ and its complementary sequence ODN_2. It has long been established that duplex formation from two DNA strands proceeds through two consecutive kinetic steps: initial formation of an unstable nucleus of three base pair followed by a "zipping" process leading to full duplex (Craig, Crothers, & Doty, 1971; Porschke & Eigen, 1971). However, according to these land-mark studies, the overall process can nevertheless be accurately described kinetically by a "global" single-step mechanism

$S_1 + S_2 \leftrightharpoons D$, because the zipping process is extremely fast in comparison of the formation of the initial nucleus.

The SPR experiments were performed at 6, 9, 14, 19, 23, 27, and 32 °C on a BiacoreTM T200 apparatus (GE Healthcare Life Sciences, Uppsala, Sweden) using a CM5 sensor chip (BiacoreTM) coated with 2600 resonance units (RU) of streptavidin (Roche Applied Sciences, Roche Diagnotics, Meylan France). The DNA oligonucleotides were purchased from Eurofins MWG Operon (Ebersberg, Germany). The DNA samples were prepared at 23 °C in a sodium phosphate buffer 20 m*M*, pH 7.2, NaCl 100 m*M*, supplemented with 0.050% Tween-20 (running buffer). Fifteen RU of the biotinylated polypurine strand were immobilized on one flow cell of the sensor chip. Another flow cell was left blank for double-referencing of the sensorgrams. The single cycle-kinetics method (Karlsson, Katsamba, Nordin, Pol, & Myszka, 2006) was used to inject the samples in duplicate over the target at 25 μL min^{-1}. The association and dissociation rate constants were determined by direct curve fitting of the sensorgrams according to the model $S_1 + S_2 \leftrightharpoons D$ with the Biacore T200 evaluation software V2.00.

The ITC experiments were performed in the same buffer (comprising 0.050% Tween-20) at 15, 20, 25, 30, and 35 °C with the parameters indicated in Fig. 8. The corresponding ETC curves are in Fig. 9 and the SPR curves with their fits are shown in Fig. 10. Again, the *kinITC–ETC* and SPR results are compared with Arrhenius plots (Fig. 11).

In agreement with Line 4 in Table 1, the ETCs are extremely sharp: the c values at 15 and 20 °C are 3147 and 1610, which lead to half-height widths (in stoichiometric ratio) of 0.11 and 0.16, respectively. Beyond 20 °C, the ETCs become featureless with favorable c values of 371 and 127, but large Θ values (Line 5 in Table 1) of 2.94 and 2.40, at 25 and 30 °C, respectively. There are thus opposite patterns (high c – low Θ) at 15 and 20 °C and (low c – high Θ) at 25 and 30 °C. The very high c values yield sharp ETCs, but prevent from obtaining accurate K_d values, whereas high Θ values prevent from distinguishing the kinetic signal from the instrument response time. On the basis of the Θ values, one would conclude that there is kinetic information only at 15 and 20 °C. However, one has to admit that the Arrhenius plots (Fig. 11) show that the systematic evolution of the k_{off} values from *kinITC–ETC* follows well that from SPR at 25 °C ($10^3/T = 3.35$ K^{-1}) and 30 °C ($10^3/T = 3.30$ K^{-1}). It is important to note that, for both *kinITC–ETC* and SPR, the experiments at the highest temperatures (32 °C for SPR ad 35 °C for ITC) were discarded as they did not provide kinetic information.

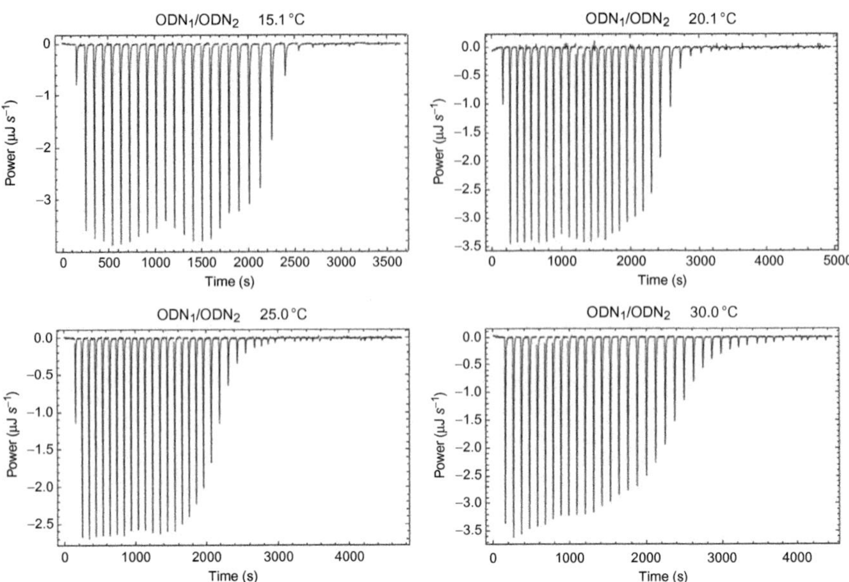

Figure 8 Example of baseline-corrected injections curves for ODN_1/ODN_2. The initial concentrations of ODN_1 (compound A) were $[A]_0 = $ **10** and 2.5 μM at 15 °C; **10**, 2.5 and 2.5 μM at 20 °C; **7.1** μM at 25 °C; **10** μM at 30 °C. The concentration of ODN_2 were τ_{ITC} 100 μM at all temperatures and the injected volumes were **1**, 0.5, 1 μL at 15 °C; **1**, 0.5 and 0.7 μL at 20 °C; **0.7** μL at 25 °C and **1** μL at 30 °C. The underlined values in bold are those for the figures shown.

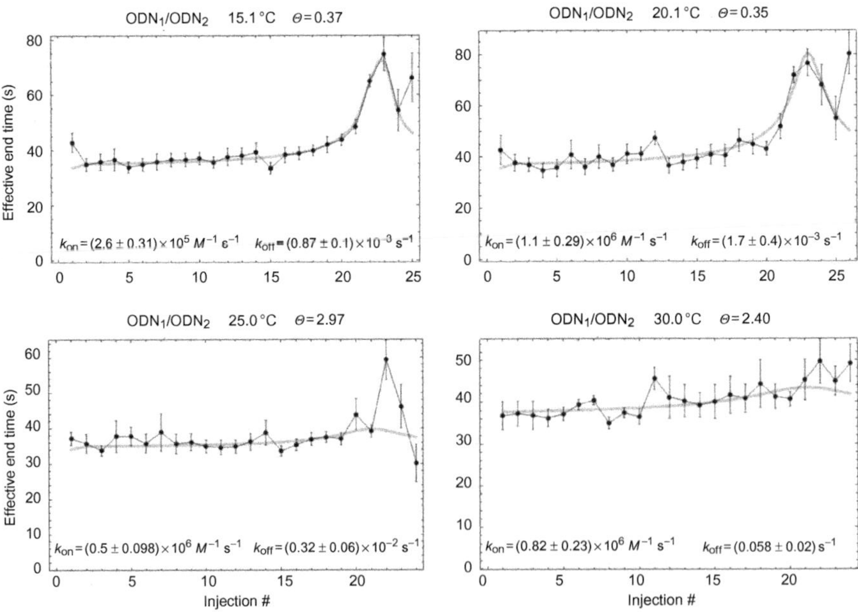

Figure 9 Fit of ODN_1/ODN_2 ETCs. Each ETC corresponds to the injection curves shown in Fig. 8. Only two injections were kept after the maximum as the effective end times were poorly determined for the following injections, which negatively affected the results. The c values (Wiseman parameter) are 3147, 1610, 371, and 127 from 15.1 °C to 30 °C. The parameter Θ indicated on top of each figure is defined in Table 1 and discussed in Appendix.

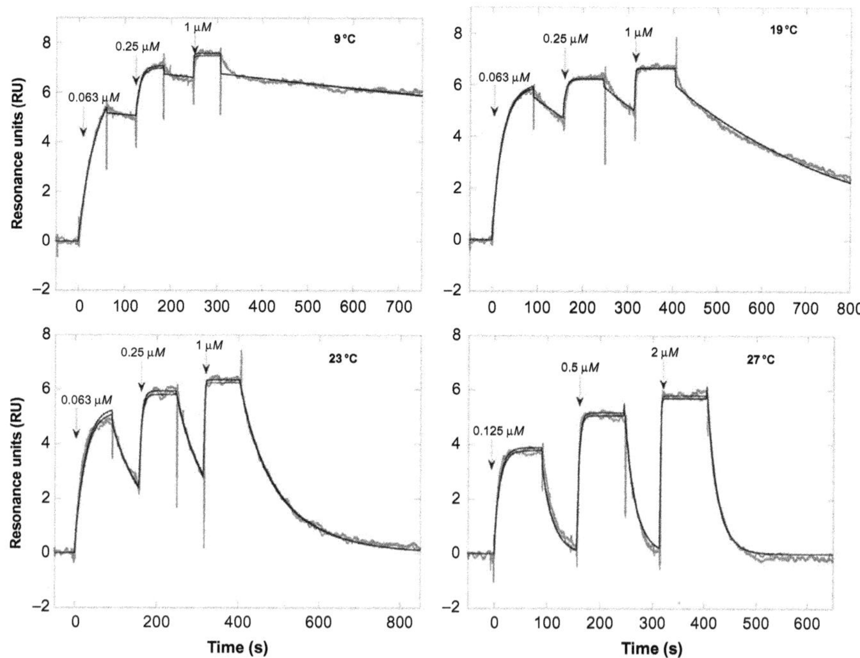

Figure 10 Kinetic analysis by SPR of duplex formation (ODN_1/ODN_2). Four experimental curves and their fits (black curves) are shown among seven. The analyte was prepared in the running buffer and injected sequentially in the order of increasing concentrations as indicated by the arrows. The regeneration of the surface was achieved with a 2-min pulse of 20 m*M* NaOH.

These results can be compared further, both on kinetic and thermodynamic grounds. First, it is striking that the activation energies of k_{on} are almost null (and even slightly negative for SPR), whereas the activation energies of k_{off} from *ITC-ETC* and SPR (Fig. 11) are large, as expected. This is in agreement with the association process not being one-step and, according to Craig et al. (1971) and Porschke and Eigen (1971), only apparent activation energies were obtained for which slightly negative values for k_{on} were indeed expected. Second, the fact that K_{d} values were obtained both by ITC and SPR, and that enthalpy values were obtained by ITC opens the way for a comparison of the experimental and van 't Hoff-determined enthalpy values (Fig. 12).

In both ITC and SPR, it appears that the van 't Hoff plots are significantly curved but also show distinct differences. As a consequence, the ΔH values (ΔH_{vH}) derived from the respective van 't Hoff plots are different

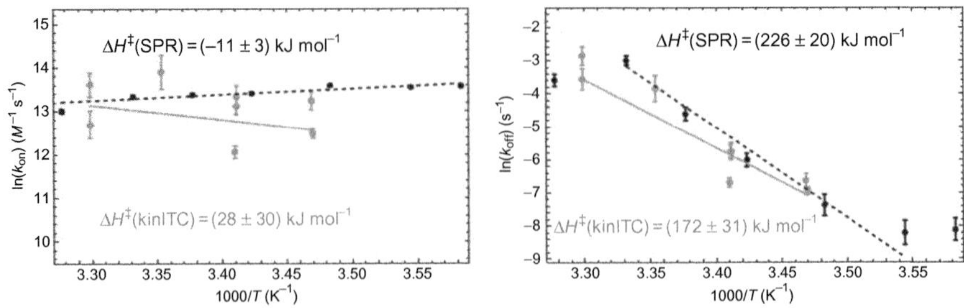

Figure 11 Arrhenius plots for the comparison of *kinITC–ETC* and SPR results for ODN_1/ODN_2. The *kinITC–ETC* results (continuous lines) are compared to the SPR results (dashed lines). The corresponding activation energies are indicated. The SPR data were not included in the Arrhenius plots at the highest (32 °C) and lowest (6 and 9 °C) temperatures due to obvious departure from linear variation for k_{off}.

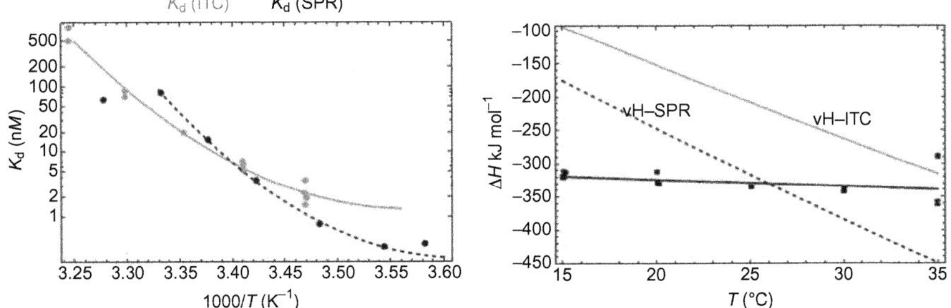

Figure 12 Comparison of van 't Hoff-derived enthalpies from ITC and SPR for ODN_1/ODN_2. Left: evolution of K_d from ITC (solid curve) and from SPR (dashed curve) versus $1/T$. The curves were obtained by a fit of the experimental K_d values (dots) with the van 't Hoff equation and a temperature-dependent ΔH according to $\Delta H = \Delta H_0 + \Delta C_p (T - T_0)$ and $1000/T_0 = 3.40\ K^{-1}$. Note that the end points were excluded from the fit for the SPR data. Right: comparison of the plots of $\Delta H = \Delta H_0 + \Delta C_p (T - T_0)$ derived from ITC and SPR van 't Hoff plots with the experimental ΔH values from ITC (solid line with squares).

(Fig. 12, right). There is another important feature: None of the ΔH_{vH} agree with ΔH_{ITC}. Even more striking, ΔH_{vH} from K_d(ITC) differs most from ΔH_{ITC}! This fact is so clear that it deserves discussion. First, recall that such differences between ΔH_{vH} and ΔH_{ITC} are not unusual (Naghibi, Tamura, & Sturtevant, 1995). The fact that association does not follow a simple one-step process (a purely kinetic feature) cannot be the cause of such a thermodynamic discrepancy. The more likely explanation is that the data do not represent merely the formation of a full duplex. The two sequences ODN_1 and

ODN_2 were designed to avoid hairpin formation but, prompted by the latter comment, one notices that two symmetric imperfect associations can form (non-WC base pairs are in bold):

```
5′-AAG AAG AGG  AG-3′                 5′-AAG AAG  AG  GAG-3′
     3′-TTC  TTC  TC CTC-5′        3′-TTC TTC  TCC TC-5′
```

(Bold in the original: the middle G of "AGG" and the middle T of "TTC" in the first association; the middle A of the second "AAG" and the middle C of "TCC" in the second.)

There is no reason why these two imperfect associations would not compete with full duplex formation, particularly at low temperature. Not only are there two possible competing structures but, in addition, dangling strands are known to stabilize terminal base pairs (Turner, 2000). Although a full study would be required, additional remarks will support this explanation. From the values of K_d and ΔH from ITC at each temperature T, the melting temperature T_m can be derived following:

$$\frac{1}{T_m}=\frac{1}{T}-\frac{R}{\Delta H}\mathrm{Ln}\frac{4K_d}{[\mathrm{ODN}]_{tot}} \tag{9}$$

with $R=8.314\,\mathrm{J\,mol^{-1}\,K^{-1}}$ being the gas constant and $[\mathrm{ODN}]_{tot}$ the total strand concentration at unit stoichiometry. Equation (9) derives from $\Delta G^0=\Delta H^0-T\Delta S^0=RT\mathrm{Ln}K_d$ and $T_m^0=\Delta H^0/\Delta S^0$, where T_m^0 is the melting temperature in the standard conditions ($[\mathrm{ODN}]_{tot}=1M$); T_m in the conditions of the experiment derives from T_m^0 by the necessary correction on concentration (the factor 4 in Eq. 9 comes from the annealing of two different strands) (Dumas, Ennifar, Disdier, & Walter, 2014; Turner, 2000). The resulting T_m values are in Fig. 13, which shows a systematic increase from 30 to 40 °C. Obviously, this melting temperature would not change with the temperature if a single species were present; the observed change is thus a strong indication of the presence of less stable species at low temperatures.

Interestingly, the theoretical T_m (Kibbe, 2007) of the imperfect association (with a G-T base pair) is either 24 or 30 °C depending on how the calculation is done, and the theoretical T_m of the full duplex is 41.9 °C, in good agreement with the maximum T_m at high temperature (Fig. 13). Also note that the ΔH_{vH} from K_d(ITC) and the measured ΔH_{ITC} are very different at low temperature but become almost equal at 35 °C (Fig. 12), which is again in agreement with a majority of the stable full duplex at that temperature.

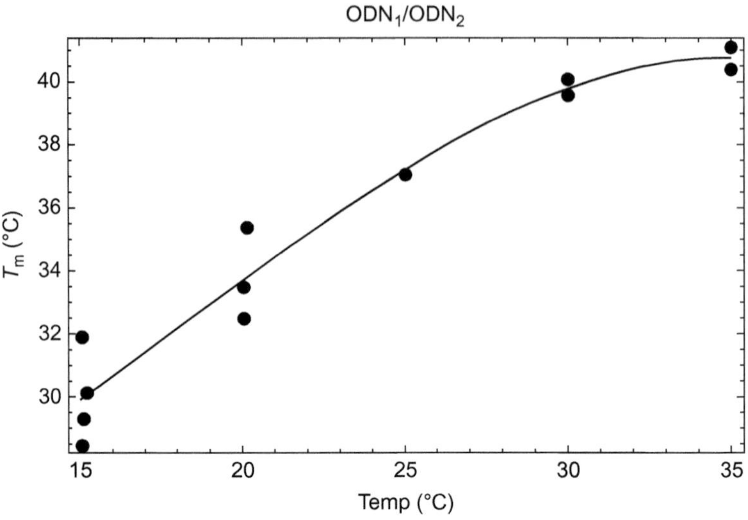

Figure 13 Evolution of the melting temperature from ITC results. The melting temperatures from Eq. (9) used with the K_d and ΔH values from all ITC experiments with ODN_1/ODN_2 are shown as dots. The curve is merely a guide for the eye.

5. INSTRUMENT RESPONSE TIME: MEASUREMENT AND IMPORTANT FACTS

Characterization of the instrument response time (for any instrument such that the response is linearly related to the excitation) is based on measuring the response to an "instantaneous" excitation, in practice to a very quick excitation. For that, one can use the dilution of a small amount of methanol into water. The differential heat of dilution of pure methanol into a large excess of water (-7.347 kJ mol^{-1}) (Bertrand, Millero, Wu, & Hepler, 1966) is too important for a modern microcalorimeter, which makes necessary to inject diluted methanol (1–2%, v/v). In view of *kinITC* experiments, the instrument must be used in the "power compensation mode," as the "heat conduction mode" would increase considerably the response time. With our ITC200 and a cell volume of 203 μL, we obtained good results with injections from 0.5 to 2 μL (at a rate of 0.5 μL s^{-1}), with a maximum sampling rate of 1 s and with a mixing speed of 1500 rpm (it may happen that 1000 rpm give better results). An estimate of the response time is obtained by an exponential fit of the end part of the power curve as shown in Fig. 14A. As a rule of thumb, one may consider as the first experimental point to be kept the one for which the power amplitude is half the maximum amplitude.

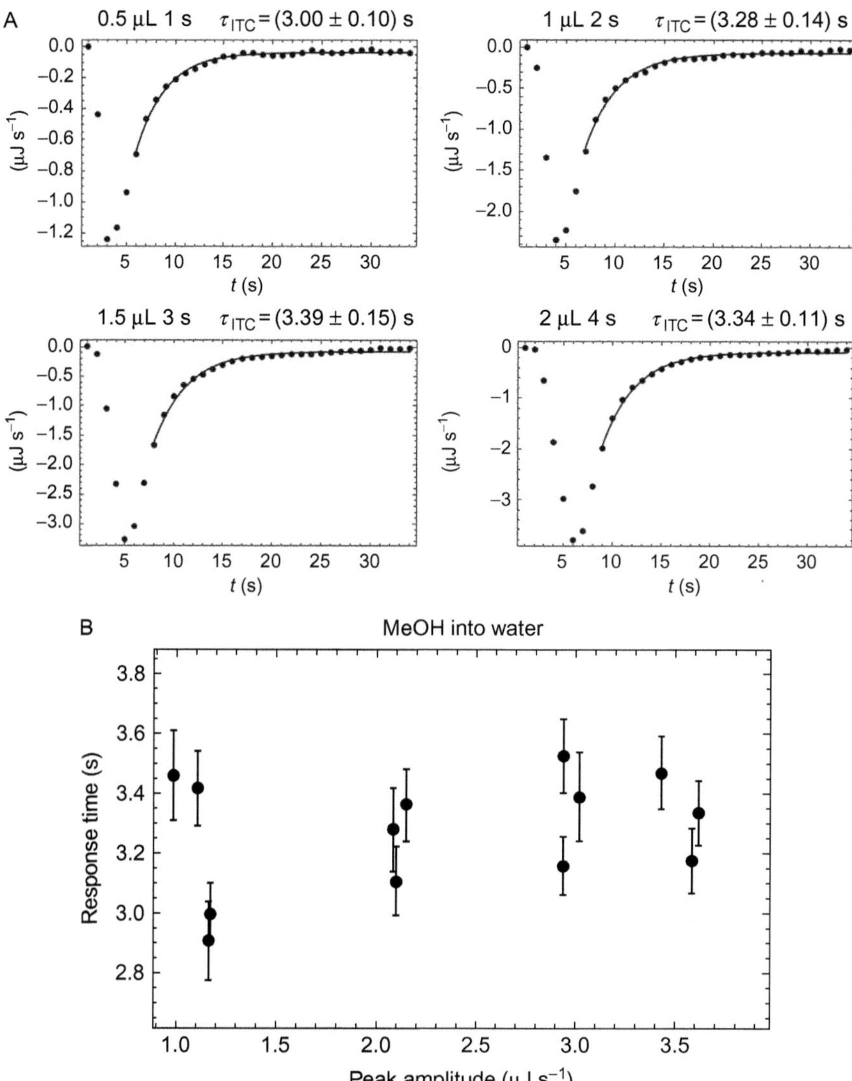

Figure 14 Response time determination. (A) Examples of data obtained with injection of methanol (1%, v/v) into water at 15 °C. The injected volumes, injection times, and response times obtained after the exponential fit (solid curves) are indicated. (B) Stability of the response times versus the amplitude of the excitation for repeated measurements.

Normally, no systematic variation of the response time with the amplitude of the signal should be seen, as in Fig. 14B. However, Fig. 14B shows significant spreading, which means that a single measurement is highly suspicious.

In addition, it should be mentioned that non-monotonous response curves with positive excursions before the return to baseline can be obtained. Such response curves should be discarded. This type of artifact results from an overshoot of the power compensation mechanism of the instrument (electronics and software), which shows that one is close to instability due to the very quick excitation following methanol dilution.

There is another fact of considerable practical importance. Contrary to a common belief, the response time is not in reality an intrinsic property of an ITC instrument since it strongly depends on the cleanness of the measurement cell. We observed, for example, that the response time obtained with the same instrument and using the same procedure (dilution of methanol) could be as high as 7 s, that is twice as much as in Fig. 14. However, there was a return to low values after extensive cleaning of the instrument with the recommended procedure. It is thus clear that frequent and quick measurement of the response time should be made. In any case, an experiment made in view of obtaining kinetic information, particularly for fast kinetics, should be preceded by such a measurement and, if necessary, by cleaning of the instrument.

6. CONCLUSIONS

The *kinITC–ETC* method, a simplified version of *kinITC* (Burnouf et al., 2012), has been examined to assess its theoretical foundation and its practical ability of obtaining kinetic information by comparison with SPR data. Apart in one case (HIV-1-RT/Nevirapine), the results from *kinITC–ETC* were noisier than those from SPR. This appeared with regression lines in Arrhenius plots being less well defined than their SPR counterparts (Figs. 5 and 11). Nevertheless, *kinITC–ETC* obtained excellent results with carbonic anhydrase/4-CBS and k_{on} and k_{off} values in the ranges 10^4–$3\times10^4\ M^{-1}\ s^{-1}$ and 5×10^{-3}–$3\times10^{-2}\ s^{-1}$. In particular, the temperature dependence was perfectly obtained. The results are reasonably good with HIV-1-RT/Nevirapine and k_{on} and k_{off} values in the ranges 10^3–$3\times10^3\ M^{-1}\ s^{-1}$ and 2×10^{-3}–$1.5\times10^{-2}\ s^{-1}$. The temperature dependence, however, was not clearly defined. It is fair to say that SPR data (at one temperature only) also showed variability with replicate experiments. The greater dispersion of the results from *kinITC–ETC* is manifest for ODN_1/ODN_2 (Fig. 11). However, the results are reasonably good at

15 and 20 °C ($k_{on} \sim (4 \pm 2) \times 10^5$ M^{-1} s^{-1} and k_{off} values in the range 10^{-3}–3×10^{-3} s^{-1}) and deviate slightly from SPR results at the higher temperatures 25 and 30 °C where the kinetics is still faster ($k_{on} \sim (4 \pm 2) \times 10^5$ M^{-1} s^{-1} and k_{off} values in the range 10^{-2}–4×10^{-2} s^{-1}). It is quite noticeable that the two techniques reached their limit at the same point: the SPR data were not usable at 32 °C and the ITC data were not usable at 35 °C.

Several important practical recommendations can be made. First, the instrument must be cleaned very thoroughly. Second, one should have a sufficient sampling of the ETC, which means that small stoichiometry increments are preferable as far as the heat power is high enough. Of course, in case of fast kinetics, one should try to lower the concentrations of compounds A and B as much as possible. In this respect, low ΔH values are obviously quite detrimental.

In conclusion, *kinITC–ETC* is a new mean of obtaining kinetic information, potentially in a wide range of k_{on} values, from 10^3 to 5×10^5 M^{-1} s^{-1}. A great interest of *kinITC–ETC* is of not requiring any sample preparation. In addition, due to its implementation in AFFINImeter (https://www.affinimeter.com/), *kinITC–ETC* results can be obtained without requiring specific knowledge or any action from the user. We therefore anticipate that *kinITC–ETC* will become a standard tool in biology. Lastly, we think it is necessary to recall that all kinetic method are model dependent, which means that the results cannot be correct if the kinetic model is not correct.

ACKNOWLEDGMENTS

We are indebted to F. Disdier for the development and to M. Zerbib for the maintenance of the web site (http://www-ibmc.u-strasbg.fr:8080/webMathematica/kinITCdemo/).

APPENDIX. GEOMETRICAL FEATURES OF AN ETC

The analysis performed in Dumas (2015) allowed to determine all geometrical characteristics of an ETC (Fig. 15). The results, obtained by making the approximation that the injection and mixing times are negligible, are given in Table 1 with reference to Fig. 15.

The parameter Θ was obtained in Burnouf et al. (2012) as a way of quantifying the possibility of discerning the true kinetic signal from the kinetic response of the instrument. With more experience, we can now state that it is safe to have $\Theta < 0.5$. However, examples obtained with ODN_1/ODN_2 showed that significantly higher values allowed obtaining kinetic information.

Table 1 Geometrical Features of an ETC (see Fig. 15) Notations: $[A]_0$, Initial Concentration in the Cell; $c=[A]_0/K_d$ (Wiseman Parameter); τ_{ITC}, Response Time of the Instrument; s is the Current Stoichiometric Ratio $[B]_{tot}/[A]_{tot}$

1. Min. equilibration time (at $s=0$)	$\alpha\left[\dfrac{1}{k_{off}(c+1)}+\tau_{ITC}\right]$	Lower horizontal line ($\alpha\approx 4.5$)
2. Max. equilibration time (at $s=1-1/c$)	$\alpha\left[\dfrac{1}{2k_{off}c^{1/2}}+\tau_{ITC}\right]=\alpha\tau_{ITC}\left[1+\dfrac{1}{2\Theta}\right]$	Upper short line (Θ defined below)
3. Max. equilibration time variation:	$\alpha k_{off}^{-1}\left[\dfrac{1}{2c^{1/2}}-\dfrac{1}{c+1}\right]$	Length of the vertical arrow (at $s=1-1/c$)
4. Half-height width	$\dfrac{4(1-c^{-1/2})\sqrt{3+2c^{1/2}+3c}}{(1+c^{1/2})^2}$	Length of the half-height dashed line
5. Parameter Θ	$\sqrt{k_{off}k_{on}[A]_0}\tau_{ITC}=c^{1/2}k_{off}\tau_{ITC}$	Defined in Burnouf et al. (2012)

It is remarkable that the maximum amplitude of equilibration time variation (arrow in Fig. 15) does not depend on τ_{ITC} and the half-height width only depends on c.

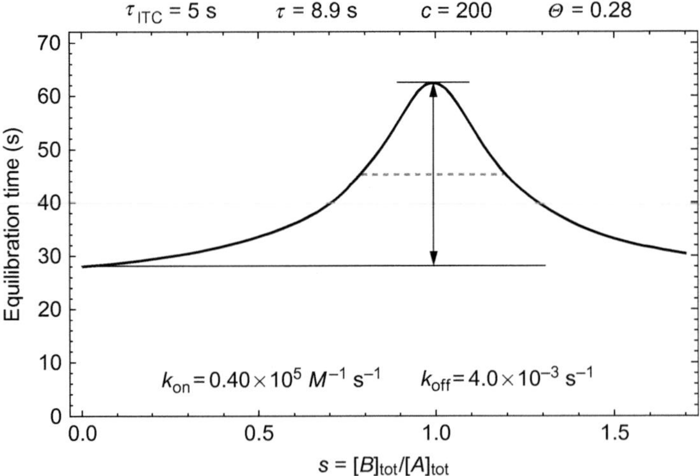

Figure 15 Geometrical features of an ETC. The minimum equilibration time at $s=0$ (lower horizontal line), the half-height width (dashed line), the maximum equilibration time (upper short line), and the maximum amplitude (vertical arrow) are highlighted. All values used to calculate the ETC are indicated. See Table 1 for equations.

REFERENCES

Bertrand, G. L., Millero, F. J., Wu, C. H., & Hepler, L. G. (1966). Thermochemical investigations of the water-ethanol and water-methanol solvent systems. I. Heats of mixing, heats of solution, and heats of ionization of water. *Journal of Physical Chemistry*, *70*, 699–705.

Burnouf, D., Ennifar, E., Guedich, S., Puffer, B., Hoffmann, G., Bec, G., et al. (2012). kinITC: A new method for obtaining joint thermodynamic and kinetic data by isothermal titration calorimetry. *Journal of the American Chemical Society*, *134*(1), 559–565.

Buzzell, A., & Sturtevant, J. M. (1951). A new calorimetric method. *Journal of the American Chemical Society*, *73*, 2454–2458.

Calvet, E., & Prat, H. (1963). *Recent progress in microcalorimetry*. Oxford: Pergamon Press.

Craig, M. E., Crothers, D. M., & Doty, P. (1971). Relaxation kinetics of dimer formation by self-complementary oligonucleotides. *Journal of Molecular Biology*, *62*, 383–401.

De Clercq, E. (1998). The role of non-nucleoside reverse transcriptase inhibitors (NNRTIs) in the therapy of HIV-1 infection. *Antiviral Research*, *38*, 153–179.

Dumas, P. (2015). Joining thermodynamics and kinetics by kinITC. In M. Bastos (Ed.), *Biocalorimetry: Foundations and contemporary approaches*.Boca Raton: Taylor & Francis. in press.

Dumas, P., Ennifar, E., Disdier, F., & Walter, P. (2014). UV melting studies with RNA. In *Handbook of RNA biochemistry* (pp. 445–480). Vol. 1. Weinheim, Germany: Wiley-VCH.

Edsall, J. T., & Gutfreund, H. (1983). *Bio-thermodynamics: The study of biochemical processes at equilibrium*. New York: John Wiley & Sons.

Egawa, T., Tsuneshige, A., Suematsu, M., & Yonetani, T. (2007). Method for determination of association and dissociation rate constants of reversible bimolecular reactions by isothermal titration calorimeters. *Analytical Chemistry*, *79*, 2972–2978.

Garcia-Fuentes, L., Baron, C., & Mayorga, O. L. (1998). Influence of dynamic power compensation in an isothermal titration microcalorimeter. *Analytical Chemistry*, *70*, 4615–4623.

Hansen, C. W., Hansen, L. D., Nicholson, A. D., Chilton, M. C., Thomas, N., Clark, J., et al. (2010). Correction for instrument time constant and baseline in determination of reaction kinetics. *International Journal of Chemical Kinetics*, *43*, 53–61.

Johnson, R. E., & Biltonen, R. L. (1975). Determination of reaction rate parameters by flow microcalorimetry. *Journal of the American Chemical Society*, *97*, 2349–2355.

Karlsson, R., Katsamba, P. S., Nordin, H., Pol, E., & Myszka, D. G. (2006). Analyzing a kinetic titration series using affinity biosensors. *Analytical Biochemistry*, *349*, 136–147.

Kibbe, W. A. (2007). OligoCalc: An online oligonucleotide properties calculator. *Nucleic Acids Research*, *35*(Web Server issue), W43–W46.

Langerman, N., & Biltonen, R. L. (1979). Microcalorimeters for biological chemistry: Applications, instrumentation and experimental design. *Methods in Enzymology*, *61*, 261–286.

Lopez-Mayorga, O., Mateo, P. L., & Cortijo, M. (1987). The use of different input signals for dynamic characterisation in isothermal microcalorimetry. *Journal of Physics E: Scientific Instruments*, *20*, 265–269.

Naghibi, H., Tamura, A., & Sturtevant, J. M. (1995). Significant discrepancies between van't Hoff and calorimetric enthalpies. *Proceedings of the National Academy of Sciences of the United States of America*, *92*, 5597–5599.

Navratilova, I., Papalia, G. A., Rich, R. L., Bedinger, D., Brophy, S., Condon, B., et al. (2007). Thermodynamic benchmark study using Biacore technology. *Analytical Biochemistry*, *364*, 67–77.

Porschke, D., & Eigen, M. (1971). Co-operative non-enzymic base recognition. 3. Kinetics of the helix-coil transition of the oligoribouridylic–oligoriboadenylic acid system and of oligoriboadenylic acid alone at acidic pH. *Journal of Molecular Biology*, *62*, 361–381.

Privalov, P. L. (1979). Stability of proteins: Small globular proteins. *Advances in Protein Chemistry*, *33*, 167–241.

Ren, J., Esnouf, R., Garman, E., Somers, D., Ross, C., Kirby, I., et al. (1995). High resolution structures of HIV-1 RT from four RT-inhibitor complexes. *Nature Structural Biology*, *2*, 293–302.

Tachoire, H., Macqueron, J. L., & Torra, V. (1986). Traitement du signal en micro-calorimétrie: Applications en cinétique et thermodynamique. *Thermochimica Acta*, *105*, 333–367.

Tian, A., & Cotie, J. (1924). Utilisation en biologie de la méthode microcalorimétrique; exemple d'application. *C.R.A.S. (Paris)*, *178*, 1390–1392.

Turner, D. H. (2000). Conformational changes. In V. A. Bloomfield, D. M. Crothers, & I. Tinocco (Eds.), *Nucleic acids, structure, properties and functions* (pp. 259–334). Sausalito, CA: University Science Books.

Vander Meulen, K. A., & Butcher, S. E. (2012). Characterization of the kinetic and thermodynamic landscape of RNA folding using a novel application of isothermal titration calorimetry. *Nucleic Acids Research*, *40*, 2140–2151.

Vander Meulen, K. A., Horowitz, S., Trievel, R. C., & Butcher, S. E. (2015). Measuring the kinetics of molecular association by isothermal titration calorimetry. In A. Feig (Ed.), *Calorimetry* (pp. 179–212). Oxford: Elsevier.

Watt, G. D., & Sturtevant, J. M. (1969). The enthalpy change accompanying the oxidation of ferrocytochrome c in the pH range 6–11 at 25 degrees. *Biochemistry*, *8*, 4567–4571.

Willson, R. J., Beezer, A. E., Mitchell, J. C., & Loh, W. (1995). Determination of thermodynamic and kinetic parameters from isothermal heat conduction microcalorimetrey: Application to long-term-reaction studies. *Journal of Physical Chemistry*, *99*, 7108–7113.

Wiseman, T., Williston, S., Brandts, J. F., & Lin, L. N. (1989). Rapid measurement of binding constants and heats of binding using a new titration calorimeter. *Analytical Biochemistry*, *179*, 131–137.

CHAPTER EIGHT

Measuring the Kinetics of Molecular Association by Isothermal Titration Calorimetry

Kirk A. Vander Meulen*,[1], Scott Horowitz†, Raymond C. Trievel‡, Samuel E. Butcher*,[1]

*Department of Biochemistry, University of Wisconsin–Madison, Madison, Wisconsin, USA
†Howard Hughes Medical Institute and Department of Molecular, Cellular, and Developmental Biology, University of Michigan, Ann Arbor, Michigan, USA
‡Department of Biological Chemistry, University of Michigan, Ann Arbor, Michigan, USA
[1]Corresponding authors: e-mail address: kirk.vandermeulen@wisc.edu; sebutcher@wisc.edu

Contents

Abstract

The real-time power response inherent in an isothermal titration calorimetry (ITC) experiment provides an opportunity to directly analyze association kinetics, which, together with the conventional measurement of thermodynamic quantities, can provide an incredibly rich description of molecular binding in a single experiment. Here, we detail our application of this method, in which interactions occurring with relaxation times ranging from slightly below the instrument response time constant (12.5 s in this case) to as large as 600 s can be fully detailed in terms of both the thermodynamics and

Methods in Enzymology, Volume 567
ISSN 0076-6879
http://dx.doi.org/10.1016/bs.mie.2015.08.012

kinetics. In a binding titration scenario, in the most general case an injection can reveal an association rate constant (k_{on}). Under more restrictive conditions, the instrument time constant-corrected power decay following each injection is simply an exponential decay described by a composite rate constant (k_{obs}), from which both k_{on} and the dissociation rate constant (k_{off}) can be extracted. The data also support the viability of this exponential approach, for k_{on} only, for a slightly larger set of conditions. Using a bimolecular RNA folding model and a protein–ligand interaction, we demonstrate and have internally validated this approach to experiment design, data processing, and error analysis. An updated guide to thermodynamic and kinetic regimes accessible by ITC is provided.

ABBREVIATIONS

ITC isothermal titration calorimetry
μJ microjoules

1. INTRODUCTION

In addition to its more traditional application in studying binding thermodynamics, the modern generation power-compensated titration microcalorimeter (Freire, Mayorga, & Straume, 1990; Wiseman, Williston, Brandts, & Lin, 1989) has been applied to the study of enzyme kinetics. The ability to measure the intrinsic heat effect for an enzyme process allows direct real-time monitoring of reaction progress in an unmodified system, circumventing the potential time, cost, or experimental difficulties associated with fluorophore attachment or other chemical modifications. Many of these studies have examined enzyme systems using Michaelis–Menten steady-state kinetics (see, for example, Bianconi, 2003; Todd & Gomez, 2001; reviewed in Bianconi, 2007; Lonhienne & Winzor, 2004; Transtrum, Hansen, & Quinn, 2015), where the substrate is injected at a sufficiently high concentration so that the initial reaction velocity (V_{init}) is relatively constant.

Here, we examine the alternative approach in which the reaction rate is determined directly from the rate of compensatory power decay following ligand injection, an approach which naturally extends to ligand binding scenarios. Freire and coworkers previously demonstrated the validity of this approach, directly monitoring the decay following substrate injection in enzyme kinetics applications (Mayorga & Freire, 1987; Morin & Freire, 1991). This method was applied sparingly and only to enzymatic systems (Franghanel, Wawra, Luke, Wildemann, & Fischer, 2006; Morin &

Freire, 1991; Williams & Toone, 1993), but more recently we adopted it to characterize reversible RNA-RNA binding (Vander Meulen & Butcher, 2012). This approach has also been applied to protein–ligand (Burnouf et al., 2012; Egawa, Tsuneshige, Suematsu, & Yonetani, 2007) and RNA–ligand (Burnouf et al., 2012) association. Burnouf et al. performed significant theoretical method development and were able to validate their findings using SPR (Burnouf et al., 2012).

This work provides a practical guide to the rate constant regimes under which time-resolved isothermal titration calorimetry (ITC) can be adopted and what information can be obtained, demonstrating these approaches with experimental data and highlighting the numerous processing steps we have developed along the way. We first examine the steps required to obtain kinetic information from the power decay following ligand injection. We then focus on the application of this methodology within a binding titration scenario, for which analysis of each injection yields a series of kinetics experiments. Minimally this reveals a robust association rate constant (k_{on}); additionally, if the dissociation constant (K_d) can be measured, the dissociation rate constant (k_{off}) can be subsequently calculated using $k_{off}=k_{on}K_d$. When the rate constants fall within a subset of values defined by reversibility throughout the full titration, both k_{on} and k_{off} can be determined simply and directly. We conclude by demonstrating the seconds–timescale association kinetics profiles accessible to titration calorimetry, which can nicely augment thermodynamics data gathered in the same experiment.

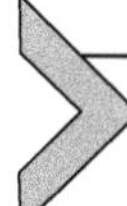

2. BACKGROUND

2.1 Extraction of Electrical Power due to Molecular Processes

As highlighted previously by Freire (Freire, Vanosdol, Mayorga, & Sanchezruiz, 1990; Mayorga & Freire, 1987), the output from a power-compensation microcalorimeter can be viewed as a convolution of the reaction "impulse" heat evolution function with the calorimeter "response" function, the latter being effectively modeled as first order with rate constant k_{ITC}. The "impulse" power function of the process of interest (E^*) can be extracted from the raw sample cell excess energy output trace (E_{XS}), perhaps most simply by way of Laplace Transform:

$$E^* = \frac{1}{k_{ITC}} \frac{\partial E_{XS}}{\partial t} + E_{XS} \tag{1}$$

The value of k_{ITC} depends on instrument design and feedback settings, among other factors (see Transtrum et al., 2015). Using a MicroCal VP-ITC in the default "high" (i.e., fast) feedback mode, we found $k_{ITC}=0.08$ ($\pm$0.02) s^{-1}, corresponding to a time constant of 12.5 s. While the listed specification for the VP-ITC is 20 s, it is noteworthy that Burnouf et al. identified a time constant of 3.5 s for the MicroCal iTC200, also significantly shorter than the corresponding specification of 10 s (Burnouf et al., 2012). With the caveats addressed by Transtrum et al. as backdrop (Transtrum et al., 2015), we simply note that the 12.5 s time constant is within range of two other experimental examinations (Demarse, Killian, Hansen, & Quinn, 2013; Garcia-Fuentes, Baron, & Mayorga, 1998); however, we are not aware of another experimental determination using a VP-ITC, which would be the appropriate comparison.

2.2 ITC Association Kinetics Analysis

E^* is approximately equal to the time-dependent rate of new complex formation, $\partial[\mathrm{C}]/\partial t$, scaled by the product of the binding enthalpy (ΔH) and the sample cell volume (V_0; for the calorimeter in our RNA experiments, $V_0=1.42$ mL)

$$E^* \approx \Delta H V_0 \frac{\partial[\mathrm{C}]}{\partial t}. \tag{2}$$

Equation (2) is approximate because the usually minor heat contributions to the cell from the mechanics of injection and ligand dilution are ignored. If the injection period is excluded, kinetics analysis only requires knowledge of the relationship between the rate of complex formation and the underlying kinetic parameters. The remainder of this section derives that relationship.

2.2.1 General Approach

Below we use a general nomenclature in which the calorimeter contains a macromolecule, M, at total concentration, $[\mathrm{M}]_{tot}$. At the start of the titration, M is at an initial total concentration, $[\mathrm{M}]_{init}$. Ligand, L, is injected from the syringe and binding generates complex, C. The overall binding process is simply described by a forward rate constant, k_{on}, and reverse rate constant, k_{off}

$$\mathrm{M}+\mathrm{L} \underset{k_{off}}{\overset{k_{on}}{\rightleftarrows}} \mathrm{C} \tag{3}$$

The equilibrium involves three species, so knowledge of $[\mathrm{M}]_{tot}$ and $[\mathrm{L}]_{tot}$ for a given injection, as dictated by the experimental setup, reduces the description of all solution concentrations to a single degree of freedom. Here, we incorporate that knowledge by using a single displacement from equilibrium variable, λ, which we define as the amount by which the equilibrium concentration of complex, $[\mathrm{C}]_{\mathrm{eq}}$, exceeds the current concentration, [C]. The reaction rate can then be written as

$$\frac{\partial[\mathrm{C}]}{\partial t} = -\frac{\partial \lambda}{\partial t} = k_{\mathrm{on}}\left([\mathrm{M}]_{\mathrm{eq}} + \lambda\right)\left([\mathrm{L}]_{\mathrm{eq}} + \lambda\right) - k_{\mathrm{off}}\left([\mathrm{C}]_{\mathrm{eq}} - \lambda\right). \quad (4)$$

Here, it is more convenient to reformulate the expression in terms of k_{on} and the dissociation constant, K_{d}. Assuming only that $K_{\mathrm{d}} = k_{\mathrm{off}}/k_{\mathrm{on}}$, Eq. (4) can be rearranged to Eq. (5)

$$-\frac{\partial \lambda}{\partial t} = k_{\mathrm{on}}\left(\lambda\left([\mathrm{M}]_{\mathrm{eq}} + [\mathrm{L}]_{\mathrm{eq}} + K_{\mathrm{d}} + \lambda\right)\right). \quad (5)$$

Integration results in the most general expression for λ

$$\ln\left(\frac{\lambda_{t=0}}{\lambda} \cdot \frac{[\mathrm{M}]_{\mathrm{eq}} + [\mathrm{L}]_{\mathrm{eq}} + K_{\mathrm{d}} + \lambda}{[\mathrm{M}]_{\mathrm{eq}} + [\mathrm{L}]_{\mathrm{eq}} + K_{\mathrm{d}} + \lambda_{t=0}}\right) = -\left([\mathrm{M}]_{\mathrm{eq}} + [\mathrm{L}]_{\mathrm{eq}} + K_{\mathrm{d}}\right) k_{\mathrm{on}} t. \quad (6)$$

In Eq. (6), $\lambda_{t=0}$ is the displacement from equilibrium following addition of ligand. More precisely in the case of an ITC experiment, it is the displacement at a hypothetical time point after ligand injection and mixing but preceding formation of any new complexes. $\lambda_{t=0}$ reflects concentration changes caused by the addition of L and the resulting displaced volume (Malvern Instruments Ltd, 2015). Equation (6) shows that in this most general case, determination of λ requires knowledge of k_{on}, in addition to K_{d} and the equilibrium concentrations. In an ITC application, the latter two term categories are known throughout a titration according to analysis of the binding curve and the experimental setup, respectively. Given these quantities and a predicted k_{on}, λ can then be determined using a root finding routine on Eq. (6). Thus, the predicted cumulative heat evolution function is given by Eq. (2), where λ and subsequently $\partial\lambda/\partial t$ are calculated from Eqs. (6) and (5), respectively.

2.2.2 Approximation Regime I: The Near-Equilibrium Approximation

The general, nonanalytical expression for λ in Eq. (6) can be simplified in two limiting cases. First, if the difference between the current and equilibrium concentration (the displacement from equilibrium) is small relative to the sum of the equilibrium concentrations of M, L, and the K_{d} ($\lambda \ll [\mathrm{M}]_{\mathrm{eq}} + [\mathrm{L}]_{\mathrm{eq}} + K_{\mathrm{d}}$), then the left-hand side of Eq. (6) is simplified and the equation can be rearranged into a first-order format. Below, we refer to this scenario as "near-equilibrium" (Bernasconi, 1976; Egawa et al., 2007):

$$\lambda = \lambda_{t=0} \cdot \mathrm{e}^{-k_{\mathrm{obs}} \times t} \tag{7}$$

(near-equilibrium)

where

$$k_{\mathrm{obs}} = k_{\mathrm{on}} \cdot \left([\mathrm{M}]_{\mathrm{eq}} + [\mathrm{L}]_{\mathrm{eq}}\right) + k_{\mathrm{off}} \tag{8}$$

(near-equilibrium)

The time derivative of Eq. (7) is required for direct analysis of E^* and can be simply calculated (see Section 3.3).

The near-equilibrium approximation holds in the following titration scenarios: (1) typically, in the first few injections, when the combined free concentration ($[\mathrm{M}]_{\mathrm{eq}} + [\mathrm{L}]_{\mathrm{eq}}$) exceeds λ largely due to the excess of M (i.e., $[\mathrm{M}]_{\mathrm{tot}} > [\mathrm{L}]_{\mathrm{tot}}$); (2) throughout a titration for sufficiently weak binding scenarios, so that the K_{d} is appreciable and the combined free reactant concentration is always in excess of λ. In Section 3.3, we show that this latter scenario holds robustly for titrations designed with a "Wiseman" c parameter ($c = [\mathrm{M}]_{\mathrm{init}}/K_{\mathrm{d}}$ (Wiseman et al., 1989)) ≤ 40.

2.2.3 Approximation Regime II: The Tight-Binding Approximation

When binding is too strong to measure accurately ($c > 200$, see Section 3.5), Eq. (6) can instead be simplified using a tight-binding approximation. K_{d} in this case is negligible; removing it from Eq. (6) and then reformulating reaction progress as the amount of unbound ligand remaining (x) results in the following kinetic equation for a forward titration:

$$\ln\left(\frac{[\mathrm{M}]_{t=0}}{[\mathrm{L}]_{t=0}} \cdot \frac{[\mathrm{L}]_{t=0} - x}{[\mathrm{M}]_{t=0} - x}\right) = -\left([\mathrm{M}]_{t=0} - [\mathrm{L}]_{t=0}\right) k_{\mathrm{on}} t \tag{9}$$

(tight binding)

Though we do not examine this scenario explicitly, it does appear to be highly robust for all injections in an experiment where $c > 200$, and it is also valid for the initial injections of weaker-binding titrations (pictorially, the "plateau region" of a titration). For example, we found that error due to this approximation is less than 10% for experiments with $c \geq 60$ if only injections are analyzed where $[\mathrm{L}]_{\mathrm{tot}}/[\mathrm{M}]_{\mathrm{tot}} < 0.5$ (unpublished results).

3. PROCEDURE AND RESULTS

Previously, we used time-resolved ITC to characterize the thermodynamics of a model RNA system along the folding reaction coordinate (Vander Meulen & Butcher, 2012). This system utilizes a ubiquitous tetraloop–receptor tertiary motif (Cate et al., 1996; Costa & Michel, 1995; Toor, Keating, Taylor, & Pyle, 2008) to drive association of two RNA helices, one containing two receptor motifs (RR) and the other containing two cognate GAAA tetraloop sequences (TT) (Fig. 1). Because RNA folding kinetics are highly dependent on both salt and temperature, the TT–RR data from our previous study provide a view of kinetics analysis across the range of accessible rate constant regimes.

3.1 Raw Thermogram Preparation

In principle, one injection of ligand into a solution containing its binding partner is sufficient to measure the forward rate constant, k_{on}. In our work, as well as that of Burnouf et al. (2012), kinetic data were obtained in a full binding titration as a "bonus," and we target that scenario in this guide. We also show in Sections 3.3, 3.4, and 4.3 that there is utility in analyzing multiple injections in a titration, because it reduces uncertainty and additionally yields a concentration-dependent kinetics dataset. However, it is our intention that a practitioner working with even a single injection could easily adapt the appropriate sections.

Thus, we begin here from a titration thermogram, as in Fig. 1. A representative raw titration power trace for the exothermic TT–RR interaction is plotted in Fig. 1A. Because the raw data can exhibit appreciable drift in the

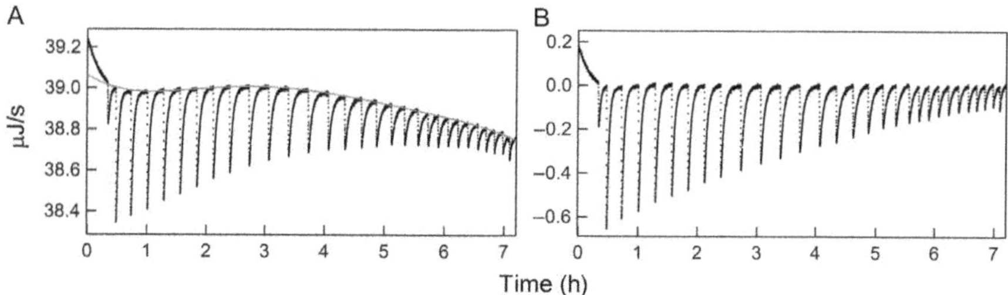

Figure 1 Demonstration of thermogram baseline subtraction (ITC power trace obtained in 0.3 m*M* $MgCl_2$ at 30 °C, using starting cell [RR] = 10 μ*M*, syringe [TT] = 171 μ*M*, and 6.5 μL injections). (A) Raw thermogram (black dots) and fitted baseline (gray curve). (B) Following baseline subtraction, the plotted data express only the additional power input in response to ligand injection and binding (referred to here as E_{XS}).

baseline power, we first perform a polynomial baseline regression and subsequent subtraction to objectively account for most of the curvature in the instrument baseline, the result of which is shown in Fig. 1B. In the regression procedure, we generally mask all data points except those $\leq$10 s prior to the next injection.

3.2 Analysis of Individual Injections

Each sharp negative deflection in the thermogram in Fig. 1 reflects the real-time compensatory reduction in electrical power following injection of an exothermic binding partner (see Section 2). Using the Laplace Transform methodology with an instrumental time constant of 12.5 s ($k_{ITC}=0.08\ s^{-1}$), the instantaneous rate of excess heat input following an injection can be easily recovered by using Eq. (1) (Mayorga & Freire, 1987; Morin & Freire, 1991).

Figure 2 plots the seventh peak from the dataset in Fig. 1 as an example with which to highlight the preparation and analysis of an individual injection using a completely general methodology. The first step illustrated is transformation to the real-time power signal; Fig. 2A shows the injection signal before and after deconvolution. This example illustrates a relatively

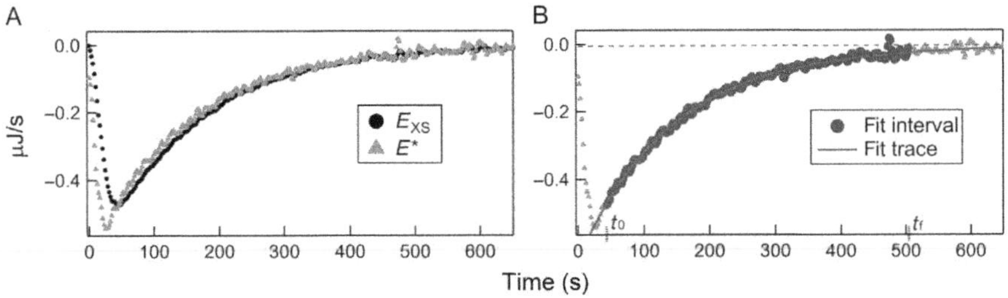

Figure 2 Demonstration of injection peak fit preparation procedure, using the seventh injection from the titration in Figure 1. (A) Deconvolution to recover the instantaneous power signal. The black (raw) data points are referred to as E_{XS} (the instrument power response) and the green (light gray in the print version) triangles (postdeconvolution) are equivalent to E^* (the instantaneous rate of heat evolution due to injection and macromolecular association). (B) Baseline, fitting regime, and curve fitting. A final baseline was not fully established in this injection, so the blue (gray in the print version) dotted line represents the linear and constant terms determined from a preliminary regression procedure. Data within the fitting region are highlighted by large blue (dark gray in the print version) circles. The starting point for fitting, t_0, corresponds to the minimum in the predeconvolution data (44 s) and, in this case, the final point, t_f, is determined from the decay in the power signal (504 s). The red (gray in the print version) curve represents the results of regression using the general set of kinetics Eqs. (2), (5), and (6).

slow association process with a half-time of approximately 105 s. Here, analysis of the raw signal results in only a slight (approximately 3%) underestimate of the rate constant. The effect of the deconvolution and thus the uncertainty related to the instrument time constant increase significantly for faster events (see Sections 3.4 and 4.2.3).

Robust kinetics analysis of each injection also requires careful consideration of both the baseline power and the fit region. The first step in baseline treatment is actually handled in the thermogram baseline subtraction (Section 3.1), which largely corrects for sloping and curvature in the baseline power signal and resets the offset to approximately 0 μJ s^{-1}. However, this procedure cannot be viewed, in general, as having completely removed baseline power from the analysis, since raw thermogram baseline drift is idiosyncratic and miscalculations due to slight imperfections in the subtraction procedure can be acute for fast association processes (see Section 4.3). Thus, we use both a linear (slope) and constant (intercept) term in the fitting procedure. Where possible, the baseline terms are determined separately and fixed during regression. Baseline selection and associated error analysis is an important but detail heavy topic, and is covered in Section 4.2.2.

The next consideration is fitting interval $[t_0, t_f]$ selection. The first time point, t_0, is the point immediately succeeding the time corresponding to the raw power trace minimum (or maximum for an endothermic process). In our implementation, this effectively excludes the injection and mixing period and the associated heat of dilution. Selection of a final time point, t_f, is also required to prevent baseline overfitting (overrepresentation of a final baseline region) in some cases. t_f is set to the minimum among the following three values: (1) the last point before the next injection, (2) a time-decay boundary point equal to 10 times the approximate half-time, and (3) a power-decay boundary point corresponding to the time at which the magnitude of the power value offset from the baseline first decreases below a threshold value, which we set to 0.001 μJ s^{-1}. Figure 2B shows the baseline and fitting range selected for the example injection peak.

The deconvolved real-time excess power input (E^*) is defined in Eq. (2). The fitting procedure includes ΔH as an unconstrained amplitude parameter (in principle, this could be related to the ΔH for the association process, but the amplitude is complicated by the time offset due to the injection/mixing period). Any obtainable kinetics quantities are buried in the decay term, which is simply the rate of complex formation, $\partial[C]/\partial t$. For a straightforward association process, this can be described in terms of the rate of change in the displacement from equilibrium, $\partial\lambda/\partial t$, as laid out in Eq. (5). The only

Table 1 Concentrations Relevant to Fig. 2

	Actual Concentrations (μ*M*)			Equilibrium Concentrations (μ*M*)			
Time Period	**[TT]**	**[RR]**	**[Complex]**	**$[TT]_{eq}$**	**$[RR]_{eq}$**	**$[Complex]_{eq}$**	**λ**
Preinjection[a]	0.37	6.90	2.80	0.37	6.90	2.80	0.00
Postinjection[a,b] ($t=0$)	0.95	6.87	2.78	0.46	6.39	3.27	0.49

[a]Concentrations include adjustments for activities as included in our previous work (Vander Meulen & Butcher, 2012; Vander Meulen, Davis, Foster, Record, & Butcher, 2008). The K_d was determined separately from the binding titration to be 907 n*M*.

[b]Theoretical state including displaced volume effect, but prior to any changes in binding in response to altered concentration.

unknown parameter in Eq. (5) is k_{on}, the rate of complex association. The concentration parameters, $[M]_{eq}$ and $[L]_{eq}$, are obtained from the titration setup and the (potentially ITC titration-derived) equilibrium dissociation constant, K_d. Table 1 displays the underlying set of concentrations ($[RR]_{eq}$ and $[TT]_{eq}$ for this specific case, along with $\lambda_{t=0}$) required for analysis of the example data in Fig. 2. The fitted curve that follows from the above-described data preparation and general fitting routine is shown in Fig. 2B.

In a titration scenario, we analyze all of the injection peaks before the stoichiometric equivalence point, where $[L]_{tot} < [M]_{tot}$, as signal to noise decreases, sometimes dramatically, beyond this point. The resulting array of k_{on} data can be averaged to increase the confidence level, although some sources of potential error are correlated and thus cannot be mitigated through averaging (see Sections 3.4 and 4.2.3).

3.3 Near-Equilibrium Approximation

While general, the approach outlined above has drawbacks in implementation (λ cannot be expressed in closed form) and in its ability to only directly calculate k_{on}. Subsequent determination of the dissociation rate constant (k_{off}) requires a single-step kinetic mechanism assumption, or at least the assumption that there is not a stable intermediate, such that $k_{off} = K_d/k_{on}$.

Where a near-equilibrium approximation can be applied to an entire titration, the kinetics analysis is both simplified and more powerful (Egawa et al., 2007). This approximation holds for titrations in which λ is always small relative to the unbound reactant concentrations ($c \leq 40$, see Section 2.2.2 and below). The near-equilibrium approximation reduces $\lambda(t)$ to the expression in Eq. (7), and subsequent substitution for $\partial[C]/\partial t$ in Eq. (2) according to Eqs. (5) and (7) (and the definition

$\partial\lambda/\partial t = -\partial[C]/\partial t$) results in a description of the rate of reaction heat evolution as pseudo first order with composite rate constant, k_{obs}

$$E^* \approx \Delta H V_0 \lambda_{t=0} k_{obs} e^{-k_{obs} t} \quad (10)$$

(near-equilibrium)

In Eq. (10), k_{obs} is a composite term that is linear in the sum of the unbound substituent concentrations, with slope k_{on} and intercept k_{off} (see Eq. 8).

Figure 3 compares rate constant determinations using both the general and near-equilibrium approach for the example titration and injection used

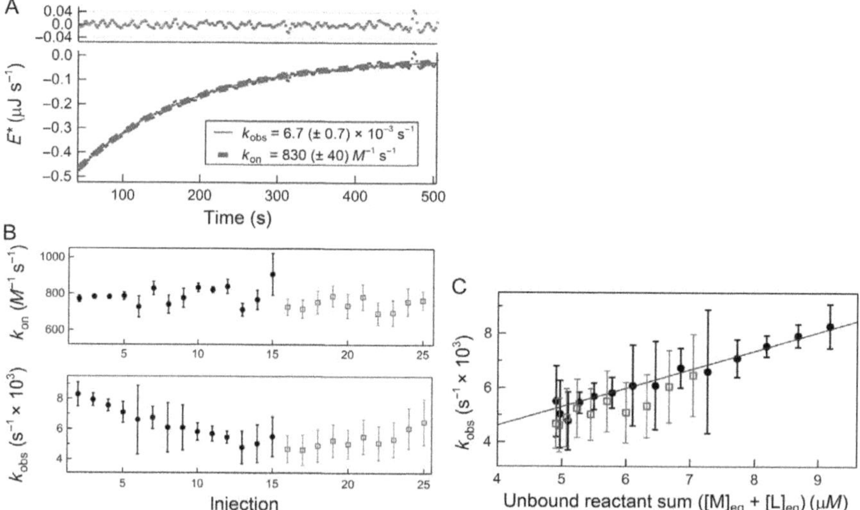

Figure 3 Demonstration and comparison of kinetics analysis using a general equation and the near-equilibrium approximation approach using the same data in Figures 1 and 2. (A) Main plot: deconvolved data within fitting region (blue (dark gray in the print version) circles) and fits using the general equation set (red (gray in the print version) solid curve) and near-equilibrium approach exponential target function (purple (dark gray in the print version) dashed curve). Top plot: corresponding fit residuals. (B) Top panel: k_{on} determinations for each injection using the general approach; bottom panel: composite rate constant (k_{obs}) determinations using the near-equilibrium approach. In both panels, closed black circles show data from the first half of the titration ($[L]_{tot} < [M]_{tot}$), which correspond to the analyzed dataset. For illustration purposes only, open gray squares show the remaining data up to injection 25 ($[L]_{tot}/[M]_{tot} \sim 2$). Error bars represent only the uncertainty that is uncorrelated with other injections (fit parameter variance and baseline uncertainty; termed "injection-level uncertainty" in Section 4.2.3). (C) k_{obs} from near-equilibrium analysis plotted against the free RNA concentration. A linear fit (closed circle data only) yields $k_{on} = 730\ (\pm 110)\, M^{-1}\, s^{-1}$ from the slope, and from the intercept $k_{off} = 1.6 \pm 0.8 \times 10^{-3}\, s^{-1}$.

in Figs. 1 and 2. Fits using these approaches are equally satisfying (Fig. 3A); neither residuals (Fig. 3A, top) nor χ^2 are useful in validating or invalidating the near-equilibrium approach. Figure 3B plots the collection of rate constant outputs from the two approaches. The microscopic rate constant k_{on}, determined using the general equation, is constant over the course of the titration, though the uncertainty increases significantly in later injections where signal to noise is reduced (Fig. 3B, top). Contrastingly, the composite rate constant k_{obs} is a strong function of the injection number (Fig. 3B, bottom). It decreases to a minimum then increases, mirroring the unbound macromolecular concentration, which decreases as bound complex increases, then subsequently increases for injections in which ligand is in excess. Figure 3C plots this linear relationship between k_{obs} and unbound reactant concentration.

Demonstration of linearity in k_{obs} versus ($[M]_{eq} + [L]_{eq}$), as seen in Fig. 3C, is necessary for validation of the near-equilibrium approach. To confirm this linearity, we examined a more extended range of k_{obs} by performing a suite of titrations using both higher and lower RNA concentrations (Fig. 4). Figure 4A displays k_{obs} data obtained as described in Fig. 3,

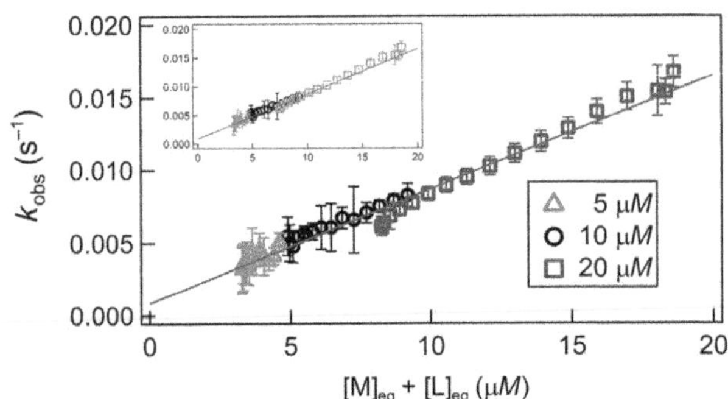

Figure 4 Demonstration of consistency in the near-equilibrium approximation, using TT–RR experiments across a range of $[M]_{init}$ at the experimental condition 0.3 m*M* $MgCl_2$, 30 °C. Main figure: plot of best-fit k_{obs} against the total unbound RNA concentration ($[M]_{eq} + [L]_{eq}$) for a set of titrations in which the starting cell concentration of a macromolecule, M (RR in this case), was varied and titrated with its ligand, L (TT). For all experiments, the syringe [TT] is 171 μ*M*; green triangles, black circles, and blue squares reflect separate experiments in which $[M]_{init}$ is 5, 10, or 20 μ*M*, respectively. The red trendline is the result of a global, linear fit; results from the individual datasets are provided in Table 2. The inset figure plots the same data except that the plot for the 20 μ*M* dataset (cyan) employed a K_d that is decreased by 10% over its best-fit value. (See the color plate.)

using the same experimental preparation with $[M]_{init}$ equal to 5 or 20 μ*M*, in addition to the 10 μ*M* dataset. Plots of k_{obs} are parallel within error and are nearly completely collinear with the exception being the offset in the 20 μ*M* dataset; a 10% increase in the K_d employed in the 20 μ*M* dataset largely alleviates this discrepancy (Fig. 4, inset). Uncertainty in the K_d should be included in a complete error analysis and our implementation of this factor is discussed more fully in Section 4.2.3. Table 2 lists the linear regression parameters (k_{on} and k_{off}) obtained from each experiment, along with the dissociation constant and kinetic data determined using the general method.

To further test methodology, we also performed a similar set of experiments in a solution containing 200 m*M* KCl at 25 °C (Fig. 5), where K_d is 360 n*M*. Relatively fast association in this condition means accuracy in the deconvolution procedure is critical, because its effect ranges from relatively minor at the lowest end of the RNA concentration ($[M]_{init}=4$ μ*M*), to much more significant at the upper end ($[M]_{init}=16$ μ*M*). Figure 5 provides representative injections corresponding to these two conditions, for which the deconvolution affects a 20% (Fig. 5B) or up to a 70% (Fig. 5C) increase in the measured rate constant relative to the raw power trace. The relative collinearity in Fig. 5A demonstrates a viable deconvolution procedure.

Table 3 lists linear regression parameters corresponding to the k_{obs} versus free reactants plots in Fig. 5. The extracted slopes are similar and in agreement with the k_{on} values obtained using the general approach, while the intercepts reflecting k_{off} are inconsistent and exhibit large uncertainties. Encouragingly however, the lowest concentration dataset, which requires the shortest extrapolation and therefore exhibits the smallest confidence interval, is within the injection-level uncertainty band (see Section 4.2.3) of the determinations using the general approach.

We compared average kinetic parameters determined via the general and near-equilibrium approaches in conditions where both kinetics and affinity measurements are robust (most experiments with $c<200$; Table 4). Examining the data according to the average titration c parameter value at that condition is instructive; for $c<43$, k_{on} and k_{off} measurements generally overlap each other within their confidence intervals (although the uncertainty levels are much higher using the near-equilibrium approach, particularly for k_{off}). This internal consistency corroborates the viability of either approach.

For $c>40$, the agreement between approaches is maintained for k_{on}, but not for k_{off}. Thus, apparently in practice, determination of k_{on} (but not k_{off}) may still be conducted according to the near-equilibrium approximation for

Table 2 Concentration-Dependent Near-Equilibrium Test for TT–RR Experiments in 0.3 m*M* $MgCl_2$, 30 °C

		Near-Equilibrium Approximation				General Method			
Dataset(s)	K_d (n*M*)	k_{on} (M^{-1} s^{-1})		k_{off} (s^{-1})$\times 10^{-5}$		k_{on} (M^{-1} s^{-1})		$k_{on}K_d$ (s^{-1})$\times 10^{-5}$	
$[M]_{init}=5$ μ*M* RR	900 (±400)	930	{±580}	30	{±220}	790	{±40}	71	{±32}
			[±630]		[±240]		[±190]		[±36]
			(±640)		(±250)		(±190)		(±36)
$[M]_{init}=10$ μ*M* RR	910 (±80)	730	{±110}	160	{±80}	790	{±10}	72	{±6}
			[±180]		[±110]		[±150]		[±15]
			(±180)		(±110)		(±150)		(±15)
$[M]_{init}=20$ μ*M* RR	1300 (±200)	940	{±50}	−110	{±60}	750	{±10}	99	{±10}
			[±170]		[±160]		[±120]		[±19]
			(±170)		(±160)		(±120)		(±19)
Avg (plotted)	1000 (±200)	840	(±140)	68	(±89)	770	(±80)	84	(±52)
Avg (all)	2000 (±100)	710	(±90)	15	(±65)	470	(±30)	96	(±7)

{} includes all injection-level uncertainty terms.
[] includes all injection-level and experiment-level uncertainty.
() includes all uncertainty contributions.

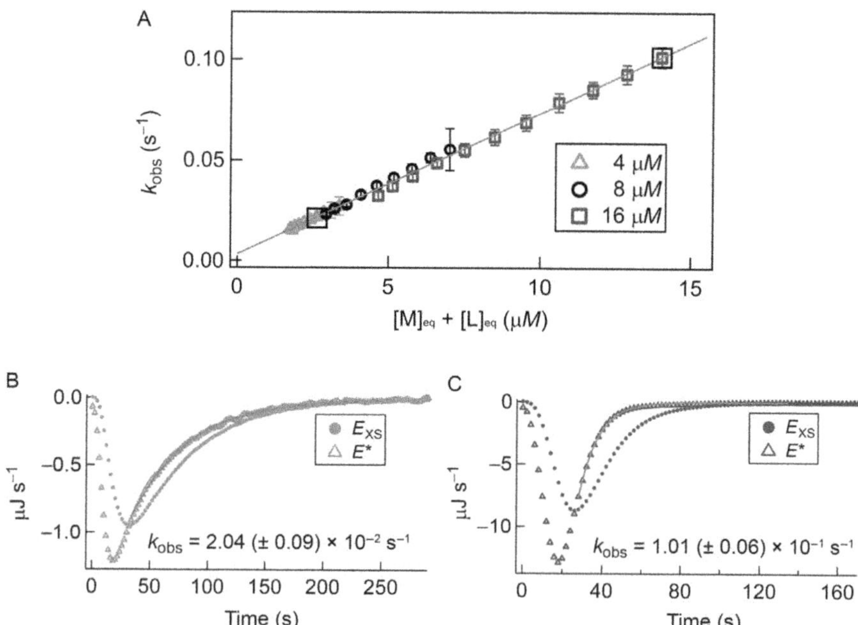

Figure 5 Demonstration of consistency in the near-equilibrium approximation at the upper end of the calorimetry-accessible kinetics window, using TT–RR experiments across a range of $[M]_{init}$ at the experimental condition 200 mM KCl, 25 °C. (A) Plot of best-fit k_{obs} against unbound RNA for a set of TT–RR titrations in which $[M]_{init}$ was varied. For all experiments, the syringe ligand (TT) concentration is 240 μM; green triangles, black circles, and blue squares reflect separate experiments in which the starting cell concentration of RR ($[M]_{init.}$) is 4, 8, or 16 μM, respectively. The red trendline is the result of a global, linear fit; results from the individual datasets are provided in Table 3. Bottom panels: demonstration of increasing importance of deconvolution for fast association kinetics. Plots depict power traces before (E_{XS}, closed circles) and after (E^*, open triangles) deconvolution using the sixth injection from the 4 μM dataset (B) and the second injection from the 16 μM dataset (C) as examples. Red curves plot best-fit exponential functions and fit intervals, and associated composite rate constants (k_{obs}) are listed. The listed uncertainties reflect only injection-level sources (see Section 4.2.3). Boxed data points in (A) correspond to the injections displayed in (B) and (C). (See the color plate.)

at least some experiments where $40 < c \leq 200$. This may simply relate to heavier weighting of the initial injections in our implementation, such that case (1), as referred to in Section 2.2.2, is effectively observed. Thus, while this finding is interesting, it may not hold for processes with rates closer to the upper limit of accessibility (see Sections 3.5 and 4.3), and it potentially might not apply at all in another adaptation. More work is required to

Table 3 Concentration-Dependent Near-Equilibrium Test for TT–RR Experiments in 200 mM KCl, 25 °C.

		Near-Equilibrium Approximation				General Method			
Dataset(s)	**K_d (nM)**	**k_{on} (M^{-1} s^{-1})**		**k_{off} (s^{-1})×10^{-5}**		**k_{on} (M^{-1} s^{-1})**		**$k_{on}K_d$ (s^{-1})×10^{-5}**	
$[M]_{init}$ = 4 μM RR	250 (±60)	7300	{±700}	280	{±140}	7300	{±100}	180	{±50}
			[±1700]		[±360]		[±1200]		[±60]
			(±1800)		(±360)		(±1200)		(±60)
$[M]_{init}$ = 8 μM RR	300 (±20)	8100	{±500}	−10	{±210}	7000	{±100}	210	{±10}
			[±1400]		[±460]		[±1000]		[±30]
			(±1900)		(±560)		(±1100)		(±40)
$[M]_{init}$ = 16 μM RR	400 (±20)	7400	{±300}	−110	{±260}	6800	{±100}	270	{±10}
			[±1100]		[±580]		[±700]		[±30]
			(±2300)		(±960)		(±1500)		(±60)
Avg (plotted)	360 (±10)	7600	(±1500)	100	(±450)	6900	(±900)	250	(±30)

{} includes all injection-level uncertainty terms.
[] includes all injection-level and experiment-level uncertainty.
() includes all uncertainty contributions.

Table 4 Comparison of Kinetic Parameters for All TT–RR Datasets with $c < 200$ and $k_{on}[M]_{init} < 0.14$

			Near-Equilibrium Approximation		General Method		
c^a	$k_{on}[M]_{init}$ $(s^{-1})^b$	K_d (nM)	k_{on} (M^{-1} s^{-1})	k_{off} (s^{-1}) × 10^{-5}	k_{on} (M^{-1} s^{-1})	$k_{off} = k_{on}K_d$ (s^{-1}) × 10^{-5}	Conditions (No. Replicates)
3	0.011	2000 (±100)	1500 (±2200)	190 (±740)	1400 (±100)	280 (±30)	0.3 mM $MgCl_2$, 50 mM KCl, 30 °C (4)
4	0.055	1700 (±100)	6700 (±2200)	1400 (±900)	7100 (±1300)	1200 (±200)	200 mM NaCl, 30 °C (4)
4	0.019	2100 (±100)	2400 (±600)	390 (±410)	2200 (±200)	480 (±40)	150 mM KCl, 25 °C (4)
5	0.007	2100 (±100)	820 (±210)	230 (±130)	690 (±50)	140 (±10)	0.3 mM $MgCl_2$, 20 mM KCl, 30 °C (7)
6	0.006	1800 (±100)	440 (±120)	83 (±56)	560 (±40)	100 (±10)	100 mM KCl, 15 °C (6)
7	0.050	1100 (±100)	6900 (±1400)	620 (±420)	6500 (±800)	720 (±90)	200 mM KCl, 30 °C (11)
8	0.022	1300 (±100)	3300 (±600)	−260 (±340)	1800 (±200)	230 (±20)	0.3 mM $MgCl_2$, 80 mM KCl, 30 °C (3)
9	0.008	2000 (±100)	710 (±90)	15 (±65)	470 (±30)	100 (±10)	0.3 mM $MgCl_2$, 30 °C (8)
11	0.011	1000 (±100)	860 (±210)	340 (±110)	1100 (±200)	110 (±20)	0.3 mM $MnCl_2$, 30 °C (2)
16	0.020	550 (±20)	2300 (±300)	250 (±110)	2400 (±200)	130 (±10)	150 mM KCl, 20 °C (3)

Continued

Table 4 Comparison of Kinetic Parameters for All TT–RR Datasets with $c < 200$ and $k_{on}[M]_{init} < 0.14$—cont'd

			Near-Equilibrium Approximation		General Method		
c	$k_{on}[M]_{init}$ (s^{-1})	K_d (nM)	k_{on} (M^{-1} s^{-1})	k_{off} (s^{-1})× 10^{-5}	k_{on} (M^{-1} s^{-1})	$k_{off}=k_{on}K_d$ (s^{-1})× 10^{-5}	Conditions (No. Replicates)
17	0.119	440 (±10)	16,000 (±5000)	790 (±980)	16,000 (±4000)	710 (±180)	250 mM NaCl, 30 °C (4)
21	0.092	350 (±10)	13,000 (±4000)	310 (±720)	12,000 (±3000)	440 (±90)	250 mM KCl, 30 °C (6)
22	0074	490 (±10)	9100 (±2300)	340 (±530)	8600 (±1600)	420 (±80)	1.0 mM $MgCl_2$, 40 °C (5)
23	0.015	420 (±20)	1800 (±100)	85 (±51)	1700 (±100)	69 (±6)	0.5 mM $MgCl_2$, 30 °C (6)
23	0.005	630 (±70)	490 (±60)	61 (±26)	500 (±70)	32 (±6)	0.3 mM $MgCl_2$, 20 mM KCl, 20 °C (2)
27	0.064	360 (±10)	7600 (±1500)	100 (±500)	6900 (±900)	260 (±30)	200 mM KCl, 25 °C (3)
29	0.015	360 (±20)	1500 (±200)	100 (±80)	1500 (±200)	52 (±8)	0.2 mM $MgCl_2$, 80 mM KCl, 20 °C (2)
36	0.005	560 (±30)	340 (±40)	22 (±32)	240 (±30)	13 (±2)	0.3 mM $MgCl_2$, 20 °C (3)
43	0.027	180 (±50)	3400 (±400)	170 (±80)	3500 (±300)	60 (±5)	0.7 mM $MgCl_2$, 30 °C (8)
45	0.056	200 (±10)	7000 (±1100)	300 (±210)	7100 (±800)	140 (±20)	1 mM $MgCl_2$, 35 °C (7)
51	0.022	180 (±10)	2300 (±200)	230 (±60)	2500 (±200)	45 (±4)	150 mM KCl, 15 °C (4)
51	0.038	190 (±10)	4000 (±500)	190 (±120)	4000 (±400)	70 (±8)	0.7 mM $MgCl_2$, 20 mM KCl, 30 °C (4)

55	0.009	170 (±10)	850 (±100)	180 (±40)	1000 (±100)	17 (±2)	0.5 m*M* $MgCl_2$, 20 °C (6)
60	0.006	190 (±10)	540 (±60)	94 (±29)	600 (±70)	12 (±1)	100 m*M* KCl, 5 °C (5)
62	0.008	110 (±10)	1100 (±100)	89 (±32)	1100 (±200)	12 (±3)	0.5 m*M* $MgCl_2$, 20 m*M* KCl, 20 °C (2)
62	0.021	160 (±10)	2000 (±200)	160 (±60)	2100 (±300)	33 (±4)	0.3 m*M* $MgCl_2$, 80 m*M* KCl, 20 °C (2)
76	0.067	130 (±10)	6400 (±1100)	290 (±230)	6600 (±800)	87 (±11)	200 m*M* KCl, 20 °C (3)
83	0.039	120 (±10)	3800 (±500)	210 (±150)	3900 (±500)	48 (±6)	0.3 m*M* $MgCl_2$, 12 m*M* KCl, 20 °C (2)
97	0.062	87 (±2)	7100 (±1100)	240 (±200)	7200 (±700)	63 (±7)	1.0 m*M* $MgCl_2$, 20 m*M* KCl, 30 °C (4)
130	0.013	78 (±4)	1200 (±100)	130 (±30)	1300 (±200)	10 (±2)	0.2 m*M* $MgCl_2$, 80 m*M* KCl, 10 °C (2)
160	0.025	62 (±4)	2400 (±300)	160 (±60)	2500 (±300)	16 (±2)	0.7 m*M* $MgCl_2$, 20 °C (2)
210	0.048	44 (±2)	5600 (±700)	250 (±100)	6000 (±700)	26 (±3)	200 m*M* NaCl, 10 °C (2)

[a] *c*: average value of $[M]_{init}/K_d$ for experiments conducted at this condition.
[b] Average "*k*-parameter" value for experiments at this condition.

determine, for each thermodynamic window, the molar ratio or binding density at which the near-equilibrium approximation is no longer valid, as well as the amount of data required to accurately determine k_{on} and k_{off}.

We add two additional comments on Tables 2–4. First, in assembling this method summary, we identified a coding error (the calculation of $\lambda_{t=0}$ prior to kinetics analysis was off by a factor of 2) that affected a uniform 10% underestimate of k_{on} in our previous work. Fortunately, this in no way affects any conclusions from that work, since the temperature dependence of k_{on} and k_{off} is completely unaffected. Second, the uncertainties reported for K_d simply reflect weighted averaging of the fitting uncertainty. Here, we are more interested in rigorous error analysis for the kinetic quantities, so we have not utilized our typical bootstrap approach, in part to simplify the presentation of nonkinetic quantities.

3.4 Maximum Accessible Rate Constant

Kinetic analyses for processes with association rates at the upper limit of the detectable range (composite rate constant $\rightarrow k_{ITC}$) are characterized by significantly increased uncertainty levels. This increased uncertainty is primarily due largely to the increased sensitivity of the measured rate constant to both the Laplace Transform procedure (Eq. 1), as well as the baseline parameters employed during fitting.

Figure 6 illustrates the importance of an accurate Laplace Transform, this time using as an example the binding between the catalytic domain of the human SET domain lysine methyltransferase SET7/9 (residues 110–366) and its substrate *S*-adenosylmethionine (AdoMet) (Horowitz, Yesselman, Al-Hashimi, & Trievel, 2011). A representative titration is shown in Fig. 6A. The rate of complex formation is close to the upper limit of the technique (see below), as shown by the fact that only near the middle of the titration do injections evidence a slightly slowed equilibration process. The combination of slightly reduced reactant concentration ($[M]_{init} = 6.0\ \mu M$ in the example), high-data quality, and extended baseline collected following each injection and decay as in these examples is crucial in examining these upper limit scenarios.

To demonstrate the effect of the deconvolution procedure, Fig. 6B displays power signals pre- and postdeconvolution for the representative example 10th injection. Clearly in this fast-associating condition, with a half-time of ~13 s, an accurate deconvolution procedure is essential to robust rate constant measurement. Using the transformed signal, a value

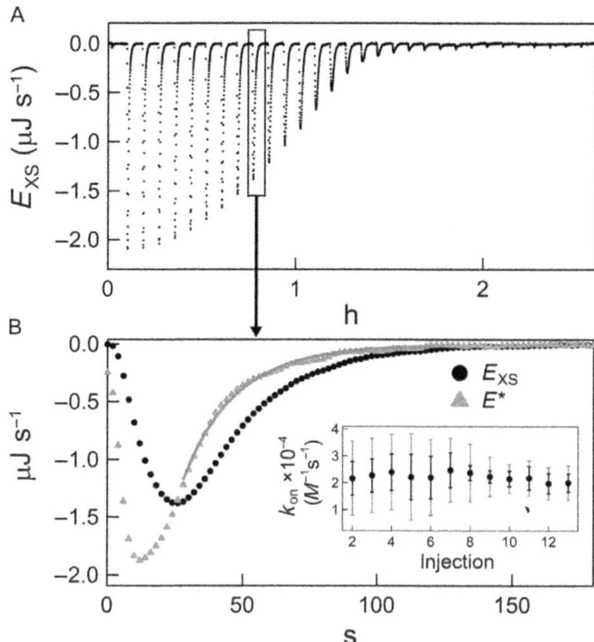

Figure 6 Characterization of the kinetics of SET7/9–AdoMet binding. (A) Thermogram following baseline subtraction for an experiment in which 6.0 μ*M* SET7/9 (residues 110–366) was titrated with 70 μ*M* AdoMet. (B) Main figure: injection 10 raw (black circles) and deconvolved (green (gray in the print version) triangles) power trace. The red (dark gray in the print version) curve denotes the graphical result of the general equation set regression procedure as well as the fitting interval. Inset: summary of all k_{on} values obtained from this titration, with error bars representing uncertainty from injection-level sources (black error bars) or from all sources (gray error bars).

$k_{on} = 22{,}000\ M^{-1}\ s^{-1}$ is obtained. Accurate transformation is even more crucial for injections earlier in the titration; for example, the decay following injection 2 exhibits a half-time of approximately 6 s (shorter than the example in Fig. 6C). In our error analysis, we utilize an empirically calculated 20% uncertainty in k_{ITC} ($0.067\ s^{-1} < k_{ITC} < 0.1\ s^{-1}$; error analysis is covered in Section 4.2.3).

There is some systematic error in the fit describing the transformed data (Fig. 6B), indicating the kinetics are approaching or slightly exceeding applicability. On the other hand, the overall picture of the dataset is mildly encouraging: over the course of the AdoMet–SET7/9 titration, the measured k_{on} values are consistent (Fig. 6B, inset). A grossly incorrect deconvolution procedure might over- or under-correct the early, faster-relaxing injection data, thus manifesting in an injection-dependent k_{on}

(not shown). For comparison, the displayed error bars reflect either the per-injection uncertainty or the total uncertainty for a single measurement, which includes uncertainty propagated through the deconvolution (see Section 4.2.3.3). These large, injection-dependent error bars underscore the massive impact of the deconvolution procedure as well as the importance of the quantity $[L]_{tot}/[M]_{tot}$ on the tractability of kinetics analysis for fast processes.

To estimate the accuracy of our measurements at the upper limit of kinetics accessibility, we utilized the generally observed logarithmic relationship between thermodynamic/kinetic quantities and salt concentration for nucleic acid processes (Record, Zhang, & Anderson, 1998). We predicted k_{on} for some of the experimental scenarios that lie on the fringe of accessibility; the results are plotted in Fig. 7 along with the actual values calculated using our general method. The dotted lines reflect extrapolations from the slower, high-confidence measurements to the conditions exhibiting fast association. Every calculation underestimates the predicted value to a remarkably similar extent, with the average underestimate being 25%. In all cases, the error is just outside of that estimated assuming a 20% uncertainty in k_{ITC}. Curvature in such plots, especially in the mixed salt experiment, is not unprecedented but the consistency of the underestimate

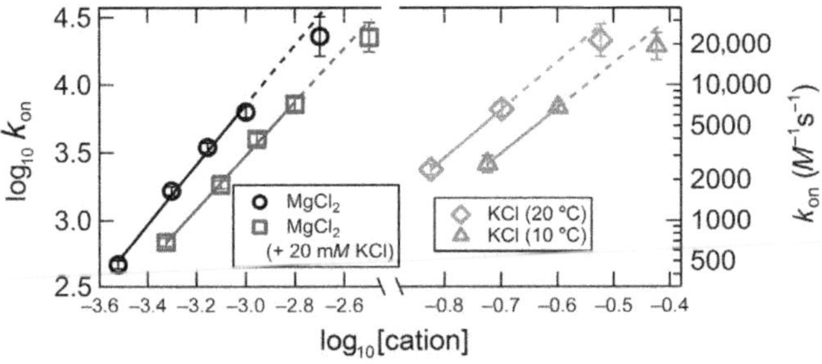

Figure 7 k_{on} measurements and predictions from extrapolation of the [salt]-dependence, for TT–RR experiments exhibiting kinetics at the upper limit of accessibility by titration calorimetry. Black circles, experiments conducted in solutions containing varying concentrations of $MgCl_2$ at 30 °C. Blue (dark gray in the print version) squares, experiments conducted in varying $[MgCl_2]$ and 20 m*M* KCl at 30 °C. Green (gray in the print version) diamonds and red (dark gray in the print version) triangles, experiments conducted in 200 m*M* KCl at 20 or 10 °C, respectively. The solid lines represent k_{on} from titration experiments within the validated region ($k_{on}[M]_{init} < k_{ITC}$), and the dotted lines extend those fits to the test data. (For visualization purposes, the blue (dark gray in the print version) and red (dark gray in the print version) datasets have been shifted to the right by 0.2 and 0.1 log units, respectively, such that they do not overlie the adjacent datasets.).

is rather convincing. Later, based on our previous observation that processes exhibiting $k_{on} \leq 8000\ M^{-1}\ s^{-1}$ (in more general framing, $k_{on}[M]_{init}$ $\leq 0.08\ s^{-1}$, comparable to k_{ITC}) showed no such underestimate for the fastest k_{on} in a trend (Vander Meulen & Butcher, 2012), we infer a logarithmic systematic uncertainty inherent in our kinetics application throughout the maximal rate regime (i.e., $k_{ITC} < k_{on}[M]_{init} <$ method maximum).

We examined correcting for this underestimate by simply decreasing k_{ITC}. A correction would place the k_{ITC} value between $0.067\ s^{-1}$ (the minimum value in our error analysis routine) and $0.05\ s^{-1}$ (which results in a >100% overestimate of k_{on} in the Fig. 7 data). However, it seems at least equally probable that we are approaching a limit where various approximations become invalid, the most important being the first-order approximation of the instrument response. Both factors may be in play, but for the purpose of this work, we use the prior-mentioned k_{ITC} of $0.08\ s^{-1}$. This value was determined simply through analysis of an injection under conditions where association is effectively instantaneous.

3.5 Guidelines for Kinetic and Thermodynamic Analysis

For an association process with appropriate kinetics, time-resolved analysis of ITC binding data can significantly enrich its biophysical characterization (Burnouf et al., 2012; Egawa et al., 2007; Vander Meulen & Butcher, 2012). Figure 8 provides an example map of ITC-measurable thermodynamic and kinetic parameters as a function of k_{on} and k_{off}. Figure 8 was generated according to the instrument specifications (e.g., VP-ITC with time constant 12.5 s) and typical experimental conditions (e.g., $[M]_{init} = 10\ \mu M$) employed in our TT–RR study (Vander Meulen & Butcher, 2012).

For a VP-ITC with $[M]_{init} = 10\ \mu M$, kinetic analysis using a titration approach requires that $1 \times 10^2\ M^{-1}\ s^{-1} \leq k_{on} \leq 1.4 \times 10^4\ M^{-1}\ s^{-1}$, though in our application exceeding $1.1 \times 10^4\ M^{-1}\ s^{-1}$ results in error of at least 10%. Above the upper limit of $1.4 \times 10^4\ M^{-1}\ s^{-1}$, we have evidence that the error is even larger than accounted for in our error analysis approach. Systematic deconvolution error inferred from our previous work and the results in Fig. 7 are plotted (and color-coded (different gray shades in the print version)) on the right.

On the low end, if the rate constant is smaller than $1 \times 10^2\ M^{-1}\ s^{-1}$, separation of the signal decay from baseline power (i.e., signal to noise) becomes problematic. We have not examined data at or below the lower bound, but we suspect a k_{on} one-half as large as the smallest we have

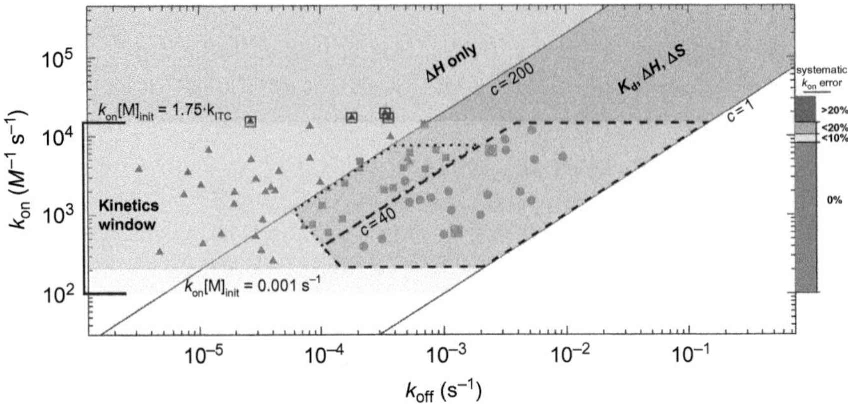

Figure 8 Relationship between the kinetics (k_{on} and k_{off}) and the attainable thermodynamic and kinetic parameters by ITC, illustrated for a titration experiment conducted with $[M]_{init} = 10$ μ*M* and instrument time constant of 12.5 s. The three main regions for consideration are denoted in blue, salmon, and yellow; corresponding overlapping regions are colored accordingly. The blue diagonal band represents the optimal thermodynamics regime for ITC in which $1 \leq c \leq 200$, and for which the full complement of thermodynamic information (K_d, ΔH, and subsequently ΔS) can be obtained. The salmon-colored region above the $c = 200$ line reflects the tight-binding regime for which, thermodynamically, only ΔH can be obtained. The horizontal yellow band represents the kinetics regime ($1 \times 10^2\ M^{-1}\ s^{-1} < k_{on} \leq 1.4 \times 10^4\ M^{-1}\ s^{-1}$). Within a subrange of the overlap of the kinetic regime with the optimal thermodynamic regime (largely consisting of $1 \leq c \leq 40$) is the near-equilibrium approximation regime, marked by dashed lines, for which this approximation is valid over the entirety of the titration. A dotted line encloses an adjacent region where, in our experience, the near-equilibrium approximation yields k_{on} values consistent with the general method. Closed symbols denote the regions within this map that we have examined using TT–RR, and the asterisk reflects the AdoMet–SET7/9 experiment. Symbol positions are "concentration-adjusted" such that they reflect the rate constants that would exhibit the experimentally observed kinetics if all experiments were conducted at $[M]_{init} = 10$ μ*M*. TT–RR closed circles and squares represent average kinetic parameters obtained at an experimental condition where k_{on} and the full complement of thermodynamic information were obtained (corresponding data are provided in Table 4). For data represented as circles ($c < 40$), k_{off} was also obtained independently of the K_d using the near-equilibrium approximation. Triangles represent conditions where we were able to obtain k_{on}, but only ΔH in terms of thermodynamics. For these experiments, a K_d extrapolated from the [salt]- or temperature dependence for TT–RR association was employed in the general equation; alternatively, for experiments in which $c > 200$, the tight-binding approximation could have been used to avoid this extrapolation requirement (additionally $\partial k_{on}/\partial K_d$ is rather negligible if $c > 200$ so precise knowledge of K_d is not required). The k_{off} for these datasets (as well as that of AdoMet–SET7/9), however, is known only as accurately as is the K_d, and thus, the associated uncertainty

(Continued)

examined is reasonable for a typical scenario with $[M]_{init} = 10\ \mu M$ and $|\Delta H| \geq 10$ kcal mol^{-1}. The figure also highlights the near-equilibrium region (dashed lines) in which, throughout the titration, the decay is exponential with composite rate constant k_{obs} (Eq. 10), allowing direct determination of both k_{on} and k_{off}.

Figure 8 overlays the kinetic data obtainable by ITC with an analogous plot illustrating the accessibility to traditional thermodynamic parameters, assuming a single-step reaction mechanism so that $K_d = k_{off}/k_{on}$. The c parameter reveals the thermodynamic parameters that can be measured in an ITC experiment. For measurement of ΔH, c must be minimally equal to one; in Fig. 8, this equates to $K_d \leq 10\ \mu M$. Measurement of K_d is additionally constrained by a lower limit; in theory, this constraint is $c \leq 1000$ (Wiseman et al., 1989), though we adhere to an upper limit of $c = 200$. For $c > 200$, K_d measurement relies on a small fraction of the titration data, meaning it may similarly rely on a small fraction of the M and/or L population. Thus, the map in Fig. 8 reflects a general case where sample homogeneity in all manners cannot be verified to ≥99%, so the upper bound of accessibility is set to $K_d \geq 50$ nM, reflecting $c \leq 200$.

There are two additional regions to note in the figure. Between the green, slightly clipped parallelogram defining the near-equilibrium region and the leftmost band defining the K_d-accessible window ($c = 200$) is a region where we have demonstrated (Table 4) consistency in applying the near-equilibrium approach to determine k_{on} (dotted lines; $40 < c \leq 200$). This region is truncated by $k_{on}[M]_{init} \leq 0.07$, because we did not collect data above this point and have some reason to doubt its generality outside of it. We also mention here the region defined by extension of the line segment clipping the aforementioned parallelogram to the $c = 200$ line, below which slow kinetics can sufficiently influence peak integration so that K_d measurements were not possible in our previous work (Vander Meulen & Butcher, 2012).

Figure 8—Cont'd would be quite high in some cases. Note the truncation at the lower left of the optimal thermodynamics regime: this denotes a region where slow kinetics prevented our accurate measurement of K_d. Boxed triangles represent data used as maximum rate constant test-cases in Fig. 7, and boxed circles were used as near-equilibrium test-cases in Figs. 4 and 5. The colored scale and labels on the right chart are inferred deconvolution-introduced errors: for $k_{on}[M]_{init} < 0.14$, the error should be less than 20% and we set this as the absolute upper limit for kinetics analysis. (See the color plate.)

For an instrument with a similar time constant, Fig. 8 can easily be adapted to an experiment using a different $[\mathrm{M}]_{\mathrm{init}}$ by noting they are directly related to the actual forward and reverse rates. A change in concentration increases the forward rate while having no effect on the reverse rate. Therefore (at least until signal to noise changes significantly), starting concentrations less than 10 μM lead to a simple, corresponding upward shift in all of the windows. For example, $[\mathrm{M}]_{\mathrm{init}}=5\ \mu M$ sets the upper limit of kinetics detection at $2.8\times10^{4}\ M^{-1}\ \mathrm{s}^{-1}$ and the lower limit $2\times10^{2}\ M^{-1}\ \mathrm{s}^{-1}$. An analog to the c parameter can be conceptualized for kinetics determination. In this respect, using a VP-ITC, kinetics are accessible as long as $0.001\ \mathrm{s}^{-1} \leq k_{\mathrm{on}}[\mathrm{M}]_{\mathrm{init}} \leq 0.14\ \mathrm{s}^{-1}$. One would expect that the fastest accessible rates will likely depend on instrumentation simply according to the difference in the instrument rate constant, so that the upper limit would be roughly defined according to $k_{\mathrm{on}}[\mathrm{M}]_{\mathrm{init}} \leq 1.75\ k_{\mathrm{ITC}}$. Similarly, differences in instrument sensitivity will affect the lower end of the kinetics window.

3.6 Summary and Outlook

We have described here our custom-built approach to extracting kinetics from an ITC titration experiment. Our focus has been describing the processing and error analysis details, and in highlighting the regimes in which different types of kinetics analyses may be performed.

Importantly, the approach described here is programmatically successful at both the upper and lower ends of the defined kinetics window. Yet, this is still early in method development, and there are several areas where it might be improved to increase the data quality and reduce the final uncertainty. Here we highlight just two possible changes. We think the most crucial improvement would be integrating the kinetics measurements into a global analysis (Zhao, Piszczek, & Schuck, 2015). Our current application requires thermodynamic analysis followed by individual analysis of each injection, first processing then fitting. Direct linking of factors such as the K_{d}, effective concentrations, and baseline analysis would likely reduce various contributors to what we refer to as injection-level and experiment-level uncertainties. Another area for improvement lies in the analysis of early time points, the accuracy of which may be improved by incorporating mixing lag as put forth by Dumas and coworkers (Burnouf et al., 2012). Along with better modeling of the instrument response time and functional form (Garcia-Fuentes et al., 1998), if possible, this increased level of sophistication might increase the upper limit of the kinetics window.

4. METHOD DETAILS

4.1 Materials

Preparation and handling of TT and RR RNA was covered in our previous kinetics work (Vander Meulen & Butcher, 2012). SET7/9 expression and enzymatic synthesis of AdoMet were performed using the methods described in Horowitz et al. (2011). Thermodynamic quantities were determined as described previously (Vander Meulen et al., 2008). All experiments used a MicroCal VP-ITC, with a high feedback setting and a 2 s filtering time. Injections were generally conducted at 0.5 μL s^{-1}. All analysis was performed using Igor Pro 6 and its inbuilt programming environment (WaveMetrics, Inc.).

4.2 Methods

4.2.1 Smoothing

We use boxcar smoothing at various stages in this method. First, we perform the Laplace Transform after minor smoothing of the power derivative term used in Eq. (1) (window size 7, 1 pass). This mild procedure is sufficient to minimize significant fluctuations in the raw data without having any measurable impact on the final measured rate constant. A more significant smoothing procedure (window size 7, 3 passes) is used prior to statistical analysis of an injection and decay (Section 3.2) in order to prevent noise-induced settings artifacts, such as an abbreviated fitting regime. Actual fitting utilizes the former, mildly smoothed data.

4.2.2 Baseline Selection and Associated Uncertainty

All kinetic functions incorporate a linear and constant term representing the underlying power baseline. An optimal approach to baseline selection uses the postrelaxation baseline period whenever possible to eliminate unnecessary fitting parameters, which carry with them inherent uncertainty that must be taken into account in characterizing k_{on} or k_{obs}. This ideal situation is most common for injections exhibiting fast kinetics such that a final baseline is quickly established, or similarly for injections that include a long post-injection (equilibration) period. In these cases, which we classify as "constrained," those predetermined baseline parameters are set and held during regression; in other cases, they are floated. Baseline classification is discussed next (subsequent propagation of baseline uncertainty is then discussed in Section 4.2.3).

We classify an injection power trace into one of three baseline groups: "constrained," "moderately constrained," and "unconstrained". These groups correspond to how reliably the baseline underlying the entirety of an injection and subsequent decay to a baseline power trace is described by that ending baseline and resultingly how rigidly baseline parameters are held during fitting. This procedure is not theoretically founded but is instead a set of boundary conditions that is based on a combination of observations of ITC power traces and manual tuning from many representative injections across the range of kinetic regimes. The goal of this procedure is to make use of a well-established baseline wherever possible so that uncertainty is not overemphasized, but in all other cases to accurately reflect the sometimes significant impact uncertainty in the baseline can have on analysis of a kinetics measurement.

To classify a baseline, we first select the final baseline region, which we consider to start at the first (boxcar-smoothed) point that is $<0.5\%$ greater in magnitude than a preliminarily determined baseline, and to conclude at the last point before the next injection. For the baseline to be considered "constrained," it must meet the following four conditions: first, the reduced $\chi 2$ ($\sigma_{\text{power}} = 0.02\ \mu\text{J s}^{-1}$ for data collected using our settings, unless a dataset is particularly noisy) must be less than 1.5; second, *F*-tests for an additional variable with either an exponential or a second-degree polynomial must be less than 2 and 3, respectively; third, extrapolation of the candidate baseline back to time zero must result in an offset not greater than either 1% of the amplitude of the peak following injection or $0.01\ \mu\text{J s}^{-1}$; and fourth, the baseline region must consist of at least 60 s. For a baseline to be considered "moderately constrained," it must meet the first condition above and a second condition that the baseline region time period is at least twice the decay half-time. Thus in our procedure, we first test for the "constrained" category, then for "moderately constrained." In either case, if the baseline meets the required conditions, it is classified as such and the procedure is concluded. Otherwise, the baseline start time is advanced to the next data point, and the tests are performed again. When the baseline candidate region is too short for the relevant categorization, the process is aborted with the default "unconstrained" baseline status maintained.

4.2.3 Error Analysis

Sources of rate constant uncertainty fall into three groups, according to their level of correlation: injection-level ($\sigma_{k,\text{inj}}$), experiment-level ($\sigma_{k,\text{exp}}$), and system-level ($\sigma_{k,\text{sys}}$). Uncertainty calculations should consider these in turn and incorporate proper nesting to obtain an accurate estimate of the uncertainty in the final measured kinetic parameters, according to the following equations

$$\sigma^2_{k,\,\mathrm{sys,\,tot}} \approx \sigma^2_{k,\,\mathrm{sys}} + \sigma^2_{k,\,\mathrm{exp\,,tot}} \tag{11}$$

where $\sigma^2_{k,\mathrm{sys,tot}}$ is the final top-level error, which includes unavoidable system-level uncertainty at a given condition, $\sigma^2_{k,\mathrm{sys}}$ (for our purposes, this is simply uncertainty related to k_{ITC}), as well as the uncertainty resulting from weighted averaging of the results from each experiment ($\sigma^2_{k,\mathrm{exp,tot}}$) at this condition

$$\sigma^2_{k,\,\mathrm{exp\,,tot}} \approx \sigma^2_{k,\,\mathrm{exp}} + \sigma^2_{k,\,\mathrm{inj,\,tot}}. \tag{12}$$

Similarly to Eq. (11), Eq. (12) obtains the uncertainty relevant to a particular experiment or titration according to the contribution from uncertainty correlated at the level of that experiment along with the uncertainty resulting from weighted averaging of the results from each injection ($\sigma^2_{k,\mathrm{inj,tot}}$) in that experiment.

4.2.3.1 Injection-Level Contributions to Uncertainty

Uncorrelated contributions to the uncertainty in kinetics measurements for each injection are related to random and systematic (i.e., drift) noise, and are classified as "fitting" uncertainty and "baseline" uncertainty. Fitting uncertainty ($\sigma_{k,\mathrm{fit}}$) comes directly from the variance in the measured k according to nonlinear regression. Baseline uncertainty results from a lack of knowledge of the true baseline underlying the power trace as discussed in Section 4.2.2. As an example, the uncertainty in k due to the slope parameter is estimated using $\Delta k/\Delta\mathrm{slope}$ in place of $\partial k/\partial\mathrm{slope}$, with $\Delta\mathrm{slope}$ being set to an estimated σ_{slope} (determined as discussed below). In this way, the combined injection-level uncertainty is calculated as

$$\sigma^2_{k,\,\mathrm{inj}} \approx \sigma^2_{k,\,\mathrm{fit}} + \Delta k^2_{\mathrm{slope}} + \Delta k^2_{\mathrm{intercept}}, \tag{13}$$

where $\Delta k_{\mathrm{slope}}$ reports the change in k affected by a one-sigma fluctuation in the baseline slope ($\Delta\mathrm{slope}$). Subsequently, $\sigma_{k,\mathrm{inj,tot}}$ is calculated through the weighted averaging of injection data.

Determination of $\Delta k_{\mathrm{slope}}$ and $\Delta k_{\mathrm{intercept}}$ for a "constrained" baseline (see Section 4.2.2) can be examined as illustrative. For this baseline scenario, $\Delta\mathrm{slope}$ is set to $\pm 0.25 \times 10^{-6}$ μJ s^{-2}. $\Delta k_{\mathrm{slope}}$ is then determined by the maximum change in k affected by either a positive or negative adjustment of this magnitude in the baseline slope. Similarly, the intercept term ($\Delta k_{\mathrm{intercept}}$) is determined by a modulation of the intercept (Δintercept) by ±1% of the injection peak's maximum magnitude. For this baseline uncertainty analysis (Eq. 13), and for the experiment-level discussion below, the covariance terms are negligible.

If the baseline is "moderately constrained," Δintercept is set to the maximum of 1% of the magnitude of the injection peak and the magnitude of the difference between the baseline's extrapolated time zero intercept and the actual value of the power trace at time zero. Δslope is the average value of the baseline slope over the course of the injection and relaxation period, as determined in the polynomial regression against the raw thermogram (Section 3.1).

For the "unconstrained" case, Δintercept is set similarly to the moderately constrained case, except that the term relative to the peak maximum is 2% instead of 1%. Δslope is the maximum of the fitted slope value and 2.0×10^{-6} μJ s^{-2}.

4.2.3.2 Experiment-Level Contributions to Uncertainty

Experiment-level contributions are those for which error is correlated among the analyses of every injection in a titration. These include uncertainty in the K_d (ΔK_d) and uncertainty in the macromolecular concentrations ($[M]_{tot}$, $[L]_{tot}$) (Δconc). Calculation of these uncertainties is conducted similarly to the previous section, and incorporates ΔK_d and Δconc values of ±10%. As an example, the uncertainty due to the K_d $\left(\Delta k_{K_d}\right)$ is determined by calculating the average k for an experiment using the best-fit K_d, then starting the analysis over using a K_d that is increased by 10%, then again for a decrease of 10%. The final experiment-level uncertainty is given by

$$\sigma_{k,\,\text{exp}}^2 \approx \Delta k_{K_d}^2 + \Delta k_{\text{conc}}^2 + \Delta k_{k,\,\text{inj},\,\text{tot}}^2 \tag{14}$$

Subsequently, $\sigma_{k,\text{exp,tot}}$ is calculated through the weighted averaging of all experiments at that condition.

4.2.3.3 System-Level Contributions to Uncertainty

Uncertainty in the instrumental rate constant (Δk_{ITC}) assumes a ±20% uncertainty (see Section 3.4) and is correlated among all experiments with a given rate constant profile. Most simply, this term is included by propagating the average affected uncertainty $\left(\Delta k_{k_{\text{ITC}}}\right)$ with the uncertainty resulting from the weighted average of the final experiment-level uncertainty calculation

$$\sigma_{k,\,\mathrm{sys,\,tot}}^2 \approx \Delta k_{k_{\mathrm{ITC}}}^2 + \sigma_{k,\,\mathrm{exp\,,tot}}^2. \tag{15}$$

For processes with $k_{\mathrm{on}}[\mathrm{M}]_{\mathrm{init}} > 0.11$, $\Delta k_{k_{\mathrm{ITC}}}$ may be insufficient, and the systematic error plotted alongside Fig. 8 could be used instead.

4.3 Example Uncertainties

Figure 9 shows an estimated uncertainty contributed by various error sources throughout titrations we performed at upper and lower limits of accessibility. Each data point in the figure represents the estimated fractional uncertainty in k_{on} contributed by each potential source of error for a single injection. As discussed above, these error sources contribute at different points in the analysis so that, for example, uncertainty due to noisy data can be minimized by averaging many injections in a titration. Similarly, uncertainty due to potential error in a K_{d} measurement can be reduced through averaging a number of experiments. However, for experiments with observed rate constants near k_{ITC}, appreciable uncertainty is unavoidably significant. The data points in the figure are an overlay of data from several experiments at a particular condition as described in the legends.

Figure 9A displays the fractional uncertainty in k_{on} from titrations with a $k_{\mathrm{on}}[\mathrm{M}]_{\mathrm{init}} = 0.006\ \mathrm{s}^{-1}$, equivalent to an experiment with $k_{\mathrm{on}} = 600\ M^{-1}\ \mathrm{s}^{-1}$ and $[\mathrm{M}]_{\mathrm{init}} = 10\ \mu M$ and significantly below the k_{ITC} value of $0.08\ \mathrm{s}^{-1}$. In

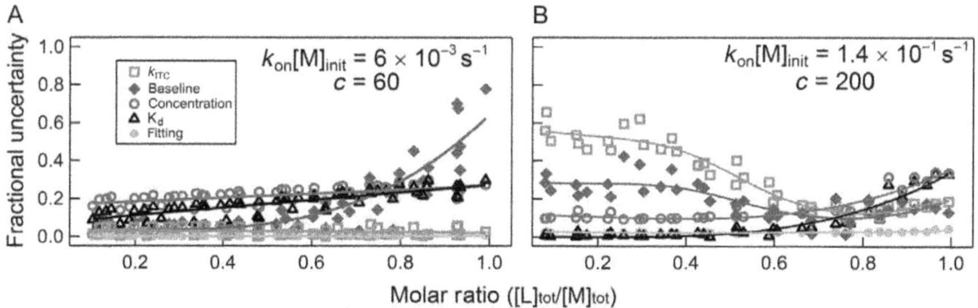

Figure 9 Estimated fractional uncertainty for k_{on} measurements at the (A) lower and (B) upper end of accessibility. (A) Summary data of five experiments collected for TT–RR binding at 5 °C, 100 mM KCl, where the average forward rate constant is $600\ M^{-1}\ \mathrm{s}^{-1}$. In each experiment, the starting cell concentration was $10\ \mu M$, so $k_{\mathrm{on}}[\mathrm{M}]_{\mathrm{init}} = 0.006\ \mathrm{s}^{-1}$. (B) Summary of data collected for the SET7/9 lysine methyltransferase, for which the average forward rate constant is $22{,}000\ M^{-1}\ \mathrm{s}^{-1}$, and the average starting cell concentration was $6.4\ \mu M$ (average $k_{\mathrm{on}}[\mathrm{M}]_{\mathrm{init}} = 0.14\ \mathrm{s}^{-1}$). Each symbol reflects the estimated contribution of each source of uncertainty to a single injection at that point in a titration. Averaging of all data points in a titration reduces the uncertainty from uncorrelated sources of error as discussed in Section 4.2.3. (See the color plate.)

this scenario, all uncertainties are relatively small in the first several injections. At later injections, the fitting-related uncertainties (random error fit variance and baseline uncertainty) begin to increase appreciably due to lowered signal to noise.

In Fig. 9B, the SET7/9–AdoMet experiment at the very upper limit of accessibility is shown, with an average $k_{on}[M]_{init} = 0.138\ s^{-1}$. In this case, the rate of association is largest at the beginning of the titration, so uncertainty related to k_{ITC} is largest here. Comparison of Figs. 7 and 9B suggests the uncertainty may be slightly underestimated in this scenario. The baseline error is also significant early on due to the small fitting range in these scenarios (see, for example, Figs. 5C and 6B). Both of these major sources of uncertainty decrease as the free macromolecular concentration is reduced and the association relaxation time is correspondingly increased; contrastingly, experiment-level sources of error do increase as the stoichiometric point is approached.

ACKNOWLEDGMENTS

This work was supported in part by grants from the University of Michigan's Biomedical Research Council and the Office for the Vice President for Research and NSF (CHE-1213484) to R.C.T., and NIH (NIGMS) grants GM065166 and GM072447 to S.E.B.

VP-ITC data for RNA experiments were obtained at the University of Wisconsin–Madison Biophysics Instrumentation Facility, which was established with support from the University of Wisconsin–Madison and grants BIR-9512577 (NSF) and S10 RR13790 (NIH).

REFERENCES

Bernasconi, C. F. (1976). *Relaxation kinetics*. New York: Academic Press.

Bianconi, M. L. (2003). Calorimetric determination of thermodynamic parameters of reaction reveals different enthalpic compensations of the yeast hexokinase isozymes. *Journal of Biological Chemistry*, *278*(21), 18709–18713.

Bianconi, M. L. (2007). Calorimetry of enzyme-catalyzed reactions. *Biophysical Chemistry*, *126*(1–3), 59–64. http://dx.doi.org/10.1016/j.bpc.2006.05.017.

Burnouf, D., Ennifar, E., Guedich, S., Puffer, B., Hoffmann, G., Bec, G., et al. (2012). kinITC: A new method for obtaining joint thermodynamic and kinetic data by isothermal titration calorimetry. *Journal of the American Chemical Society*, *134*(1), 559–565. http://dx.doi.org/10.1021/ja209057d.

Cate, J. H., Gooding, A. R., Podell, E., Zhou, K. H., Golden, B. L., Kundrot, C. E., et al. (1996). Crystal structure of a group I ribozyme domain: Principles of RNA packing. *Science*, *273*(5282), 1678–1685.

Costa, M., & Michel, F. (1995). Frequent use of the same tertiary motif by self-folding RNAs. *The EMBO Journal*, *14*(6), 1276–1285.

Demarse, N. A., Killian, M. C., Hansen, L. D., & Quinn, C. F. (2013). Determining enzyme kinetics via isothermal titration calorimetry. *Methods in Molecular Biology*, *978*, 21–30. http://dx.doi.org/10.1007/978-1-62703-293-3_2.

Egawa, T., Tsuneshige, A., Suematsu, M., & Yonetani, T. (2007). Method for determination of association and dissociation rate constants of reversible bimolecular reactions by isothermal titration calorimeters. *Analytical Chemistry, 79*(7), 2972–2978. http://dx.doi.org/10.1021/ac062183z.

Franghanel, J., Wawra, S., Luke, C., Wildemann, D., & Fischer, G. (2006). Isothermal calorimetry as a tool to investigate slow conformational changes in proteins and peptides. *Analytical Chemistry, 78*(13), 4517–4523. http://dx.doi.org/10.1021/Ac052040x.

Freire, E., Mayorga, O. L., & Straume, M. (1990). Isothermal titration calorimetry. *Analytical Chemistry, 62*(18), A950–A959.

Freire, E., Vanosdol, W. W., Mayorga, O. L., & Sanchezruiz, J. M. (1990). Calorimetrically determined dynamics of complex unfolding transitions in proteins. *Annual Review of Biophysics and Biophysical Chemistry, 19*, 159–188.

Garcia-Fuentes, L., Baron, C., & Mayorga, O. L. (1998). Influence of dynamic power compensation in an isothermal titration microcalorimeter. *Analytical Chemistry, 70*(21), 4615–4623.

Horowitz, S., Yesselman, J. D., Al-Hashimi, H. M., & Trievel, R. C. (2011). Direct evidence for methyl group coordination by carbon-oxygen hydrogen bonds in the lysine methyltransferase SET7/9. *Journal of Biological Chemistry, 286*(21), 18658–18663. http://dx.doi.org/10.1074/jbc.M111.232876.

Lonhienne, T. G., & Winzor, D. J. (2004). A potential role for isothermal calorimetry in studies of the effects of thermodynamic non-ideality in enzyme-catalyzed reactions. *Journal of Molecular Recognition, 17*(5), 351–361. http://dx.doi.org/10.1002/jmr.706.

Malvern Instruments Ltd. (2015). MicroCal ITC analysis software using Origin.

Mayorga, O. L., & Freire, E. (1987). Dynamic analysis of differential scanning calorimetry data. *Biophysical Chemistry, 27*(1), 87–96.

Morin, P. E., & Freire, E. (1991). Direct calorimetric analysis of the enzymatic activity of yeast cytochrome c oxidase. *Biochemistry, 30*(34), 8494–8500.

Record, M. T., Zhang, W. T., & Anderson, C. F. (1998). Analysis of effects of salts and uncharged solutes on protein and nucleic acid equilibria and processes: A practical guide to recognizing and interpreting polyelectrolyte effects, Hofmeister effects, and osmotic effects of salts. *Advances in Protein Chemistry, 51*, 281–353.

Todd, M. J., & Gomez, J. (2001). Enzyme kinetics determined using calorimetry: A general assay for enzyme activity? *Analytical Biochemistry, 296*(2), 179–187. http://dx.doi.org/10.1006/abio.2001.5218.

Toor, N., Keating, K. S., Taylor, S. D., & Pyle, A. M. (2008). Crystal structure of a self-spliced group II intron. *Science, 320*(5872), 77–82.

Transtrum, M. K., Hansen, L. D., & Quinn, C. (2015). Enzyme kinetics determined by single-injection isothermal titration calorimetry. *Methods, 76*, 194–200. http://dx.doi.org/10.1016/j.ymeth.2014.12.003.

Vander Meulen, K. A., & Butcher, S. E. (2012). Characterization of the kinetic and thermodynamic landscape of RNA folding using a novel application of isothermal titration calorimetry. *Nucleic Acids Research, 40*(5), 2140–2151. http://dx.doi.org/10.1093/nar/gkr894.

Vander Meulen, K. A., Davis, J. H., Foster, T. R., Record, M. T., Jr., & Butcher, S. E. (2008). Thermodynamics and folding pathway of tetraloop receptor-mediated RNA helical packing. *Journal of Molecular Biology, 384*(3), 702–717.

Williams, B. A., & Toone, E. J. (1993). Calorimetric evaluation of enzyme-kinetic parameters. *Journal of Organic Chemistry, 58*(13), 3507–3510.

Wiseman, T., Williston, S., Brandts, J. F., & Lin, L. N. (1989). Rapid measurement of binding constants and heats of binding using a new titration calorimeter. *Analytical Biochemistry, 179*(1), 131–137.

Zhao, H. Y., Piszczek, G., & Schuck, P. (2015). SEDPHAT—A platform for global ITC analysis and global multi-method analysis of molecular interactions. *Methods, 76*, 137–148. http://dx.doi.org/10.1016/j.ymeth.2014.11.012.

CHAPTER NINE

Isothermal Titration Calorimetry to Characterize Enzymatic Reactions

Luca Mazzei, Stefano Ciurli[1], Barbara Zambelli
Laboratory of Bioinorganic Chemistry, Department of Pharmacy and Biotechnology, University of Bologna, Bologna, Italy
[1]Corresponding author: e-mail address: stefano.ciurli@unibo.it

Contents

Abstract

Isothermal titration calorimetry (ITC) is a technique that measures the heat released or absorbed during a chemical reaction as an intrinsic probe to characterize any chemical process that involves heat changes spontaneously occurring during the reaction. The general features of this method to determine the kinetic and thermodynamic parameters of enzymatic reactions (k_{cat}, K_M, ΔH) are described and discussed here together with some detailed applications to specific cases. ITC does not require any modification or labeling of the system under analysis, can be performed in solution, and needs only small amounts of enzyme. These properties make ITC an invaluable, powerful, and unique tool to extend the knowledge of enzyme kinetics to drug discovery.

Methods in Enzymology, Volume 567
ISSN 0076-6879
http://dx.doi.org/10.1016/bs.mie.2015.07.022

1. INTRODUCTION

Understanding biological processes requires a complete determination of the thermodynamic and kinetic parameters of the biochemical reactions at the basis of life. Most biochemical reactions occur spontaneously at rates not compatible with life. Therefore, living organisms exploit enzymes as biological catalysts to accelerate the speed of substrate consumption or product formation. In a typical enzymatic reaction scheme (Eq. 1), substrate (S) reversibly binds the enzyme (E) forming the enzyme–substrate (ES) complex, which subsequently proceeds irreversibly towards the formation of the product (P) and the enzyme itself:

$$E + S \underset{k-1}{\overset{k_1}{\rightleftharpoons}} ES \xrightarrow{k_{cat}} E + P \quad (1)$$

In Eq. (1), k_1 is the rate constant for the formation of the ES complex, k_{-1} is the rate constant for the dissociation of the ES complex, while k_{cat} is the catalytic rate constant.

Even though this reaction mechanism is a simplification that assumes that the overall enzyme-catalyzed reaction is irreversible and has only one intermediate complex, it is the starting point to interpret enzyme kinetics (Seibert & Tracy, 2014). This approach is usually described by the Michaelis–Menten theory (Michaelis & Menten, 1913), which shows that enzymes can be studied by measuring the initial rate of product formation or substrate consumption (v) as a function of active enzyme concentration, substrate concentration, and two kinetic parameters, namely, the Michaelis constant (K_M) and the catalytic rate constant (k_{cat}):

$$v = \frac{k_{cat} \cdot [E] \cdot [S]}{K_M + [S]} \quad (2)$$

In Eq. (2), $K_M = (k_{-1} + k_{cat})/k_1$ and $k_{cat} = v_{max}/[E]$, v_{max} being the maximal velocity reached when all enzyme is bound to the substrate. The k_{cat}/K_M ratio is referred to as the catalytic efficiency of an enzyme. In practice, determination of K_M and k_{cat} for a specific reaction provides a complete description of the catalysis. These kinetic parameters are usually obtained through a set of continuous or discontinuous time course experiments, in which an observable event accompanying the transformation of substrate into product is measured. A plot of initial velocity versus substrate concentration is constructed (Fig. 1) and fitted according to the Michaelis–Menten equation (Eq. 2). Often, the substrate or product concentration is measured

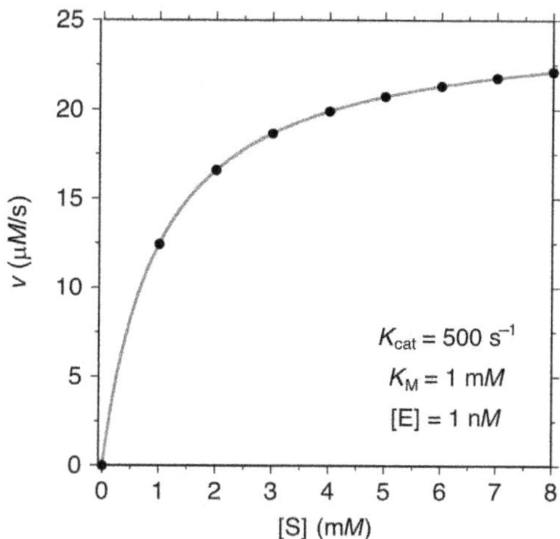

Figure 1 Representative Michaelis–Menten plot describing the catalytic activity of a 1 n*M* enzyme, showing $K_M = 1$ m*M* and $k_{cat} = 500\ s^{-1}$ for its substrate.

over time by light absorbance or fluorescence at characteristic wavelengths. However, most substrates and/or products are not spectroscopically active, and therefore either a chromophore or fluorophore or an optically active substrate analogue must be included to follow the reaction. Alternatively, the use of coupled assays, with the product of the catalysis being a substrate for another fast reaction yielding a product with measurable concentration, resolves the problem of product detection. However, all these methods introduce additional variables that reduce the accuracy of the determined values of K_M and k_{cat}. In addition, the characterization of enzyme inhibition by small ligands using spectroscopic methods could be complicated by the absorbance or fluorescence of the small ligand itself, which may alter the reliability of the spectroscopic assay. Alternative methods to measure kinetic parameters involve stopped flow techniques, in which the amount of product is quantified using mass spectrometry or chromatography. Even though these techniques are precise and reliable, they are time-consuming and expensive for routine analysis.

Unlike spectroscopic methods, calorimetry is a label-free methodology because it uses heat released or absorbed as a probe to monitor the reaction that occurs over time. In particular, isothermal titration calorimetry (ITC) measures the heat that is exchanged with the surroundings at a constant temperature (Freyer & Lewis, 2008). As almost all chemical reactions produce or absorb heat, ITC is a universal detection system and a very convenient methodology to quantify the amount of reacting molecules (i.e., binding thermodynamics), as well as to measure the reaction rate (i.e., kinetics). In particular, ITC has been adopted as a method of choice to characterize the thermodynamics of biomolecular equilibria, involving protein–ligand, protein–protein, protein–metal ions, and protein–DNA interactions (Duff, Grubbs, & Howell, 2011; Ghai, Falconer, & Collins, 2012; Ladbury, 2004; Leavitt & Freire, 2001; Zambelli, Bellucci, Danielli, Scarlato, & Ciurli, 2007; Zambelli et al., 2008). In addition, the ability of ITC to provide kinetic information makes it a very powerful system to measure enzyme catalysis, even though the potential of this application is still underestimated (Bianconi, 2007; Demarse, Killian, Hansen, & Quinn, 2013; Mazzei, Ciurli, & Zambelli, 2014; Olsen, 2006; Todd & Gomez, 2001; Transtrum, Hansen, & Quinn, 2015). Indeed, calorimetric measurements can be done with species that are spectroscopically silent, as well as on heterogeneous or turbid solutions, and over a range of biologically relevant conditions (temperature, salt concentration, pH, etc.) (Freyer & Lewis, 2008). The versatility of this method is demonstrated by its possible use to measure enzyme activity in crude tissue extracts (Sica, Gilli, Briand, & Sari, 1987) and in the presence of insoluble substrates (Lonhienne, Baise, Feller, Bouriotis, & Gerday, 2001). These studies indicate the capability of calorimetry to determine enzyme kinetics in complex solutions. Moreover, this technique is straightforward and fast, requires small amount of material, and can be performed in solution (Baumann et al., 2011; Noske, Cornelius, & Clarke, 2010).

2. INSTRUMENTAL SETUP

The basic instrument to carry out ITC measurements is a calorimeter. This is made of an adiabatic shield containing two coined-shaped cells, connected to the outside with narrow access tubes (Fig. 2A). The sample cell is loaded with the enzyme solution, while the reference cell is generally filled with water or with the solvent used for the analysis. A rotating syringe with a long needle and an attached stirring paddle is mounted on the sample cell and dispenses the substrate solution. A thermoelectric device measures the difference of temperature between the sample and the reference cell, and using a cell feedback network it maintains this difference at zero by adding or subtracting heat to the sample cell. During a standard experiment, the substrate is injected by the syringe into the enzyme solution at a constant temperature set up by the user. When the enzymatic reaction takes place, the amount of heat released or absorbed is proportional to the number of substrate molecules

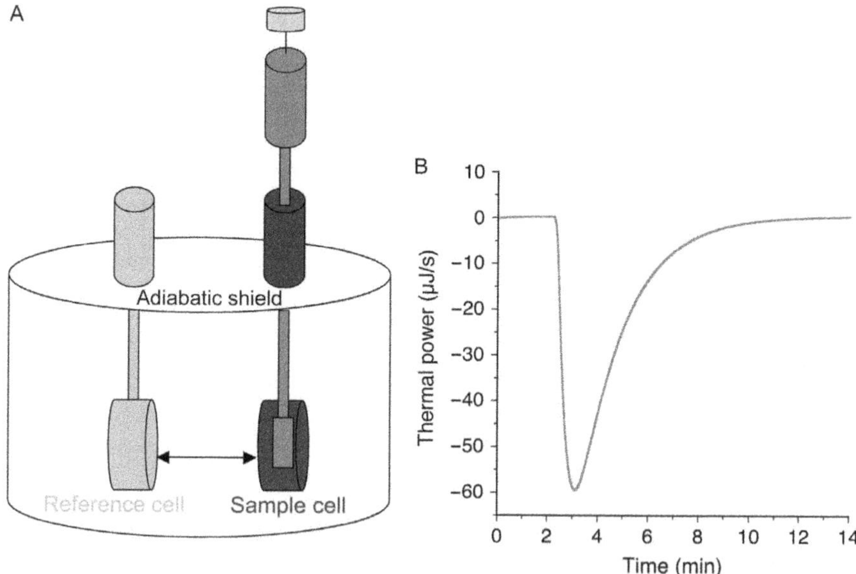

Figure 2 (A) Schematic representation of the isothermal titration calorimeter to study enzymatic reactions. The reference cell and the sample cell are indicated in green (light gray in the print version) and blue (dark gray in the print version), respectively, and the titration syringe is colored in red (gray in the print version). (B) The resulting change of the thermal power released by the calorimeter, needed to keep the difference of temperature between the sample cell and the reference cell constant, is represented as a deviation of the baseline in the produced raw data.

that are converted into product molecules. In addition, the rate of heat flow is directly related to the rate of the reaction. The measured data, appearing as a deviation of the heat trace from initial baseline (Fig. 2B), represent the thermal power (μJ/s) supplied by the instrument to the sample cell, which is proportional to the heat flow occurring in the sample cell over time.

3. METHODS AND THEORY

The rate of a reaction is directly proportional to the thermal power, defined as the heat (Q) produced as a function of time ($\mathrm{d}t$): thermal power $= \mathrm{d}Q/\mathrm{d}t$. Overall, the heat change (Q) is proportional to the molar enthalpy measured for the reaction in the established conditions (ΔH_{app}) and the number of moles of product generated (n), which in turn is given by the total volume times the concentration of product:

$$Q = n \cdot \Delta H_{\mathrm{app}} = V \cdot [\mathrm{P}] \cdot \Delta H_{\mathrm{app}} \tag{3}$$

The product formation over time (d[P]/dt), which corresponds to the reaction rate, can thus be related to the amount of heat generated over the same time (dQ/dt) through the relation:

$$v = \frac{d[P]}{dt} = \frac{1}{V \cdot \Delta H_{app}} \cdot \frac{dQ}{dT} \tag{4}$$

According to this equation, in order to obtain a Michaelis–Menten plot and to derive the kinetic parameters K_M and k_{cat}, it is necessary to measure (i) the total apparent molar enthalpy ΔH_{app} and (ii) the heat flow dQ/dt at different substrate concentrations. The apparent molar enthalpy is generally measured in experiments in which all the injected substrate is converted into the product, in a given time period. In these experiments, the baseline response returns to the initial value after the substrate injection (Fig. 2B), and integration of the peak with respect to time yields the total heat produced by the reaction. Dividing this value by the known amount of substrate added and converted gives the total enthalpy change. Two different methods have been described to determine the power generated over time (dQ/dt). The first method, called *multiple-injection method*, involves subsequent injections of substrate concentrations using the normal titration mode of ITC. As each injection provides a different concentration of substrate, multiple rate determinations at different substrate concentrations under steady-state conditions can be performed within a single experiment, producing a plot of heat rate versus substrate concentration, typically with 15–30 data points. An alternative strategy, named *single-injection method*, involves continuous rate measurements following a single injection of substrate at concentrations higher than K_M. In these experiments, the thermal power dQ/dt is monitored as the substrate is completely depleted, and the reaction rate is determined using Eq. (4). The concentration of substrate at any given time can be determined from the integral of the heat evolved. Therefore, plots of rate versus substrate concentration can be obtained to give a continuous kinetic curve. While in the multiple-injection method, two separate experiments, one to measure the ΔH_{app} and the other to determine dQ/dt, should be performed, the single-injection method provides the Michaelis–Menten curve in a single experiment, because ΔH_{app} and dQ/dt can be derived using the same curve that describes substrate consumption over time.

4. EXPERIMENTAL CONDITIONS

4.1 Preparing Samples

Enzyme and substrate must be diluted in buffer with identical composition to minimize the heat of dilution and mixing during the substrate addition. This can be achieved by dialyzing the enzyme solution and by using the dialysis buffer for substrate dilution or, alternatively, using a desalting or size-exclusion chromatography column. If organic solvents (e.g., DMSO) are included in the enzyme solution, they must be added into the substrate solution at exactly the same concentration. Buffer concentration must be adequate to ensure good pH buffering of the system in order to prevent pH change during the experiment. Before running the experiment, the pH mismatch of both the enzyme and the substrate solutions should be verified to be minimal under the experimental conditions (±0.05 pH units). Specific effects of the buffer or additives molecules on the system under analysis have to be taken into account. Enzyme and substrate concentrations must be carefully determined in order to obtain accurate and reliable calculation of the thermodynamic and kinetic parameters because concentration errors in either [S] or [E] will result in incorrect values of ΔH_{app}, K_M, and k_{cat}. The concentration of the solutes in the final solutions should be accurately determined with a reliable analytical procedure (e.g., absorbance, colorimetric assays, BCA assays) after appropriate dilution of the stock samples. The experimental temperature for enzymatic catalysis falls in the 25–37 °C range, even though ITC generally allows working temperatures between 2 and 80 °C. ITC samples are typically equilibrated to a temperature that is 2–3 °C below the desired experimental temperature, to avoid long equilibration periods before the experiment begins, because of the passive instrumental cooling device. Number and volume of substrate additions, as well as the time spacing between injections, must be chosen to allow a measurable heat signal and a good signal-to-noise ratio. Given the intrinsic difference of every enzymatic system and the different heat produced by different reactions, which cannot be quantified a priori, the determination of the optimal experimental conditions (i.e., enzyme and substrate concentrations, number and volume of injections, spacing time, etc.) for a specific system often requires few exploratory experiments.

4.2 Measuring the Enthalpy Change

In order to obtain the value of ΔH_{app} of the reaction, the total heat generated by substrate consumption must be measured. This is carried out in a single-injection experiment, performed at enzyme concentrations (usually in the nM–μM range) that are adequate to achieve the total substrate conversion into the product over the time of experiment (1000–2000 s for most systems). When the reaction has proceeded to completion, the heat signal returns to the original baseline. In principle, a single injection is enough to determine the ΔH_{app} of the reaction. However, performing 2–5 additions with relatively large injected volumes (5–100 μL) is recommended to improve the precision of the measurement and to explore possible product inhibition effects. In this case, the spacing time between each injection should be sufficient to ensure that the heat signal returns to the baseline before the next addition.

4.3 Multiple-Injection Method

The multiple-injection method provides the rate of heat change ($\mathrm{d}Q/\mathrm{d}t$) at different substrate concentrations. Subsequent small injections (generally 15–30 injections of 2–10 μL) are performed. The interval between the injections should allow the system to stabilize the thermal power to the new baseline, but must be short enough to avoid the conversion of a significant amount of substrate, allowing the measurement to be performed under steady-state conditions. The usual spacing time between injections is 120–180 s and requires the user to watch the titration closely to optimize it to the specific reaction and concentrations being studied. As each data point takes 2–3 min, a single experiment usually takes 1–2 h. A displacement of the baseline occurs after each injection, indicative of a change of heat rate in the sample cell. A further addition of substrate increases the substrate concentration and, therefore, the rate of the reaction, which results in a new displacement of the baseline. The value of $\mathrm{d}Q/\mathrm{d}t$ is determined by measuring the difference between the original baseline and the new baseline after each injection. In these conditions, the injected substrate is marginally (less than 5%) consumed, and the enzyme reaction proceeds at the steady state. For these considerations, the enzyme concentration used for this experiment is generally lower (usually in pM–nM range) than in the single-injection experiments. The substrate concentration in the syringe should be enough to assure that the final substrate concentration in the sample cell is higher than K_M, in order to obtain a reliable kinetic curve. The main advantage of this method is that the product inhibition is minimal, due to the low amount of product formed over time. However, the heat flow is quite low, especially for low values of ΔH_{app}, rendering this assay sensitive to baseline drifts. The latter should take into account the effect of enzyme dilution occurring upon substrate sequential additions because the initial rate of the reaction, reflected in the baseline position after the substrate injection, depends on the enzyme concentration. In this instrumental scheme, the inverse titration (enzyme added from the syringe to the substrate solution in the sample cell) is not possible due to the potential leakage of the syringe after each injection.

4.4 Single-Injection Method

In this experimental scheme, the substrate solution is injected into the cell containing the enzyme solution, producing a heat response that eventually returns to baseline after all the substrate has reacted, similarly to the method to calculate enthalpy (Section 4.2). Following the single injection, a continuous rate measurement is performed, while the concentration of substrate decreases over the time of reaction. A second injection of substrate can be made to collect additional information about the reaction (e.g., the presence of product inhibition). The second injection should occur after the time necessary to allow the first reaction event to be complete and the baseline to go back to the initial value. The concentration of substrate in the reaction cell should be adjusted to assure that the substrate is consumed in 30–60 min and that the initial concentration of substrate is higher than K_{M}. Even though this experimental setup is faster than the multiple-injection case because it requires a single experiment and a single injection, it is not often used because the data treatment is considerably more challenging than the multiple-injection data, and the instrument time constant can significantly affect the data (Transtrum et al., 2015). The single-injection assay is however less prone to errors related to baseline drifts over time, because it shows a postreaction baseline that allows for correction effects from side reactions, such as possible enzyme aggregation over time, or external temperature changes over the course of the experiment. In any case, both in the single- and multiple-injection methods, baseline levels should be checked by performing a control experiment that consists in titrating the substrate in the buffer solution without the enzyme.

5. DATA ANALYSIS

Data are typically analyzed using widely used spreadsheet software. The primary experimental data are the heat flow, measured in μJ/s, which is directly proportional to the reaction rate. This can be obtained either from the multiple-injection or from the single-injection data.

The data from the multiple-injection experiment are converted to reaction rate according to Eq. (4) and using the enthalpy value obtained as described below. These data are plotted against the substrate concentration, derived by dividing the molar amount of substrate added during the titration by the sample cell volume (taking into account dilution effects). This operation yields a typical Michaelis–Menten plot that can be fitted to Eq. (2) to obtain the values of K_M and k_{cat}.

In the single-injection method, the substrate concentration at any time is calculated using the integrated heat as a function of time, according to Eq. (5):

$$[S]_t = [S]_{t=0} - \frac{\int_{t=0}^{t} \frac{dQ}{dt} dt}{\Delta H_{app} V} \tag{5}$$

The value of ΔH_{app}, needed in both Eqs. (4) and (5), is calculated by integrating the peak resulting from baseline deviation in the thermogram of the single-injection experiment and dividing the result by the total substrate concentration according to Eq. (6):

$$\Delta H = \frac{\int_{t=0}^{\infty} \frac{dQ}{dt} dt}{V[S]_{t=0}} \tag{6}$$

The total heat change measured in this way represents the sum of all events occurring during the reaction under analysis and depends on the molar enthalpy of all the processes involved, including, if applicable, proton release or uptake from the buffer. In a generic process in which the reacting system acquires protons from a buffered solution, the buffer itself releases protons in order to maintain the pH constant, thus producing additional heat within the reaction cell. Therefore, the measured enthalpy change is apparent and not absolute (ΔH_{app}) and includes the intrinsic enthalpy change of the reaction (ΔH_{int}) as well as the ionization enthalpy of the buffer (ΔH_{ion}) (Fukada & Takahashi, 1998):

$$\Delta H_{app} = \Delta H_{int} + n\Delta H_{ion} \tag{7}$$

Using Eq. (7), it is possible to determine *n* (the number of exchanged protons) and ΔH_{int} by performing the same experiment in buffers with different ionization enthalpies, and plotting the measured ΔH_{app} as a function of the known ΔH_{ion}, specific for each buffer. From the resulting linear regression, *n* is derived from the slope: a negative slope reflects protons acquired by the buffer, while a positive slope represents protons released from the buffer. ΔH_{int} is derived from the intercept on the *Y*-axis. Equation (7) also indicates that the value of ΔH_{app} is strongly influenced by the ionization enthalpy of the buffer. Therefore, buffer with small ionization enthalpies, such as phosphate, gives a little contribution to the observed heat of reaction. On the other hand, buffers with large ionization enthalpies, such as Tris, provide larger heat of reaction. This characteristic is useful when systems with low ΔH_{int} are analyzed because using a buffer with a higher ionization enthalpy increases the signal-to-noise ratio. Chapters "Avoiding buffer interference in ITC experiments: A case study from the analysis of entropy driven reactions of glucose-6-phosphate dehydrogenase" by Bianconi and "Isothermal titration calorimetry measurements of metal ions binding to proteins" by Quinn et al. also describe the relationship between parameters obtained from ITC and the proton exchange effects of the buffer.

6. REPRESENTATIVE RESULTS

6.1 Ni(II)-Dependent Urease

Urease (EC 3.5.1.5; urea amidohydrolase) is a nickel-containing enzyme found in archaea, bacteria, unicellular eukaryotes, and plants (Maroney & Ciurli, 2014; Zambelli, Musiani, Benini, & Ciurli, 2011). This protein acts in the last step of organic nitrogen mineralization, catalyzing the hydrolysis of urea to ammonia and carbamate, which spontaneously decomposes to give a second molecule of ammonia and bicarbonate (Eq. 8):

$$H_2N{-}C({=}O){-}NH_2 + H_2O \xrightarrow{\text{urease}} H_2N{-}C({=}O){-}O^- + NH_4^+ \xrightarrow{H_3O^+} HO{-}C({=}O){-}O^- + 2NH_4^+ \quad (8)$$

This reaction causes an overall increase of the pH of the environment, which is responsible of negative effects for human health and agriculture (Maroney & Ciurli, 2014; Zambelli et al., 2011). The ureolytic activity of *Canavalia ensiformis* (jack bean) urease (JBU) was chosen as a representative reaction to demonstrate the applicability of ITC in quantitatively determining enzymatic catalysis (Mazzei et al., 2014). A value of $\Delta H_{app} = -43.9$ kJ/mol was measured in a single injection (Fig. 3A), by

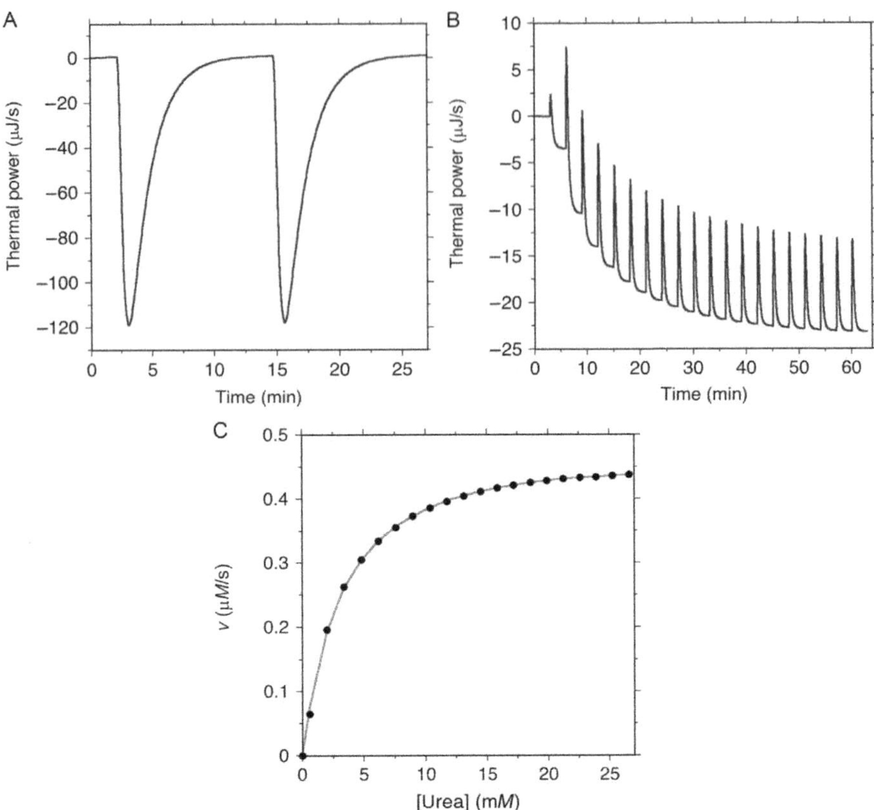

Figure 3 Urea hydrolysis by jack bean urease (JBU) determined by ITC. JBU (type C3) was purchased as lyophilized enzyme and diluted in 20 m*M* HEPES, pH 7, to the final concentration. The concentration of the enzyme in the reaction mixture was calculated by comparing the determined activity of the purchased product to the activity of the pure enzyme, reported to be 6200 U/mg (Hausinger & Karplus, 2001). Molar mass of 545kDa corresponds to the homohexameric protein. (A) Thermal power observed in the single-injection experiment performed at 25 °C with an injection of 5 μL (repeated twice to verify reproducibility) of 20 m*M* urea (in the injection syringe) into 1 n*M* JBU (in the sample cell). A spacing of 750 s between the injections was applied to allow the baseline returning to the initial value. (B) Thermal power observed in the multiple-injection experiment, obtained by titrating 400 m*M* urea (20 × 5 μL injections) into 15 p*M* JBU. A spacing of 180 s was applied between injections, in order to allow the system to reach and maintain the steady-state level. dQ/dt at different substrate concentrations is obtained as displacement of the baseline after each addition. (C) The baseline displacement in panel (B) was converted to the reaction rate using Eq. (4). Fit of the obtained data (filled circles) was performed using Eq. (2) and is represented as a solid red (gray in the print version) line.

integrating the thermal power resulting by injecting urea substrate into the urease solution in 20 m*M* HEPES at pH 7, and allowing the reaction to proceed to completion. Subsequently, the thermal power registered in a multiple-injection experiment (Fig. 3B) was converted to the reaction rate using Eq. (4) (Fig. 3C). Fit of the obtained data to the Michaelis–Menten equation provided the kinetic parameters for JBU as $k_{cat}=30{,}200\ s^{-1}$ and $K_M=3.45$ m*M*, in agreement with previously reported data (Callahan, Yuan, & Wolfenden, 2005; Krajewska, van Eldik, & Brindell, 2012; Mazzei et al., 2014). A similar experiment (Benini, Cianci, Mazzei, & Ciurli, 2014) performed on urease from *Sporosarcina pasteurii* (SPU), a widespread and highly ureolytic soil bacterium, in 50 m*M* HEPES, pH 7, provided a $\Delta H_{app}=41.86$ kJ/mol, $K_M=16.5$ m*M*, and $k_{cat}=9010\ s^{-1}$.

The total heat change measured in the single-injection experiment, representing the sum of all events occurring during the reaction and depending on the molar enthalpy of all the processes involved, including proton acquisition or release from the buffer, can be used to calculate the intrinsic heat of reaction ΔH_{int} and the number of exchanging protons (*n*), according to Eq. (7). Three single-injection experiments, performed for JBU in different buffers (HEPES, Tris–HCl, phosphate) characterized by three different ΔH_{ion} (Fig. 4), yielded ΔH_{int} (−61.5 kJ/mol) and the number of protons (0.98) released from the buffer, calculated from the intercept and the slope of the linear fit of the experimental data according to Eq. (7) (Mazzei et al., 2014). These results are in full agreement with data previously reported for the same reaction catalyzed by *Helicobacter pylori* urease (Todd & Gomez, 2001).

6.2 Lactate Dehydrogenase

Lactate dehydrogenase (EC 1.1.1.27; LDH) is an enzyme found in nearly all living organisms, playing a key role in glycolysis by catalyzing the reversible transformation of pyruvate to L-lactate with simultaneous oxidation of reduced nicotinamide adenine dinucleotide (NADH) to oxidized nicotinamide adenine dinucleotide (NAD^+), the latter being essential to maintain the glycolytic flow (Everse & Kaplan, 1973). LDH exists either as a homotetramer or a heterotetramer enzyme, made of different combinations of two different subunits (named A and B, or M and H), forming five possible isoenzymes in mammals (Everse & Kaplan, 1973).

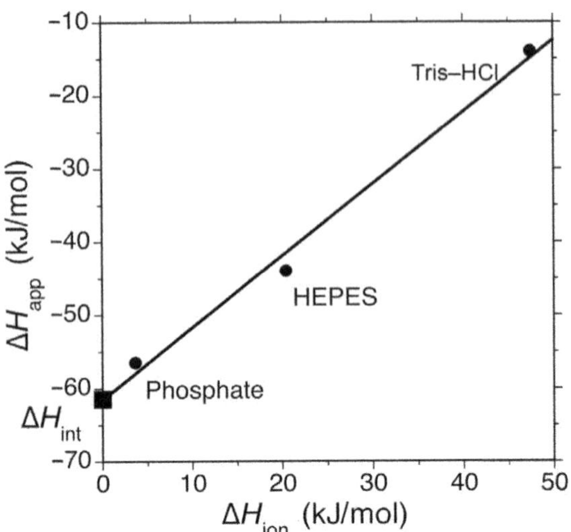

Figure 4 Determination of ΔH_{int} of urease reaction. Values of ΔH_{app} in different buffers (filled circles) were determined by performing the M1 experiments in 20 m*M* HEPES ($\Delta H_{ion}=20.3$ kJ/mol), 20 m*M* Tris–HCl ($\Delta H_{ion}=47.5$ kJ/mol), and 20 m*M* phosphate ($\Delta H_{ion}=3.6$ kJ/mol) (Goldberg, Kishore, & Lennen, 2002). Linear regression analysis (solid line) of the data allowed determining ΔH_{int} of the reaction (filled square), as well as the number of protons exchanged during urea turnover.

NADH + H^+ NAD$^+$ O$^-$ O$^-$ LDH OH (9)

The enzymatic activity of LDH-A was tested by ITC. First, the value of $\Delta H_{app}=-198$ kJ/mol was determined in a single-injection experiment (Fig. 5A), according to Eq. (6), integrating the thermal power resulting by injecting the substrate (pyruvate) into the LDH-A solution (in 50 m*M* phosphate buffer, pH 7.5 containing 250 μ*M* NADH), and allowing the reaction to proceed to completion. Subsequently, the thermal power registered in a multiple-injection experiment (Fig. 5B) was converted to the reaction rate using Eq. (4) (Fig. 5C). Fit of the obtained data to the Michaelis–Menten Eq. (2) provided the kinetic parameters for LDH-A as $k_{cat}=100\ s^{-1}$ and $K_M=0.16$ m*M*, which are similar to the ones measured previously by spectroscopic methods, following the decrease of the optical

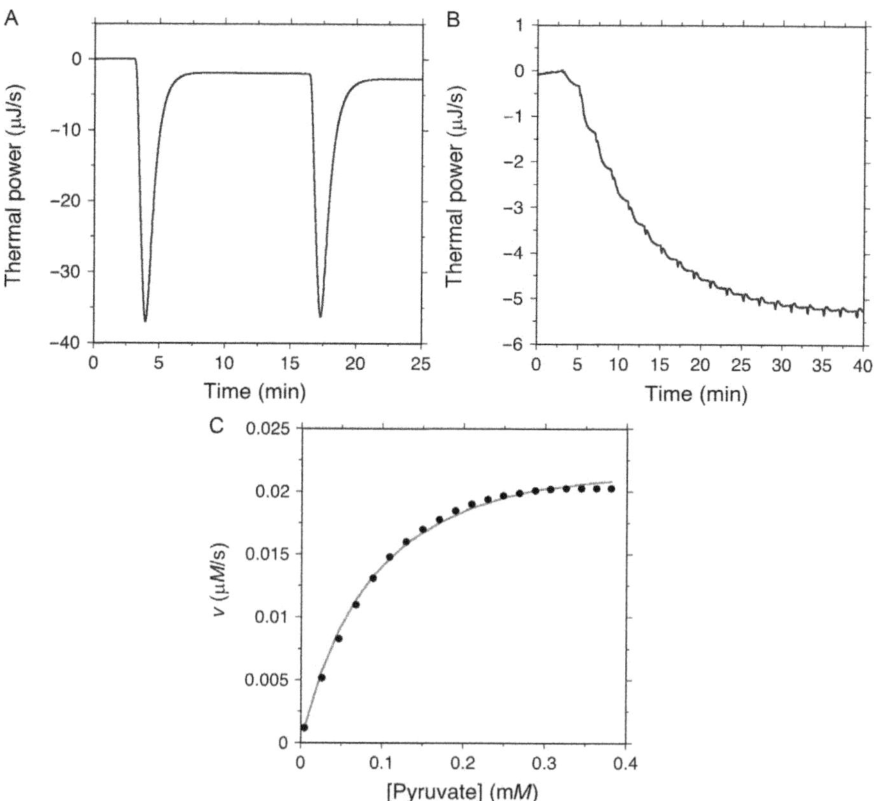

Figure 5 L-Lactate formation by lactate dehydrogenase-A (LDH-A) determined by ITC. Solution of human LDH-A was purchased from Lee Biosolutions (St. Louis, MO, USA) and diluted in 50 m*M* phosphate buffer, pH 7.5, to the final concentration. The concentration of the enzyme in the stock solution was 8.5 mg/mL according to the manifacturer's instructions, and the molar mass of the tetramer is 135 kDa. (A) Thermal power observed in a single-injection experiment, performed at 25 °C, with an injection of 80 μL (repeated twice to verify reproducibility) of 300 μ*M* pyruvate (in the injection syringe) into 28 n*M* LDH (in the sample cell), in the presence of 250 μ*M* of NADH cofactor. A spacing of 800 s was applied between injections. (B) Thermal power observed in the multiple-injection experiment, obtained by titrating 3 m*M* pyruvate (20 × 10-μL injections) into 280 p*M* LDH-A, in the presence of 250 μ*M* NADH. A spacing of 120 s was applied between injections, in order to allow the system to reach and maintain the steady-state level. dQ/dt at different substrate concentrations is obtained as displacement of the baseline after each addition. (C) The baseline displacement in panel (B) was converted to the reaction rate using Eq. (4). Fit of the obtained data (filled circles) was performed using Eq. (2) and is represented as a solid red (gray in the print version) line.

activity of NADH (Eszes, Sessions, Clarke, Moreton, & Holbrook, 1996; Pettit, Nealon, & Henderson, 1981).

6.3 Enzyme Inhibition Studies

Enzyme inhibition by small molecules serves as a major control mechanism of biological systems. This is often used as a strategy for drug discovery and can provide insight into the mechanism of enzyme activity, for example, by identifying residues critical for catalysis. Enzyme inhibition can be reversible or irreversible. The latter occurs when the inhibitor binds tightly to the enzyme, often covalently, and dissociates very slowly from the target. The reversible inhibition, on the other hand, is characterized by a rapid dissociation of the enzyme–inhibitor complex. Different types of reversible inhibitions can take place as described below (Cornish-Bowden, 2014; Segel, 1993):

1. *Competitive inhibition*: the inhibitor binds to the enzyme independently of the presence of substrate, in a reversible equilibrium of the type $E + I \rightleftarrows EI$. The competitive inhibitor resembles the substrate and binds to the active site of the enzyme, preventing substrate binding thus diminishing the proportion of enzyme molecules bound to a substrate. Under these conditions, the substrate can outcompete the inhibitor, and inhibition can be relieved at high substrate concentrations. For the competitive inhibition, the dissociation constant is expressed as K_{ic}, and the rate of enzymatic reaction is given in Eq. (10):

$$v = \frac{v_{max}[S]}{[S] + K_M \times \left(1 + \frac{[I]}{K_{ic}}\right)} \quad (10)$$

2. *Uncompetitive inhibition*: the inhibitor binds to the enzyme in the presence of substrate in an inhibition equilibrium of the type $ES + I \rightleftarrows ESI$. An uncompetitive inhibitor acts by decreasing the turnover number (corresponding to $v_{max}/[E]$) while maintaining the proportion of enzyme molecules that are bound to substrate. Therefore, uncompetitive inhibition cannot be overcome by increasing the substrate concentration.

   The dissociation constant of uncompetitive inhibition is expressed as K_{iu} and the rate of enzymatic reaction is given by Eq. (11):

$$v = \frac{v_{\max}[S]}{K_M + [S] \times \left(1 + \frac{[I]}{K_{iu}}\right)} \tag{11}$$

3. *Mixed inhibition*: the inhibitor binds to the enzyme both in the presence and absence of substrate, so that both equilibria above are operating. In this case, the rate of enzymatic reaction is given by Eq. (12):

$$v = \frac{v_{\max}[S]}{[S]\left(1 + \frac{[I]}{K_{iu}}\right) + K_M\left(1 + \frac{[I]}{K_{ic}}\right)} \tag{12}$$

Once the kinetics of enzymatic reaction is determined by calorimetry, the reaction can be repeated in the presence of different types and concentrations of enzyme inhibitors in the sample cell, in order to measure the different inhibition constants. In particular, the kinetic parameters can be obtained from a general fit of the experimental reaction rates obtained by calorimetry at different inhibitor concentrations, using the general inhibition (Eq. 12) and nonlinear regression analysis. Alternatively, it is possible to perform the experiments at variable inhibitor concentrations while maintaining the substrate concentration constant. However, the multiple-injection scheme is built up to vary the substrate concentration upon the subsequent injections, therefore, maintaining this experimental scheme, with the enzyme–inhibitor complex preformed in the sample cell and the substrate added in multiple injections from the syringe, is more convenient to obtain comparable data. Competitive (K_{ic}) and uncompetitive (K_{iu}) inhibition constants can be determined, thus allowing us to distinguish among different types of inhibition. In conclusion, this method can be used as a fast and economic way to test a large number of enzyme inhibitors for pharmaceutical and industrial applications, such as in drug discovery and optimization.

6.3.1 Fluoride-Induced Urease Inhibition

An example of the use of calorimetry to determine inhibition constants and mechanism is the kinetic study of the fluoride-induced inhibition of SPU (Benini et al., 2014). Urease inhibitors have both medical and agricultural applications because the high activity of urease triggers an overall increase in pH, which causes negative consequences on human and animal health, as well as in plant crop production (Maroney & Ciurli, 2014). Fluoride is a known inhibitor for urease activity (Saboury & Moosavi-Movahedi, 1997). The activity of SPU was measured using the multiple-injection experiment at fluoride concentrations ranging from 0 to 800 μ*M* and at pH values of 6.5 (Fig. 6A), 7.0 (Fig. 6B), and 8.0 (Fig. 6C) (Benini et al., 2014). The competitive (K_{ic}) and uncompetitive (K_{iu}) inhibition constants were obtained at the diverse pHs by regression analysis of Eq. (12). For this enzymatic inhibition, K_{iu} is generally smaller than K_{ic}, indicating a predominance of the uncompetitive inhibition within the explored pH range (Fig. 6). In addition, an increase of the uncompetitive inhibition mechanism with pH is reported, and the competitive inhibition becomes negligible at pH 8.0 (Fig. 6). The analysis of the inhibition data, coupled to the structure of the enzyme–inhibitor complex that revealed the presence of two fluoride anions coordinated to the Ni(II) ions in the active site, allowed inferring the occurrence of a direct interaction of fluoride with the metal ions in the active site: one fluoride binds competitively to the urea-binding site, while the other fluoride uncompetitively substitutes the Ni(II)-bridging hydroxide, which is the key nucleophile of the hydrolysis reaction.

6.3.2 Oxamate-Induced Inhibition of LDH

Research on LDH inhibition has been quite active in the last years, as this approach could constitute a valid therapeutic strategy for diseases so different as malaria and cancer (Granchi, Bertini, Macchia, & Minutolo, 2010). Indeed, the isoform expressed by the malaria parasite *Plasmodium falciparum* is a key enzyme for energy generation of this organism (Gomez et al., 1997). In addition, LDH-A expression is constantly upregulated in tumors and is a critical factor in tumorigenesis. Therefore, inhibition of LDH-A may constitute a way to interfere with tumor proliferation (Levine & Puzio-Kuter, 2010). Oxamate is considered to be the LDH-A reference inhibitor, and it is competitive with the pyruvate substrate. Oxamate-induced LDH inhibition was measured using the multiple-injection experiment at oxamate concentrations ranging from 0 to 100 μ*M* (Fig. 7), under the same experimental conditions described in Section 6.2. Data were fitted by regression analysis with Eq. (12). This enzymatic inhibition results fully competitive, with $K_i = 80$ μ*M* (Fig. 7), coherent to that measured previously (Read, Winter, Eszes, Sessions, & Brady, 2001).

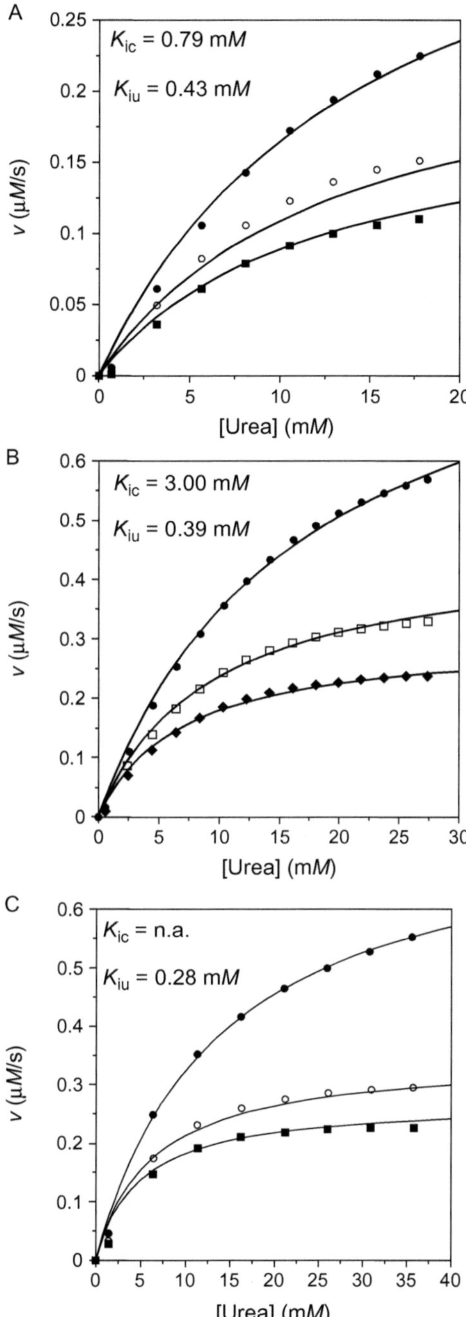

Figure 6 Michaelis–Menten plots of SPU activity at pH 6.5 (A), pH 7.0 (B), and pH 8.0 (C), in the presence of 0 μ*M* (filled circles), 300 μ*M* (empty circles), 400 μ*M* (empty squares), 500 μ*M* (filled squares), 800 μ*M* (solid diamonds) of fluoride inhibitor. The lines represent calculated nonlinear regression analysis fits carried out using Eq. (12). The calculated inhibition constants are indicated.

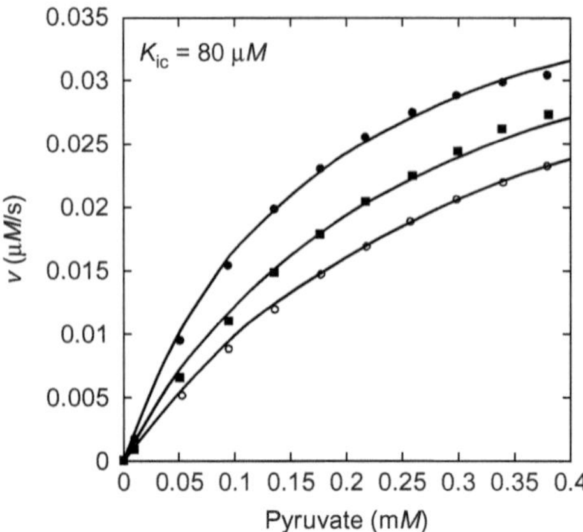

Figure 7 Michaelis–Menten plots of LDH-A activity at oxamate concentration of 0 μ*M* (filled circles), 50 μ*M* (filled squares) and 100 μ*M* (empty circles). The lines represent calculated nonlinear regression analysis fits carried out using Eq. (12). The calculated competitive inhibition constant is indicated.

7. CONCLUSIONS

In addition to its classical application to study binding equilibria, ITC provides a reliable and fast method to characterize enzymatic reactions in solution, using the heat of reaction as a probe and without requiring system modification or labeling. The analysis of the process is generally performed in two different experiments, which provide the total molar enthalpy of the reaction and the heat change over time at different substrate concentrations. Alternatively, a single-injection scheme can be applied, when a single-substrate injection can provide concentrations higher than the K_M of the reaction. Therefore, this method can be used as a fast and economic way to test enzyme inhibitors, with potentially useful pharmaceutical and industrial applications, such as in drug screening and development.

REFERENCES

Baumann, M. J., Murphy, L., Lei, N., Krogh, K. B., Borch, K., & Westh, P. (2011). Advantages of isothermal titration calorimetry for xylanase kinetics in comparison to chemical-reducing-end assays. *Analytical Biochemistry*, *410*(1), 19–26. http://dx.doi.org/10.1016/j.ab.2010.11.001.

Benini, S., Cianci, M., Mazzei, L., & Ciurli, S. (2014). Fluoride inhibition of *Sporosarcina pasteurii* urease: Structure and thermodynamics. *Journal of Biological Inorganic Chemistry*, *19*(8), 1243–1261. http://dx.doi.org/10.1007/s00775-014-1182-x.

Bianconi, M. L. (2007). Calorimetry of enzyme-catalyzed reactions. *Biophysical Chemistry*, *126*(1–3), 59–64. http://dx.doi.org/10.1016/j.bpc.2006.05.017.

Callahan, B. P., Yuan, Y., & Wolfenden, R. (2005). The burden borne by urease. *Journal of the American Chemical Society*, *127*(31), 10828–10829. http://dx.doi.org/10.1021/ja0525399.

Cornish-Bowden, A. (2014). Current IUBMB recommendations on enzyme nomenclature and kinetics. *Perspectives on Science*, *1*, 74–87.

Demarse, N. A., Killian, M. C., Hansen, L. D., & Quinn, C. F. (2013). Determining enzyme kinetics via isothermal titration calorimetry. *Methods in Molecular Biology*, *978*, 21–30. http://dx.doi.org/10.1007/978-1-62703-293-3_2.

Duff, M. R., Jr., Grubbs, J., & Howell, E. E. (2011). Isothermal titration calorimetry for measuring macromolecule-ligand affinity. *Journal of Visualized Experiments*, *55*, 2796. http://dx.doi.org/10.3791/2796.

Eszes, C. M., Sessions, R. B., Clarke, A. R., Moreton, K. M., & Holbrook, J. J. (1996). Removal of substrate inhibition in a lactate dehydrogenase from human muscle by a single residue change. *FEBS Letters*, *399*(3), 193–197.

Everse, J., & Kaplan, N. O. (1973). Lactate dehydrogenases: Structure and function. *Advances in Enzymology and Related Areas of Molecular Biology*, *37*, 61–133.

Freyer, M. W., & Lewis, E. A. (2008). Isothermal titration calorimetry: Experimental design, data analysis, and probing macromolecule/ligand binding and kinetic interactions. *Methods in Cell Biology*, *84*, 79–113. http://dx.doi.org/10.1016/S0091-679X(07)84004-0.

Fukada, H., & Takahashi, K. (1998). Enthalpy and heat capacity changes for the proton dissociation of various buffer components in 0.1 M potassium chloride. *Proteins*, *33*(2), 159–166.

Ghai, R., Falconer, R. J., & Collins, B. M. (2012). Applications of isothermal titration calorimetry in pure and applied research—Survey of the literature from 2010. *Journal of Molecular Recognition*, *25*(1), 32–52. http://dx.doi.org/10.1002/jmr.1167.

Goldberg, R., Kishore, N., & Lennen, R. (2002). Thermodynamic quantities for the ionization reactions of buffers. *Journal of Physical and Chemical Reference Data*, *31*(2), 231–370.

Gomez, M. S., Piper, R. C., Hunsaker, L. A., Royer, R. E., Deck, L. M., Makler, M. T., et al. (1997). Substrate and cofactor specificity and selective inhibition of lactate dehydrogenase from the malarial parasite *P. falciparum*. *Molecular and Biochemical Parasitology*, *90*(1), 235–246.

Granchi, C., Bertini, S., Macchia, M., & Minutolo, F. (2010). Inhibitors of lactate dehydrogenase isoforms and their therapeutic potentials. *Current Medicinal Chemistry*, *17*(7), 672–697.

Hausinger, R. P., & Karplus, P. A. (2001). Urease. In A. Messerschmidt, R. Huber, T. Poulos, & K. Wieghardt (Eds.), *Handbook of metalloproteins* (pp. 867–879). Chichester, UK: John Wiley & Sons.

Krajewska, B., van Eldik, R., & Brindell, M. (2012). Temperature- and pressure-dependent stopped-flow kinetic studies of jack bean urease. Implications for the catalytic mechanism. *Journal of Biological Inorganic Chemistry*, *17*(7), 1123–1134. http://dx.doi.org/10.1007/s00775-012-0926-8.

Ladbury, J. E. (2004). Application of isothermal titration calorimetry in the biological sciences: Things are heating up! *Biotechniques*, *37*(6), 885–887.

Leavitt, S., & Freire, E. (2001). Direct measurement of protein binding energetics by isothermal titration calorimetry. *Current Opinion in Structural Biology*, *11*(5), 560–566.

Levine, A. J., & Puzio-Kuter, A. M. (2010). The control of the metabolic switch in cancers by oncogenes and tumor suppressor genes. *Science*, *330*(6009), 1340–1344. http://dx.doi.org/10.1126/science.1193494.

Lonhienne, T., Baise, E., Feller, G., Bouriotis, V., & Gerday, C. (2001). Enzyme activity determination on macromolecular substrates by isothermal titration calorimetry: Application to mesophilic and psychrophilic chitinases. *Biochimica et Biophysica Acta*, *1545*(1–2), 349–356.

Maroney, M. J., & Ciurli, S. (2014). Nonredox nickel enzymes. *Chemical Reviews*, *114*(8), 4206–4228. http://dx.doi.org/10.1021/cr4004488.

Mazzei, L., Ciurli, S., & Zambelli, B. (2014). Hot biological catalysis: Isothermal titration calorimetry to characterize enzymatic reactions. *Journal of Visualized Experiments*, *86*, e51487. http://dx.doi.org/10.3791/51487.

Michaelis, L., & Menten, M. (1913). Die kinetik der invertinwirkung. *Biochemische Zeitschrift*, *49*, 333–369.

Noske, R., Cornelius, F., & Clarke, R. J. (2010). Investigation of the enzymatic activity of the Na^+, K^+-ATPase via isothermal titration microcalorimetry. *Biochimica et Biophysica Acta*, *1797*(8), 1540–1545. http://dx.doi.org/10.1016/j.bbabio.2010.03.021.

Olsen, S. N. (2006). Applications of isothermal titration calorimetry to measure enzyme kinetics and activity in complex solutions. *Thermochimica Acta*, *448*(1), 12–18. http://dx.doi.org/10.1016/J.Tca.2006.06.019.

Pettit, S. M., Nealon, D. A., & Henderson, A. R. (1981). Purification of lactate dehydrogenase isoenzyme-5 from human liver. *Clinical Chemistry*, *27*(1), 88–93.

Read, J. A., Winter, V. J., Eszes, C. M., Sessions, R. B., & Brady, R. L. (2001). Structural basis for altered activity of M- and H-isozyme forms of human lactate dehydrogenase. *Proteins*, *43*(2), 175–185.

Saboury, A. A., & Moosavi-Movahedi, A. A. (1997). A simple novel method for determination of an inhibition constant by isothermal titration microcalorimetry. The effect of fluoride ion on urease. *Journal of Enzyme Inhibition*, *12*(4), 273–279.

Segel, I. H. (1993). *Enzyme kinetics: Behavior and analysis of rapid equilibrium and steady-state enzyme systems*. New York: John Wiley & Sons, Inc.

Seibert, E., & Tracy, T. S. (2014). Fundamentals of enzyme kinetics. *Methods in Molecular Biology*, *1113*, 9–22 (2014/02/14 ed.).

Sica, L., Gilli, R., Briand, C., & Sari, J. C. (1987). A flow microcalorimetric method for enzyme activity measurements: Application to dihydrofolate reductase. *Analytical Biochemistry*, *165*(2), 341–348.

Todd, M. J., & Gomez, J. (2001). Enzyme kinetics determined using calorimetry: A general assay for enzyme activity? *Analytical Biochemistry*, *296*(2), 179–187. http://dx.doi.org/10.1006/abio.2001.5218.

Transtrum, M. K., Hansen, L. D., & Quinn, C. (2015). Enzyme kinetics determined by single-injection isothermal titration calorimetry. *Methods*, *76*, 194–200. http://dx.doi.org/10.1016/j.ymeth.2014.12.003.

Zambelli, B., Bellucci, M., Danielli, A., Scarlato, V., & Ciurli, S. (2007). The Ni^{2+} binding properties of Helicobacter pylori NikR. *Chemical Communications*, *35*, 3649–3651. http://dx.doi.org/10.1039/b706025d.

Zambelli, B., Danielli, A., Romagnoli, S., Neyroz, P., Ciurli, S., & Scarlato, V. (2008). High-affinity Ni^{2+} binding selectively promotes binding of *Helicobacter pylori* NikR to its target urease promoter. *Journal of Molecular Biology*, *383*(5), 1129–1143. http://dx.doi.org/10.1016/j.jmb.2008.08.066.

Zambelli, B., Musiani, F., Benini, S., & Ciurli, S. (2011). Chemistry of Ni^{2+} in urease: Sensing, trafficking, and catalysis. *Accounts of Chemical Research*, *44*(7), 520–530. http://dx.doi.org/10.1021/ar200041k.

Avoiding Buffer Interference in ITC Experiments: A Case Study from the Analysis of Entropy-Driven Reactions of Glucose-6-Phosphate Dehydrogenase

M. Lucia Bianconi[1]
Instituto de Bioquímica Médica Leopoldo de Meis, Universidade Federal do Rio de Janeiro, Rio de Janeiro, Brazil
[1]Corresponding author: e-mail address: bianconi@bioqmed.ufrj.br

Contents

Abstract

Isothermal titration calorimetry (ITC) is a label-free technique that allows the direct determination of the heat absorbed or released in a reaction. Frequently used to determining binding parameters in biomolecular interactions, it is very useful to address enzyme-catalyzed reactions as both kinetic and thermodynamic parameters can be

Methods in Enzymology, Volume 567
ISSN 0076-6879
http://dx.doi.org/10.1016/bs.mie.2015.08.025

obtained. Since calorimetry measures the total heat effects of a reaction, it is important to consider the contribution of the heat of protonation/deprotonation that is possibly taking place. Here, we show a case study of the reaction catalyzed by the glucose-6-phosphate dehydrogenase (G6PD) from *Leuconostoc mesenteroides*. This enzyme is able to use either NAD^+ or $NADP^+$ as a cofactor. The reactions were done in five buffers of different enthalpy of protonation. Depending on the buffer used, the observed calorimetric enthalpy (ΔH^{cal}) of the reaction varied from −22.93 kJ/mol (Tris) to 19.37 kJ/mol (phosphate) for the $NADP^+$-linked reaction, and −11.67 kJ/mol (Tris) to 7.32 kcal/mol or 30.63 kJ/mol (phosphate) for the NAD^+ reaction. We will use this system as an example of how to extract proton-independent reaction enthalpies from kinetic data to ensure that the reported accurately represent the intrinsic heat of reaction.

1. INTRODUCTION

Isothermal titration calorimetry (ITC) is a very powerful technique for the thermodynamic study of intermolecular interactions, such as ligand binding to proteins or partitioning of solutes into membranes, as well as catalysis. Isothermal calorimetry has been used to study chemical reactions since the late 1930s (Sturtevant, 1937), but the application of ITC to biological systems was first described by Wiseman, Williston, Brandts, and Lin (1989).

Although frequently related to the study of biomolecular interactions (Ladbury & Chowdhry, 1996; Leavitt & Freire, 2001; Limo & Perry, 2015; Malesinski, Tsvetkov, Kruczynski, Peyrot, & Devred, 2015; Velazquez-Campoy, Leavitt, & Freire, 2004; Wafer, Streicher, McCallum, & Makhatadze, 2012), ITC is a very useful technique to determining both thermodynamic and kinetic properties of enzymes (Bianconi, 2003, 2007; Cai, Cao, & Lai, 2001; Olsen, 2006; Todd & Gomez, 2001).

Enzyme activity is usually determined by measuring the amount of product formed or substrate consumed in a reaction by spectrophotometric methods, such as UV–visible spectrophotometry and fluorescence that, in many cases, require labeling, which, in turn, introduces undesirable artifacts. On the contrary, ITC allows the direct determination of the heat absorbed or released in a reaction, avoiding those artifacts.

In the case of enzymes, reaction kinetics can be directly studied even if reactants and/or products do not present optical activity. Furthermore, turbid solutions do not represent a limitation to the method since the measurements are based in the heat developed during the reaction. In Chapter 9, "ITC to characterize enzyme reactions", Ciurli gives a description of the determination of enzyme kinetics parameters by ITC.

Briefly, ITC is a technique based on the measurement of the heat evolved in a reaction started by the injection of a small amount of a reactant into a cell containing the other reactant(s). The heat measured in the ITC is actually the sum of all heat effects taking place, thus the observed calorimetric enthalpy (ΔH^{cal}) includes the net heat of the reaction plus the heat due to all other events such as conformational changes, dilution, changes in solvation, and/or extent of burial of polar/apolar surface area as well as protonation/deprotonation effects. If one (or more) of these effects is (are) neglected, one certainly will reach misleading conclusions.

1.1 Protonation/Deprotonation Heat Effects

Many binding or kinetic reactions proceed with the exchange of protons. ITC can be a powerful technique to measure any linked protonation effects by repeating the measurement in the presence of different buffers. Baker and Murphy (1996) described a theoretical approach to evaluate proton linkage in protein binding reactions by ITC. The methodology is based on the determination of the enthalpies of protonation and pK_a changes upon binding (Baker & Murphy, 1996).

The same effect can be observed in enzyme-catalyzed reactions as previously shown for the yeast hexokinase (HK) isozymes (Bianconi, 2003). This enzyme was studied in five buffer systems with enthalpies of protonation varying from −5.1 kJ/mol (phosphate) to −48.2 kJ/mol (Tris:HCl). The observed enthalpy of the HK reaction (ΔH^{cal}) varies according to the heat of buffer protonation: for the PI isozyme, ΔH^{cal} varies from −26.4 kJ/mol in phosphate buffer to −69.5 kJ/mol (Tris:HCl) and for the PII HK, from −14.8 kJ/mol (phosphate) to −65.0 kJ/mol (Tris). This shows how important it is to take the heat of buffer protonation into account when determining the reaction enthalpy (Bianconi, 2003).

If proton release/uptake is involved, in order to calculate the intrinsic enthalpy of reaction (ΔH^{R}), one must know the enthalpy of protonation (or ionization) of the buffer used. In a very extensive work, using a flow calorimeter, Fukada and Takahashi (1998) studied the heat of deprotonation of 18 biological buffers in 0.1 *M* KCl. The experiments, performed between 5 and 45 °C, allowed the determination of the enthalpy and the heat capacity (ΔCp) changes for the deprotonation of many common buffers. Nevertheless, it is interesting to determine the heat of protonation of the buffer used in a particular reaction medium as the experimental conditions can affect ΔH^{P} (see Section 2.1).

The linear dependence between the observed enthalpy (ΔH^{cal}) and the buffer protonation enthalpy (ΔH^{P}) allows the calculation of ΔH^{R} and the number of protons (n) released in the reaction by Eq. (1):

$$\Delta H^{cal} = \Delta H^{R} + n\Delta H^{P} \tag{1}$$

The same equation with opposite sign is used in the case of buffer deprotonation, i.e., when there is uptake of proton by the reaction, as $\Delta H^{P} = -\Delta H^{D}$.

Typically, ΔH^{cal} is determined in three or more buffer systems; thus, Eq. (1) can be used to determine ΔH^{R} from the intercept and the number of protons, from the slope.

In this chapter, the buffer protonation effect in ΔH^{cal}, as well as the determination of ΔH^{R} and the number of protons, will be exemplified with the reaction catalyzed by the glucose-6-phosphate dehydrogenase (G6PD) from *Leuconostoc mesenteroides*.

1.2 G6PD from *L. mesenteroides*

G6PD (EC 1.1.1.49) is the first enzyme of the anabolic pentose phosphate pathway that catalyzes the oxidation of glucose 6-phosphate (G6P) into 6-phosphoglucoo-δ-lactone by reducing $NADP^+$ or NAD^+. In mammals, G6PD is specific for $NADP^+$ and it is activated by various extracellular oxidants that lead to the decrease in NADPH level (Stanton, 2012). There is no evidence for *in vivo* regulation of G6PD by the NADPH/NADP ratio, although this can be observed *in vitro* (Holten, Procsal, & Chang, 1976; Stanton, 2012).

The G6PD from *L. mesenteroides* is a very interesting enzyme that is able to use either NAD^+ or $NADP^+$ as coenzyme, depending on the metabolic state of the bacterium. $NADP^+$ is used in biosynthetic pathways while NAD^+ participates in the catabolic metabolism when G6P concentrations are high (Rowland, Basak, Gover, Levy, & Adams, 1994). This unusual dual coenzyme specificity is important as *L. mesenteroides* lacks a complete glycolytic pathway and uses the heterolactic (phosphoketolase) pathway to metabolize sugars (Cogan & Jordan, 1994).

The G6PD from *L. mesenteroides* is a homodimer of 54 kDa per subunit. Each subunit presents two domains: a large $\alpha+\beta$ domain at the interface of the dimer and a coenzyme classic domain with the structural $\beta\alpha\beta$ motif characteristic of Rossman fold (Adams, Basak, Gover, Rowland, & Levy, 1993). Although NAD^+ and $NADP^+$ bind to the same site of the enzyme, the kinetic mechanism of the reaction is quite different depending

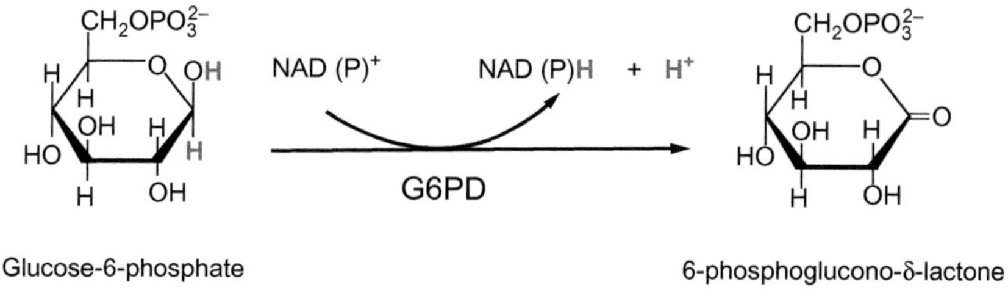

NAD (P)$^+$ → NAD (P)H + H$^+$

G6PD

Glucose-6-phosphate → 6-phosphoglucono-δ-lactone

Scheme 1 Reaction catalyzed by glucose-6-phosphate (G6PD) from *Leuconostoc mesenteroides*. NAD(P)$^+$ indicates that the enzyme is able to use either NAD$^+$ and NADP$^+$ as coenzyme.

on the coenzyme. The NADP$^+$-linked reaction is ordered and sequential with the coenzyme binding first and presenting a strong NADPH competitive inhibition and noncompetitive inhibition toward G6P (Levy, 1989; Levy, Christoff, Ingulli, & Ho, 1983). On the other hand, the NAD$^+$-linked reaction has a random-order mechanism with noncompetitive inhibition as regards to both NADH and G6P (Levy, 1989; Levy et al., 1983).

G6PD catalyzes the reduction of NAD(P)$^+$ to NAD(P)H through the oxidation of G6P into 6-phosphoglucono-δ-lactone in a reaction that occurs with a release of a proton (Scheme 1).

As mentioned before, in calorimetry, release and/or uptake of protons lead to interferences that mask the heat effect due to the reaction only. Since ΔH^{cal} comprises all heat effects taking place, when studying an enzyme-catalyzed reaction such as that of the G6PD, one should be aware that the proton released in the reaction protonates the buffer, and depending on the buffer used, different values of ΔH^{cal} are determined (see Bianconi, 2003). Here, we will discuss the heat effect of the buffer protonation and the way to deal with that in order to calculate the reaction enthalpy.

2. CORRECTING FOR PROTONATION EFFECTS

2.1 Determination of Heat of Buffer Protonation

As described before, in order to correct the calorimetric enthalpy for protonation effects, it is necessary to determine the heat of protonation/deprotonation of the buffer used. Usually, ΔH^{cal} is determined in, at least, three buffers with different enthalpies of protonation (ΔH^{P}). Then, Eq. (1) is applied, correlating the experimental ΔH^{cal} with ΔH^{P} to obtain the enthalpy of the reaction (intercept). This correlation also gives the number of protons involved in the reaction.

Depending on the buffer and the temperature used in the experiments, the interference of heat of protonation/deprotonation effects could mask the actual enthalpy of the reaction. Although Fukada and Takahashi (1998) published a comprehensive work determining the heat of deprotonation of 18 biological buffers in 0.1 *M* KCl, it is important to determine ΔH^{P} of the buffer system if working under different conditions. It is important to notice that for some buffers, heat capacity (ΔCp) changes for the protonation/deprotonation effect can be very large (Fukada & Takahashi, 1998). Therefore, the experiment described here must be done across the desired temperature range chosen for the determination of the reaction enthalpy.

The enthalpy of buffer protonation (ΔH^{P}) was determined from thermograms (heat flux, given in μcal/s as a function of time, given in s) obtained in a VP-ITC from MicroCal (Malvern Instruments, Northampton, MA) at 25 °C. The experiment consists of injection of 1 m*M* HCl into the calorimetric cell ($V = 1.422$ mL) loaded with the desired buffer or Milli-Q water. For purpose of this example, the same buffers chosen to study the G6PD reaction were used: 50 m*M* buffer at pH 7.6 (Tris:HCl, MOPS, imidazole, and HEPES) or 10 m*M* phosphate buffer, as follows for the VP-ITC:

1. Begin by degassing the buffers and Milli-Q water for ca. 5 min.
2. Load the reference cell with Milli-Q water.
3. Load the sample cell with buffer.
4. Load the injection syringe with 0.1 m*M* HCl and carefully insert it into the sample cell access tube.
5. Choose the desired parameters: the setup used here was for four identical injections of 10 μL each with 20 s duration (time for each injection) and 300 s spacing (time between injections). The reference power was 15 μcal/s (which corresponds to 62.8 kJ/s), and the stirring speed was 310 rpm. The initial delay was 60 s.
6. Repeat this procedure with Milli-Q water in the sample cell, in order to determine the heat of HCl dilution, which is significative.
7. After doing the experiment with all the buffers and with water, data analysis was done with the Origin 5.0 software provided by MicroCal. First, the heat of HCl dilution obtained from four separated experiments (raw data) was averaged. The same was done with the four independent experiments with each buffer. The heat of dilution was, then, subtracted from the raw data obtained with the buffers. After baseline correction, the heat of buffer protonation was calculated by integrating each peak, giving the total heat for each injection (Q_T). The enthalpy of buffer protonation (ΔH^{P}) was calculated by dividing Q_T by the HCl concentration in the cell after the injection (see Fig. 1 and Table 1).

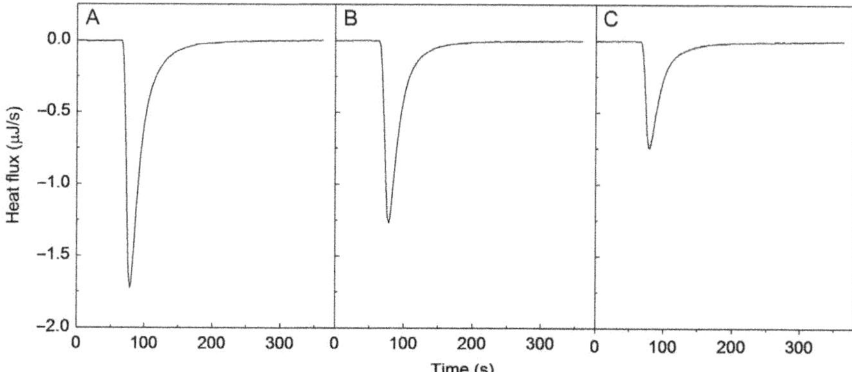

Figure 1 Determination of the heat of buffer protonation. Raw data showing the heat flux (μJ/s) as a function of time (in s) measured at 25 °C for the injection of 10 μL of 1.0 m*M* HCl into (A) Tris:HCl, (B) imidazole, and (C) MOPS, after subtraction of the heat of HCl dilution, which accounted for −5.9 ± 0.2 kJ/mol of the total heat. All buffers were 50 m*M*, pH 7.6.

Table 1 Enthalpy of Buffer Protonation at 25 °C

Buffer	**pK_a**	**ΔH^P (kJ/mol)**
Phosphate	6.81	−5.22 ± 0.1[a]
		−5.12 ± 0.03[b]
HEPES	7.45	−21.1 ± 0.1[a]
		−21.01 ± 0.07[b]
MOPS	7.09	−22.2 ± 0.6[a]
		−21.82 ± 0.03[b]
Imidazole	6.95	−35.31 ± 0.1[a]
		−36.59 ± 0.06[b]
Tris:HCl	8.10	−48.5 ± 0.9[a]
		−47.49[c]
		−47.57[d]

[a]This work: 50 m*M* buffer, pH 7.6.
[b]Fukada and Takahashi (1998): the original data refers to the enthalpy of deprotonation, which has the same value, but opposite sign, of the protonation enthalpy. The experiments were done in 10 m*M* buffer with 0.1 *M* KCl.
[c]Datta, Grzybowski, and Weston (1963).
[d]Ramette, Culberson, and Bates (1977): with 30‰ synthetic seawater prepared in Tris:HCl and containing NaCl, KCl, Na_2SO_4, $CaCl_2$, and $MgCl_2$.

Care is needed when loading the cells and the injection syringe to avoid bubbles that can interfere in the quality of the experiment—bubbles cause a high (and irregular) noise level that can distort the peaks and/or result in unstable baselines.

Figure 1 shows the raw data obtained for three of the five buffers used here, after subtraction of the heat of HCl dilution. Each experiment was repeated four times, and ΔH^P obtained here for five buffers are shown in Table 1 and are similar to those determined by Fukada and Takahashi (1998) and Datta et al. (1963). Increasing the ionic strength by adding KCl (Fukada & Takahashi, 1998) or synthetic seawater at 30–45‰ salinity (Ramette et al., 1977) did not change ΔH^P.

2.2 Determination of Reaction Enthalpy and Number of Protons Release/Uptake by ITC

In order to calculate the reaction enthalpy (ΔH^R) and the number of protons released/uptake in an enzyme-catalyzed reaction, ΔH^{cal} should be determined in different buffers, as follows:

1. After proper degassing, load the sample cell with the reaction medium containing the reactant(s). It is important that the concentration of the reactant is such that it will be completely consumed during the reaction. If more than one reactant is present in the reaction medium, only one should be in a limited concentration and all the others should be in excess.
2. Load the injection syringe with the enzyme solution. To avoid denaturation, degasification is not recommended for the enzyme solution. Therefore, the enzyme should be diluted in degassed Milli-Q water.
3. Choose the desired parameters: the setup used for the G6PD catalyzed reaction was for two identical injections of 10 μL each with 20 s duration (time for each injection). The time between the injections was set at 1200 s for the first injection and 300 s for the second one, and the stirring speed was 310 rpm. The initial delay was 120 s. Note that in the case studied here, 20 min was enough to have the return of the heat effect to the baseline level, indicating the end of the reaction. The second injection is interesting to determine the heat of the enzyme dilution, which was negligible in the case studied here.
4. Integrate the peaks to determine the total heat (Q_T), subtract the heat of the second injection from Q_T obtained in the first one, and divide by the amount of product formed, which is the same as the concentration of the reactant used in limited amount in the reaction medium. This will give

the calorimetric enthalpy (ΔH^{cal}) of the reaction. ΔH^{cal} corresponds to the sum of all heat effects taking place in the reaction. In the case where more than one reactant is present, it is interesting to check if ΔH^{cal} is the same when one or the other is used as a limiting reactant.

5. According to Eq. (1), the plot of ΔH^{cal} determined in buffer systems with different protonation enthalpies as a function of the buffer protonation/deprotonation enthalpy (ΔH^{P}) allows the determination of the reaction enthalpy (ΔH^{R}; intercept) as well as the number of protons involved in the reaction (slope). This will be exemplified in Fig. 4.

3. MATERIALS AND METHODS

3.1 Materials

G6PD (EC 1.1.1.49) from *L. mesenteroides* was purchased from Sigma-Aldrich (St. Louis, MO) as crystalline suspensions in ammonium sulfate. Prior to use, the ammonium sulfate was removed by centrifugation and dialysis as follows: small volumes of G6PD (usually 10 μL) were centrifuged at 15,000 × *g* for 10 min at 4 °C using a Sorvall Legend Micro 17R centrifuge. After resuspension of the pellet in the buffer of the reaction medium, the enzyme was dialyzed for 2 h against the same buffer in the microdialyzer system 100 from Pierce to eliminate traces of ammonium sulfate. NAD^+, $NADP^+$, G6P, and the buffers (Tris, Mops, HEPES, and imidazole) were also from Sigma-Aldrich and were used without further purification. Monosodium phosphate and disodium phosphate were from Merck Indústrias Químicas (Brazil). All reagents were analytical grade.

3.2 Determination of Reaction Enthalpy and Number of Protons Released by ITC

The observed calorimetric enthalpies (ΔH^{cal}) and the number of protons released of the G6PD reaction were determined from thermograms (heat flux as a function of time) obtained in a VP-ITC from MicroCal (Malvern Instruments, Northampton, MA) at 25 °C. The reactions started by the injection of 10 μL of G6PD solution into the sample cell (V = 1.422 mL) loaded with the reaction medium, which contained 5 m*M* G6P and 0.1 m*M* NAD^+ or 0.1 m*M* $NADP^+$. The reaction medium (pH 7.6) was prepared with five different buffers (Tris, imidazole, MOPS, HEPES, or phosphate). Buffer concentration was 50 m*M* with the exception of phosphate (10 m*M*). In the presence of KCl or NaCl (up to 200 m*M*), the concentration of all buffers was lowered to 10 m*M*. The

G6PD solution was prepared in the same buffer as the reaction medium in order to minimize the heat effects during the enzyme dilution. For the determination of the reaction enthalpy, the reaction was followed for 20–25 min to allow the return of the heat flux to the baseline level, which indicated the end of the reaction, i.e., when one of the substrates was completely consumed; a second injection of G6PD was done in order to determine the heat of dilution of the enzyme solution, which was subtracted from the total heat of reaction. The heat of G6PD dilution was negligible (0.3–0.7% of the total heat) in the reaction media containing buffer only and corresponded to 2–3% of the total heat when the salt (KCl or NaCl) concentration was 120 m*M* and higher. The duration of the 10 μL injection was 20 s, and the reactions were followed with stirring speed of 310 rpm. The thermograms were analyzed with the ORIGIN 5.0 software provided by MicroCal.

3.3 Determination of Reaction Rate by ITC

The determination of the initial rate of reaction was done in a similar way as described above for the enthalpy of the reaction but using 5 m*M* G6P and 1 m*M* $NAD(P)^+$ to assure a 10-min pseudo-first-order kinetics.

3.4 Spectrophotometric Determination of G6P

Spectrophotometric assays were run in parallel to the ITC experiments by measuring the concentration of NAD(P)H at 340 nm, which is proportional to the G6P concentration formed in the reaction. The same reaction medium was used in the ITC and in the spectroscopic experiments. For the latter, a falcon tube containing the same volume of reaction medium ($V = 1.422$ mL) was left in a water bath (HAAKE DC1 Circulating Bath) at 25.0 ± 0.3 °C. The reaction started at the same time as the reaction in the ITC cell by adding the same amount of G6PD as that injected into the cell, and an aliquot (700 μL) was used in the spectrophotometric assay in a UV/Vis spectrophotometer T-70 (PG Instruments) equipped with a constant temperature cell changer and Peltier module for temperature control. The molar extinction coefficient of NAD(P)H is 6220 M^{-1}/cm at 340 nm.

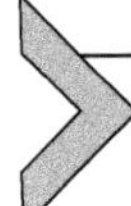

4. RESULTS AND DISCUSSION

4.1 Determination of the Enthalpy of the G6PD Reaction

The enthalpy of the reaction catalyzed by G6PD from *L. mesenteroides* was determined at 25 °C in five buffers with different heat of protonation.

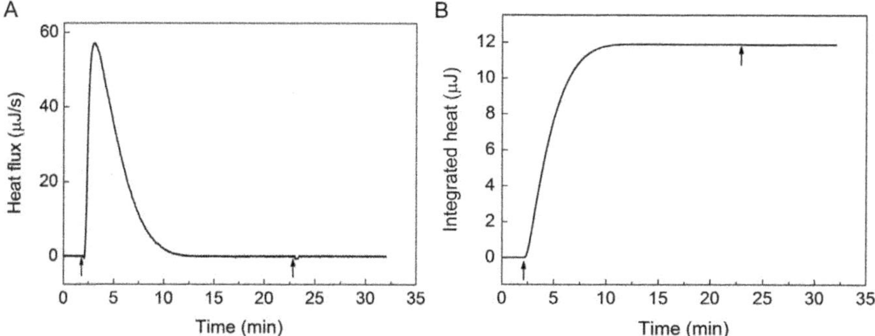

Figure 2 Calorimetric thermograms of the G6PD reaction. (A) Raw data showing the heat flux (μJ/s) as a function of time (in s) measured at 25 °C in 10 m*M* phosphate buffer, pH 7.6, containing 5 m*M* glucose-6-phosphate and 0.1 m*M* $NADP^+$. The reaction started after equilibration at the desired temperature by the injection of 10 μL of the G6PD solution into the calorimeter cell. The upward displacement of the baseline indicates the endothermic nature of the reaction. The return to the baseline level indicates the complete consumption of $NADP^+$. A second injection of 10 μL of the G6PD solution was obtained after completion of the reaction. The arrows indicate the time of each injection. (B) Integrated heat from the thermogram shown in (A).

In order to determine ΔH^{cal}, the calorimeter cell was filled with the reaction medium. After the injection of the G6PD solution, the heat flux (μcal/s) as a function of time (in s) was measured for 20 min (Fig. 2A). The integrated heat (Fig. 2B) shows the kinetic profile of the reaction.

The final concentration of G6PD in the cell was 0.1 U/mL. The reaction medium had an excess of G6P but limited amounts of NAD^+ or $NADP^+$. Therefore, the coenzyme was entirely consumed in the reaction when the heat flux returned to the baseline level, which was confirmed by spectrophotometric data.

As pointed before, the G6PD reaction occurs with a concomitant proton release to the solution, and the observed enthalpy depends on the buffer used as the proton absorption by the buffer is responsible for part of the total heat developed. The ITC determination of the enthalpy of buffer protonation was done in a VP-ITC (MicroCal/Malvern) by a single injection of 10 μL of 1 m*M* HCl into the cell containing 50 m*M* buffer, pH 7.6. The area under the peak corresponds to the total heat (Q_T) released in the protonation reaction. ΔH^P was calculated by dividing Q_T by the amount of HCl (in moles) in the cell. ΔH^P values determined for the five buffers used in this work were very similar to those from Fukada and Takahashi (1998).

Figure 3 shows the calorimetric traces obtained in four buffers: phosphate, MOPS, imidazole, and Tris:HCl. It is very clear that the reaction

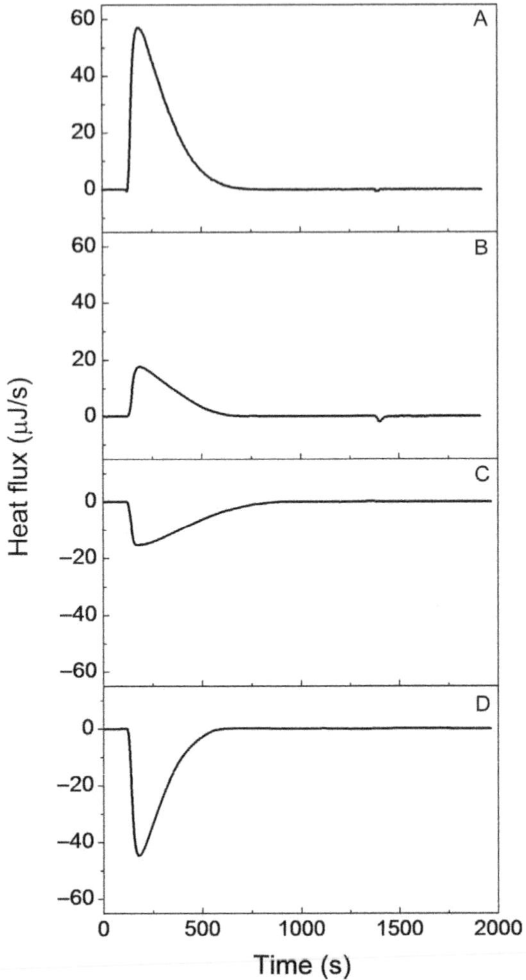

Figure 3 Calorimetric thermograms of the G6PD reaction obtained in different buffers. The reaction of G6PD was followed at 25 °C in similar conditions as described in Fig. 1A. The buffers used were 10 m*M* phosphate (A) or 50 m*M* MOPS (B), imidazole (C), and Tris:HCl (D), pH 7.8.

can be exothermic in Tris ($\Delta H^{P} = -47.9$ kJ/mol) and in imidazole ($\Delta H^{P} = -36.61$ kJ/mol) or endothermic in MOPS ($\Delta H^{P} = -21.94$ kJ/mol) or phosphate ($\Delta H^{P} = -5.09$ kJ/mol). By using HEPES ($\Delta H^{P} = -21.1$ kJ/mol), the reaction is also endothermic (data not shown). A second injection was done in order to calculate the heat of the enzyme dilution, which was very small as compared to the total heat of the reaction (see Fig. 3).

As described before, the area under the peak corresponds to the total heat of the reaction (Q_T), and the calorimetric enthalpy was calculated by dividing Q_T by the amount of NADH or NADPH formed (determined by spectrophotometry). The linear correlation between ΔH^{cal} and ΔH^{P} (Fig. 4) is described by Eq. (1) in which ΔH^{R} can be obtained by the intercept. For both coenzymes, ΔH^{R} was positive. However, ΔH^{R} was significantly different ($p<0.0001$) depending on the coenzyme used. For the $NADP^{+}$-linked reaction, ΔH^{R} was 24.5 ± 1.5 kJ/mol and for the NAD^{+}-linked reaction, ΔH^{R} was 35.7 ± 1.3 kJ/mol. In both cases, as expected there was one proton released: $n=0.9$ and 1.1 for the reactions with $NADP^{+}$ and NAD^{+}, respectively.

The results are very interesting for two reasons: (i) the reaction is endothermic, which is not the expected behavior for spontaneous reactions, and (ii) there is a significant difference in ΔH^{R} depending on the coenzyme used.

An endothermic reaction is spontaneous if the entropy of the system increases and the entropic term, $-T\Delta S$, is negative. This means that an entropic compensation is taking place. The nature of the increase in entropy during the reaction can be related to an increase in the conformational freedom of the enzyme and/or the release of water upon binding of the

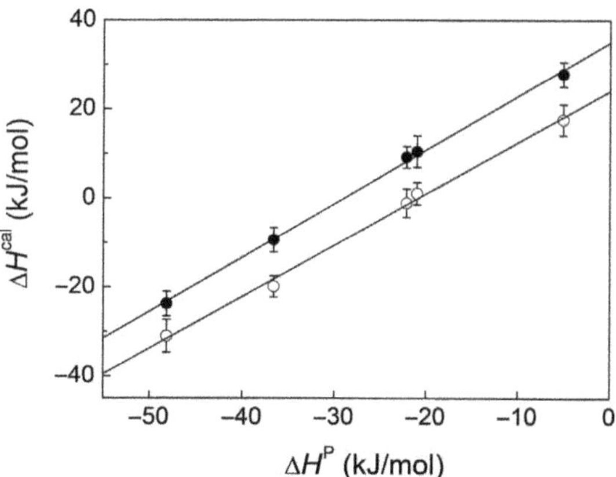

Figure 4 Correlation between the calorimetric enthalpy of the G6PD reaction and the enthalpy of buffer protonation. ΔH^{cal} was calculated from the area under the peaks for the reaction with NAD^{+} (○) or $NADP^{+}$ (•) in 50 m*M* Tris ($\Delta H^{P}=-48.5$ kJ/mol), imidazole ($\Delta H^{P}=-35.3$ kJ/mol), MOPS ($\Delta H^{P}=-22.2$ kJ/mol) or HEPES ($\Delta H^{P}=-21.1$ kJ/mol), and 10 m*M* phosphate ($\Delta H^{P}=-5.22$ kJ/mol). The symbols represent the mean and the standard deviation of the mean of at least five experiments. The curves were obtained by linear regression of the experimental data with linear correlation $r=0.9998$ for the $NADP^{+}$-linked reactions and $r=0.9982$ for the $NADP^{+}$-linked reactions.

substrates or during the reaction. In fact, by comparing the crystal structure of the G6PD in the presence and in the absence of G6P, the number of water molecules is different, especially near to the G6P binding site (Fig. 5). However, this effect needs to be further explored, and both differential scanning calorimetry and molecular dynamic simulation of the enzyme in the presence of the substrates are in progress.

The difference in ΔH^R depending on the coenzyme used is also intriguing. It has been shown that both coenzymes bind in the same site on the enzymes (Levy, 1989). Yet, the catalytic mechanism is not the same. Furthermore, it has been suggested that either coenzyme cause different conformational changes in the G6PD as seen by treatments with protease, urea, and temperature (Kurlandsky, Hilburger, & Levy, 1988). These authors showed that the protective effect of both G6P and NAD^+ is substantially greater than $NADP^+$, suggesting a greater conformational change when NAD^+ is bound. This difference together with the different K_D for NAD^+ or $NADP^+$ binding to G6PD probably plays an important role in the dual coenzyme specificity as either nucleotide binds in the same site.

It is important to notice that the rate of the reaction did not change with most of the buffer used. However, phosphate buffer caused an inhibitory effect in the enzyme. In order to minimize this effect, the concentration of the phosphate was 10 mM instead the 50 mM used with the other buffers.

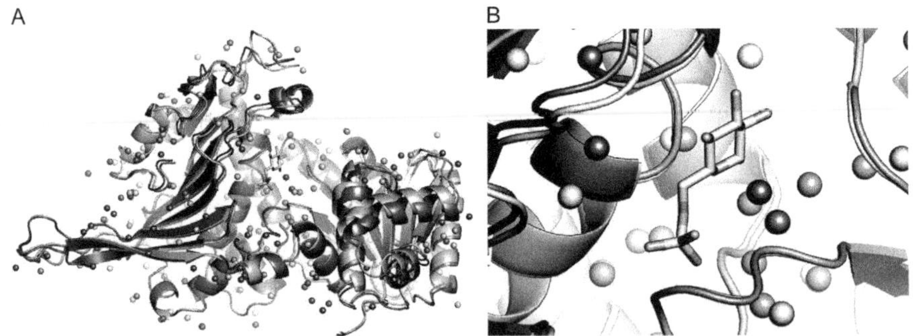

Figure 5 Comparison between the crystal structures of the apo- and G6P-bound enzyme. The crystal structures of the apo-enzyme (1H9B.pdb, in green) and the G6P-bound enzyme (1E77.pdb, in blue) were superimposed. Upon G6P binding, no significant structural change was observed, but the amount of water molecules in the solvation shell as well as in the active site significantly decreases. On the right, a zoom of the active site with a G6P molecule bound. (See the color plate.)

4.2 Effect of Ionic Strength on the Rate and the Enthalpy of the Reaction

The G6PD reaction was studied with NaCl and KCl (up to 200 m*M*). Since very similar results were found with both salts, the results are exemplified with KCl for the reaction with both coenzymes (Fig. 6). The rate of the

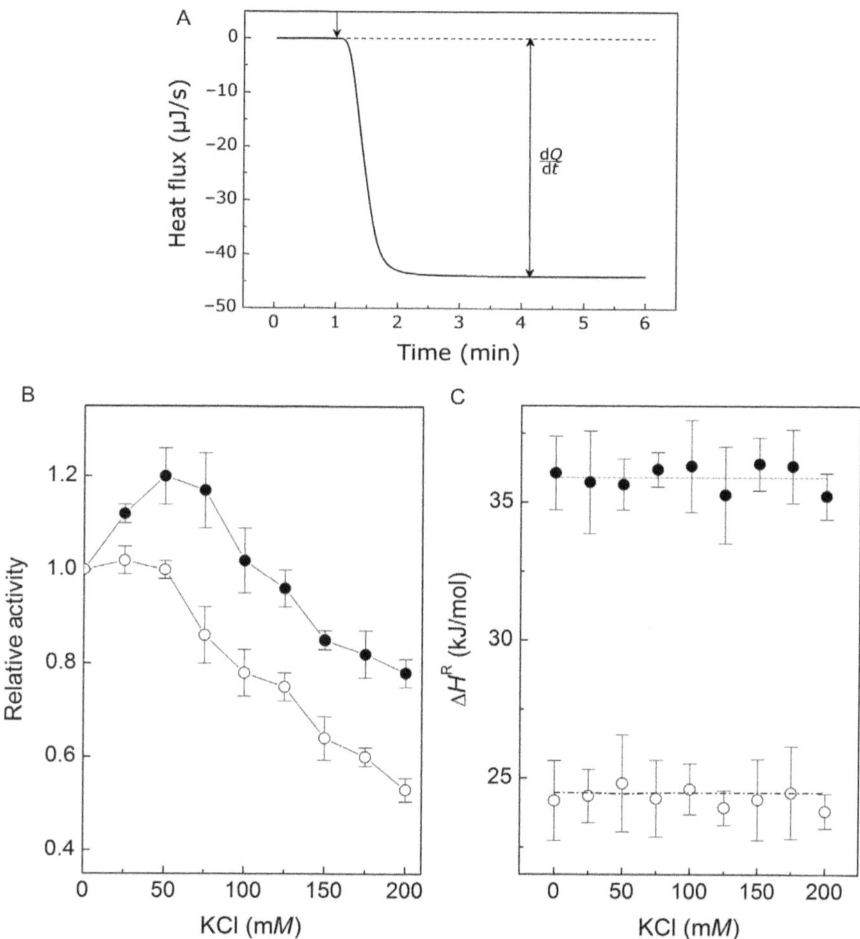

Figure 6 Effect of ionic strength on the rate of the reaction and in ΔH^R for the G6PD from *Leuconostoc mesenteroides*. (A) Calorimetric trace of the reaction in steady-state conditions (5 m*M* G6P and 1 m*M* NAD^+) in 50 m*M* Tris:HCl. (B) The rate of the reaction was calculated as shown in (A), for the $NADP^+$-linked reaction (•) and the NAD^+-linked reaction (○) as a function of KCl concentration where the symbols represent the mean and the standard deviation of the mean of at least five experiments. (C) The enthalpy of the reaction as a function of KCl concentration. The symbols are the same as in (B).

reaction was measured by calorimetry in steady-state conditions as shown in Fig. 6A. In this case, the heat flux (μJ/s) is proportional to the rate of the reaction (Bianconi, 2003, 2007; Todd & Gomez, 2001).

For the $NADP^+$-linked reaction, the rate increases with KCl or NaCl up to 75 m*M* and decreases with higher salt concentrations (Fig. 6B). For the NAD^+-linked reaction, there is no significant change in the rate with up to 50 m*M* KCl (Fig. 6B), which decreases with higher KCl concentrations. Interestingly, even though the salt effect on the reaction rate is not the same, ΔH^{cal} and, consequently ΔH^{R}, was not affected by KCl or NaCl (Fig. 6).

It is very intriguing that ΔH^{cal} is not affected by the ionic strength. It was already shown that ΔH^{P} is not affected by the ionic strength since the data obtained here were very similar to those obtained by Fukada and Takahashi (1998) and Ramette et al. (1977). In the latter, the authors used synthetic seawater at high salinity level (30–45‰). Table 1 shows some of these data as compared to those obtained here with low ionic strength.

Nevertheless, this result is very similar to those obtained by Morin and Freire (1991) for cytochrome *c* oxidase and Bianconi (2003) for the yeast HK isozymes in which ΔH^{R} is a parameter independent of ionic strength. Similar to that found with the $NADP^+$-linked reaction of the G6PD, the dependence of the rate of the cytochrome *c* oxidase reaction on ionic strength has a bell shape with a maximum at 50 m*M* KCl. In all cases, the decrease on the rate of the reaction with NaCl or KCl suggests an entropic nature of these effects.

5. TAKING ADVANTAGE OF THE HEAT EFFECT OF BUFFER PROTONATION/DEPROTONATION

It is very clear that the choice of the buffer will affect the calorimetric enthalpy of a reaction in such a way that can lead to a misinterpretation of the data. As shown here for the particular case of the reaction catalyzed by the G6PD from *L. mesenteroides*, the calorimetric enthalpy can be positive or negative depending on the buffer system used. Therefore, if the heat of protonation/deprotonation of the buffer used is not taking into account, the reaction can be explained as being exothermic or endothermic. This can be very deceiving as the conclusions will certainly be misled.

Furthermore, although ΔH^{P} does not change with the ionic strength, it is very important to determine ΔH^{P} in different temperatures as the heat capacity changes (ΔCp) significantly varies depending on the buffer system (see Fukada and Takahasi for details). ΔCp can be positive, as in the case of MOPS (39 J/K/mol), or negative, as observed with imidazole (−16 J/K/mol), for instance; ΔCp can also be very small, as in the case of BES and Bicine (2 J/K/mol for both), or very large, as in the case of phosphate (−187 J/K/mol) and acetate (−128 J/K/mol).

However, in order to determine the rate of the reactions, one can take advantage of the heat of buffer protonation. By using a buffer with high ΔH^{P}, the signal can be intensified and the signal to noise ratio can be improved, so the rate of the reaction is more accurately determined.

The best approach to use for any unknown reaction, even for ligand binding studies, is to do the first experiments in two buffers of very different ΔH^{P}. If there is no change in the observed enthalpy, no proton release/uptake is involved. If ΔH^{cal} changes, a more detailed studied is necessary, as described in this chapter.

6. FINAL REMARKS

In this work, G6PD from *L. mesenteroides* was used as a model to illustrate buffer interference in ITC experiments. In the literature, we can find some examples of the effect of the heat of buffer protonation/deprotonation in the enthalpy of ligand binding to proteins. This was also shown in the thermodynamic study of the reaction catalyzed by the two isozymes, PI and PII, of the yeast HK (Bianconi, 2003). In this case, independent of the buffer used, the reactions were exothermic, with negative ΔH^{cal} and ΔH^{P}.

However, depending on the buffer used, the reaction catalyzed G6PD from *L. mesenteroides* was apparently exothermic (Tris and imidazole) or endothermic (MOPS, HEPES, and phosphate). By determining ΔH^{cal} in buffers of different heat of protonation, it is possible to calculate ΔH^{R} as well as the number of protons involved in the reaction, by the linear correlation between both enthalpic parameters. It was interesting to note that, once ΔH^{P} is accounted for, it is clear that the G6PD reaction is endothermic indicating that it is entropy driven. The reason for that remains unclear at this point, although we can postulate that an increase in conformational freedom occurs during the reaction. This result was unexpected as exothermic ΔH^{R} values are usually observed—substrates tend to increase the stability of the enzymes due to the decrease in conformational freedom. For instance, the yeast HKs present two independent domains in each subunit. Upon glucose binding, there is a great conformational change, in which the two domains move about 8 Å, closing the cleft between the lobes (Bennett & Steitz, 1978), increasing the stability and the cooperativity of the thermal transition of the enzyme (Catanzano, Gambuti, Graziano, & Barone, 1997; Takahashi, Casey, & Sturtevant, 1981).

Another possibility is the release of bound water molecules from the active site and/or the surface of the enzyme. Indeed, when we compare the crystal structures of the apo- and G6P-bound enzyme, the number of water molecules is higher in the apo-enzyme, especially in active site (Fig. 5). Molecular dynamics simulation of the G6PD in aqueous environment as well as differential scanning calorimetry studies is in progress in order to elucidate this finding.

It was also interesting to find different thermodynamic behaviors when comparing the coenzymes. Even though the reaction is endothermic in both cases, the difference in ΔH^{R} suggests that a phenomenon other than the reaction itself is taking place. The differences found here for both the ΔH^{R} values and the effect of salts in the rate of the reaction can be related to the difference in the binding affinity of each coenzyme to the G6PD.

In short, the G6PD from *L. mesenteroides* is a good example that the expected behavior is not always observed. It is important to check if this peculiar thermodynamic behavior is also found in G6PDs from other organisms.

ACKNOWLEDGMENTS

The author thanks Dr. M.G. D'Andrea for the critical reading of the manuscript and Dr. P.H.M. Torres for the drawings in Fig. 5.

REFERENCES

Adams, M. J., Basak, A. K., Gover, S., Rowland, P., & Levy, R. (1993). Site-directed mutagenesis to facilitate X-ray structural studies of Leuconostoc mesenteroides glucose 6-phosphate dehydrogenase. *Protein Science*, *2*, 859–862.

Baker, B. M., & Murphy, K. P. (1996). Evaluation of linked protonation effects in protein binding reactions using isothermal titration calorimetry. *Biophysical Journal*, *71*, 2049–2055.

Bennett, W. S., Jr., & Steitz, T. A. (1978). Glucose-induced conformational change in yeast hexokinase. *Proceedings of the National Academy of Sciences of the United States of America*, *75*, 848–4852.

Bianconi, M. L. (2003). Calorimetric determination of thermodynamic parameters of reaction reveals different enthalpic compensations of the yeast hexokinase isozymes. *The Journal of Biological Chemistry*, *278*, 18709–18713.

Bianconi, M. L. (2007). Calorimetry of enzyme-catalyzed reactions. *Biophysical Chemistry*, *126*, 59–64.

Cai, L., Cao, A., & Lai, L. (2001). An isothermal titration calorimetric method to determine the kinetic parameters of enzyme catalytic reaction by employing the product inhibition as probe. *Analytical Biochemistry*, *299*, 19–23.

Catanzano, F., Gambuti, A., Graziano, G., & Barone, G. (1997). Interaction with D-glucose and thermal denaturation of yeast hexokinase B: A DSC study. *Journal of Biochemistry*, *121*, 568–577.

Cogan, T. M., & Jordan, K. N. (1994). Metabolism of *Leuconostoc* bacteria. *Journal of Dairy Science*, *77*, 2704–2717.

Datta, S. P., Grzybowski, A. K., & Weston, B. A. (1963). The acid dissociation constant of the protonated form of tri(hydroxymethyl)methylamine. *Journal of the Chemical Society*, 792–796. http://dx.doi.org/10.1039/jr9630000792.

Fukada, H., & Takahashi, K. (1998). Enthalpy and heat capacity changes for the proton dissociation of various buffer components in 0.1 M potassium chloride. *Proteins*, *33*, 159–166.

Holten, D., Procsal, D., & Chang, H. L. (1976). Regulation of pentose phosphate pathway dehydrogenases by NADP+/NADPH ratios. *Biochemical and Biophysical Research Communications*, *68*, 436–441.

Kurlandsky, S. B., Hilburger, A. C., & Levy, H. R. (1988). Glucose-6-phosphate dehydrogenase from Leuconostoc mesenteroides: Ligand-induced conformational changes. *Archives of Biochemistry and Biophysics*, *264*, 93–102.

Ladbury, J. E., & Chowdhry, B. Z. (1996). The application of isothermal titration calorimetry to thermodynamics studied of biomolecular interaction. *Chemistry & Biology*, *3*(10), 791–801.

Leavitt, S., & Freire, E. (2001). Direct measurement of protein binding energetics by isothermal titration calorimetry. *Current Opinion in Structural Biology*, *11*, 560–566.

Levy, H. R. (1989). Glucose-6-phosphate dehydrogenase from *Leuconostoc mesenteroides*. *Biochemical Society Transactions*, *17*, 313–315.

Levy, H. R., Christoff, M., Ingulli, J., & Ho, E. M. (1983). Glucose-6-phosphate dehydrogenase from *Leuconostoc mesenteroides*: Revised kinetic mechanism and kinetics of ATP inhibition. *Archives of Biochemistry and Biophysics*, *222*, 473–488.

Limo, M. J., & Perry, C. C. (2015). Thermodynamic study of interactions between ZnO and ZnO binding peptides using isothermal titration calorimetry. *Langmuir*, *31*, 6814–6822.

Malesinski, S., Tsvetkov, P. O., Kruczynski, A., Peyrot, V., & Devred, F. (2015). Stathmin potentiates vinflunine and inhibits paclitaxel activity. *PloS One*, *10*(6), e0128704. http://dx.doi.org/10.1371/journal.pone.0128704.

Morin, P. E., & Freire, E. (1991). Direct calorimetric analysis of the enzymic activity of yeast cytochrome c oxidase. *Biochemistry*, *30*, 8494–8500.

Olsen, S. N. (2006). Applications of isothermal titration calorimetry to measure enzyme kinetics and activity in complex solution. *Thermochimica Acta*, *448*, 12–18.

Ramette, R. W., Culberson, C. H., & Bates, R. G. (1977). Acid-base properties of tris(hydroxymethyl)aminomethane (tris) buffers in seawater from 5 to 40 °C. *Analytical Chemistry*, *49*, 867–870.

Rowland, P., Basak, A. K., Gover, S., Levy, H. R., & Adams, M. J. (1994). The three-dimensional structure of glucose 6-phosphate dehydrogenase from *Leuconostoc mesenteroides* refined at 2.0 Å resolution. *Structure*, *15*, 1073–1087.

Stanton, R. C. (2012). Glucose-6-phosphate dehydrogenase, NADPH, and cell survival. *IUBMB Life*, *64*, 362–369.

Sturtevant, J. M. (1937). Calorimetric investigations of organic reactions. I. Apparatus and method. The inversion of sucrose and the decomposition of diacetone alcohol. *Journal of the American Chemical Society*, *59*, 1528–1537.

Takahashi, K., Casey, J. L., & Sturtevant, J. M. (1981). Thermodynamics of the binding of D-glucose to yeast hexokinase. *Biochemistry*, *20*, 4693–4697.

Todd, M. J., & Gomez, J. (2001). Enzyme kinetics determined using calorimetry: A general assay for enzyme activity? *Analytical Biochemistry*, *296*, 179–187.

Velazquez-Campoy, A., Leavitt, S. A., & Freire, E. (2004). Characterization of protein-protein interactions by isothermal titration calorimetry. *Methods in Molecular Biology*, *261*, 35–54.

Wafer, L. N., Streicher, W. W., McCallum, S. A., & Makhatadze, G. I. (2012). Thermodynamic and kinetic analysis of peptides derived from CapZ, NDR, p53, HDM2, and HDM4 binding to human S100B. *Biochemistry*, *51*, 7189–7201.

Wiseman, T., Williston, S., Brandts, J. F., & Lin, L. N. (1989). Rapid measurement of binding constants and heats of binding using a new titration calorimeter. *Analytical Biochemistry*, *179*, 131–137.

CHAPTER ELEVEN

ITC Methods for Assessing Buffer/Protein Interactions from the Perturbation of Steady-State Kinetics: A Reactivity Study of Homoprotocatechuate 2,3-Dioxygenase

Kate L. Henderson, Delta K. Boyles, Vu H. Le, Edwin A. Lewis[1], Joseph P. Emerson
Department of Chemistry, Mississippi State University, Mississippi State, Mississippi, USA
[1]Corresponding authors: e-mail address: elewis@chemistry.msstate.edu

Contents

Abstract

Isothermal titration calorimetry (ITC) can be used to study the thermodynamics of enzyme substrate binding or the kinetics of substrate turnover (or both). Substrate-binding interactions are observed in a typical ITC titration experiment in which the heat change for the addition of an aliquot of substrate to a solution containing the enzyme is determined for a number of titrant (i.e., substrate) injections and the data fit for the thermodynamic parameters (ΔG, ΔH, and $-T\Delta S$) for substrate binding. Of course, these measurements must be made under conditions where the substrate binds but does not turnover. In the ITC "kinetics" experiment, the power change observed after injection of an excess of substrate into a solution of the enzyme is a direct measure of the rate

Methods in Enzymology, Volume 567
ISSN 0076-6879
http://dx.doi.org/10.1016/bs.mie.2015.08.034

at which substrate is converted to product, and the ITC data can be analyzed for the kinetic parameters (V_{max}, k_{cat}, K_M, and k_{cat}/K_M). The ITC technique is particularly versatile in that it can be applied to systems where there might not be a change in a spectroscopic signal for either substrate binding or the reaction of the substrate to form product. A complication is that if there are competing reactions, for example, buffer protonation, or product binding, to name just two, the enthalpy change measured for either substrate binding or for substrate turnover will be a summation of all of the reaction heats. Enzyme studies are typically done in buffered solutions at constant pH. The general, and often incorrect, assumption is that the buffer components are simply spectators and not participants in either substrate binding or substrate turnover. This chapter describes how we have used ITC measurements to identify problem buffers that impact the kinetics for an enzyme catalyzed reaction. Herein, we show the effects of several buffers on the steady-state kinetics for the conversion of the substrate, 3,4-dihydroxyphenyl acetate (homoprotocatechuate), to the ring-opened product, 5-carboxymethyl-2-hydroxymuconic semialdehyde by the nonheme iron(II) metalloenzyme, homoprotocatechuate 2,3-dioxygenase. Several buffers were observed to engage in buffer/enzyme interactions within the active site pocket. These enzyme–buffer interactions were shown to inhibit substrate turnover and to contribute additional enthalpy terms to the overall heat of reaction observed for substrate turnover (and for substrate binding).

1. INTRODUCTION

The use of buffers to maintain the pH of a solution with the expectation that the buffering components (both the conjugate weak acid and conjugate base) will be inert (unreactive) in solution is a common laboratory practice in biochemistry. In reality, this may not be the case for many commonly used buffers, particularly when studying metalloenzymes (Desmarais et al., 2002; Harutynunyan et al., 1996; Kirsch, Lomonosova, Korth, Sustmann, & de Groot, 1998; Lucarini & Kilikian, 1999; McPhail & Holt, 1999; Taha, Gupta, Khoiroh, & Lee, 2011; Welch, Davis, & Aust, 2002; Yu, Kandegedara, Xu, & Rorabacher, 1997). Metalloenzymes and bioinorganic complexes can be particularly sensitive to common buffer components which have a proclivity for binding to available metal ions. These competing interactions must be accounted for and where neglected have led to erroneous interpretation of the thermodynamic and or kinetic behavior of metalloproteins such as homoprotocatechuate 2,3-dioxygenase (HPCD) (Whiting, Boldt, Hendrich, Wackett, & Que, 1996).

Isothermal titration calorimetry (ITC) can be used to directly to follow enzyme kinetics by simply following the rate of heat production after introduction of the substrate. If the only reaction occurring is substrate turnover, changes in heat rate (ITC power) will be a direct measure of the turnover rate and the heat of reaction for substrate conversion to product. In the event that substrate–buffer or enzyme–buffer reactions are occurring simultaneously with substrate turnover, the measured change in power will be proportional to the rate of the substrate turnover and to the sum of the enthalpy changes for substrate turnover and all of the complicating reactions. In a typical ITC kinetic experiment, an excess amount of a small molecule (substrate) is injected into a solution of the macromolecule (enzyme) (Freyer & Lewis, 2008). As heat (Q) is produced or consumed by the reaction, the calorimeter adjusts the compensating power, dQ/dt (μW, μJ/s) supplied to the cell as required to maintain a constant cell temperature.

The addition of a large excess of substrate to the enzyme results in a steady-state rate of heat change, all of the enzyme is saturated with substrate, and the heat produced (or consumed) will be equal to the rate of substrate conversion to product and the sum of the enthalpy changes for substrate conversion and any other reactions that may be taking place simultaneously (Fig. 1).

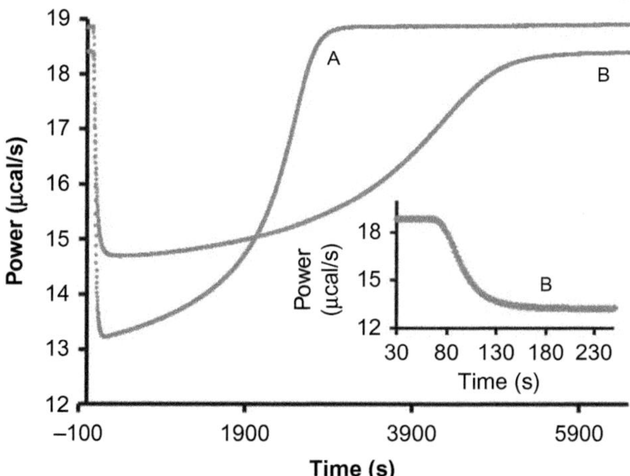

Figure 1 The raw heat for the ring-opening reaction of HPCA by HPCD in 100 m*M* MOPS (blue (Line A in the print version)) and TRIS (red (Line B in the print version)) buffers, pH 7.2. The reaction in TRIS buffer takes significantly longer to return to baseline, indicating a dramatically slower rate of reaction in TRIS buffer (inset). The enlarged view of the steady-state region for the blue (dark gray in the print version) trace in this experiment.

The steady-state production of heat from substrate turnover reaction will remain unchanged until substrate is depleted (or until product inhibition slows the rate of substrate conversion). When substrate conversion is complete and all of the injected substrate has been converted to product, the calorimeter signal will return to baseline since $\mathrm{d}Q/\mathrm{d}t=0$ at this point. The rate of the overall reaction at any point in time is simply determined from the measured $\mathrm{d}Q/\mathrm{d}t$ at that point in time and the value of ΔH_{obs}, as shown in Eq. (1) (Bianconi, 2007; Cai, Cao, & Lai, 2001; Todd & Gomez, 2001).

$$\frac{\mathrm{d}Q}{\mathrm{d}t}=\frac{\mathrm{d}nP}{\mathrm{d}t}V\Delta H_{\mathrm{obs}} \tag{1}$$

where $\mathrm{d}nP/\mathrm{d}t$ is the number of moles of product formed over time, t, V is the volume in the reaction cell, and ΔH_{obs} is the observed reaction enthalpy for the system, which will depend on the ΔH for substrate conversion, and if applicable to the ΔH for buffer protonation and the number of protons released (or taken up) per mol of substrate turned over, and the ΔH for any other complicating reactions (e.g., buffer/substrate or buffer/enzyme interactions and/or product binding). By rearranging Eq. (1), the rate of reaction can be determined if ΔH_{obs} is known or can be estimated.

$$\mathrm{rate}=\frac{\mathrm{d}P}{\mathrm{d}t}=\frac{\mathrm{d}Q}{\mathrm{d}t}\frac{1}{(V\Delta H_{\mathrm{obs}})} \tag{2}$$

Integrating $\mathrm{d}Q/\mathrm{d}t$ as a function of time over the entire experiment, i.e., from $t=0$ to $t=t_{\mathrm{complete}}$ where all of the injected substrate has been consumed and the calorimetric signal has returned to baseline yields an experimental value for ΔH_{obs}. At the steady-state maximum, observed at or near ($t=0$), a V_{max} can be directly determined for the reaction. Additionally, by utilizing various buffers and plotting the buffer ionization enthalpy ($\Delta H_{\mathrm{ionization}}$) against the observed enthalpy (ΔH_{obs}), the reaction enthalpy (ΔH_{rxn}) can be obtained, along with the total number of protons released/consumed by the reaction, according to the following equation:

$$\Delta H_{\mathrm{obs}}=\Delta H_{\mathrm{rxn}}+n(\Delta H_{\mathrm{ionization}}) \tag{3}$$

Herein, we describe how a series of ITC experiments were used as a method to measure the influence of complicating buffer interactions on the kinetics of substrate turnover for a metalloenzyme. The system we have chosen to illustrate the use of ITC methods to evaluate whether enzyme–buffer interactions are confounding the kinetics of substrate turnover is the nonheme iron(II) HPCD catalyzed ring opening of the substrate, 3,4-dihydroxyphenyl acetate (homoprotocatechuate or HPCA), to form the

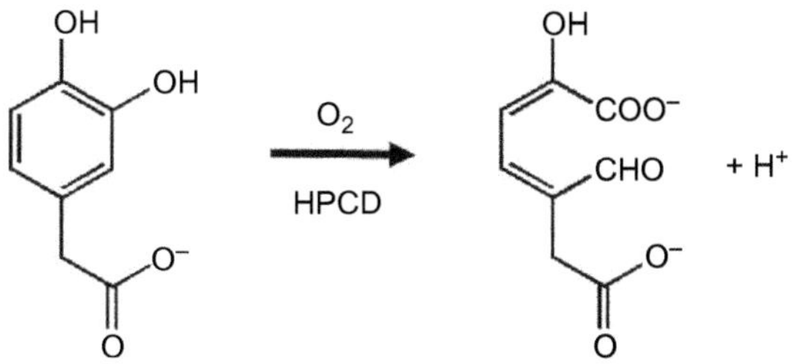

Scheme 1 The homoprotocatechuate 2,3-dioxygenase (HPCD) catalyzed ring opening of the substrate, 3,4-dihydroxyphenyl acetate (HPCA), to form the product, 5-carboxymethyl-2-hydroxymuconic semialdehyde (5CHMSA).

product, 5-carboxymethyl-2-hydroxymuconic semialdehyde (5CHMSA). The structures of the substrate and product of the HPCD reaction are shown in Scheme 1.

2. REQUIRED MATERIALS

- VP-ITC, ITC_{200}, TA instruments Nano ITC or equivalent isothermal titration calorimeter
- Loading syringe
- Enzyme solutions (titrate), prepared and dialyzed in the buffer of choice
- Substrate solutions (titrant), prepared and dialyzed in the buffer of choice
- Buffer components
- ITC detergent cleaning solution
- 18 MΩ or ddH_2O

3. INSTRUMENTATION AND SAMPLE CONSIDERATIONS

When preparing for a steady-state kinetics experiment by ITC, it is important to consider the concentrations of the reactants (both the substrate titrant concentration and the enzyme titrate concentration). A nominally high titrant concentration is required to deliver the excess substrate needed to establish the zero-order reaction condition at $t=0$. The high substrate concentration may result in higher than normal heat of dilution effects for titrant addition and will certainly require careful blank titrations and heat corrections for the heat of substrate dilution in the ITC kinetics experiment.

The enzyme concentration should be approximately 1000-fold lower than the initial substrate concentration ($[E] \ll [S]_0$) to enable observation of the steady-state heat rate (or power). However, for some enzymatic reactions which proceed at very fast rates (e.g., carbonic anhydrase) it may be difficult to add an excess of substrate in a small volume injection making it difficult to achieve a steady-state over a reasonable length of time. This problem can be overcome through a "reverse titration" reaction, in which a low concentration of enzyme is used as the titrant, and the substrate is placed in the cell. This technique also works well for difficult substrates that have either low solubility or exhibit large heats of dilution.

Because the ITC measures the sum of the heats produced or consumed in the cell during the experiment, solutions should be matched as accurately as possible with respect to all solute concentrations in order to minimize heats from solute dilution and other nonspecific processes. This is successfully done through exhaustive dialysis of the enzyme solution to remove low-molecular-weight contaminants and then using the dialysate from the last buffer change as the solvent for preparation of the titrant (substrate) solutions. Exhaustive dialysis typically involves dialyzing a small volume of the enzyme stock solution against a 1000-fold larger volume of buffer and at least three changes of the buffer solution. The substrate titrant solutions are then made by dissolution of a weighed amount of the substrate into the dialysate (from the last enzyme dialysis step). It is important to degas the enzyme titrate (or substrate titrate in the case of a reverse titration experiment) before filling the calorimeter cell.

The ITC "kinetic" experiment parameters are setup for a single injection followed by compensation power data collection over an extended period of time. In this case, a fast burst of titrant is added to the cell at $t=0$, and then power (or heat rate) data collected until the reaction has gone to completion (i.e., the time required for the ITC signal to return to baseline). Additional injections can be performed for reproducibility if product inhibition is not present.

4. STEADY-STATE ITC METHOD APPLIED TO HPCD

HPCD is a nonheme iron(II) enzyme that catalyzes the ring-opening step in the degradation of aromatic compounds (Lipscomb, 2008; Vaillancourt, Bolin, & Eltis, 2006). In the monomeric form, HPCD is made up of two domains, where two copies of a βαβββ motif can be found in each domain (Armstrong, 2000; Gerlt & Babbitt, 2001). It is this motif that forms the funnel-shaped active site through hydrophobic interactions between motifs (Fig. 2A and B). At the center of this active site sits a six-coordinate

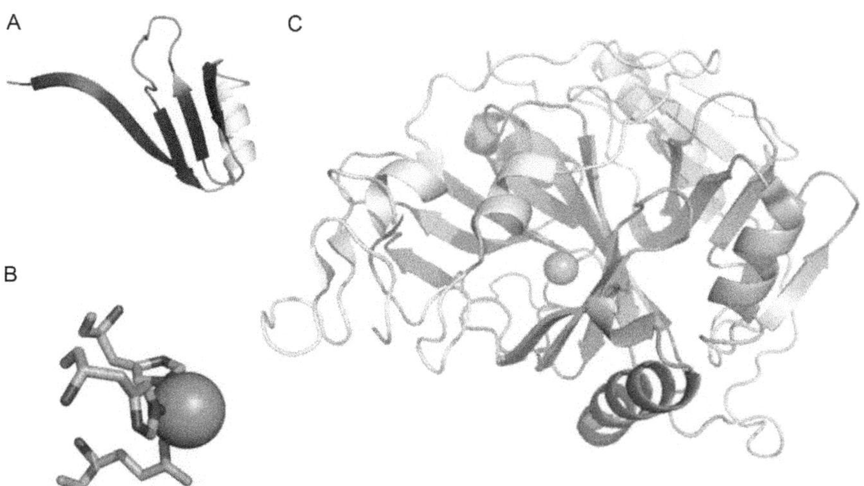

Figure 2 The structure of HPCD. (A) The βαβββ motif, of which HPCD contains multiple copies to form the active site cavity. (B) The globular structure of the HPCD monomeric unit, which shows the N-terminal domain (blue), the C-terminal domain (green) which holds the 2H1C facial triad, and the lid domain (yellow). (C) The iron(II) center ligated to the 2H1C (green residues) of HPCD. (See the color plate.)

nonheme iron(II) ion, ligated by two histidine and one glutamate residues on one face of its octahedral coordination sphere. These ligands make up what is referred to as the 2-His-1-carboxylate facial triad, a common metal-binding motif among many of the enzymes in the extradiol dioxygenase family (Fig. 2C) (Hegg & Que, 1997; Leitgeb & Nidetzky, 2008). This facial triad allows for binding of substrates and dioxygen activation to occur in close proximity on the same face of the iron(II) octahedron, resulting in efficient catalysis for the enzyme (Kovaleva & Lipscomb, 2007). Additionally, HPCD contains a loop with a conserved arginine residue which interacts with the distal end of the substrate. This loop serves as a lid to close in the active site when a substrate molecule is bound. HPCD functions as a homotetramer and initial substrate-binding studies along with crystal structures of the substrate–enzyme complex seem to suggest some degree of cooperativity between the four units of HPCD in the absence of oxygen (Henderson, Le, Lewis, & Emerson, 2012; Kovaleva & Lipscomb, 2012).

Previously, we reported the thermodynamic properties of native substrate binding to HPCD (Henderson et al., 2012). For these binding reactions, the heats of ionization of various commonly used buffers were exploited to develop a linear correlation plot of the observed enthalpy

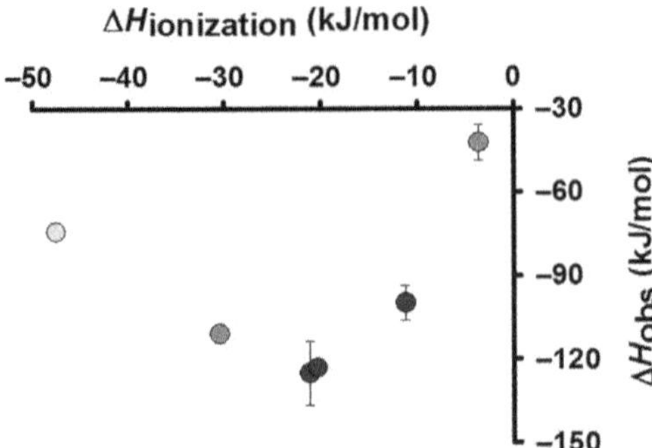

Figure 3 Enthalpy plot of the HPCA-binding reaction to HPCD. The substrate-binding reaction was performed in various buffers to utilize their heats of ionization. A linear correlation was found with HEPES, MOPS, and PIPES buffers (plotted as blue (dark gray in the print version) circles), whereas ACES and potassium phosphate buffers (pink (dark gray in the print version)), lie off the line 38 kJ/mol. The data associated with TRIS buffer (green (light gray in the print version)) is low by 117 kJ/mol. Error bars represent one standard deviation of the mean.

change for substrate binding versus the enthalpy change for buffer ionization (Goldberg, Kishore, & Lennen, 2002; Grossoehme, Spuches, & Wilcox, 2010). During these ITC substrate-binding experiments, it was observed the $\Delta H_{\text{binding}}$ data collected in several buffers did not follow the expected linear trend of ΔH_{obs} versus buffer ionization enthalpy. Specifically, the HPCA/HPCD-binding enthalpy changes observed in ACES (*N*-(2-acetamido)-2-aminoethanesulfonic acid) and in potassium phosphate buffer deviated from the trendline by approximately 38 kJ/mol (Fig. 3, pink (dark gray in the print version)), while the observed binding enthalpy change in TRIS (*tris*-(hydroxymethyl) aminomethane) buffer (green (light gray in the print version)) was 117 kJ/mol lower than predicted. In all three cases, enthalpy change values for the binding reaction were lower than expected, suggesting that additional processes are occurring during HPCA/HPCD binding in these particular buffering systems that are not accounted for in the simple linear relationship shown in Eq. (3). We believe that the explanation for this observation is that in the case of the three buffers which exhibit the anomalous heats effects, strong enzyme–buffer interactions are occurring which might be reflected in the inhibition of the HPCD catalytic activity under turnover conditions in these buffers.

4.1 HPCD Samples and Preparations

All reagents and buffers were purchased as the highest grade commercially available and used without further purification. The following buffer stock solutions: MOPS, PIPES, HEPES, and ACES were prepared by dissolution of a weighed amount of the free acid form of the selected buffer into 18 MΩ water with the final solution pH adjusted to 7.2 by the addition of small volumes of concentrated NaOH solution. TRIS buffer solutions were prepared by dissolution of a weighed amount of the free TRIS base into 18 MΩ water with the final solution pH adjusted to 7.2 by the addition of small volumes of concentrated HCl. Phosphate buffer solutions were prepared by dissolution of weighed amounts of KH_2PO_4 into 18 MΩ water with the final solution pH adjusted to 7.2 by the addition of small volumes of concentrated NaOH solution. All solutions and media were prepared by using 18 MΩ water obtained from a Millipore ultrapurification system and all ITC experiments were done at 25 °C, unless otherwise noted.

HPCD was overexpressed in *E. coli* BL21(DE3) and purified as previously reported with a small modification (Wang & Lipscomb, 1997). Briefly, cells were lysed and HPCD was initially purified as previously reported. Fractions were then collected from a diethylaminoethyl (DEAE) strong anion exchange column and further separated using size exclusion chromatography. HPCD relevant fractions were collected, pooled, and concentrated. HPCD stock solutions were finally prepared in each of the study buffers by performing a buffer exchange by exhaustive dialysis into each of the study buffers, 100 m*M* (MOPS, PIPES, HEPES, ACES, TRIS, and phosphate) at pH 7.2. The final dialysate solution for each buffer system was kept for subsequent dilution of the HPCD stock solutions, preparation of the matched substrate titrant solutions, or for use in the appropriate ITC blank injection experiments.

4.2 Steady-State Kinetics by ITC

For the HPCD catalyzed turnover of HPCA to 5CHMSA (see Scheme 1), HPCD enzyme titrate solutions, having a nominal concentration of 8 n*M*, were prepared in each of the study buffers by dilution of the appropriate HPCD stock solution with the final buffer dialysate solution, while the matching HPCA substrate titrant solutions, having a nominal concentration of 14 m*M*, were prepared using the same final buffer dialysate solution as the solvent. Approximately, 1.5 mL of the 8 n*M* HPCD enzyme solution was loaded into the ITC cell and the matching 14 m*M* HPCA solution loaded into the ITC syringe. A single injection of 10 μL of the HPCA titrant was made into the HPCD solution in the ITC cell, and the microcalorimeter power signal collected for approximately 6500 s. Blank titrant injection experiments were performed in which 10 μL of the 14 m*M* HPCA titrant solution was added in a single injection into the ITC cell filled with the matched buffer solution. The titrant solution buffer and the buffer solution in the cell must be matched to minimize blank heat effects other than the dilution of the titrant, in these experiments, the approximate 1500-fold decrease in the HPCA concentration from 14 mM in the syringe to approximately 9000 nM in the ITC cell (at $t=0$). In determining the value for ΔH_{rxn} (or ΔH_{obs}) for the kinetic experiment, any detected heat effect for titrant (substrate) dilution observed in the blank experiment must be subtracted from the overall heat measured in the kinetic experiment. Data analysis was performed using software written by the Lewis Biophysics Laboratory at Mississippi State University. This software contains a graphical interface developed in MATLAB and was developed to assist in data visualization and extraction of steady-state parameters for enzymes exhibiting large k_{cat} values, e.g., HCPD and carbonic anhydrase. This software is available for free download at http://lewis.chemistry.msstate.edu. All experiments were reproduced in triplicate and the reported associated error is one standard deviation from the mean.

4.3 Analysis of the ITC Data

The reaction enthalpy plot for the HPCD catalyzed conversion of HPCA to 5CHMSA is seen in Fig. 4. The HPCD catalyzed steady-state reaction was carried out in the six buffers of interest, yielding a buffer-independent ΔH_{rxn}

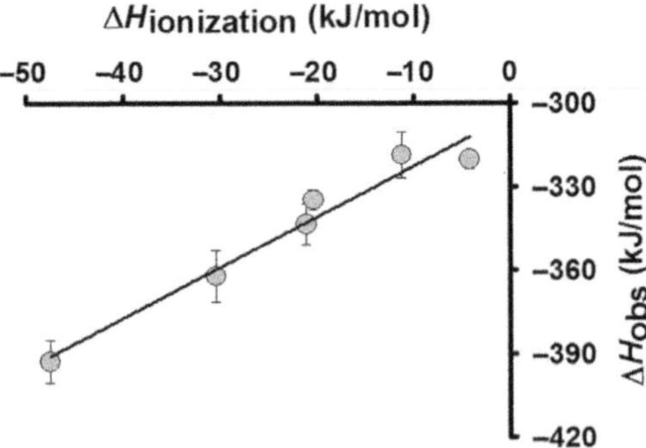

Figure 4 Reaction enthalpy plot for the ITC kinetic data. The enthalpies for the HPCA–HPCD reaction in various buffers are plotted against the enthalpy of ionization of the buffer to yield a linear relationship, where the *y*-intercept indicates the reaction enthalpy independent of the buffer (-305 ± 5 kJ/mol), and the slope is equal to the number of protons released by the system (1.7 ± 0.2 protons).

of -305 ± 5 kJ/mol and a release of 1.7 ± 0.2 protons, which is in good agreement with the accepted mechanism for the reaction (Emerson, Kovaleva, Farquhar, Lipscomb, & Que, 2008).

A comparison of the raw data for the reaction in MOPS and TRIS buffers (Fig. 1), it becomes immediately apparent that the HPCD catalytic reaction is slowed in TRIS buffer, when compared to the same reaction in MOPS. Kinetic analysis of the steady-state data indicates that the V_{max} for the reaction varies in the six different buffers (Table 1). Clearly, the reaction kinetics have the highest rates in MOPS buffer. Curiously, the HPCD reaction in potassium phosphate, and TRIS buffers have significantly lower V_{max} values, and the ring-cleavage kinetics in the remaining three buffers fall somewhere in the middle and are within error of each other.

By plotting the data in a traditional Lineweaver–Burk plot, visualization of the effect of the buffering environment on the catalytic activity of HPCD becomes more apparent (Fig. 5). Although this method is often used in more traditional kinetic analyses, our method of V_{max} determination allows for a more direct measurement of the maximum rate which we believe to be more accurate. However, by plotting all of the rates contained within the ITC experiment in a Lineweaver–Burk fashion, the observation of trends in the changing V_{max} and K_M values becomes more obvious. From this plot, we can see that the $1/V_{max}$ (y-intercept) for the reaction in HEPES, ACES, phosphate, and TRIS are very similar. Moreover, the $-1/K_M$ value (x-intercept) for phosphate, HEPES, and TRIS buffers are very similar to MOPS buffer, suggesting that these buffers could be acting as noncompetitive inhibitors, as classically described by a changing V_{max} while K_M is unchanged. A K_I value can then be calculated using the Michaelis–Menten equation in the presence of an inhibitor:

$$\text{app}\, V_{max} = \frac{(V_{max}[S])}{(\alpha K_M + \alpha'[S])} \tag{4}$$

For noncompetitive inhibitors (i.e., ACES), $\alpha = \alpha'$ and the equation can be simplified at high concentrations of substrate to:

$$\text{app}\, V_{max} = \frac{V_{max}}{\alpha} \tag{5}$$

where

$$\alpha = 1 + \frac{[I]}{K'_I} \tag{6}$$

Table 1 Kinetic Parameters for the Ring-Opening Reaction of HPCA by HPCD

	MOPS	**PIPES**	**HEPES**	**ACES**	**Phosphate**	**TRIS**
V_{max}	2.77 ± 0.09	2.34 ± 0.17	1.97 ± 0.45	1.83 ± 0.16	1.75 ± 0.02	1.65 ± 0.11
k_{cat}	336 ± 11	283 ± 19	243 ± 50	228 ± 11	215 ± 6	200 ± 13
K_M	7 ± 2	13 ± 16	5 ± 9	12 ± 8	9 ± 3	11 ± 4
k_{cat}/K_M	8×10^5	4×10^5	8×10^5	3×10^5	4×10^5	3×10^5

The kinetic terms are reported in the following units: V_{max} is in μ*M*/min, k_{cat} is in min^{-1}, and K_M is in μ*M*.

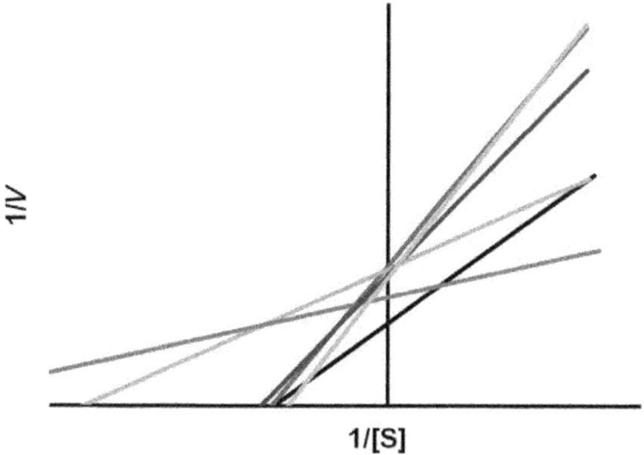

Figure 5 Lineweaver–Burk plots of the steady-state kinetic data associated with HPCA ring opening by HPCD in different buffer systems at pH 7.2. Legend: MOPS (black), TRIS (red), ACES (green), PIPES (purple), phosphate (blue), and HEPES (orange). (See the color plate.)

Rearrangement of the equation results in calculation of K_I:

$$K_I = \frac{[I]}{\left(\frac{V_{max}}{\text{app}\,V_{max}} - 1\right)} \tag{7}$$

For example, when we use the kinetic data collected, a K_I value for TRIS is calculated to be approximately 147 mM. Other K_I values can be calculated to yield a similar result in the high mM concentration range.

4.4 Confirming the Steady-State Measurement

A flask containing 100 mL of 5 μM HPCD and 5 mM HPCA in 100 mM MOPS buffer pH 7.2 was stirred for 48 h to ensure that all HPCA was converted to product. The reaction was quenched with 3 mL of 2 N HCl and placed in a separatory funnel. 3 × 100 mL of dichloromethane (DCM) were added to the flask and gently shaken. The two layers were allowed to separate and the DCM layer was extracted and evaporated, leaving pure product. The product was weighed by mass difference with the flask, and placed in 3 mL of water. The characteristic UV spectrum of the product was observed, which has a λ_{max} of 380 nm at pH 8.0 ($\varepsilon_{380} = 36{,}000\ M^{-1}\ \text{cm}^{-1}$), in agreement with the literature value for the product of the HPCD reaction (Miller & Lipscomb, 1996). A pH titration was then performed on 5CHMSA to determine the λ_{max} at acidic pH, for use in HPLC analysis. A spectral shift to λ_{max} of 320 nm was observed and the resultant molar absorptivity was calculated to be 13,914 M^{-1} cm^{-1}. A mass spectrum was obtained for the compound and identified with a mass of 199.0214 m/z in negative mode, in excellent agreement with the $[M+H]^-$ theoretical value (199.028 m/z), indicating successful separation of 5CHMSA.

A standard line of 5CHMSA varying in concentration from 50 to 750 μM was made in water and spiked with 30 μL of 2 N HCl. A 5CHMSA control was prepared in 25 mM MOPS and TRIS buffers, pH 7.2, and also spiked with 2 N HCl for HPLC analysis. A 1 μM HPCD sample was prepared in each buffer, and HPCA was added to the HPCD samples for final concentration of 1 mM substrate in 1 mL total volume. The samples were briefly vortexed and allowed to react for 5 min, at which point the samples were spiked with 30 μL of 2 N HCl to quench the reaction. All samples and standards were centrifuged at 13,000 rpm for 2 min and transferred to HPLC vials. A mobile phase of 1% acetic acid (A) and methanol (B) was used for the elution of 5CHMSA on a C18 column using the following gradient: 0–5 min 100% A; 5–6 min 100–70% A; 6–9 min 70% A; 9–10 min 70–0% A; 10–11 min 0% A. An injection delay period of 5 min was used between samples to reequilibrate the column in 100% solvent A. The product was detected using a UV–vis diode array detector at a wavelength of 320 nm with a reference wavelength of 600 nm. Signature spectra were collected on all chromatographic peaks as shown in Fig. 6. Product

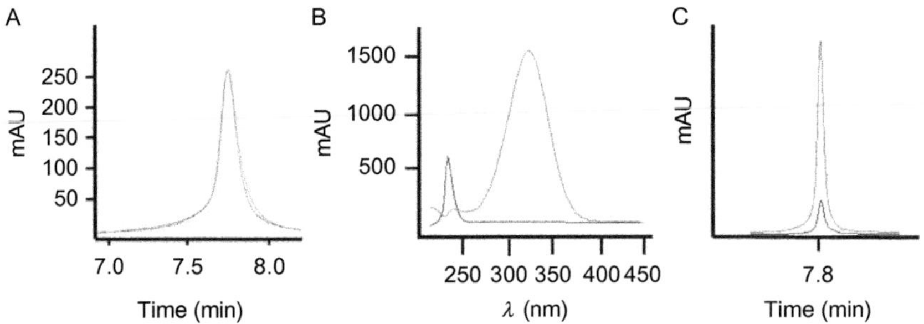

Figure 6 HPLC data for the separation and detection of 5CHMSA. (A) An overlay trace of a 100 μM 5CHMSA control in MOPS (red (light gray in the print version)) and TRIS (blue (dark gray in the print version)) buffers. (B) The UV–vis signature spectrum of the 5CHMSA standard (red (light gray in the print version)) against the reference (blue (dark gray in the print version)). (C) The elution peak of 5CHMSA from the catalytic reaction by HPCD in MOPS (red (light gray in the print version)) and TRIS (blue (dark gray in the print version)) buffers. The reaction was allowed to proceed for 5 min and quenched with 2 N HCl to denature the enzyme.

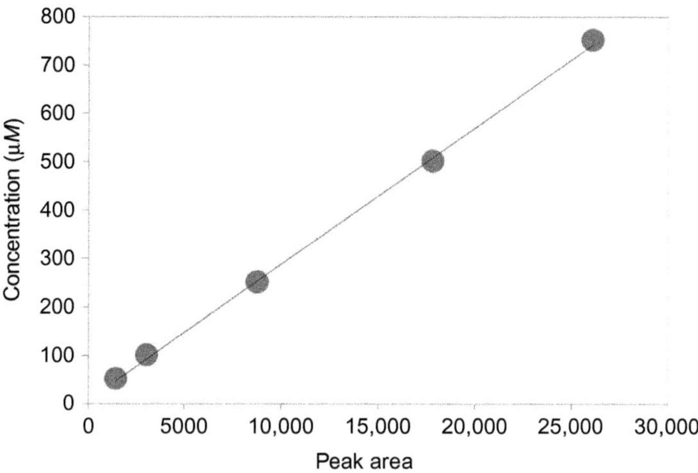

Figure 7 5CHMSA standard line for the quantitation of reaction product. The standard line ranged in concentration from 50 to 750 μ*M* 5CHMSA in water. Elution of 5CHMSA from the HPLC column occurred at approximately 7.8 min. A plot of the peak area against the standard concentration provided a linear relationship in which to quantify reaction products in MOPS and TRIS buffers.

samples were collected for the HPCD catalyzed reaction in each buffer, with the final 5CHMSA concentration quantified using the 5CHMSA standard line (see Fig. 7). These substrate conversion results in each buffer were compared to further support the ITC kinetic results.

A brief quantification study was performed using HPLC to further support the ITC kinetics result. A standard line of 5CHMSA was created to then assess the amount of product formed after 5 min reaction time in both MOPS and TRIS buffers. The excellent linearity of the standard line yielded an R^2 value of 0.999 and gave the means to quantify the reaction product by integration of the chromatographic peak at an elution time of 7.8 min. Additionally, a UV–vis signature spectrum from 200 to 500 nm was collected on the chromatographic peak, providing a second means of identification for 5CHMSA.

Control samples 100 μ*M* 5CHMSA in each buffer were separated by HPLC and analyzed to ensure that the peak areas and the resulting calculated concentrations were identical in each buffering system, and that the buffers were not forming complexes with the product, effectively lowering the molar extinction coefficient of the product as suggested in previous reports (Whiting et al., 1996). The results of this analysis yielded an average peak area for the controls of 3218 ± 123, which corresponds to an average concentration of 98 ± 11 μ*M* across the buffers tested. A representative overlay of the 100 μ*M* controls in TRIS and MOPS buffers can be seen in Fig. 6A, whereas the absorption spectra of this product are given in Fig. 6B.

The HPLC chromatogram of 5CHMSA for the reaction that was quenched after 5 min clearly shows a difference in the amount of product produced for the reaction in MOPS and TRIS buffers (Fig. 6C). Calculation of the concentration based on the 5CHMSA standard line indicates that after 5 min, 204 ± 24 μ*M* was produced from the reaction in MOPS buffer, while the reaction in TRIS buffer only produced 131 ± 4 μ*M*. This indicates that the reaction is dramatically slowed in TRIS buffer. Varying concentrations of 5CHMSA were produced for the reaction in the additional buffers (data not shown), which corroborates with the kinetic data obtained by ITC.

5. DISCUSSION

The use of buffering systems in biochemical reactions is necessary and routine for the study of biological macromolecules *in vitro*. These buffering systems maintain the pH of the reaction solution where the buffering capacity is imperative in the realm of biochemical experiments; as the stability and reactivity of most enzymes are pH dependent. However, over the years it has been demonstrated that this central function of buffers is not necessarily the only reaction they perform in solution (Desmarais et al., 2002; Kirsch et al., 1998; McPhail & Holt, 1999; Taha et al., 2011; Welch et al., 2002; Yu et al., 1997). For example, several inorganic salts (e.g., phosphate buffers) can have an effect on the conformational stability of an enzyme or protein, which has led to the increased usage of Good's buffering systems (McPhail & Holt, 1999). These commonplace zwitterionic buffers were developed to be highly soluble in water and have low solubility in organic solvents (Ferguson et al., 1980; Good & Izawa, 1972; Good et al., 1966). In addition to these properties, Good's buffers were considered to be relatively inert when used in the concentration range of 10–50 m*M*. The presence of the bulky tertiary amine groups was supposed to limit the formation of ion complexes in solution. However, 17 of the 20 Good's buffers have been shown to form complexes with biologically important metal ions (Yu et al., 1997). TRIS, one of the most commonly used biological buffers, is also well known to form complexes with a number of metal ions (Desmarais et al., 2002). The formation of buffer metal ion complexes can be a serious complication when studying metalloproteins. This is the case whether the metal ions are playing a catalytic or structural role. It has also been shown that several common biochemical assays are greatly affected by the choice of the buffering system (Lucarini & Kilikian, 1999). In this report, we recount the effects of several buffering systems on the structure and function of the HPCD enzyme and attempt to deconvolute the steady-state kinetics for HPCA turnover with respect to the effects that enzyme–buffer interactions may have on the intrinsic HPCD kinetic parameter values.

Previously, we reported the thermodynamic parameters for binding the native substrate, HPCA (3,4-dihydroxyphenylacetic acid) to HPCD as determined in ITC experiments done under anaerobic conditions (Henderson et al., 2012). These experiments were performed in a number of different buffers having well known enthalpy changes for buffer ionization (or protonation). These ITC experiments were analyzed to provide insight into the number of protons released from the enzyme or substrate upon formation of the enzyme–substrate complex. The buffering systems chosen for these binding experiments were based upon their commonplace use in the laboratory, their pK_a's or buffering capacity at pH 7.2, and the differences in their heats of ionization. However, during the course of these substrate-binding studies, additional endothermic enthalpies were observed in three of the buffering systems, defying the predicted linear trend. Specifically, the overall ΔH values measured for HPCA binding to HPCD in ACES, potassium phosphate, and TRIS buffers were observed to deviate from the ΔH values predicted from Eq. (3) and an estimated loss of 1.7 protons by up to 117 kJ/mol. The proposed explanation for this behavior was that additional complicating reactions, for example, a buffer–protein interaction, was responsible for the additional enthalpy change term in the overall enthalpy change observed for substrate binding in the three noninert buffer systems.

In order to assess this interaction, our intentions were to observe any changes in structure and stability, as well as determine whether the rate of the catalytic reaction by HPCD was also affected by the proposed buffer–enzyme interactions. For this purpose, we employed the ITC kinetic experiment method to monitor the steady-state kinetics for the catalytic turnover of the native substrate HPCA by the HPCD enzyme. The ITC kinetic experiment method has been well established as a practical means for the measurement of enzyme catalyzed rates of reaction under steady-state conditions, particularly for following the rates of substrate conversion in systems at low (n*M*) concentrations (Cai et al., 2001; Lonhienne, Baise, Feller, Bouriotis, & Gerday, 2001; Todd & Gomez, 2001). The steady-state kinetic parameters for the turnover of HPCA by HPCD have been reported by

Groce et al., in MOPS buffer at pH 8.0, and by Emerson et al. in MOPS buffer at pH 7.5 Groce, Miller-Rodeberg, & Lipscomb, 2004; Emerson et al., 2008. However, the spectrum of the 5CHMSA product changes with pH and requires adjustments to the molar extinction coefficient dependent on the pH of the reaction. Additionally, Whiting et al. reported a change in extinction coefficient of 5CHMSA when reactions in MOPS and TRIS were compared (Whiting et al., 1996). This idea is contradictory to our HPLC analysis that indicates there is no significant complexation between the product, 5CHMSA and either of these buffers as the UV spectrum for the product 5CHMSA in these buffers is identical to the UV spectrum for pure 5CHMSA, and to the UV spectra for control samples at similar 5CHMSA concentrations.

The common practice of measuring Michaelis–Menten kinetics by spectroscopy is limited in that the reported V_{max} is a theoretical value based on data fitting, and not a true, directly measured term. While employing the ideas of Lineweaver–Burk provides a means of obtaining a value for V_{max}, the Lineweaver–Burk relationship can result in large errors in the estimates obtained for the LB slope at low concentrations of substrate and inaccurate values for V_{max}. The ITC kinetic method provides at least two benefits over the typical spectroscopic experiment: (1) the heat of reaction is monitored and observation of the reaction is no longer restricted to substrates (and or products) that are spectroscopically active (possess a chromophore with a characteristic absorbance) and (2) by initially saturating the enzyme with substrate, the heat rate measured at $t=0$ provides a direct measurement of the maximum reaction rate, V_{max} without any need for data fitting or extrapolation.

The kinetic parameters obtained here by ITC are in agreement with the findings presented in previously published work of Emerson et al. and Groce et al. In both cases, the kinetic analyses were performed in MOPS buffer at higher pH using spectroscopic techniques, and it was found that the optimum pH for the reaction is at pH 8.0. Experiments performed at pH 8.0 have the highest turnover of substrate ($k_{cat}=600$ min^{-1}) (Groce et al., 2004), whereas the experiments performed at pH 7.8 report a k_{cat} value of 470 min^{-1} (Emerson et al., 2008), illustrating the reduction in speed at lower pH. Our experiments resulted in a k_{cat} value of 336 min^{-1} in MOPS buffer at pH 7.2, following this trend.

Interestingly, the V_{max} and subsequent calculation of k_{cat} for HPCD do drastically change in phosphate and TRIS buffers, in direct agreement with our thermodynamic ITC-binding data that suggest an additional process taking place during the binding event. Using the Lineweaver–Burk plots to analyze the kinetic data, it becomes visually obvious that some of the buffering systems could be acting as weak, noncompetitive or uncompetitive inhibitors. Structurally speaking, the Good's buffers used in these experiments all contain either a piperazine or morpholine ring. ACES, along with TRIS contain primary amines in their structures, allowing them to be privy to various molecular interactions with the enzyme. In addition, TRIS is a known strong metal-binding ligand, while potassium phosphate has been implicated in binding with arginine and lysine residues, which could be crucial to the function of the lid domain which contains an arginine residue suspected of providing favorable contacts with the bound substrate (Desmarais et al., 2002; Harutynunyan et al., 1996; McPhail & Holt, 1999). It could then be inferred that for the monitoring of the catalytic reaction of HPCD *in vitro*, Good's buffers containing the piperazine or morpholine ring structures (such as PIPES, HEPES, and MOPS) may be more inert, and less likely to participate in enzyme–buffer interactions, which makes them more suitable for use in the binding studies of HPCD.

By performing experiments on the steady-state kinetics of HPCD using ITC, the ΔH_{obs} was also obtained in each of the buffering systems. When these enthalpies are plotted against the $\Delta H_{ionization}$, a correlation graph is obtained and linearly fit by Eq. (7). From this, we obtained a ΔH_{rxn} of -305 ± 5 kJ/mol. The $\Delta H_{binding}$ for the binding reaction of HPCA to HPCD was reported to be -71 kJ/mol, leaving -163 kJ/mol of enthalpic energy for O_2 binding and activation, incorporation into the ring, ring cleavage, and product release. Additionally, we observe 1.7 protons released from the turnover reaction. This proton loss is in agreement with our reported reaction mechanism, where both protons on the catechol substrate are lost during substrate binding and prior to O_2 binding, and the resulting muconic semialdehyde species is then released to the bulk solvent and reprotonated (Henderson et al., 2012). The additional loss of 0.7 protons could be attributed to the partial loss of a proton from the formation of an enolate complex in the 5CHMSA product. Curiously, when the $\Delta H_{binding}$ plot (Fig. 3) is compared to the ΔH_{rxn} plot (Fig. 4), no additional endothermic processes are noticeable in the ΔH_{rxn} plot, where the ΔH_{obs} in various buffers all fall within 12 kJ/mol of the linear fit. This could be due to a reversible process, such as a specific enzyme–buffer interaction that is removed for the binding reaction to take place (i.e., within the active site). When oxygen is present and turnover results, completion of the reaction allows for the enzyme–buffer interaction to once again occur, which would be in line with the assessment that the buffers are providing a noncompetitive inhibition-binding interaction. This interaction would be equal in enthalpic magnitude but opposite in sign, thus unobservable during reaction enthalpy measurements but visible through kinetic studies. However, when just the binding reaction is performed in the absence of O_2, only the removal of the enzyme–buffer interaction is observed with substrate binding, since the turnover reaction is not allowed to proceed. Therefore, the additional enthalpic process confounds the $\Delta H_{\text{binding}}$ data.

If these interactions are truly noncompetitive in nature, we can assess the K_I for each of the buffers, which indicates high concentration (>100 m*M*) values. While large K_I values are generally disregarded as inconsequential, they become relevant and of concern when dealing with buffer systems, as many biochemical reactions are performed in buffer systems with concentrations within a range of 10–150 m*M*. In some cases, buffer concentrations can be many orders of magnitude higher than the biological molecules under consideration.

6. CONCLUSIONS

To summarize, the data shown here suggest that buffer components and buffering systems may not be innocent bystanders in enzyme catalyzed reactions. Weak enzyme–buffer interactions that can confound the measurement of both the thermodynamic parameters for substrate binding and the kinetic parameters for the steady-state conversion of substrate to product. In the case of the HPCD–HPCD system, the anomalous or confounding buffer–enzyme interactions appear to hinder the substrate-binding process. The offending buffers, TRIS, phosphate, and ACES, should probably be avoided when studying substrate-binding reactions in 2-His-1-carboxylate metalloenzymes. As illustrated in the HPCD experiments described here, the favored buffers for the nonheme iron(II) HPCD enzyme are the Good's buffers, which contain morpholine or piperazine rings. These systems generally use a sterically bulky tertiary amine as the ionizable group that provides the buffering capacity. While the avoidance of buffers limits the number of buffering systems that can be utilized for proton release analysis, it ultimately allows for a more accurate determination of the enthalpy of substrate binding and for a more accurate determination of the heat of reaction for substrate conversion and the kinetic parameters which define the efficiency of the enzyme.

Lastly, it is clear that ITC kinetic methods have a place in the biochemists toolbox for the purpose of characterizing the thermodynamics of protein substrate (and/or inhibitor or product) interactions as well as a place in characterizing the kinetic parameters for an enzyme catalyzed reaction. Perhaps even more importantly, ITC kinetic methods have been shown to be useful in deconvoluting the participation of buffer components (or perhaps even other noninert solute species) in the overall process of substrate binding and hence substrate conversion.

REFERENCES

Armstrong, R. N. (2000). Mechanistic diversity in a metalloenzyme superfamily. *Biochemistry, 39*, 13625–13632.

Bianconi, M. L. (2007). Calorimetry of enzyme-catalyzed reactions. *Biophysical Chemistry, 126*, 59–64.

Cai, L., Cao, A., & Lai, L. (2001). An isothermal titration calorimetric method to determine the kinetic parameters of enzyme catalytic reaction by employing the product inhibition as probe. *Analytical Biochemistry, 299*, 19–23.

Desmarais, W. T., Bienvenue, D. L., Bzymek, K. P., Holz, R. C., Petsko, G. A., & Ringe, D. (2002). The 1.20 A resolution crystal structure of the aminopeptidase from Aeromonas proteolytica complexed with tris: A tale of buffer inhibition. *Structure, 10*, 1063–1072.

Emerson, J. P., Kovaleva, E. G., Farquhar, E. R., Lipscomb, J. D., & Que, L., Jr. (2008). Swapping metals in Fe- and Mn-dependent dioxygenases: Evidence for oxygen activation without a change in metal redox state. *Proceedings of the National Academy of Sciences of the United States of America, 105*, 734–752.

Ferguson, W. J., Braunschweiger, K. I., Braunschweiger, W. R., Smith, J. R., McCormick, J. J., Wasmann, C. C., et al. (1980). Hydrogen ion buffers for biological research. *Analytical Biochemistry, 104*, 300–310.

Freyer, M. W., & Lewis, E. A. (2008). Isothermal titration calorimetry: Experimental design, data analysis, and probing macromolecule/ligand binding and kinetic interactions. *Methods in Cell Biology, 84*, 79–113.

Gerlt, J. A., & Babbitt, P. C. (2001). Divergent evolution of enzymatic function: Mechanistically diverse superfamilies and functionally distinct suprafamilies. *Annual Review of Biochemistry, 70*, 209–246.

Goldberg, R. N., Kishore, N., & Lennen, R. M. (2002). Thermodynamic quantities for the ionization reactions of buffers. *Journal of Physical and Chemical Reference Data, 31*, 231–370.

Good, N. E., & Izawa, S. (1972). Hydrogen ion buffers. *Methods in Enzymology, 24*, 53–68.

Good, N. E., Winget, G. D., Winter, W., Connolly, T. N., Izawa, S., & Singh, R. M. M. (1966). Hydrogen ion buffers for biological research. *Biochemistry, 5*, 167–177.

Groce, S. L., Miller-Rodeberg, M. A., & Lipscomb, J. D. (2004). Single-turnover kinetics of homoprotocatechuate 2,3-dioxygenase. *Biochemistry, 43*, 15141–15153.

Grossoehme, N. E., Spuches, A. M., & Wilcox, D. E. (2010). Application of isothermal titration calorimetry in bioinorganic chemistry. *Journal of Biological Inorganic Chemistry, 15*, 1183–1191.

Harutynunyan, E. H., Kuranova, I. P., Vainshtein, B. K., Hohne, W. E., Lamzin, V. S., Dauter, Z., et al. (1996). X-ray structure of yeast inorganic pyrophosphatase complexed with manganese and phosphate. *European Journal of Biochemistry, 239*, 220–228.

Hegg, E. L., & Que, L., Jr. (1997). The 2-His-1-carboxylate facial triad—An emerging structural motif in mononuclear non-heme iron(II) enzymes. *European Journal of Biochemistry, 250*, 625–629.

Henderson, K. L., Le, V. H., Lewis, E. A., & Emerson, J. P. (2012). Exploring substrate binding in homoprotocatechuate 2,3-dioxygenase using isothermal titration calorimetry. *Journal of Biological Inorganic Chemistry, 17*, 991–994.

Kirsch, M., Lomonosova, E. E., Korth, H. G., Sustmann, R., & de Groot, H. (1998). Hydrogen peroxide formation by reaction of peroxynitrite with HEPES and related tertiary amines. Implications for a general mechanism. *The Journal of Biological Chemistry, 273*, 12716–12724.

Kovaleva, E. G., & Lipscomb, J. D. (2007). Crystal structures of Fe^{2+} dioxygenase superoxo, alkylperoxo, and bound product intermediates. *Science, 316*, 453–457.

Kovaleva, E. G., & Lipscomb, J. D. (2012). Structural basis for the role of tyrosine 257 of homoprotocatechuate 2,3-dioxygenase in substrate and oxygen activation. *Biochemistry, 51*, 8755–8763.

Leitgeb, S., & Nidetzky, B. (2008). Structural and functional comparison of 2-His-1-carboxylate and 3-His metallocentres in non-haem iron(II)-dependent enzymes. *Biochemical Society Transactions, 36*, 1180–1186.

Lipscomb, J. D. (2008). Mechanism of extradiol aromatic ring-cleaving dioxygenases. *Current Opinion in Structural Biology, 18*, 644–649.

Lonhienne, T., Baise, E., Feller, G., Bouriotis, V., & Gerday, C. (2001). Enzyme activity determination on macromolecular substrates by isothermal titration calorimetry: Application to mesophilic and psychrophilic chitinases. *Biochimica et Biophysica Acta, 1545*, 349–356.

Lucarini, A. C., & Kilikian, B. V. (1999). Comparative study of Lowry and Bradford methods: Interfering substances. *Biotechnology Techniques, 13*, 149–154.

McPhail, D., & Holt, C. (1999). Effect of anions on the denaturation and aggregation of β-lactoglobulin as measured by differential scanning microcalorimetry. *International Journal of Food Science and Technology, 34*, 477–481.

Miller, M. A., & Lipscomb, J. D. (1996). Homoprotocatechuate 2,3-dioxygenase from Brevibacterium fuscum. A dioxygenase with catalase activity. *The Journal of Biological Chemistry, 271*, 5524–5535.

Taha, M., Gupta, B. S., Khoiroh, I., & Lee, M. J. (2011). Interactions of biological buffers with macromolecules: The ubiquitous "smart" polymer PNIPAM and the biological buffers MES, MOPS, and MOPSO. *Macromolecules, 44*, 8575–8589.

Todd, M. J., & Gomez, J. (2001). Enzyme kinetics determined using calorimetry: A general assay for enzyme activity? *Analytical Biochemistry, 296*, 179–187.

Vaillancourt, F. H., Bolin, J. T., & Eltis, L. D. (2006). The ins and outs of ring-cleaving dioxygenases. *Critical Reviews in Biochemistry and Molecular Biology, 41*, 241–267.

Wang, Y. Z., & Lipscomb, J. D. (1997). Cloning, overexpression, and mutagenesis of the gene for homoprotocatechuate 2,3-dioxygenase from Brevibacterium fuscum. *Protein Expression and Purification, 10*, 1–9.

Welch, K. D., Davis, T. Z., & Aust, S. D. (2002). Iron autoxidation and free radical generation: Effects of buffers, ligands, and chelators. *Archives of Biochemistry and Biophysics, 397*, 360–369.

Whiting, A. K., Boldt, Y. R., Hendrich, M. P., Wackett, L. P., & Que, L., Jr. (1996). Manganese(II)-dependent extradiol-cleaving catechol dioxygenase from Arthrobacter globiformis CM-2. *Biochemistry, 35*, 160–170.

Yu, Q., Kandegedara, A., Xu, Y., & Rorabacher, D. B. (1997). Avoiding interferences from Good's buffers: A contiguous series of noncomplexing tertiary amine buffers covering the entire range of pH 3-11. *Analytical Biochemistry, 253*, 50–56.

SECTION II

DSC

CHAPTER TWELVE

Modern Analysis of Protein Folding by Differential Scanning Calorimetry

Beatriz Ibarra-Molero*, Athi N. Naganathan†, Jose M. Sanchez-Ruiz*,[1], Victor Muñoz‡,§,[1]

*Facultad de Ciencias, Departamento de Química-Física, Universidad de Granada, Granada, Spain
†Department of Biotechnology, Bhupat & Jyoti Mehta School of Biosciences, Indian Institute of Technology Madras, Chennai, India
‡Centro Nacional de Biotecnología, Consejo Superior de Investigaciones Científicas, Madrid, Spain
§School of Engineering, University of California, Merced, California, USA
[1]Corresponding authors: e-mail address: sanchezr@ugr.es; vmunoz@cnb.csic.es

Contents

Methods in Enzymology, Volume 567
ISSN 0076-6879
http://dx.doi.org/10.1016/bs.mie.2015.08.027

Abstract

Differential scanning calorimetry (DSC) is a very powerful tool for investigating protein folding and stability because its experimental output reflects the energetics of all conformations that become minimally populated during thermal unfolding. Accordingly, analysis of DSC experiments with simple thermodynamic models has been key for developing our understanding of protein stability during the past five decades. The discovery of ultrafast folding proteins, which have naturally broad conformational ensembles and minimally cooperative unfolding, opens the possibility of probing the complete folding free energy landscape, including those conformations at the top of the barrier to folding, via DSC. Exploiting this opportunity requires high-quality experiments and the implementation of novel analytical methods based on statistical mechanics. Here, we cover the recent exciting developments in this front, describing the new analytical procedures in detail as well as providing experimental guidelines for performing such analysis.

1. INTRODUCTION

A differential scanning calorimetry (DSC) run involves the transfer of energy to a sample cell that contains the protein solution and a reference cell containing the same buffer performed in such a way so that their temperature increases at a constant rate (the scanning rate). Since the two cells differ in composition, ensuring that temperature increases at the same rate for both cells implies that slightly different amounts of energy need to be transferred to the reference and sample cells. The difference in heating energy can be used to calculate an apparent value for the heat capacity of the protein. "Apparent" is used here because, as we discuss in detail further below, unprocessed DSC heat capacity data are always distorted by a water displacement effect (a distortion that can be corrected) and often distorted by the time dependence of protein folding–unfolding and/or protein irreversible denaturation (distortions that may be impossible to correct). The experimental outcome is a profile of protein heat capacity versus temperature that shows "peaks" signaling the occurrence of protein unfolding/denaturation. This makes sense because such processes are endothermic and thus require the transfer of energy to keep the sample cell temperature equal to that of the reference cell.

DSC is, arguably, the most informative methodology that can be used to study protein thermal denaturation. Furthermore, modern instrumentation allows DSC experiments to be performed efficiently and reliably. There is, however, a drawback. Proper analysis is neither trivial nor straightforward, and requires concepts and tools from equilibrium thermodynamics, chemical kinetics, and statistical mechanics. The need to develop creative approaches for data analysis explains the limited popularity of DSC in these times of high-throughput screening and instant gratification. It must be recognized, nevertheless, that DSC studies have contributed greatly to our current understanding of protein stability and energetics. For instance: (1) they have led to the currently accepted description of the temperature dependence of protein stability, including the existence of heat and cold denaturation (Becktel & Schellman, 1987; Privalov, 1990); (2) they have provided a basis for much of our understanding of structure–energetics relationships in proteins, and its application in protein design (Freire, 2001; Makhatadze & Privalov, 1995; Robertson & Murphy, 1997); (3) they have demonstrated that protein stability is often controlled by kinetic factors (Sanchez-Ruiz, 1992, 2010), thus providing compelling evidence for the biophysical and evolutionary relevance of protein kinetic stability; (4) they have led to a connection between bulk experimental folding/unfolding data and ensemble views that take into account protein conformational heterogeneity and folding energy landscapes.

The fourth application mentioned above is the actual focus of this chapter. Experimental data on protein folding/denaturation are widely analyzed with models that assume a small number of discrete macrostates, such as the native and unfolded states of the two-state model. Such analyses are useful and convenient in many cases, but it is important to recognize their phenomenological character and their limitations. Strictly speaking, proteins in their native state are not unique conformations, as naive interpretations derived from static, X-ray 3D structures might seem to imply. In fact, it is now widely accepted that native proteins in solution are ensembles of conformations with interconversion dynamics that are key to many phenomena relevant for biological function, including catalysis, promiscuity, and molecular recognition (Bahar, Lezon, Yang, & Eyal, 2010; Henzler-Wildman et al., 2007; Khersonsky & Tawfik, 2010; Leone, Marinelli, Carloni, & Parinello, 2010; Sikosek & Chan, 2014; Vogt & Di Cera, 2012; Zou, Risso, Gavira, Sanchez-Ruiz, & Ozkan, 2015). The ensemble nature of protein macrostates is even more germane to unfolded or partially unfolded states, which are inherently heterogeneous. It naturally follows that the conformational heterogeneity of partially folded ensembles, such as those controlling the overall folding/unfolding kinetics (the transition state ensemble; Onuchic, Socci, Luthey-Schulten, & Wolynes, 1996), must be intermediate between those of the end states. Analyzing protein folding/unfolding experiments using statistical mechanical models that go beyond discrete macrostates is thus desirable, but challenging (Muñoz, 2001). However, such challenges can be best met when the available experimental data includes DSC. This is so because the equilibrium DSC profile is equivalent to the relevant partition function of the system, as was demonstrated by Biltonen and Freire many years ago (Freire & Biltonen, 1978). Provided that equilibrium is maintained during the scan, protein heat capacity profiles contain information about all the microstates that become significantly populated during thermal denaturation (Muñoz & Sanchez-Ruiz, 2004; Sanchez-Ruiz, 2011).

In recent years, a variety of such statistical mechanics models have been developed and applied to the analysis of protein unfolding. These efforts have been motivated by the discovery of many proteins capable of folding to completion in the microsecond timescale (Gruebele, 2008; Muñoz, 2007; Prigozhin & Gruebele, 2013). The folding rate for these proteins approaches the folding speed limit (Kubelka, Hofrichter, & Eaton, 2004), and thus, these are proteins that must fold by crossing very small free energy barriers or no barrier at all. Their folding free energy surfaces are so shallow that all the relevant conformational microstates may become significantly populated during unfolding and are thus amenable to detection by thermodynamic means (Muñoz, 2002). This idea has sparked the development of multiple equilibrium unfolding methods to study fast-folding proteins, including the combination of multiple spectroscopic probes (Garcia-Mira, Sadqi, Fischer, Sanchez-Ruiz, & Muñoz, 2002; Naganathan & Muñoz, 2014), and the analysis at atomic resolution using NMR alone (Sadqi, Fushman, & Munoz, 2006) or in combination with atomistic computer simulations (Sborgi et al., 2015).

Among those approaches, DSC analysis has stood out on its own, demonstrating capacity for estimating the very small free energy barriers of fast and ultrafast folding proteins (Sanchez-Ruiz, 2011). But, the reader must not conclude from this discussion that a DSC experiment automatically provides a value for the folding/unfolding free energy barrier. Determination of marginal folding free energy barriers from DSC is involved and must meet the following criteria: (1) the experimental DSC data in question must reflect the true equilibrium thermodynamics; (2) the data must be of high quality, and in absolute heat capacity units for comparison with the heat capacity levels expected for the native and fully unfolded protein; (3) statistical mechanics analyses should provide a more accurate and physically reasonable description of the data than conventional thermodynamic analyses that assume discrete macrostates; and (4) the analysis should ideally be carried out with different statistical mechanics models to obtain consistent, model-independent, results.

In this chapter, we deal with what we term modern analysis of protein denaturation by DSC, discussing the four requirements one by one. Our motivation is to provide detailed guidelines for researchers interested in performing DSC experiments that are conducive to such modern analysis as well as specific instructions for carrying out the theoretical analysis of DSC data in terms of conformational ensembles and free energy surfaces.

2. THERMODYNAMICS AND THE DSC EXPERIMENT

Determining whether an equilibrium thermodynamics analysis of DSC profiles is acceptable depends on the results obtained on the two basic, yet essential, experimental tests: calorimetric reversibility and scan rate effects. The possible outputs obtained on these tests define various thermodynamic scenarios that need to be tackled in specific ways. Figure 1 shows a flowchart with the four different experimental scenarios highlighted in color (red (thick

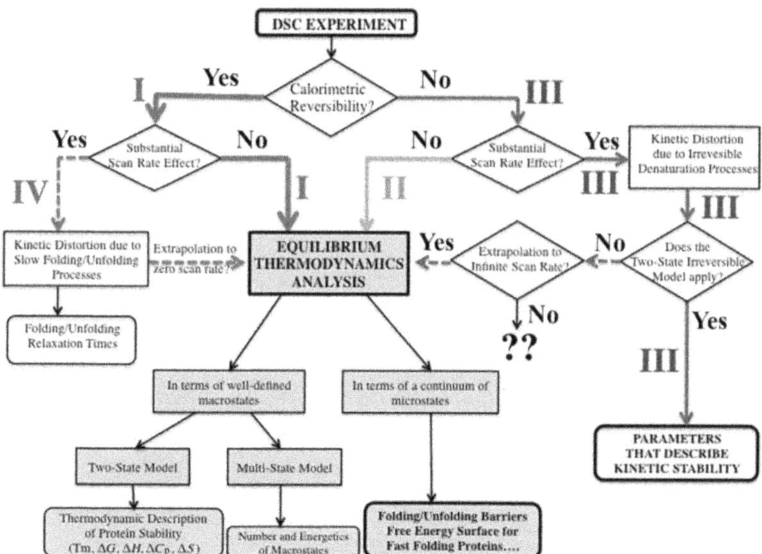

Figure 1 Flowchart showing the different scenarios for analysis of experimental DSC data on the basis of the results provided by tests of calorimetric reversibility and scan rate effects.

dark gray in the print version), orange (light gray in the print version), blue (thin gray in the print version), and dashed red (dashed gray in the print version) pathways in flowchart) and the procedures one must follow in each case to make the equilibrium thermodynamics analysis feasible. Common scenarios are shown with continuous lines, while dashed lines are used for situations that occur less often or have not been fully explored in the literature.

This section is not intended to provide an exhausive description of DSC methods, but rather highlight the key experimental issues to be considered when DSC experiments aim at determining the general properties of the thermodynamic free energy surface for protein folding. The reader will be walked through the flowchart in a concise and practical manner, including references to previous methodological works when required for further information. Note that the "equilibrium thermodynamics analysis" occupies a central position. This is so because those conditions are sought after by most DSC practitioners and are also conducive to the modern theoretical analysis that is the main topic of this chapter. It is important to keep in mind, however, that most DSC experiments on protein denaturation do not permit an equilibrium thermodynamics analysis, but are controlled by kinetic effects. In other words, the validity of performing an equilibrium thermodynamics analysis of DSC profiles needs to be explicitly demonstrated in each particular experimental case.

2.1 The DSC Experiment

Detailed descriptions on methodological aspects of protein denaturation calorimetry have been provided in previous works (Johnson, 2013; Lopez & Makhatadze, 2002; Makhatadze, 2001). Here, we only discuss some crucial points:

1. Preparing a protein sample of ultra-high purity is essential. The sample must be exhaustively dialyzed against the proper buffer (three buffer changes every 8 h or more, depending on membrane cut-off) to ensure that protein and reference solutions have identical pH and composition. This is a key issue because DSC instruments operate in differential mode. The actual DSC output is the difference in heat capacity between the reference and the sample cells, which implies that any minute difference in buffer compositions results in large baseline artifacts.
2. The selected buffer must have low ionization enthalpy to ensure a low temperature dependence of the pH value (Goldberg, Kishore, & Lennen, 2002).

3. Several buffer–buffer baselines (three or four typically) must be recorded to establish the thermal history of the instrument and ensure baseline repeatability according to the manufacturer specifications. Baseline repeatability is key for obtaining absolute heat capacity profiles. For manual calorimeters, baseline repeatability could be a tricky issue as it depends on the filling technique. Before running the sample, it is advisable for beginners to check baseline repeatability with and without a refilling cycle.
4. Protein concentrations used for DSC experiments are within 0.1–2 mg mL^{-1}. In terms of volume, 400 μL to 1 mL of protein solution will be needed for automated or manual calorimeters, respectively. Prior to protein concentration measurements, insoluble aggregates and other particles must be removed by sample centrifugation (15 min, 12,000 × *g*). The buffer solution must be also filtered. Finally, accurate protein concentration determination is key for data analysis, especially when seeking absolute heat capacity measurements. Any error in sample concentration will systematically distort the DSC profile.

2.2 Testing Calorimetric Reversibility

Following the protein–buffer scan, a reheating of the sample after cooling to room temperature must be collected in order to check for calorimetric reversibility. If more than ~80% of the DSC transition is recovered in the rescan (after baseline subtraction), then the unfolding process is typically considered calorimetrically reversible. The degree of reversibility of protein denaturation often depends on the final temperature achieved during the first scan because the rate of irreversible processes such as protein aggregation increases.

The incidence of irreversible alterations can be minimized by stopping the heating ramp of the first scan immediately after the unfolding transition (peak) is observed, to then quickly proceed to the cooling cycle followed by the reheating run. Nevertheless, it must be noted that if high reversibility is obtained when stopping immediately after the transition but low reversibility is observed when stopping a few degrees above, the high-temperature baseline is likely to be distorted. This is an important point because the analyses described in Sections 4 and 5 require reliable experimental baselines.

Calorimetric reversibility is usually taken as an indicator for thermodynamic reversibility meaning that the actual DSC peak reflects a true equilibrium process. However, calorimetric reversibility does not necessarily imply thermodynamic reversibility. In some cases, there is calorimetric reversibility, but the DSC profile is distorted from slow folding/unfolding kinetic processes (compared with the time scale of the DSC experiment). Therefore, a determination of scan rate effects must also be carried out to fully ascertain the validity of an equilibrium thermodynamic analysis.

2.3 Substantial Scan Rate Effects?

Current commercial calorimeters can perform DSC experiments at heating (scanning) rates between $\sim 10^{-1}$ and $\sim 10^{2}$ degrees per hour. The Fastest scan rates may require correction for the calorimeter time response (dynamic correction), whereas the slowest scan rates are time-consuming and lead to low signal-to-noise. In practice, scan rate effects are checked performing several experiments at scan rates within the 15–200 K h^{-1} range. Significant scan rate effects indicate that the state of the system (the protein solution) at a given temperature depends on the time required to reach that temperature, and, therefore, that the DSC transition is kinetically distorted. For calorimetrically irreversible transitions, the same sources of irreversibility, such as protein aggregation, are also most likely responsible for the scan rate effect. For calorimetrically reversible transitions, slow folding–unfolding kinetics relative to the scan rate produce distortions due to the inability to reach proper equilibration. When is a scan rate effect considered significant? A rule of thumb is that scan rate effects are significant when the shift in transition temperature (temperature of the peak) is comparable to or larger than the width of the transition, which implies that the denaturation process is over at the fastest scan rate while it is barely starting at the slowest scan rate used.

Once calorimetric reversibility and scan rate effects have been characterized, the DSC profile can be further analyzed following the procedures outlined in Fig. 1 for the specific scenario that applies to the data.

2.4 Scenario I (Red (Thick Dark Gray in the Print Version)): Calorimetric Reversibility and No Substantial Scan Rate Effect

This scenario is the most straightforward from the viewpoint of performing an equilibrium thermodynamics analysis. Figure 2A shows an archetypical example of a scenario I DSC profile in which the conventional thermodynamic analysis with a two-state model (i.e., all-or-none transition, see Section 4) renders an excellent fit, and meaningful thermodynamic parameters. This scenario, however, is more the exception than the rule, in particular when working with large and complex protein systems, which

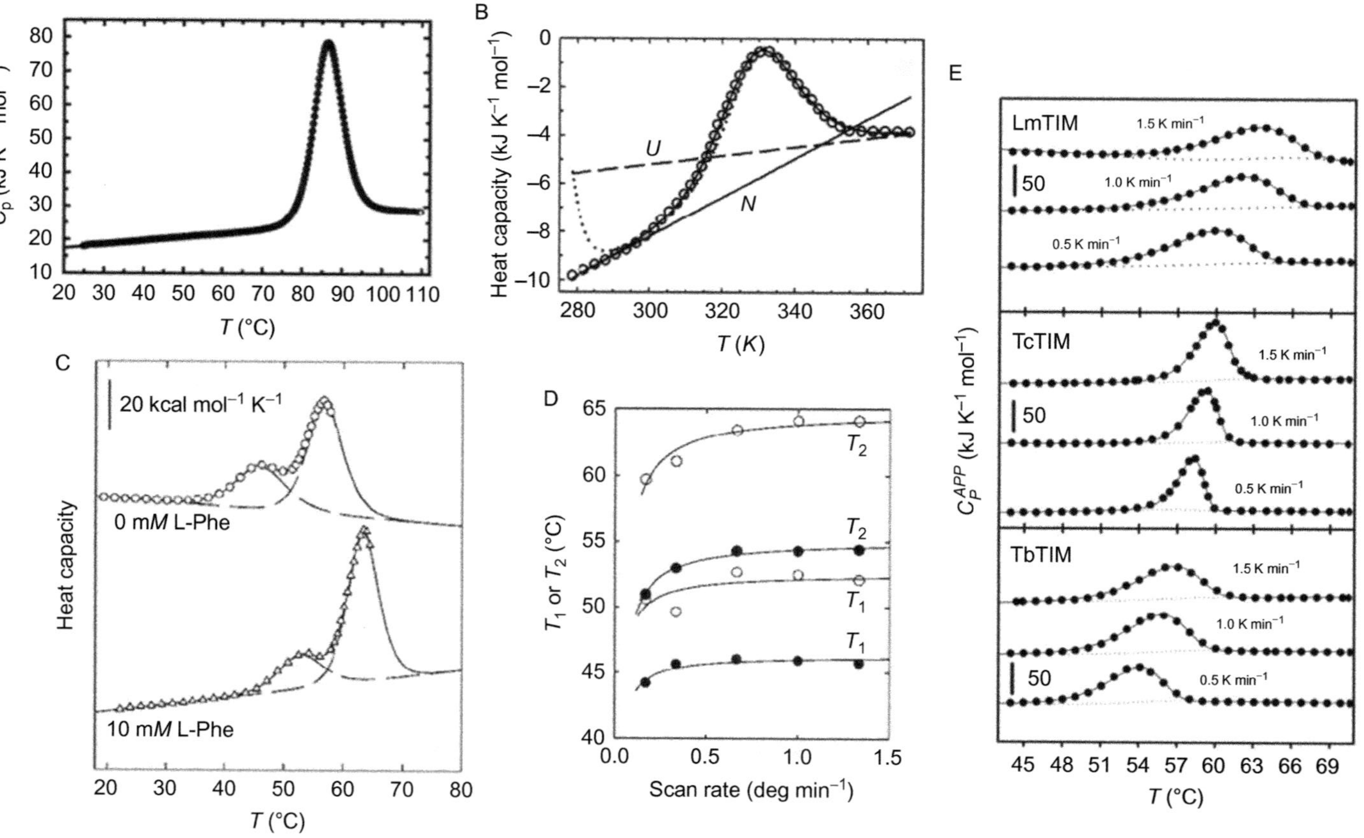

Figure 2 See legend on next page.

are highly prone to irreversible alteration processes. On the other hand, small fast-folding proteins have kinetics that are much faster than the timescales for DSC experiments and are also less likely to undergo irreversible denaturation during the DSC run. Nevertheless, in contrast to the example shown in Fig. 2A, DSC profiles of fast-folding proteins are intrinsically broad and produce unrealistic baselines when analyzed with the thermodynamic two-state model (Fig. 2B; see also Sections 3 and 4), which makes them excellent candidates for the application of the theoretical methods described in Section 5 (Fung, Li, Godoy-Ruiz, Sanchez-Ruiz, & Muñoz, 2008; Godoy-Ruiz et al., 2008; Muñoz & Sanchez-Ruiz, 2004; Naganathan, Li, Perez-Jimenez, Sanchez-Ruiz, & Muñoz, 2010). The latter analysis can be also successfully applied to scenario I DSC profiles that are unusually broad even if the protein is not a fast folder (Halskau et al., 2008).

Figure 2—Cont'd Representative examples of the experimental DSC scenarios described in Fig. 1. (A) Thermal denaturation of thioredoxin. DSC transitions are calorimetrically reversible and scan rate independent. Circles represent the experimental data and the continuous line is the best fit to the two-state model. The fit is excellent and permits to determine the thermodynamic parameters describing the unfolding energetics. (B) Thermal denaturation of a variant of the ultrafast folding protein BBL. DSC transitions are calorimetrically reversible and scan rate independent. Unlike the example in panel A, the DSC transition is very broad and the pre- and posttransition baselines are not well defined. A two-state model produces a visually excellent fit (continuous line), but the native and unfolded baselines cross in the middle of the experimental temperature range. (C–D) Thermal denaturation of human phenylalanine hydroxylase with and without L-phenylalanine. DSC transitions were calorimetrically irreversible, but showed little scan rate dependence (panel D), supporting the equilibrium thermodynamics analysis despite the calorimetric irreversibility. The analysis was performed after eliminating the precipitous drop in high capacity observed at high temperature. PAH is a multidomain protein that produces a DSC profile with two partially overlapping transitions, none of each adhering to the two-state model (continuous lines are fits to a pseudo-two-state model with different values for calorimetric and van't Hoff enthalpies). (E) Thermal denaturation of triosephosphate isomerases (Tc: *Trypanosoma cruzi*; Tb: *Trypanosoma brucei*; Lm: *Leishmania mexicana*). DSC transitions are calorimetrically irreversible and strongly scan rate dependent, indicating that thermal stability of these proteins is kinetically controlled. The profiles could be fitted to the two-state irreversible (Lumry–Eyring) model. *Panel A: Taken with permission from Georgescu RE et al. (2001)* Eur. J. Biochem. *268:1477–1485. Panel B: Taken with permission from: Naganathan AN et al. (2005)* Biochemistry *44:7435–7449. Panels C–D: Taken with permission from: Thórólfsson M et al. (2002)* Biochemistry *41:7573–7585. Panel E: Taken with permission from: Costas M et al. (2009)* J. Mol. Biol. *385:924–937.*

2.5 Scenario II (Orange (Light Gray in the Print Version)): Calorimetric Irreversibility and No Scan Rate Effects

This scenario has been rarely reported. Calorimetric irreversibility is normally accompanied by the observation of scan rate effects that indicate kinetic distortions caused by the same processes responsible for irreversibility (e.g., aggregation). The unusual combination of calorimetric irreversibility and lack of scan rate dependence is most easily interpreted by assuming that the processes that cause irreversibility are only taking place at very high temperatures where the protein denaturation process is already completed (Fig. 2C and D). In such case, an equilibrium thermodynamics analysis may be acceptable, at least up to the transition temperature or a few degrees above the transition temperature. However, there are two important points that must be considered when dealing with this scenario:

- The lack of significant scan rate effects must be clearly and convincingly demonstrated using the widest possible scan rate range (Thorolfsson et al., 2002). This key control guarantees that the source of irreversibility is not causing distortions within the denaturation transition, which would invalidate an equilibrium thermodynamics analysis.
- The high-temperature baseline of the DSC profile is likely to be distorted by the irreversible process even in the absence of scan rate effects. A distorted baseline precludes the reliable determination of the change in unfolding heat capacity, an important parameter for defining the temperature dependence of protein stability.

2.6 Scenario III (Blue (Thin Gray in the Print Version)): Calorimetric Irreversibility and Substantial Scan Rate Effects

This is a very common scenario, in particular when dealing with large, or complex, protein systems (Fig. 2E). The combination of calorimetric irreversibility and substantial scan rate effect indicates that the DSC transitions are distorted by the kinetics of the irreversible processes, which can include aggregation, proteolysis, chemical modification of amino-acidic residues, and irreversible cofactor loss. This scenario is often discussed in the literature in terms of the Lumry–Eyring model (Sanchez-Ruiz, 1992), which extends the standard two-state equilibrium unfolding by adding a step in which the unfolded state is modified irreversibly leading to a dead end or "final" state,

$$N \Leftrightarrow U \rightarrow F$$

This model can be further generalized including any number of intermediate states in chemical equilibrium prior to the irreversible step.

The two-state irreversible model predicts that under conditions in which the irreversible step is significantly faster than all prior equilibria the overall process consists on the progressive conversion of the native state into the final denatured state without significant population of any other species. This simplifying condition permits an elaborate mathematical analysis that leads to specific quantitative predictions of the DSC profile as well as the scan rate effect. Such predictions provide straightforward tests of the applicability of the model (Costas et al., 2009; Pey, 2013). Over the years, a large number of proteins have been found that follow the two-state irreversible kinetic model (Sanchez-Ruiz, 2010). In such cases, the analysis of the DSC profile provides information about the kinetics of protein denaturation, including an assessment of the magnitude of the kinetic free energy barrier to unfolding (continuous blue (thin gray in the print version) lines in Fig. 1). Moreover, the widespread applicability of the two-state irreversible model suggests that kinetic stability is a major natural selective pressure for many proteins (Sanchez-Ruiz, 2010). Natural selection for kinetic stability requires the emergence of a high free energy barrier for unfolding that facilitates the protein remaining functional during the physiologically relevant times in a harsh *in vivo* environment.

Some irreversible and strongly scan rate dependent DSC transitions do not follow the simple two-state irreversible kinetic model (dashed blue (thin gray in the print version) lines in Fig. 1). One possibility is that in such cases the unfolded or partially unfolded states have nonnegligible populations during the denaturation process. This is an interesting case because theoretical analyses suggest the possibility of obtaining equilibrium information from an extrapolation of the transition parameters to an infinite scan rate. However, we are getting here into uncharted territory, as there are very few examples in the literature in which this extrapolation approach has been explored (Vogl, Jatzke, Hinz, Benz, & Huber, 1997).

2.7 Scenario IV (Dashed Red (Dashed Gray in the Print Version): Calorimetric Reversibility and Substantial Scan Rate Effects

Calorimetric reversibility indicates the absence of irreversible alterations during the thermally induced protein denaturation process. The only additional factor that could give rise to the observation of scan rate effects is the presence of slow folding–unfolding kinetics relative to the scan rate. Under such conditions, there is not enough time to reach equilibrium during temperature scanning producing different DSC profiles depending on the rate. This slow equilibrium condition was first described many years ago (Mayorga & Freire, 1987), and it is likely to be encountered in the following two situations:

- DSC experiments performed on proteins with intrinsically very slow folding–unfolding kinetics (Kaushik, Ogasahara, & Yutani, 2002) or on slow folding proteins in the presence of chemical denaturants (urea, guanidine) at concentrations close to the chemical denaturation midpoint, which corresponds to their minimal folding/unfolding relaxation rate (the bottom of the kinetic chevron plot) (Plaza del Pino, Pace, & Freire, 1992).
- DSC experiments that show cold denaturation. Cold denaturation takes place at very low temperatures at which the folding/unfolding kinetics tends to become much slower (Romero-Romero, Ingles-Prieto, Ibarra-Molero, & Sanchez-Ruiz, 2011).

Regardless of the origin of the observation, the lower the scan rate the closer will be the system to attain equilibrium conditions during the temperature scan. The operating scan rate range of the calorimeter may preclude the use of sufficiently slow scan rates that ensure proper equilibration. An alternative is to collect thermograms at various scan rates and extrapolate the parameters to zero scan rate as an estimate of the true equilibrium transition. Furthermore, proper kinetic analysis of the slow equilibrium data permits the accurate determination of folding/unfolding relaxation times in timescales of minutes to hours (Freire, van Osdol, Mayorga, & Sanchez-Ruiz, 1990; Sanchez-Ruiz, 1995). However, this kind of analysis has been seldom explored in the literature (Romero-Romero et al., 2011). On the other hand, the slow equilibrium condition is unlikely to be of concern for the modern analysis described in Section 5 given that its best protein candidates are small fast folding single domains (Muñoz & Sanchez-Ruiz, 2004; Naganathan, Doshi, & Muñoz, 2007). A potential exception could come from proteins within that group that happen to have very rugged energy landscapes that produce very sluggish kinetics at low temperatures, such as some recently studied *de novo* designed proteins (Sadqi, de Alba, Pérez-Jiménez, Sanchez-Ruiz, & Muñoz, 2009).

3. ABSOLUTE HEAT CAPACITY AND DSC BASELINES

The scenarios discussed in Section 2 provide an overview of the limiting cases faced when performing DSC experiments. It should be clear by now that most experimental DSC scenarios do not permit an equilibrium thermodynamic analysis, which is in turn the main requirement for applying the methods described in following sections. However, for those cases that do satisfy the thermodynamic equilibrium tests from Section 2, the next step is to produce DSC data in absolute heat capacity units. Having access to the absolute values of the heat capacity is a very important requirement for an in depth thermodynamic equilibrium analysis of protein denaturation because there is significant amount of information to be extracted from the native (low temperature) and unfolded (high temperature) baselines of the DSC thermogram.

3.1 Determining Absolute Heat Capacities by DSC

The differential nature of the DSC experiment makes it not trivial to determine the absolute heat capacity of protein unfolding. The calorimeter measures the difference between the heat capacities of a protein solution and the pure solvent (buffer) placed in the reference cell. The result is typically expressed per mole of protein and is an apparent heat capacity value contaminated by a water displacement effect. This is so because the concentration of protein is high enough to occupy a significant fraction of the sample cell volume and thus to change the concentration of water relative to the reference cell. Apparent heat capacities may even be negative at some temperatures given that the heat capacity of water is larger than that of the protein. The "true" partial molar heat capacity of the protein, commonly known as the absolute heat capacity, is only accessed after an accurate water displacement correction. Water displacement can be estimated from a single thermogram, but the most accurate procedure involves performing several DSC experiments at different protein concentrations (Guzman-Casado, Parody-Morreale, Robic, Marqusee, & Sanchez-Ruiz, 2003; Kholodenko & Freire, 1999). For each temperature, the slope obtained after plotting the apparent heat capacity values against protein concentration includes varying contributions from the absolute protein heat capacity and from water displacement, which permits correcting for the latter. This procedure not only improves the accuracy of the absolute heat capacity determination, but also serves as diagnostics for several common problems in DSC experiments. Among those problems, instrumental baseline irreproducibility and lack of proper calorimeter equilibration are observed as scatter in the plots of apparent heat capacity versus protein concentration, whereas protein association–aggregation results in curved plots.

3.2 Analysis of Low- and High-Temperature Baselines

The availability of reliable absolute heat capacity profiles permits direct comparison with empirical predictions for the heat capacity values expected for the protein in its native and unfolded states (i.e., the expected native and unfolded baselines). Heat capacities for unfolded proteins are estimated from equations based on model compound data (Häckel, Hinz, & Hedwig, 1999; Makhatadze & Privalov, 1990), and heat capacities of native proteins are estimated from empirical equations based on the statistical analysis of available calorimetric data from a collection of globular proteins (Gomez, Hilser, Xie, & Freire, 1995). We shall refer to the latter as the Freire baseline in the following discussion. Comparisons with the predicted native baseline are particularly informative. For instance, if the low temperature absolute heat capacity data are substantially above the predicted native baseline, the protein is likely to be partially unfolded in native-like conditions, a common occurrence for proteins that bind nucleic acids and which experience folding coupled to DNA binding (Spolar & Record, 1995).

Such information highlights the presence of conformational heterogeneity and/or partial unfolding under conditions in which the protein is native like. The test is useful to ascertain the validity of a conventional thermodynamic analysis based on well-defined macroscopic states (e.g., the two-state model; see Section 4) and is key for the implementation of the modern statistical mechanical analysis of Section 5. For instance, comparison between the absolute heat capacity values (C_p) recorded for the protein at temperatures below the denaturation transition and the native heat capacity baseline predicted from the Freire equation can result in four different scenarios (Fig. 3): (A) experimental low temperature C_p is in good agreement with the Freire native baseline both in magnitude and slope (temperature dependence); (B) the magnitude of the experimental C_p at low temperature is comparable to that predicted by the Freire equation, but the slope is steeper; (C) the experimental heat capacity profile is upshifted with respect to the native baseline but the slope is similar to that expected from the Freire equation; (D) both the magnitude and slope of the low temperature C_p are higher than Freire's prediction.

Baseline scenario A indicates that the enthalpic fluctuations experienced by the protein in the temperature range prior to the denaturation transition are commensurate with those expected for the protein in a well-folded globular state. Therefore, baseline scenario A is in principle consistent with a conventional thermodynamic equilibrium analysis based on two well-defined macrostates, provided, of course, that other criteria for the validity of the two-state model hold (see next section for more details). The other three baseline scenarios indicate conformational heterogeneity and/or progressive melting of protein structure taking place even at the lowest temperatures. The existence of conformational fluctuations at temperatures below the denaturation transition suggests that a conventional analysis with well-defined macrostates is inappropriate.

A simple empirical recipe to determine whether a DSC profile includes structural disorder at low temperature is to inspect the baselines obtained for the native and unfolded states when the DSC data are fitted to a two-state model. This simple test can be performed even with apparent heat capacity data. The presence of conformational fluctuations will result in fitted native and unfolded baselines that cross within the experimental temperature range (Muñoz, 2007; Naganathan, Doshi, Fung, Sadqi, & Muñoz, 2006; Naganathan, Perez-Jimenez, Sanchez-Ruiz, & Munoz, 2005; Naganathan et al., 2010; e.g., Fig. 2B). The closer is the baseline crossing to the protein transition temperature (DSC peak) the less realistic the obtained native and unfolded baselines. In general, the observation of baseline crossing rules out scenario A because it implies that an inversion of the change in heat capacity upon unfolding is taking place in the middle of the denaturation process. Such heat capacity inversion is physically implausible given that the changes in unfolding heat capacity mostly reflect the difference in accessible surface area between the native and unfolded states, which are always positive.

Baseline scenarios B–D are not uncommon, having been described for multiple DNA-binding domains (Privalov & Dragan, 2007) and microsecond-folding proteins (Naganathan, Perez-Jimenez, Muñoz, & Sanchez-Ruiz, 2011; Naganathan, Perez-Jimenez, et al., 2005). DSC profiles with signs of structural fluctuations in the native baseline should be analyzed with modern methods that rely on statistical mechanical models with ensembles of microstates that account for both the wells and barriers on the folding free energy surface (see Section 5).

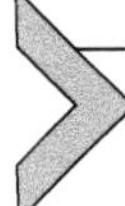

4. EQUILIBRIUM THERMODYNAMICS ANALYSIS: MICROSTATES VERSUS MACROSTATES

There are basically two procedures for the thermodynamic equilibrium analysis of DSC data: in terms of discrete macrostates and in terms of ensembles of microstates. The two procedures are not incompatible

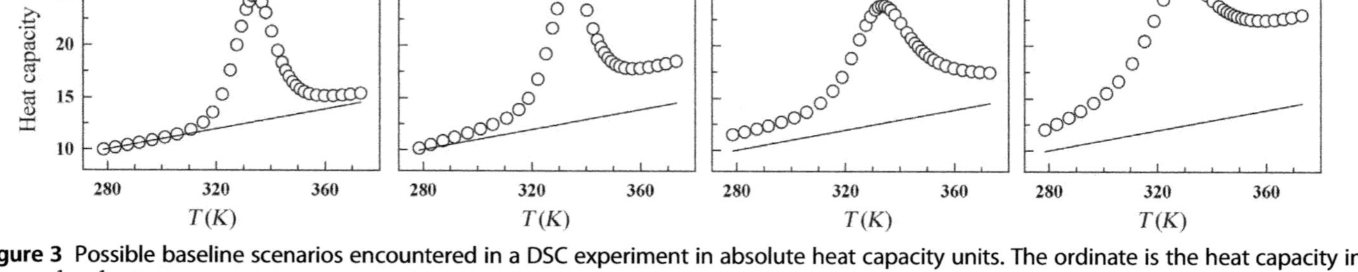

Figure 3 Possible baseline scenarios encountered in a DSC experiment in absolute heat capacity units. The ordinate is the heat capacity in kJ mol^{-1} K^{-1} and the straight line corresponds to the native baseline estimated from the Freire equation.

because models of the first type are particular cases (or limiting cases) of models that describe protein unfolding using ensembles of conformational microstates.

What we term the unfolded state of a protein is indeed an ensemble of microstates that share the lack of defined structure. Native states are also ensembles since folded proteins exhibit conformational dynamics that reflect interconversion between different structured conformations (James & Tawfik, 2003; Palmer, 1997). When we analyze protein denaturation as an equilibrium between native and unfolded states (the two-state model), we are assuming that these two ensembles are thermodynamically distinct. From statistical mechanics, we know that this condition implies that their interconversion involves visiting highly unstable microstates with intermediate conformational properties, which in terms of free energy projections implies that the native and unfolded ensembles are wells on the folding free energy surface separated by a high free energy barrier.

The application of the two-state model involves *de facto* the assumption that there is a high free energy barrier separating the two macrostates. In fact, one of the original applications of DSC was to provide a straightforward test for the validity of the two-state approximation to protein unfolding. The test is based on the capability of DSC experiments for determining the unfolding enthalpy change in two different ways: from the area under the peak (calorimetric enthalpy) and from the width of the transition (van't Hoff enthalpy). The calorimetric enthalpy is model free, whereas the van't Hoff enthalpy is obtained from the two-state analysis. Disagreement between the two values indicates that the two-state model does not hold. Agreement is a necessary condition for the applicability of the two-state model, but it is by itself not sufficient. Typically, the test is performed by fitting the DSC profile to an equation based on a pseudo-two-state model that includes the two enthalpies as independent fitting parameters (Risso et al., 2015). The fit must be visually good and render a calorimetric to van't Hoff enthalpy ratio close to unity, but it is also critical that the fitted native and unfolded baselines are physically plausible and do not cross at temperatures within the denaturation transition (see previous section).

When the three conditions are met, we may then conclude that the two-state model provides an acceptable description of the thermal unfolding process. In this case, the relevant thermodynamic parameters for the protein unfolding process (ΔG, ΔH, ΔC_p, ΔS) are conveniently determined (Becktel & Schellman, 1987; Schellman, 1987). This kind of analysis has been performed for a number of model proteins (Robertson & Murphy,

1997), providing the basis for a consistent description of the relationships existing between protein structure and energetics that have proven very useful in molecular design.

Here, we are interested in the procedures for analyzing DSC profiles from scenario I that cannot be properly described with the two-state model (baseline scenarios B–D). The conventional approach in such case has been to assume that additional intermediate macrostates become populated during the thermally induced process, including as many as is necessary to obtain a good fit. However, one must consider whether such analysis is physically realistic, because a model with N macrostates is implicitly assuming the existence of $N-1$ high free energy barriers separating them. Whereas this approach may be justifiable for complex proteins with multiple domains unfolding more or less independently, it does not make much sense for small single-domain proteins (e.g., fast folders). For the latter, one should resource to a statistical mechanics analysis in which the protein conformational space is described in terms of ensembles of microstates (see Section 5). The rationalization for this assertion comes from the realization that protein folding is not a chemical reaction that involves formation or breakage of a single covalent bond, but a conformational reorganization of a polymer driven by the cooperation of thousands of very weak interactions. There is no physical reason to assume *a priori* the existence of high free energy barriers in protein folding/unfolding, and in fact, energy landscape theory postulates that folding free energy barriers are caused by entropic bottlenecks and are intrinsically small (Onuchic, Luthey-Schulten, & Wolynes, 1997; Portman, Takada, & Wolynes, 2001). Moreover, there is now ample empirical evidence coming from work on ultrafast folding proteins (Prigozhin & Gruebele, 2013) confirming that free energy barriers for single-domain folding are indeed rather small (Akmal & Muñoz, 2004) and often times marginal or nonexisting (downhill folding) (Garcia-Mira et al., 2002; Muñoz, 2007; Naganathan et al., 2007; Sadqi et al., 2006).

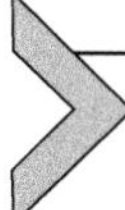

5. STATISTICAL MECHANICS ANALYSIS OF DSC EXPERIMENTS

5.1 Heat Capacity and the Partition Function

As discussed above, the DSC profile contains information about all of the protein conformational microstates that become minimally populated during the thermal denaturation process. This key statement can be easily demonstrated using a generic statistical thermodynamics model that defines the protein conformational space as an arbitrary series of N states in equilibrium with one another:

$$I_1 \rightleftharpoons I_2 \cdots I_{i-1} \rightleftharpoons I_i \rightleftharpoons I_{i+1} \cdots I_{N-1} \rightleftharpoons I_N. \tag{1}$$

The N species can be either microstates or ensembles of microstates (macrostates), and their number can vary from just two (reducing to the two-state model) to the nearly infinite number of species that are implicitly considered in a continuum of microstates. The temperature (T) dependent total partition function (Q) of such a model is defined as,

$$Q(T) = \sum_{i=1}^{N} w_i = \sum_{i=1}^{N} \exp\left(-\frac{G_i}{RT}\right) = \sum_{i=1}^{N} \exp\left(\frac{S_i}{R}\right) \exp\left(-\frac{H_i}{RT}\right) \tag{2}$$

where, for each species I_i, w_i is the statistical weight and $\exp(S_i/R)$ is the density of microstates (Ω_i) with an enthalpy value H_i in the conventional statistical mechanical representation of the partition function (Freire & Biltonen, 1978). The temperature-dependent probability (p_i) of each of the species can be calculated from,

$$p_i(T) = \frac{w_i}{Q}.$$

The effective heat capacity of the system $\langle C_p \rangle$ is expressed as,

$$\langle C_p \rangle = \frac{\mathrm{d}\langle H \rangle}{\mathrm{d}T} = \frac{\mathrm{d}\left(\sum_{i=1}^{N} H_i P_i\right)}{\mathrm{d}T}.$$

Differentiating, we get

$$\begin{aligned} \langle C_p \rangle &= \sum_{i=1}^{N} \left(p_i \frac{\mathrm{d}(H_i)}{\mathrm{d}T} + H_i \frac{\mathrm{d}p_i}{\mathrm{d}T} \right) \\ &= \sum_{i=1}^{N} \left(p_i C_p^i + \frac{H_i^2 p_i - (H_i p_i)^2}{RT^2} \right). \\ &= \left\langle C_p^i \right\rangle + \frac{\langle H_i^2 \rangle - \langle H_i \rangle^2}{RT^2} \\ &= \left\langle C_{p,i}^{\mathrm{int}} \right\rangle + C_p^{\mathrm{ex}} \end{aligned} \tag{3}$$

The first term is called the intrinsic heat capacity of the system (or chemical baseline), whereas the second part is the excess heat capacity resulting from the unfolding transition. The intrinsic heat capacity corresponds to changes in the probability weighted heat capacity values of the different species as a function of temperature, and thus it is intimately related to the low- and high-temperature baselines discussed in Section 3. The excess heat capacity arises from the temperature dependence of the denaturation equilibrium, i.e., the contribution to the heat capacity arising from enthalpic fluctuations in the system. The area between the chemical baseline and the experimental heat capacity curve is the total enthalpy realized during the transition (i.e., the calorimetric enthalpy; ΔH_{Cal}). The determination of ΔH_{Cal} does not depend on the number of species involved, but is sensitive to the definition of baselines (see Sections 3 and 4). For a continuum of states, Eq. (2) can be written as,

$$Q(T) = \int \Omega_i \exp(-H_i/RT)\mathrm{d}H. \tag{4}$$

Equation 4 is in fact analogous to the Laplace transform of the density of states $\boldsymbol{\Omega}$, which conversely indicates that the density of states can be extracted from the partition function (and thus from the DSC profile) by an inverse Laplace transform. Unfortunately, the inverse Laplace transformation of experimental (inherently noisy) DSC data is a mathematically ill determined procedure that can result in physically unreasonable results (Kaya & Chan, 2000). Moreover, the procedure requires that is performed for a varying number of states ($N = 1, 2$, etc.) to determine the minimum number of states required to explain the experimental data.

A more practical alternative to the "model-free" procedure is to employ the partition function from a physically reasonable statistical mechanical model of protein folding and estimate the probabilities for all the microstates defined in the model that best reproduce the experimental DSC profile using an iterative procedure to fit the model parameters. The probabilities for microstates can also be lumped together as a function of a suitable order parameter i, thus providing the partial partition functions (Q_i) from which a temperature-dependent one-dimensional free energy surface (G) is constructed (projected free energy surface),

$$G_i(T) = -RT\ \ln(Q_i(T)). \tag{5}$$

This model-dependent procedure provides a physically reasonable set of microstates and their probabilities (a thermodynamic folding free energy surface) that reproduce the DSC profile. The obtained surface (Eq. 5) provides estimates of the position and height of the thermodynamic free energy barrier separating the native and unfolded ensembles (β), which can be compared with the folding and unfolding rate constants (k) obtained from kinetic experiments (Naganathan, Sanchez-Ruiz, & Muñoz, 2005). If such comparison shows a linear trend for data from multiple proteins, it becomes possible to equate the thermodynamic (from DSC) and kinetic (from kinetic experiments) folding free energy barriers and thus estimate the pre-exponential term (k_0) to the folding rate equation,

$$k = k_0 \exp\left(-\beta/RT\right). \tag{6}$$

5.2 Free Energy Surface Models of Protein Folding

The main caveat of the general procedure outlined above is its dependence on the particular theoretical model of choice. There is not an obvious way to eliminate this requirement given the intrinsic limitations of the analytical procedures to calculate the inverse Laplace transform. However, one can minimize the model dependence and obtain statistically significant general results by performing the analysis with various microstate models of protein folding (each with its own assumptions) and then devising a procedure to rank their results using Bayesian inference.

In particular, four different statistical mechanical models of protein folding have been applied to the analysis of DSC data. The four models have been widely and successfully used in the literature to analyze and interpret a vast array of protein folding experiments. The models include different definitions of protein conformational space and energetics and involve different assumptions, but they can all be adapted to the analysis discussed in Section 5.1. In the following paragraphs, we introduce the general properties of the four models to the reader, who is referred to the original articles for the mathematical details. We also illustrate the performance of the models through the analysis of the DSC profiles of two exemplary proteins: gpW and SH3. These two proteins are single domains with essentially the same size (57 and 56 residues) and melting temperature (~340 K), but they fold at very different rates (gpW folds with a rate of ~40,000 s^{-1} at 310 K (Fung et al., 2008) and SH3 with a rate of ~10 s^{-1} at the same temperature (Viguera, Martinez, Filimonov, Mateo, & Serrano, 1994)), indicating that they fold by crossing very different free energy barriers.

5.2.1 Variable-Barrier Model

The Variable-Barrier (VB) model is inspired by the Landau theory of phase transitions and was developed specifically for analyzing DSC data (Muñoz & Sanchez-Ruiz, 2004). The VB model assumes a continuous distribution of conformational microstates using the enthalpy as order parameter. The free energy as a function of the order parameter (the free energy surface) is defined as a Landau quartic polynomial (see Muñoz and Sanchez-Ruiz (2004) for the whole derivation),

$$G_0(H) = -2\beta\left(\frac{H}{\alpha}\right)^2 + |\beta|\left(\frac{H}{\alpha}\right)^4 \tag{7}$$

where β and α are the two basic model parameters (Table 1). For $\beta > 0$, the free energy surface has two minima at $H = \pm\alpha$ separated by a maximum at $H = 0$ that defines a barrier of height β (Muñoz & Sanchez-Ruiz, 2004). If the barrier is higher than three times the thermal energy, the resulting surface is two-state like and the change in unfolding enthalpy is simply 2α. For $\beta < 0$, the free energy surface presents a single minimum at $H = 0$ thus mimicking a downhill folding (one-state) free energy surface (Garcia-Mira et al., 2002). To account for the fact that native states have fewer structural fluctuations (and thus fewer enthalpy fluctuations since H is used as order parameter) than the unfolded state, a parameter α_N was introduced for $H < 0$ and α_P for $H > 0$. This modification results in surfaces with wells of different width, which allows reproducing the typically asymmetric protein DSC profiles (high-temperature baseline above the low temperature one).

The VB model makes no assumptions about folding mechanisms and thus it provides a purely phenomenological approach. The model is also highly constrained, requiring only six fitting parameters (the same number required for a trivial two-state analysis) (Table 1). It is important to note that analysis with the VB model requires that all nonstructural contributions to the heat capacity of the native state (solvent contributions) are eliminated by subtracting the native Freire baseline (see Section 3) from the experimental DSC data prior to fitting. From a practical standpoint, it is recommended that several slight modifications of the Freire baseline (in slope and magnitude) are attempted during the fitting procedure to identify the best fit, and thus minimize artifacts from the experimental uncertainty in the determination of the absolute heat capacity.

Despite its simplicity, the thermodynamic barrier height estimated from the VB model for a collection of 17 proteins of large spread in folding rates

Table 1 Statistical Mechanical Models for the Analysis of DSC Profiles

Models	Features	Order Parameter	Thermodynamic Parameters[a]	Outputs
Variable-Barrier (VB)	Inspired by the Landau theory of phase transitions	Enthalpy (continuous)	$\boldsymbol{\beta}$, $\Sigma\alpha$, T_0, f	FE profiles, barriers, native ensemble heterogeneity
Muñoz–Eaton Single Sequence Approximation (ME-SSA)	Ising like, ensemble with only single stretches of native-like residues, mean-field energetics	Number of native-like peptide bonds (discrete)	$\Delta H_{res}^{T_{ref}}/\varepsilon$, a_{ASA}, $\Delta S_{conf,res}^{385}$	FE profile, barriers; residue probabilities
Muñoz–Eaton Exact Solution (ME-ES)	Ising like, ensemble with all combinations of native like and unfolded residues, energetics weighted by contact map	Number of native-like peptide bonds (discrete)	ΔH_{cont}, $\Delta S_{conf,res}$	FE profile, barriers, residue probabilities, structural–energetic connection
Mean-field (MF)	Functional based, mean-field energetics, empirical size-scaling of thermodynamic parameters	Nativeness (continuous)	$\Delta S_{res}^{n=0}$, ΔH_{res}^{385}, $\kappa_{\Delta H}$, $\Delta C_{p,res}$	FE profile, barriers

[a]A grid-based search approach can be employed on the parameters in bold to eliminate possible interparameter correlations.
All models require the experimental data in absolute heat capacity units. ME models additionally require a PDB file with the native structure.

correlated very well with the folding rates determined experimentally using kinetic experiments (Muñoz, 2007; Naganathan, Sanchez-Ruiz, et al., 2005). As means of example, the VB model analysis of the experimental DSC profiles of gpW and SH3 (Figs. 4A and 5A) reveal vastly different properties, with thermodynamic barriers of only 0.5 kJ mol^{-1} and of 16.3 kJ mol^{-1}, respectively (Figs. 4B and 5B). Native state heterogeneity is estimated to be much higher (a broad native well) for gpW than for SH3, consistent with the gpW DSC profile being significantly upshifted from the Freire baseline.

5.2.2 Muñoz–Eaton Ising-Model with Single Sequence Approximation (ME-SSA)

The general ME model is a statistical mechanical model of protein folding similar to the Ising theory of ferromagnetism in which atomic spins can take two configurations: +1 or −1. From a protein perspective, the fundamental units are the peptide bonds whose conformation is defined by specific pairs of dihedral angles (Φ, ψ). The model is binary, so each peptide bond is allowed to be either in the native state (dihedral angles of the native structure; represented as a 1) or in the unfolded state (any other combination of dihedral angles; represented as a 0). A N-residue protein ($N-1$ peptide bonds) can thus have 2^{N-1} different microstates (or 2^N when the residue is used as unit rather than the peptide bond), which are defined by all the possible combinations of native and unfolded units (Muñoz & Eaton, 1999; Wako & Saito, 1978a). Because the unfolded state for each unit includes many conformations (combinations of dihedral angles), the model defines a fundamental parameter $\Delta S_{conf} = S_U - S_F$ that accounts for the entropic penalty of fixing a given unit into the native conformation. From a statistical mechanical perspective $\Delta S_{conf} = R\ln(\Omega_U/\Omega_F)$, where $\Omega_U/\Omega_F \gg 1$. The model is also native-centric and thus only considers interactions that are present in the protein 3D structure. Therefore, the energetics of each specific protein microstate (any string of $N-1$ units combining native and unfolded conformations) is determined by how many native interactions are structurally consistent with the particular conformation.

In practice, the ME model has been implemented calculating the whole partition function (see next section) or invoking various simplifying approximations (Muñoz & Eaton, 1999). The simplest version assumes that only single stretches of native peptide bonds are formed simultaneously (the single sequence approximation, SSA). This approximation is justified by the fact

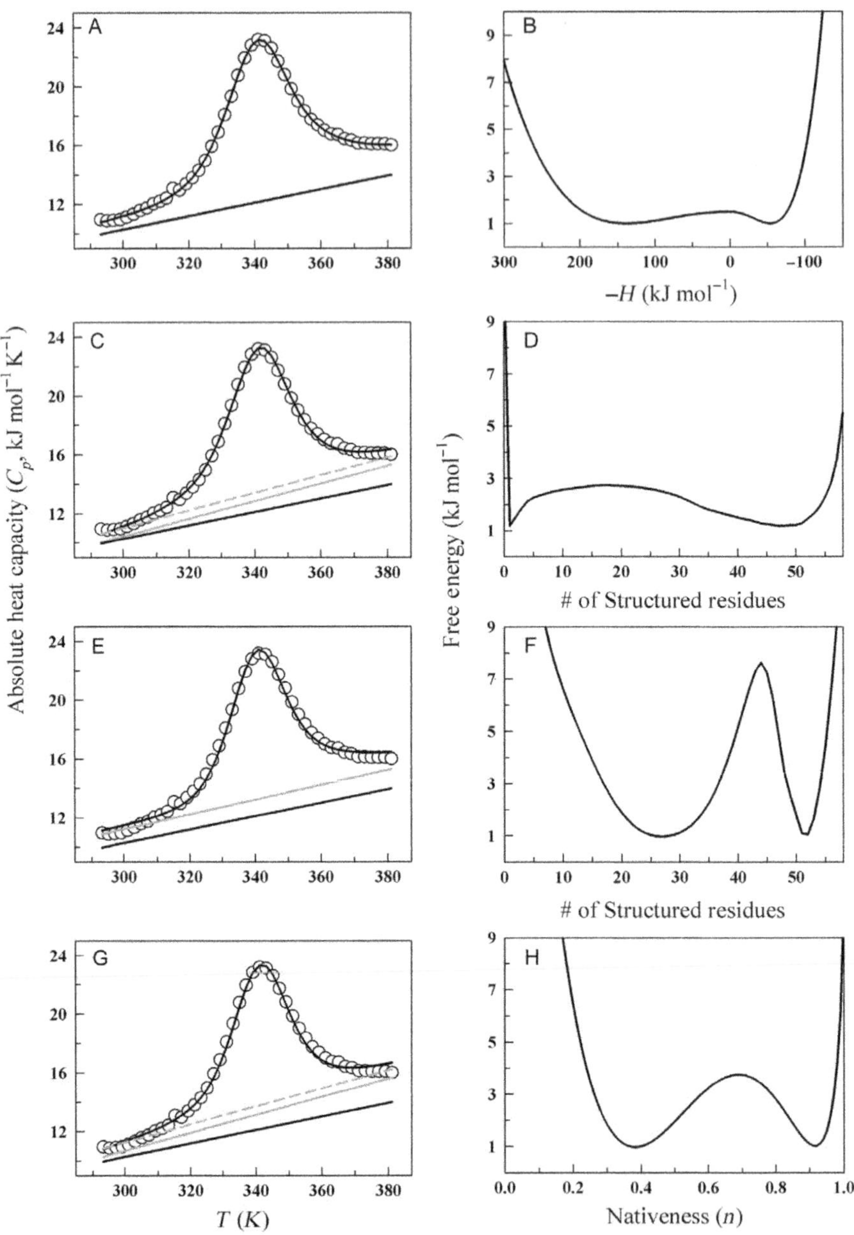

Figure 4 Representative fits (curves) from different models to the experimental DSC profile of gpW (Fung et al., 2008) (circles: panels A, C, E, and G) and the corresponding free energy surfaces as a function of the relevant order parameter (panels B, D, F, and H). In panels A, C, E, and G, the black continuous line corresponds to the Freire native baseline, whereas the gray continuous and dashed lines indicate the fitted folded and unfolded baselines, respectively, from each models. The models are: VB (panels A and B), ME-SSA (C and D), ME-ES (E and F), and MF model (G and H).

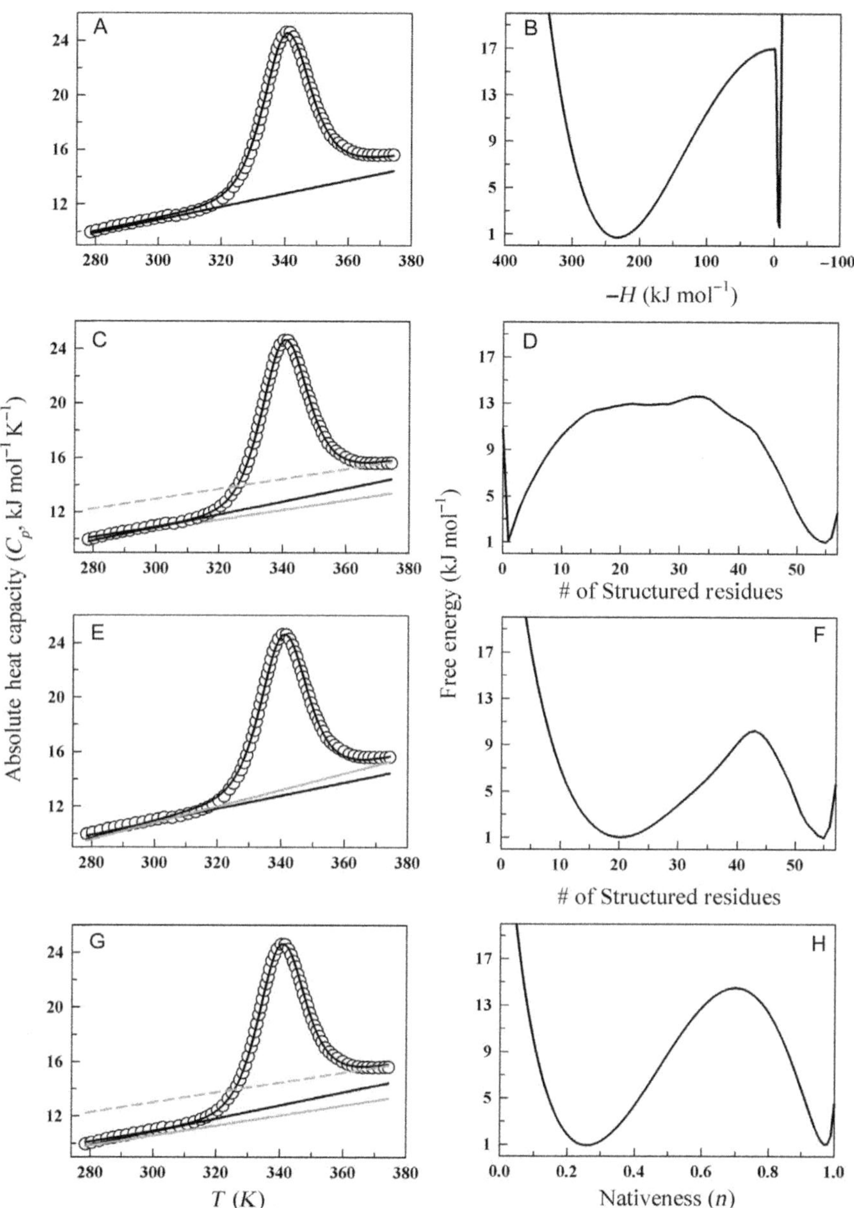

Figure 5 Same as in Fig. 4 but for the experimental DSC profile of the α-spectrin SH3 domain (Viguera et al., 1994).

that ΔS_{conf} is large and that stabilizing interactions are only formed when all peptide bonds connecting the interacting pair of residues are simultaneously in native conformation. It defines a nucleation–propagation mechanism for structure formation that is similar to helix-coil transition models (Muñoz & Serrano, 1995). For the ME-SSA version, the number of microstates is drastically reduced to only $N(N+1)/2+1$ (Muñoz & Eaton, 1999). The treatment of protein energetics is mean-field like and has been implemented in two different ways. The first implementation defines the free energy (ΔG) of every microstate with a simple Gibbs free energy expression,

$$\begin{aligned}\Delta G(T) = {} & n \cdot \Delta H_{\mathrm{res}}^{T_{\mathrm{ref}}} + \Delta C_p(T - T_{\mathrm{ref}}) \\ & - T\left[n \cdot \Delta S_{\mathrm{conf,res}}^{385} + \Delta C_p \ln(T/T_{385})\right]\end{aligned} \tag{8}$$

with $\Delta C_p = a_{\mathrm{ASA}}\Delta\mathrm{ASA}$, and where n is the number of peptide bonds in native conformation, ΔC_p is the heat capacity change calculated from the difference in accessible surface area (ΔASA), $\Delta S_{\mathrm{conf,res}}^{385}$ is the entropic penalty associated with fixing a unit (or residue) in the native state at 385 K, $\Delta H_{\mathrm{res}}^{T_{\mathrm{ref}}}$ is the mean gain in enthalpy per residue at a reference temperature T_{ref}. The entropic cost is defined at 385 K since at this temperature the solvation contribution to the entropic free energy tends to zero (convergence temperature) (Robertson & Murphy, 1997). In the second implementation, the enthalpy contribution is weighted by the number of atomic contacts within a specified cut-off (spherical radius of typically 5 or 6 Å in an all-heavy-atom representation) and with a mean van der Waals interaction energy ε or ΔH_{cont} (Muñoz & Eaton, 1999). In practical terms, the analysis of DSC data with the ME-SSA model requires five parameters: three determining the thermodynamics $\left(\Delta H_{\mathrm{res}}^{T_{\mathrm{ref}}}/\varepsilon, a_{\mathrm{ASA}}, \Delta S_{\mathrm{conf,res}}^{385}\right)$ and two defining the native baseline (Table 1). The heat capacity curves are calculated using equations such as those given in Section 5.1.

The natural reaction coordinate of this model is the number of peptide bonds in native conformation (n). The ME-SSA model has proven very effective in predicting the folding rates (and hence the relative free energy barriers) of a large database of proteins (Muñoz & Eaton, 1999). An ME-SSA analysis of the DSC thermograms of gpW and SH3 produces distinct free energy surfaces, but the quality of fits, the estimated free energy barriers, and structural heterogeneity of the native ensemble are all similar to those obtained with the VB model (Figs. 4C, D and 5C, D).

5.2.3 Exact Solution of the Muñoz–Eaton Ising-Model (ME-ES)

Whereas computing the statistical weights of every one of the 2^N microstates of the general ME model in residue space is impractical, it is still possible to calculate the total partition function either from a transfer-matrix formalism or from iterative algorithmic enumerations, thus giving rise to the ME-ES. The former approach is extremely rapid and is feasible assuming that all of the peptide bonds connecting two residues must be native for a stabilizing interaction to be made. Following the transfer-matrix formalism of Wako and Saitô, the partition function (Q) of a protein with N residues can be calculated by matrix multiplication of N N-by-N square matrices (Wako & Saito, 1978a,1978b). Each of the N square matrices carries information of the interactions made by that residue with subsequent residues (see the original works Wako & Saito (1978a,1978b) for the specific details of the calculation).

The interaction energy is divided in quanta of energy per contact (ΔH_{cont}), and thus the strength of the interaction between two residues depends on ΔH_{cont} and the number of atomic contacts between the interacting pair that are found in the native 3D structure (obtained from a contact map with a cut-off of 5 or 6 Å). In its simplest version, the MS-ES model requires just four parameters: two thermodynamic (ΔH_{cont}, $\Delta S_{\mathrm{conf,res}}$) and two for the native baseline. Practically, additional variables, such as the distance cut-off for defining a native contact and the number of nearest neighbors to exclude from the calculation, need to be considered as well (Table 1).

This model links the observation of structural contacts in the 3D structure and the energetics of the protein. The use of structure–energy relationships is interesting from a mechanistic standpoint, but makes the calculations highly dependent on the quality of the available 3D structure and produces fits of poorer quality compared to those from more phenomenological models (Naganathan, 2012). For example, the analysis of the fast folder gpW with the ME-ES model produces a theoretical DSC profile that is sharper than the experimental one and with a native baseline that has lower temperature dependence (Fig. 4E). Consequently, the fitted free energy surface has a too sharp native well and overestimates the free energy barrier (~6.5 kJ mol^{-1}) compared to analyses with VB and ME-SSA models (Fig. 4F). The situation is the exact opposite for the slow folder SH3 in which the native baseline slope estimated by the model is higher than the Freire baseline thus producing a poorer fit (Fig. 5E) and an underestimated free energy barrier (Fig. 5F) compared to estimates from VB and ME-SSA models.

5.2.4 Mean-Field Free Energy Surface Model

The mean-field (MF) model uses the continuous variable nativeness (n) as the reaction coordinate. Nativeness is defined as the average probability of finding any residue in native-like conformation (De Sancho, Doshi, & Muñoz, 2009; Naganathan et al., 2007). Therefore, $n=1$ defines the state in which all the peptide bonds are native, corresponding to the fully folded conformation, whereas $n=0$ defines the fully unfolded state in which the probability for any given peptide bond to be native is 0. Each intermediate value of n defines an ensemble of microstates that is consistent with a probability of n for any given peptide bond to be in native conformation. In this model, there is no explicit enumeration of individual protein microstates and no folding mechanism is set *a priori*.

The MF model defines specific functional forms for the enthalpy, entropy, and heat capacity as a function of the reaction coordinate n (see De Sancho and Muñoz (2011) for the complete mathematical formulation). The conformational entropy as a function of n ($\Delta S^{\text{conf}}(n)$) is calculated using the formula,

$$\Delta S_{\text{conf}}(n) = N\left(-R[n\ln(n) + (1-n)\ln(1-n)] + (1-n)\Delta S_{\text{res}}^{n=0}\right) \tag{9}$$

where the native state is the reference state, $\Delta S_{\text{res}}^{n=0}$ is the cost in conformational entropy of fixing one residue in native conformation, and N is the number of residues. The second term is the difference in entropy between the fully native and fully unfolded states, whereas the first is the combinatorial entropy of mixing residues in native and nonnative conformation. The enthalpy at the reference temperature (again 385 K) and the heat capacity are defined as Markov chains, and thus they have an exponential dependence on n with curvature determined by parameter $\kappa_{\Delta H}$ (De Sancho & Muñoz, 2011). The combination of the three thermodynamic functions results in the free energy as a function of n (order parameter) and temperature,

$$\Delta G(T,n) = \Delta H(T,n) - T\Delta S(T,n). \tag{10}$$

The shape of the free energy surface arises from the different curvature of the entropic and enthalpic contributions. The curvature of the entropic contribution is constant (Eq. 9) and thus the steepness of the enthalpic exponential decay solely determines the height of the free energy barrier (purely entropic barrier like in all previous models). A steeper enthalpy function (higher $\kappa_{\Delta H}$), which would represent a larger contribution from nonlocal interactions to protein stability (De Sancho & Muñoz, 2011), results in a larger entropy–enthalpy mismatch, and thus in free energy surfaces with two wells separated by a high free energy barrier. Smoother enthalpy functions (lower $\kappa_{\Delta H}$), which would represent proteins with native states largely stabilized by local interactions (De Sancho & Muñoz, 2011), result in little or no mismatch producing one-state downhill free energy surfaces. Therefore, the structural information of the protein is fully encased in the sole parameter $\kappa_{\Delta H}$. The model requires three additional thermodynamic parameters $\left(\Delta S_{\text{res}}^{n=0}, \Delta H_{\text{res}}, \Delta C_{p,\text{res}}\right)$ plus two more for defining the native heat capacity baseline. As in the other structure-free models (VB and ME-SSA), the MF model fits to the gpW and SH3 thermograms are of very high quality, capturing both the pretransition slope and the width of the thermogram (Figs. 4G, H and 5G, H).

5.3 Bayesian Approach for Global Analysis with Multiple Models

The objective of employing multiple statistical mechanical models in the DSC analysis was to extract key information from the experiment in a model-independent manner. This obviously requires procedures for comparing the models and ranking their performance. The first important issue is that the models vary in the order parameter and assumptions, and thus result in free energy surfaces of very different shape (position and width of the wells and peaks) (see Figs. 4 and 5). The performance of the models is also very different, depending on their implicit assumptions, the number of parameters they use, and the particular experimental data that is being analyzed. For instance, an extensive analysis with various versions of the four models to the gpW and SH3 DSC profiles (21–23 independent fits) produced predicted barrier heights (β) in the range −0.04 and 7.0 kJ mol^{-1} for gpW and in the range 3–22 kJ mol^{-1} for SH3 (Naganathan et al., 2011). Likewise, the quality of fits (evaluated from the sum of least squares, SLS) differs by more than an order of magnitude for both proteins. The agreement of the fitted native baseline with the Freire prediction varies widely as well (Naganathan et al., 2011).

Although the direct comparison of free energy surfaces from various models is not straightforward, the property of most interest from an experimental viewpoint is the height of the thermodynamic barrier (which defines the width of the ensemble at the denaturation midpoint), which can be calculated in a simple way from the free energy surfaces produced by all models.

To extract a model-independent estimate of the thermodynamic barrier from the DSC profile, one thus only need to define a statistical procedure to compare the physical reasonableness of the different fits. For this purpose, we defined a simple Bayesian approach that ranks the fits based on the SLS and the degree of deviation of the native baseline slope form that expected from the Freire equation. This approach results in simple expressions for calculating the statistically best estimate of the free energy barrier and its associated standard error (see Naganathan et al. (2011) for full derivation of the procedure and details on its implementation).

Figure 6 summarizes the results of the Bayesian analysis of the gpW and SH3 DSC profiles. This figure shows that the fits to the gpW profile consistently produce low barriers and those to the SH3 profile high barriers. Obtaining such a clear-cut result requires that fit quality (SLS) and physical reasonableness (magnitude of the native baseline slope) are combined as criteria for the Bayesian approach (especially for fast-folding proteins with

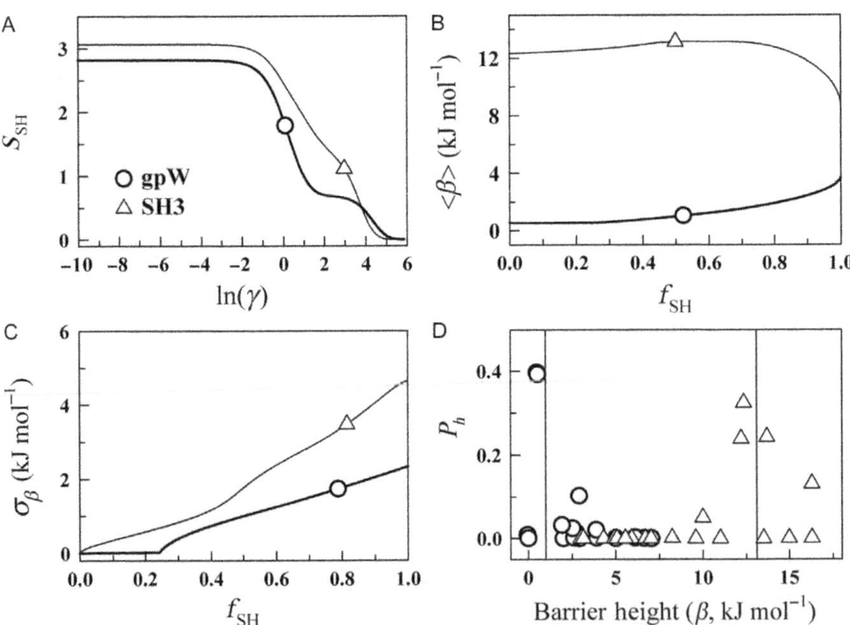

Figure 6 Summary of the results from the multimodel Bayesian analysis of the DSC profiles of gpW (circles) and α-spectrin SH3 domain (triangles). (A) Shannon entropy (see Naganathan et al. (2011) for details) as a function of the regularization constant. (B and C) Estimated mean barrier height and associated standard deviation as a function of the relative Shannon entropy. (D) The probability of a given barrier height obtained from 21 (gpW) and 23 (α-spectrin SH3 domain) different fits.

marginal barriers as gpW). Ultimately, the Bayesian analysis estimates a barrier of 13.3 ± 1.8 kJ mol^{-1} for SH3, which is consistent with its previous adscription to the two-state folding regime and with its slow folding rate (Viguera et al., 1994). The barrier estimated for gpW is 1.0 ± 1.0 kJ mol^{-1}, which is lower than thermal energy thus placing this protein in the downhill folding regime, consistently with its fast microsecond folding and observed dispersion in melting temperatures in atomic resolution nuclear magnetic resonance experiments and long-timescale molecular dynamics simulations (Sborgi et al., 2015).

6. CONCLUDING REMARKS

It is now widely accepted that proteins in solution populate ensembles of different conformations. However, including conformational heterogeneity in the analysis of protein denaturation experiments was for a very long time, simply out of the question. Consequently, protein thermal denaturation was (and still is, for the most part) interpreted using models that assume discrete macrostates, and most researchers would see little point in going beyond this simple, phenomenological description. Justification for such analysis has usually come from the outcome of thermodynamic protein denaturation experiments, which typically show a simple sigmoidal change in the physical properties of interest.

Recent developments in multiple fronts have challenged this conventional wisdom. Here, we have reviewed those developments in the context of the analysis of protein denaturation DSC experiments. As discussed, DSC experiments can be nowadays interpreted in terms of ensembles of protein microstates provided that certain conditions hold: (i) the DSC thermogram must accurately describe an equilibrium unfolding process, thus guaranteeing that the heat capacity versus temperature profile contains accurate information about all the significantly populated states; (ii) The DSC profile must be in absolute heat capacity units, which is essential to distinguish the signs of conformational heterogeneity from the intrinsic solvation heat capacity of the native protein; (iii) the analysis of the experimental results with a variety of ensemble models should provide physically reasonable and consistent results. We refer to these developments as "modern DSC analysis" of protein folding. These modern methods offer new opportunities for DSC, but most certainly do not invalidate or replace more traditional approaches, such as the two-state adherence of protein unfolding from the calorimetric to van't Hoff enthalpy ratio, the interpretation of experiments on large complex proteins as the combination of unfolding events from individual domains, the determination of kinetic stability effects from scan rate dependent transitions for protein irreversible denaturation, or the possibility of screening thermal stability for multiple protein variants.

Finally, we want to emphasize that the modern approaches reviewed here are not of general applicability, but they break new ground by providing direct links between conventional folding–unfolding experiments and the conformational dynamics features of proteins that are essential for their biological function.

ACKNOWLEDGMENTS

This work was funded through grants CSD2009-00088 (V.M., J.M.S.R.), BIO2011-28092 (V.M.), and BIO2012-34937 (J.M.S.R.) from the Spanish Ministry of Economy and Competitiveness, grant P09-CVI-5073 (B.I.M.) from the Andalucian Regional Government, and grant ERC-2012-ADG-323059 (V.M.) from the European Research Council.

REFERENCES

Akmal, A., & Muñoz, V. (2004). The nature of the free energy barriers to two-state folding. *Proteins*, *57*(1), 142–152.

Bahar, I., Lezon, T. R., Yang, L. W., & Eyal, E. (2010). Global dynamics of proteins: Bridging between structure and function. *Annual Review of Biophysics*, *39*, 23–42.

Becktel, W. J., & Schellman, J. A. (1987). Protein stability curves. *Biopolymers*, *26*(11), 1859–1877.

Costas, M., Rodriguez-Larrea, D., De Maria, L., Borchert, T. V., Gomez-Puyou, A., & Sanchez-Ruiz, J. M. (2009). Between-species variation in the kinetic stability of TIM proteins linked to solvation-barrier free energies. *Journal of Molecular Biology*, *385*(3), 924–937.

De Sancho, D., Doshi, U., & Muñoz, V. (2009). Protein folding rates and stability: How much is there beyond size. *Journal of the American Chemical Society*, *131*(6), 2074–2075.

De Sancho, D., & Muñoz, V. (2011). Integrated prediction of protein folding and unfolding rates from only size and structural class. *Physical Chemistry Chemical Physics*, *13*(38), 17030–17043.

Freire, E. (2001). The thermodynamic linkage between protein structure, stability and function. *Methods in Molecular Biology*, *168*, 37–68.

Freire, E., & Biltonen, R. L. (1978). Statistical mechanical deconvolution of thermal transitions in macromolecules. 1. Theory and application to homogeneous systems. *Biopolymers*, *17*(2), 463–479.

Freire, E., van Osdol, W. W., Mayorga, O. L., & Sanchez-Ruiz, J. M. (1990). Calorimetrically determined dynamics of complex unfolding transitions in proteins. *Annual Review of Biophysics and Biophysical Chemistry*, *19*, 159–188.

Fung, A., Li, P., Godoy-Ruiz, R., Sanchez-Ruiz, J. M., & Muñoz, V. (2008). Expanding the realm of ultrafast protein folding: gpW, a midsize natural single-domain with alpha + beta topology that folds downhill. *Journal of the American Chemical Society*, *130*(23), 7489–7495.

Garcia-Mira, M. M., Sadqi, M., Fischer, N., Sanchez-Ruiz, J. M., & Muñoz, V. (2002). Experimental identification of downhill protein folding. *Science*, *298*(5601), 2191–2195.

Godoy-Ruiz, R., Henry, E. R., Kubelka, J., Hofrichter, J., Munoz, V., Sanchez-Ruiz, J. M., et al. (2008). Estimating free-energy barrier heights for an ultrafast folding protein from calorimetric and kinetic data. *The Journal of Physical Chemistry. B*, *112*(19), 5938–5949.

Goldberg, R. N., Kishore, N., & Lennen, R. M. (2002). Thermodynamic quantities for the ionization reactions of buffers. *Journal of Physical and Chemical Reference Data*, *31*(2), 231–370.

Gomez, J., Hilser, V. J., Xie, D., & Freire, E. (1995). The heat capacity of proteins. *Proteins*, *22*, 404–412.

Gruebele, M. (2008). Fast protein folding. In V. Muñoz (Ed.), *Protein folding, misfolding and aggregation: Classical themes and novel approaches*. Cambridge: Royal Society of Chemistry.

Guzman-Casado, M., Parody-Morreale, A., Robic, S., Marqusee, S., & Sanchez-Ruiz, J. M. (2003). Energetic evidence for formation of a pH-dependent hydrophobic cluster in the denatured state of Thermus thermophilus ribonuclease H. *Journal of Molecular Biology*, *329*, 731–743.

Häckel, M., Hinz, H.-J., & Hedwig, G. R. (1999). A new set of peptide-based group heat capacities for use in protein stability calculations. *Journal of Molecular Biology*, *291*, 197–213.

Halskau, O., Perez-Jimenez, R., Ibarra-Molero, B., Underhaug, J., Munoz, V., Martinez, A., et al. (2008). Large-scale modulation of thermodynamic protein folding barriers linked to electrostatics. *Proceedings of the National Academy of Sciences of the United States of America*, *105*(25), 8625–8630.

Henzler-Wildman, K. A., Lei, M., Thai, V., Kerns, S. J., Karplus, M., & Kern, D. (2007). A hierarchy of timescales in protein dynamics is linked to enzyme catalysis. *Nature*, *450*, 913–916.

James, L. C., & Tawfik, D. S. (2003). Conformational diversity and protein evolution—A 60-year-old hypothesis revisited. *Trends in Biochemical Sciences*, *28*(7), 361–368.

Johnson, C. M. (2013). Differential scanning calorimetry as a tool for protein folding and stability. *Archives of Biochemistry and Biophysics*, *531*(1–2), 100–109.

Kaushik, J. K., Ogasahara, K., & Yutani, K. (2002). The unusually slow relaxation kinetics of the folding-unfolding of pyrrolidone carboxyl peptidase from a hyperthermophile, *Pyrococcus furiosus*. *Journal of Molecular Biology*, *316*(4), 991–1003.

Kaya, H., & Chan, H. S. (2000). Polymer principles of protein calorimetric two-state cooperativity. *Proteins*, *40*(4), 637–661.

Khersonsky, O., & Tawfik, D. S. (2010). Enzyme promiscuity: A mechanistic and evolutionary perspective. *Annual Review of Biochemistry*, *79*, 471–505.

Kholodenko, V., & Freire, E. (1999). A simple method to measure the heat capacity of proteins. *Analytical Biochemistry*, *270*, 336–338.

Kubelka, J., Hofrichter, J., & Eaton, W. A. (2004). The protein folding 'speed limit'. *Current Opinion in Structural Biology*, *14*(1), 76–88.

Leone, V., Marinelli, F., Carloni, P., & Parinello, M. (2010). Targeting biomolecular flexibility with metadynamics. *Current Opinion in Structural Biology*, *20*, 148–154.

Lopez, M. M., & Makhatadze, G. I. (2002). Differential scanning calorimetry. *Methods in Molecular Biology*, *173*, 113–119.

Makhatadze, G. I. (2001). Measuring protein thermostability by differential scanning calorimetry. *Current Protocols in Protein Science*. Chapter 7: p. 7.9.1–7.9.14.

Makhatadze, G. I., & Privalov, P. L. (1990). Heat capacity of proteins. I. Partial molar heat capacity of individual aminoacid residues in aqueous solution; hydration effect. *Journal of Molecular Biology*, *213*, 375–384.

Makhatadze, G. I., & Privalov, P. L. (1995). Energetics of protein structure. *Advances in Protein Chemistry*, *5*, 507–510.

Mayorga, O. L., & Freire, E. (1987). Dynamic analysis of differential scanning calorimetry data. *Biophysical Chemistry, 27*(1), 87–96.
Muñoz, V. (2001). What can we learn about protein folding from Ising-like models? *Current Opinion in Structural Biology, 11*(2), 212–216.
Muñoz, V. (2002). Thermodynamics and kinetics of downhill protein folding investigated with a simple statistical mechanical model. *International Journal of Quantum Chemistry, 90*(4–5), 1522–1528.
Muñoz, V. (2007). Conformational dynamics and ensembles in protein folding. *Annual Review of Biophysics and Biomolecular Structure, 36*, 395–412.
Muñoz, V., & Eaton, W. A. (1999). A simple model for calculating the kinetics of protein folding from three-dimensional structures. *Proceedings of the National academy of Sciences of the United States of America, 96*(20), 11311–11316.
Muñoz, V., & Sanchez-Ruiz, J. M. (2004). Exploring protein folding ensembles: A variable barrier model for the analysis of equilibrium unfolding experiments. *Proceedings of the National Academy of Sciences of the United States of America, 101*, 17646–17651.
Muñoz, V., & Serrano, L. (1995). Elucidating the folding problem of helical peptides using empirical parameters. 2. Helix macrodipole effects and rational modification of the helical content of natural peptides. *Journal of Molecular Biology, 245*(3), 275–296.
Naganathan, A. N. (2012). Predictions from an Ising-like statistical mechanical model on the dynamic and thermodynamic effects of protein surface electrostatics. *Journal of Chemical Theory and Computation, 8*(11), 4646–4656.
Naganathan, A. N., Doshi, U., Fung, A., Sadqi, M., & Muñoz, V. (2006). Dynamics, energetics, and structure in protein folding. *Biochemistry, 45*(28), 8466–8475.
Naganathan, A. N., Doshi, U., & Muñoz, V. (2007). Protein folding kinetics: Barrier effects in chemical and thermal denaturation experiments. *Journal of the American Chemical Society, 129*(17), 5673–5682.
Naganathan, A. N., Li, P., Perez-Jimenez, R., Sanchez-Ruiz, J. M., & Muñoz, V. (2010). Navigating the downhill protein folding regime via structural homologues. *Journal of the American Chemical Society, 132*(32), 11183–11190.
Naganathan, A. N., & Muñoz, V. (2014). Thermodynamics of downhill folding: Multiprobe analysis of PDD, a protein that folds over a marginal free energy barrier. *The Journal of Physical Chemistry. B, 118*(30), 8982–8994.
Naganathan, A. N., Perez-Jimenez, R., Muñoz, V., & Sanchez-Ruiz, J. M. (2011). Estimation of protein folding free energy barriers from calorimetric data by multi-model Bayesian analysis. *Physical Chemistry Chemical Physics, 13*(38), 17064–17076.
Naganathan, A. N., Perez-Jimenez, R., Sanchez-Ruiz, J. M., & Munoz, V. (2005). Robustness of downhill folding: Guidelines for the analysis of equilibrium folding experiments on small proteins. *Biochemistry, 44*(20), 7435–7449.
Naganathan, A. N., Sanchez-Ruiz, J. M., & Muñoz, V. (2005). Direct measurement of barrier heights in protein folding. *Journal of the American Chemical Society, 127*(51), 17970–17971.
Onuchic, J. N., Luthey-Schulten, Z., & Wolynes, P. G. (1997). Theory of protein folding: The energy landscape perspective. *Annual Review of Physical Chemistry, 48*, 545–600.
Onuchic, J. N., Socci, N. D., Luthey-Schulten, Z., & Wolynes, P. G. (1996). Protein folding funnels: The nature of the transition state ensemble. *Folding & Design, 1*(6), 441–450.
Palmer, A. G., 3rd. (1997). Probing molecular motion by NMR. *Current Opinion in Structural Biology*, 7(5), 732–737.
Pey, A. L. (2013). Protein homeostasis disorders of key enzymes of amino acids metabolism: Mutation-induced protein kinetic destabilization and new therapeutic strategies. *Amino Acids, 45*(6), 1331–1341.

Plaza del Pino, I. M., Pace, C. N., & Freire, E. (1992). Temperature and guanidine hydrochloride dependence of the structural stability of ribonuclease T1. *Biochemistry*, *31*(45), 11196–11202.

Portman, J. J., Takada, S., & Wolynes, P. G. (2001). Microscopic theory of protein folding rates. I. Fine structure of the free energy profile and folding routes from a variational approach. *The Journal of Chemical Physics*, *114*(11), 5069–5081.

Prigozhin, M. B., & Gruebele, M. (2013). Microsecond folding experiments and simulations: A match is made. *Physical Chemistry Chemical Physics*, *15*(10), 3372–3388.

Privalov, P. L. (1990). Cold denaturation of proteins. *Critical Reviews in Biochemistry and Molecular Biology*, *25*, 281–305.

Privalov, P. L., & Dragan, A. I. (2007). Microcalorimetry of biological macromolecules. *Biophysical Chemistry*, *126*, 16–24.

Risso, V. A., Manssour-Triedo, F., Delgado-Delgado, A., Arco, R., Barroso-DelJesus, A., Ingles-Prieto, A., et al. (2015). Mutational studies on resurrected ancestral proteins reveal conservation of site-specific amino acid preferences throughout evolutionary history. *Molecular Biology and Evolution*, *32*(2), 440–455.

Robertson, A. D., & Murphy, K. P. (1997). Protein structure and the energetics of protein stability. *Chemical Reviews*, *97*(5), 1251–1267.

Romero-Romero, M. L., Ingles-Prieto, A., Ibarra-Molero, B., & Sanchez-Ruiz, J. M. (2011). Highly anomalous energetics of protein cold denaturation linked to folding-unfolding kinetics. *PloS One*, *6*, e23050.

Sadqi, M., de Alba, E., Pérez-Jiménez, R., Sanchez-Ruiz, J. M., & Muñoz, V. (2009). A designed protein as experimental model of primordial folding. *Proceedings of the National Academy of Sciences of the United States of America*, *106*, 4127–4132.

Sadqi, M., Fushman, D., & Munoz, V. (2006). Atom-by-atom analysis of global downhill protein folding. *Nature*, *442*(7100), 317–321.

Sanchez-Ruiz, J. M. (1992). Theoretical analysis of Lumry-Eyring models in differential scanning calorimetry. *Biophysical Journal*, *61*(4), 921–935.

Sanchez-Ruiz, J. M. (1995). Differential scanning calorimetry of proteins. *Sub-Cellular Biochemistry*, *24*, 133–176.

Sanchez-Ruiz, J. M. (2010). Protein kinetic stability. *Biophysical Chemistry*, *148*(1–3), 1–15.

Sanchez-Ruiz, J. M. (2011). Probing free-energy surfaces with differential scanning calorimetry. *Annual Review of Physical Chemistry*, *62*, 231–255.

Sborgi, L., Verma, A., Piana, S., Lindorff-Larsen, K., Cerminara, M., Santiveri, C. M., et al. (2015). Interaction networks in protein folding via atomic-resolution experiments and long-time-scale molecular dynamics simulations. *Journal of the American Chemical Society*, *137*, 6506–6516.

Schellman, J. A. (1987). The thermodynamic stability of proteins. *Annual Review of Biophysics and Biophysical Chemistry*, *16*, 115–137.

Sikosek, T., & Chan, H. S. (2014). Biophysics of protein evolution and evolutionary protein biophysics. *Journal of the Royal Society, Interface*, *11*, 20140419.

Spolar, R. S., & Record, M. T. (1995). Coupling of local folding to site-specific binding of proteins to DNA. *Science*, *263*, 777–784.

Thorolfsson, M., Ibarra-Molero, B., Fojan, P., Petersen, S. B., Sanchez-Ruiz, J. M., & Martinez, A. (2002). L-phenylalanine binding and domain organization in human phenylalanine hydroxylase: A differential scanning calorimetry study. *Biochemistry*, *41*(24), 7573–7585.

Viguera, A. R., Martinez, J. C., Filimonov, V. V., Mateo, P. L., & Serrano, L. (1994). Thermodynamic and kinetic-analysis of the Sh3 domain of spectrin shows a 2-state folding transition. *Biochemistry*, *33*(8), 2142–2150.

Vogl, T., Jatzke, C., Hinz, H. J., Benz, J., & Huber, R. (1997). Thermodynamic stability of annexin V E17G: Equilibrium parameters from an irreversible unfolding reaction. *Biochemistry*, *36*(7), 1657–1668.

Vogt, A. D., & Di Cera, E. (2012). Conformational selection or induced fit? A critical appraisal of the kinetic mechanism. *Biochemistry*, *51*, 5894–5902.

Wako, H., & Saito, N. (1978a). Statistical mechanical theory of protein conformation. 2. Folding pathway for protein. *Journal of the Physical Society of Japan*, *44*(6), 1939–1945.

Wako, H., & Saito, N. (1978b). Statistical mechanical theory of protein conformation. 1. General considerations and application to homopolymers. *Journal of the Physical Society of Japan*, *44*(6), 1931–1938.

Zou, T., Risso, V., Gavira, J. A., Sanchez-Ruiz, J. M., & Ozkan, S. B. (2015). Evolution of conformational dynamics determines the conversion of a promiscuous generalist into a specialist enzyme. *Molecular Biology and Evolution*, *32*, 132–143.

CHAPTER THIRTEEN

A Guide to Differential Scanning Calorimetry of Membrane and Soluble Proteins in Detergents

Zhengrong Yang*, Christie G. Brouillette*,†,1
*Center for Structural Biology, University of Alabama at Birmingham, Birmingham, Alabama, USA
†Department of Chemistry, University of Alabama at Birmingham, Birmingham, Alabama, USA
[1]Corresponding author: e-mail address: christie@uab.edu

Contents

Abstract

Differential scanning calorimetry (DSC) detects protein thermal unfolding by directly measuring the heat absorbed. Simple DSC experiments that require relatively small amounts of pure material can provide a wealth of information related to structure, especially with respect to domain architecture, without the need for a complete

Methods in Enzymology, Volume 567
ISSN 0076-6879
http://dx.doi.org/10.1016/bs.mie.2015.08.014

thermodynamic analysis. Thus, DSC is an ideal additional tool for membrane protein characterization and also offers several advantages over indirect thermal unfolding methods. Integral membrane proteins (IMPs) that comprise both large multitopic transmembrane domains (TMDs) and extramembranous domains (EMDs) are differentially affected by detergent interactions with both domains. In fact, in some cases, destabilization of the EMD by detergent may dominate overall IMP stability. This chapter will (1) provide a perspective on the advantages of DSC for membrane protein characterization and stability measurements, including numerous examples spanning decades of research; (2) introduce models for the interaction and destabilization of IMPs by detergents; (3) discuss two case studies from the authors' lab; and (4) offer practical advice for performing DSC in the presence of detergents.

ABBREVIATIONS

CD circular dichroism
CFTR cystic fibrosis transmembrane conductance regulator
CMC critical micelle concentration
DSC differential scanning calorimetry
EMD extramembranous domain
IMP integral membrane protein
NBD1 first nucleotide-binding domain
PDC protein–detergent complex
Pgp P-glycoprotein
TMD transmembrane domain

DETERGENT ABBREVIATIONS (I.E., SHORT NAMES)

CYMAL-5 5 cyclohexylpentyl β-D-maltoside
CHAPS 3-[(3-cholamidopropyl)-dimethylammonio]-1-propane sulfonate
DDM *n*-dodecyl-β-D-maltopyranoside
FC-14 *n*-tetradecyl-phosphocholine
LPG14 1-myristoyl-2-hydroxy-*sn*-glycero-3-phospho-(1′-rac-glycerol)
OG *n*-octyl2b2D-glucopyranoside
SDS sodium dodecyl sulfate
TTAB trimethyl(tetradecyl) ammonium bromide
also see Table 3 for additional detergents

1. INTRODUCTION

All membrane proteins produced from recombinant expression systems will come in contact with one or more detergents in the purification protocol. Consequently, as a matter of convenience or necessity, a large fraction of biophysical and structural studies on membrane proteins are conducted in detergent. Generally, the considerations and pitfalls are similar for conducting any biophysical experiment on a protein in detergent solution, whether it is membrane derived or soluble. Moreover, an investigator familiar with using detergents for such experiments probably will find that most of the methodology and precautions particular to differential scanning calorimetry (DSC) are intuitively obvious. This chapter will also devote some discussion to various models, which may not be widely recognized, for the destabilization of proteins by detergents, and begin with a brief background on relevant advantages to DSC of membrane proteins.

We believe this chapter will be most helpful to structural biologists interested in learning more about the title subject. However, we expect that detergent aficionados with little experience using DSC will also find the chapter informative, and we hope that calorimetrists with little prior experience working with detergents will also find utility in understanding the impact of detergents on biophysical analysis. We also hope to convince the structural biologist, who is considering crystallography as their first line of study, that DSC may be a better initial foray. It is common for multiple milligram quantities of purified membrane protein, likely obtained from heroic laboratory efforts, to be surrendered merely to find and optimize crystallization conditions before data collection for structure determination is even attempted. Devoting only a milligram or less of the same purified preparation for a couple of DSC experiments will typically yield useful, perhaps crucial, information on the physical properties and inferred structural features. The information obtained on a membrane protein from DSC analysis, as described in this chapter, will undoubtedly also aid in interpretation of the eventual high-resolution structure obtained by NMR or crystallography.

The selection of publications given in Table 1 contains excellent discussions of the pertinent background information on detergents with which the reader should be familiar. This list is merely representative, and by no means exhaustive. In this chapter, we will only touch on some of these topics to aid in clarity and continuity. If interested in a historical perspective, one may delve back into the Methods in Enzymology archives to discover that the principles of, and obstacles to, membrane protein purification and stability in detergents have not changed much (see Table 1). On the other hand, today, there is an impressive array of chemically diverse detergents, and the commercial availability of detergents has grown at least 10-fold from about 30 years ago.

Interest in the thermodynamics of membrane protein stability led to their immediate study by DSC upon the introduction of high sensitivity DSC instruments (Jackson, Kostyla, Nordin, & Brandts, 1973; Jackson & Sturtevant, 1978). For both theoretical and technical reasons, DSC offers

Table 1 Background Information on Detergents

Comment	Title, Authors	Publication Date
Detergent properties and their selection for membrane protein purification and structural studies	Methods in enzymology Vol. 557, Membrane proteins—Engineering, purification and crystallization. Shukla (Ed.) Chapter 3: Membrane preparation and solubiization. Roy Chapter 4: Amphipathic agents for membrane protein study. Sadaf, Cho, Byrne, and Chae Chapter 17: Inducing two-dimensional crystallization of membrane proteins by dialysis for electron crystallography. Uddin and Schmidt-Krey Chapter 18: Crystallization of membrane proteins by vapor diffusion. Delmar, Bolla, Su, and Yu	2015
Protein purification	Methods in enzymology Vol. 463, Guide to protein purification, 2nd ed. Burgess and Deutscher (Eds.)	2009
Detergent structures and properties; selection criteria for membrane protein solubilization	Chapter 34: Detergents: An overview. Guide to protein purification. Linke	
Selection criteria for membrane protein solubilization; biophysical studies in detergents	Chapter 35: Purification of membrane proteins. Lin and Guidotti	

Case studies and general guidelines to membrane protein functional and structural studies	Methods Vol. 55(4), Membrane protein technologies for structural biology. Stevens (Ed.)	2011
New detergent classes	Review article found in Vol. 55: New amphiphiles for membrane protein structural biology. Zhang, Tao, and Hong	
Membrane protein functional and structural studies	Detergents for the stabilization and crystallization of membrane proteins. Privé	2007
	A pedestrian guide to membrane protein crystallization. Wiener	2004
	Detergents as tools in membrane biochemistry. Garavito and Ferguson-Miller	2001
Historical perspective	Methods in enzymology Vol. 182, Guide to protein purification. Deutscher (Ed.)	1990
	Chapter 18: Detergents: An overview. Neugebauer	
	Chapter 19: Solubilization of native membrane proteins. Hjelmeland	
	Methods in enzymology Vol. 104(Pt. C), Enzyme purification and related techniques. Jakoby (Ed.)	1984

some distinct advantages over other methods aimed at measuring protein stability. DSC stands alone as a tool that directly measures the heat of a cooperative protein conformational change, e.g., unfolding. It thereby provides model-independent thermodynamic parameters, namely, the molar enthalpy and heat capacity change upon thermal unfolding, from which the entropy and free energy changes can be obtained. The inexperienced reader is referred to Cooper, Nutley, & Wadood (2000) and/or Johnson (2005) for introductory guides to DSC theory, instrumentation, experimentation, and data analysis. A theoretical foundation for DSC and a statistical thermodynamics description of unfolding transitions can be found in Freire (1995), and earlier reviews of the theory and instrumentation, as well as DSC experimental examples, are found in Privalov (1979) and Sturtevant (1987).

By assuming a model for unfolding, only the van't Hoff enthalpy can be obtained from all other indirect measurements of unfolding, e.g., thermal unfolding monitored by changes in a spectral property. However, analysis of the DSC transition can provide both the model-independent calorimetric enthalpy and a van't Hoff enthalpy. These two parameters provide insight into the mechanism of unfolding not achievable with indirect methods (see Freire, 1995 for an in-depth discussion). From extrapolation of this concept, one can see that information on complex domain architecture can be obtained. While we focus on the secondary and tertiary structure of proteins in this chapter, it is worth noting that DSC can detect and identify changes in the quaternary structure as well. Some examples of DSC of multidomain membrane proteins, including oligomeric structures, in detergent and/or lipid are compiled in Table 2.

Related to the greater information content of DSC curves is the fact that technical difficulties limit the number of data points one can obtain through indirect measures of stability, whether determined by thermal or isothermal denaturation. For instance, at each discrete temperature, a measurement of the observable (e.g., fluorescence, absorbance, or ellipticity at a specific wavelength(s)) is most commonly made as essentially a unique experiment, while measurement of heat change in a DSC experiment is practically a continuous measurement providing at least hundreds of data points through the unfolding transition, depending on the data sampling rate and scan rate. Complex integral membrane proteins (IMPs) with an extensive extramembranous structure are common. Consequently, it can be expected that two or more domains with different stabilities exist within these structures. Partly due to the limited data collected from indirect techniques, it is often not possible to detect the presence of multiple domains, let alone obtain reliable or statistically significant parameters for the stability of each domain from

Table 2 Thermal Unfolding Studies on Multidomain/Oligomeric Membrane Proteins and/or Their Isolated Subunits/Domains[a]

Protein(s) Studied	Oligomeric State	DSC Included?	Detergent(s) used	Complementary Methods Used	Citation	Section(s) in Text Where Cited
Human erythrocyte membranes	N/A	Yes	In membrane, not solubilized	Far-UV CD Enzyme Activity	Jackson et al. (1973)	Section 1
Purple membrane	Trimer and higher order	Yes	In membrane, not solubilized	Absorbance	Jackson and Sturtevant (1978)	Section 1
Bacteriorhodopsin in purple membrane	Trimer and higher order	Yes	In membrane, not solubilized	Absorbance, fluorescence, near- and far-UV CD	Brouillette, Muccio, and Finney (1987)	Sections 1, 3, and 4
Bacteriorhodopsin	Monomer	Yes	CHAPS/lipid mixed micelles	Absorbance, fluorescence, near- and far-UV CD	Brouillette, McMichens, Stern, and Khorana (1989)	Sections 1 and 3
Ca^{2+} ATPase of sarcoplasmic reticulum	Monomer	Yes	In membrane, not solubilized	Functional assays	Anteneodo et al. (1994)	Section 1
Cytochrome c oxidase (from *Paracoccus denitrificans*)	3-Subunit	Yes	DDM, Triton X-100	Infrared spectroscopy	Haltia, Semo, Arrondo, Goñi, and Freire (1994)	Section 1
Cytochrome c oxidase (bovine)	13-Subunit	Yes	DDM with or without added lipids	Far-UV CD	Sedlák, Varhač, Musatov, and Robinson (2014)	Section 1

Continued

Table 2 Thermal Unfolding Studies on Multidomain/Oligomeric Membrane Proteins and/or Their Isolated Subunits/Domains—cont'd

Protein(s) Studied	Oligomeric State	DSC Included?	Detergent(s) used	Complementary Methods Used	Citation	Section(s) in Text Where Cited
Mannitol permease (from *E. coli*)	Monomer	Yes	Reconstituted in DMPC	Far-UV CD	Meijberg, Schuurman-Wolters, Boer, Scheek, and Robillard (1998)	Sections 1 and 4
Transferrin receptor complex (from *Neisseria meningitidis*)	2-Subunit	Yes (on the soluble subunit)	Elugent	Far-UV CD	Krell et al. (2003)	Section 1
Na^+-K^+ ATPase (from rabbit kidney)	3-Subunit	Yes	$C_{12}E_8$, and reconstituted in liposomes	Enzyme activity	Yoneda, Rigos, and Ciancaglini (2013)	Sections 1 and 3
Light-harvesting antenna complex	Multisubunit	Yes	In lipid vesicles	Fluorescence and visible CD	Krumova et al. (2014)	Sections 1 and 4
P-glycoprotein (murine)	Monomer	Yes	DDM, and reconstituted in liposomes	Fluorescence and far-UV CD	Bai et al. (2011) and Yang et al. (2015)	Sections 1, 3, 6, and 7
NaCHBac sodium channel (from *Bacillus halodurans*)	Tetramer	No	CYMAL-5	Far-UV CD (synchotron radiation)	Powl, Miles, and Wallace (2012)	Sections 3 and 4
Bovine photoreceptor protein rhodopsin	Monomer	Yes	DDM	Far-UV CD	Reyes-Alcaraz, Martínez-Archundia, Ramon, and Garriga (2011)	Section 4

Glycerol facilitator (from *E. coli*)	Tetramer	No	DDM, LPC14, SDS	Near- and far-UV CD and fluorescence	Galka, Baturin, Manley, Kehler, and O'Neil (2008)	Sections 3 and 6
NBD1 of human CFTR	Monomer	Yes	None	Near- and far-UV CD and fluorescence	Protasevich et al. (2010)	Sections 4 and 6
NBD1 of human CFTR	Monomer	Yes	20 Detergents from three structural classes	Far-UV CD	Yang et al. (2014)	Sections 3, 5, and 6

[a]Citations listed in order of their first appearance in the Chapter.

deconvolution, unless the transitions do not overlap significantly. This is usually not an issue for DSC, and frequently a seemingly single unfolding transition can be reliably deconvoluted to reveal unfolding intermediates. Further technical advantages of DSC are it does not require covalent labeling or extrinsic probes, and while the spectroscopic techniques commonly used to monitor membrane protein folding, especially in lipid vesicles, can be susceptible to light scattering artifacts (Kelly, Jess, & Price, 2005; Mao & Wallace, 1984; Moon & Fleming, 2011; Wallace & Mao, 1984), DSC has no analogous problem. While not related to artifact, it is worth noting a thorough review of static light scattering theory and practical applications for characterizing protein–detergent mixed micelles by Slotboom, Duurkens, Olieman, and Erkens (2008).

Regarding disadvantages, thermal denaturation of membrane proteins does not lead to complete unfolding; therefore, one may conclude that any estimate of the free energy change will be an underestimate (Harris & Booth, 2012; Hong, Joh, Bowie, & Tamm, 2009; Minetti & Remeta, 2006). A larger issue is related to the fact that thermal denaturation of membrane proteins, whether in detergent, the native membrane or reconstituted into lipid vesicles, is irreversible and sometimes kinetically controlled (Sedlák et al., 2014). Consequently, a thermodynamic analysis of thermal unfolding data can be problematic or impossible. On the other hand, under certain conditions, and for a limited number of cases, reversible *isothermal* unfolding by SDS or chaotrope/detergent mixtures has been achieved and a thermodynamic analysis performed (Booth & Curnow, 2009; Harris & Booth, 2012; Hong, 2014; Roman & Gonzalez Flecha, 2014). However, caveats to this type of analysis are also described by Broecker and Keller (2013). Since thermodynamic equilibrium rarely will be achieved for complex IMP unfolding, the authors believe the advantages of DSC, especially with respect to defining multidomain architecture, outweigh the disadvantages. The reader is referred to the following additional published examples that illustrate the unique insights DSC can provide into domain and subunit architecture (Brandts, Hu, Lin, & Mas, 1989; Chamani, 2010; Griko, Rogov, & Privalov, 1992; Piszczek, D'Auria, Staiano, Rossi, & Ginsburg, 2004; Uchiyama, Ohshima, Yoshida, Ohkubo, & Kobayashi, 2013; Yang et al., 2004).

2. DETERGENT PROPERTIES

In this section, we give a brief overview of the detergent properties that are pertinent to conducting DSC experiments and analyzing DSC data, which include the chemical structure, micellization properties, and the effect of temperature on micellization.

2.1 Chemical Structure

Detergents are amphipathic molecules that, in the most general description, contain two mostly segregated molecular surfaces—one comprising hydrophilic moieties and the other hydrophobic moieties. The classical structure is a linear molecule containing a hydrophilic "headgroup" and a hydrophobic "tail." Structural classes which differ from this general description will be introduced in Section 5.3. This functional group segregation facilitates aggregation into micelles in aqueous media via the hydrophobic surfaces, when the detergent concentration is sufficiently high. Please refer to Linke (2009) and Lin and Guidotti (2009) for more in-depth discussions of relevant subject matter. The most important factor that determines whether a detergent is harsh (may denature protein) or mild (nondenaturing) is its overall charge or lack thereof. Detergents are broadly classified into four classes based on their ionic properties: anionic, cationic, zwitterionic, and nonionic (see Table 3). In general, charged detergents, anionic or cationic, are more denaturing than zwitterionic detergents, which in turn are more denaturing than nonionic detergents. Not surprisingly, the most popular detergents used in IMP structural determination are nonionic (Moraes, Evans, Sanchez-Wetherby, Newstead, & Stewart, 2014; Newstead, Ferrandon, & Iwata, 2008).

2.2 Micellization Properties: Critical Micelle Concentration and Aggregation Number

The minimum or threshold detergent concentration at which micelles form is called the CMC (critical micelle concentration). Any detergent solution that contains micelles also contains the detergent monomer at a concentration that is approximately equal to the CMC. The Aggregation Number (N_{agg}) is the number of monomeric detergents in the micelle and therefore determines the size of the micelle. The values for CMC and N_{agg} are both required to calculate the actual concentrations of detergent monomers and micelles in solution. Since either or both species may interact with the protein and their effect will be observed by DSC, it is necessary to know their molar concentrations before analyzing the data.

Several empirical correlations between the detergent structure and micellization properties have been observed. For example, within the same structural class, detergents with longer tails have lower CMCs and larger N_{agg}'s. These properties also correlate with the harshness of the detergents, which will be discussed in Section 6.

For a given detergent, its CMC and N_{agg} are dependent on the environmental variables, such as pH, ionic strength, and organic solvent content.

Table 3 Representatives From Each Detergent Class

Short Name	Full Name	Structure	MW	Ionic Property	CMC in Water (m*M*)	N_{agg} in Water
SDS	Sodium dodecyl sulfate	$Na^+ \ {}^-O-S(=O)_2-O-(CH_2)_{11}-CH_3$	289	Anionic	7–10	62
TTAB	Trimethyl(tetradecyl) ammonium bromide	$CH_3-(CH_2)_{12}-CH_2-N^+(CH_3)_3 \ Br^-$	337	Cationic	4–5	80
DiC_6PC	1,2-Dihexanoyl-*sn*-glycero-3-phosphocholine	$(CH_3)_3N^+-CH_2CH_2-O-P(=O)(O^-)-O-CH_2-CH(O-C(=O)-(CH_2)_5-CH_3)-CH_2-O-C(=O)-(CH_2)_5-CH_3$	454	Zwitterionic	14.3	30
FC-14	*n*-Tetradecyl-phosphocholine	$(CH_3)_3N^+-CH_2CH_2-O-P(=O)(O^-)-O-(CH_2)_{13}-CH_3$	380	Zwitterionic	0.12	108
DDM	*n*-Dodecyl-β-D-maltopyranoside	maltose–O–$(CH_2)_{11}-CH_3$	511	Nonionic	0.15	98
$C_{10}E_6$	Hexaethylene glycol monodecyl ether	$CH_3-(CH_2)_9-O-(CH_2CH_2O)_5CH_2CH_2OH$	423	Nonionic	0.9	40

CMC and N_{agg} obtained from www.anatrace.com or http://www.sigmaaldrich.com/life-science/core-bioreagents/learning-center/biological-detergents-guide.html.

Usually, the micellization of uncharged detergents is not influenced by pH or salt concentration as much as the charged detergents. Therefore, the CMC and N_{agg} of uncharged detergents in water may be used instead of values obtained under the actual experimental conditions. However, this approximation is not suitable for charged detergents. Methods for CMC and N_{agg} determination include light scattering (Gracia, Gómez-Barreiro, González-Pérez, Nimo, & Rodríguez, 2004), small-angle neutron scattering (Berr & Jones, 1989), and fluorescent dye binding or quenching (Jumpertz et al., 2011; Tummino & Gafni, 1993). The CMC and N_{agg} in water of each detergent, as those shown in Table 3, can usually be obtained from the detergent manufacturers' Websites, along with other helpful information about the detergents such as their water solubility and spectral properties.

2.3 Effect of Temperature on Micellization Properties

For some detergents, it is important to note the Cloud Point (T_C) and Krafft Point (T_K), which are two unique temperatures where the solubility of the detergent changes drastically. Detergents aggregate and precipitate out of solution if the temperature exceeds their T_C. Although the T_C of most detergents is higher than 100 °C, several popular nonionic detergents, such as Triton X-100 and Tween-20, have T_C that fall in the DSC experimental temperature range (typically 5–100 °C). If it is necessary to use these detergents in DSC, proper control and careful analysis are needed to obtain meaningful data. That is, the T_C should be carefully determined, and any heat effect associated with the precipitation should be measured by recording a DSC scan on the detergent in buffer. In addition, if any DSC transition is observed near the T_C for a protein sample in this detergent, then non-calorimetric techniques (see Section 4) are required to confirm that the transition originates from a structural change in the protein. The Krafft point is the minimum temperature at which the detergent will form micelles. If the temperature falls below T_K at or above the detergent CMC, the detergent will precipitate due to insolubility. Some ionic detergents, such as long-chained alkane sulfonates, have a T_K that is higher than RT and so have limited solubility at room temperature. T_K may be lowered by adding salt (Chu & Feng, 2012) or by mixing with another detergent (Tsujii, Saito, & Takeuchi, 1980). For detergents whose T_C and T_K are outside of the typical experimental temperature range of DSC, the CMC dependence on temperature is not very pronounced. Therefore, the detergent monomer and micelle concentrations remain relatively constant throughout the DSC experiment.

3. DETERGENT EFFECTS ON PROTEINS AND DETECTION BY DSC

The issue of membrane protein stability commonly arises during the process of detergent selection for membrane protein extraction and/or purification and is often discussed in terms of a protein's monodispersity and its propensity to aggregate (e.g., Privé, 2007; Tate, 2010; see detergent-related references given in Table 1). These physical properties are related, and strongly influenced, if not dictated, by the folding/unfolding thermodynamics. The unfolding kinetics also factor into protein stability, e.g., when aggregation accompanies or follows unfolding, rendering unfolding irreversible.

For any given membrane protein, considerable experimentation must go into choosing the best detergent(s) for purification and downstream structural studies (see e.g., Volumes 556 and 557 of Methods in Enzymology, Shukla (Ed.) 2015a, 2015b). In the presence of detergent, protein function and, by inference, native structure must be maintained, i.e., stabilized. In fact, for the purpose of detergent selection, many screening assays are designed to either detect temperature-dependent loss of function (Serrano-Vega, Magnani, Shibata, & Tate, 2008; Slowik & Henderson, 2015) or detect thermal destabilization of native structure (Alexandrov, Mileni, Chien, Hanson, & Stevens, 2008; Heydenreich, Vuckovic, Matkovic, & Veprintsev, 2015; Mancusso, Karpowich, Czyzewski, & Wang, 2011; Sonoda et al., 2011).

With respect to thermal stability, which is related to thermodynamic stability, the effect of a given detergent depends on both its structure and concentration. Once the membrane protein is solubilized from the membrane with a given detergent, a comparison among different detergents, or at different concentrations of the same detergent, may exhibit one of any possible outcomes, i.e., stabilization relative to another detergent; continual destabilization at increasing concentrations of the same detergent; or little to no change in thermal stability at increasing concentrations (Yang et al., 2014). Stability as measured by changes in protein size or spectral properties displays similar trends (Chou, Krishnamurthy, Randolph, Carpenter, & Manning, 2005; Lundahl, Mascher, Kameyama, & Takagi, 1990; Otzen, Sehgal, & Westh, 2009; Vajdos et al., 2001). The unfolding temperature (T_m) of the protein, measured by DSC, is a direct indication of the protein's thermal stability. As noted above, DSC also measures the molar unfolding enthalpy change (ΔH_c), the amount of heat required to unfold 1 mol of protein. Either a decrease in T_m or a decrease in ΔH_c (beyond that expected from the denaturational heat capacity change) or both suggests

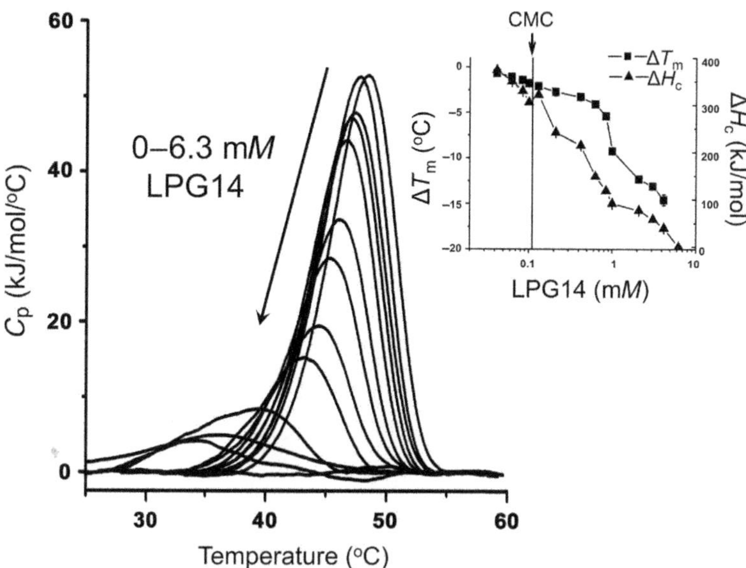

Figure 1 DSC of NBD1 from human CFTR in the presence of increasing amount of LPG14 from 0 to 6.3 m*M*. 0.5 mg/ml NBD1 in 20 m*M* HEPES, pH 7.5, 150 m*M* NaCl, 1 m*M* TCEP, 10% glycerol and 10% ethylene glycol, 20 μ*M* ATP, 3 m*M* $MgCl_2$ with LPG14 added at 0, 0.042, 0.063, 0.084, 0.125, 0.2, 0.4, 0.64, 1.0, 2.0, 3.0, 4.2, and 6.3 m*M*. DSC scan rate was 2 °C/min. Inset: ΔH_c and change in T_m (ΔT_m) as a function of LPG14 concentration. *Reproduced with permission from Yang et al. (2014) with change in the unit of the Y-axis.*

destabilization. If ΔH_c is reduced to zero, then the protein lacks a well-defined cooperative tertiary structure and, practically speaking, is denatured. Figure 1 shows the example of destabilization and denaturation of the isolated first nucleotide-binding domain (NBD1) derived from the human cystic fibrosis transmembrane conductance regulator (CFTR) by the anionic detergent LPG14 (Yang et al., 2014). Spectroscopic evidence from circular dichroism (CD) supported the interpretation of the DSC that NBD1 was denatured (see Section 4 for an overview of indirect thermal unfolding methods and Section 6.1 for a Case Study on NBD1).

In addition to the qualitative conclusions that are drawn based on the T_m and ΔH_c, a clue to the molecular mechanism for these changes is provided by thermodynamic models. The simplest model to represent a folding/unfolding process is the two-state model, where under equilibrium thermal unfolding, the protein exists in either the native state (N) or the unfolded state (U) (Fig. 2). Binding of detergents to either N or U or both states may shift the equilibrium toward either side, which causes the change in the measured T_m and ΔH_c.

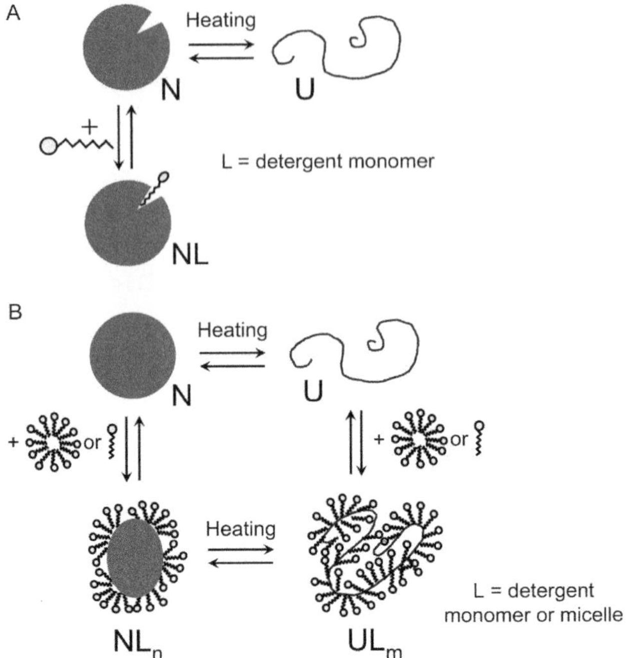

Figure 2 Three types of detergent-binding mechanisms to a soluble protein. (A) Stoichiometric binding of a detergent monomer to the native state. N, the native state; U, the unfolded state, the thermal unfolding is two-state; NL, detergent-bound native state without significant conformational change. (B) Detergent binding to the unfolded state (right leg) and the native state (left leg) with possible conformational change in the native state. NL_n, the native state with n detergent monomers or micelles bound, the conformation of the protein may or may not be the same as N; UL_m, the unfolded state with m detergent monomers or micelles bound; the thermal unfolding of NL_n is two-state. *Adapted from the thermodynamic models in Yang et al. (2014).*

As shown in Fig. 2, three different types of detergent-binding mechanisms can be envisaged: (1) Stoichiometric binding of detergent monomers to specific sites on the native protein (Fig. 2A). This type of binding is similar to the binding of natural ligands to a protein and usually does not cause a significant structural change. Because the detergent-binding site(s) only exist in N, N is favored in the presence of detergents, and the protein is stabilized. Some serum albumins are stabilized in the presence of extremely low concentrations of SDS, which is likely due to the specific binding of the detergent monomer (Kragh-Hansen, Hellec, de Foresta, le Maire, & Mller, 2001). (2) Detergent binding to the unfolded state (the right leg in Fig. 2B). The exposure of the protein's hydrophobic core in U provides many nonspecific sites for interaction with both detergent monomers and micelles. This type of binding destabilizes the protein because the unfolding equilibrium is shifted to U (Cooper et al., 2000; Waldron & Murphy, 2003). (3) Detergent binding to the native state with accompanied structural changes that result in destabilization or denaturation (the left leg in Fig. 2B). SDS denaturation belongs to this type of binding (Otzen, 2011), as does the example shown in Fig. 1. In reality, mechanism (3) is always accompanied by mechanism (2) because the detergent-bound N thermally unfolds into the detergent-bound U, which completes the thermodynamic cycle depicted in Fig. 2B. Even if the detergent-bound N is already denatured, further unfolding in the secondary structure may occur as the temperature increases.

We have also shown that the same mechanisms apply to detergent interactions with lysozyme, a soluble protein that reversibly unfolds via a two-state mechanism, thus raising the possibility that these mechanisms are generally applicable and not confined to membrane proteins and/or their extramembranous domains (EMDs). We have used this thermodynamic cycle to globally fit NBD1 DSC data obtained in the presence of detergents at various concentrations, as those shown in Fig. 1, to obtain the apparent binding parameters of the detergents–protein interaction. These studies will be discussed in more detail in Section 6.1. A significant observation from those studies was the similar effects detergents had on this isolated EMD in comparison to the intact IMP (Yang et al., 2014).

Detergent interactions with IMPs can be modeled from the same point of view, i.e., there exist four states, the detergent-free N and U, and the detergent-bound N and U. However, because detergent extraction of IMPs from the membrane is an irreversible process, the thermodynamic cycle is broken (Fig. 3, the ligand-binding equilibrium is replaced by the irreversible reaction). Let us first consider the unfolding equilibrium in the native membrane environment. Due to the high stability of transmembrane helices, the transmembrane domain (TMD) is usually more thermally stable than the EMD (Haltia & Freire, 1995; Powl et al., 2012). In addition, if the EMD interacts with the TMD, the EMD is expected to be more stable within the full-length IMP than in isolation (Bhaskara & Srinivasan, 2011; Di Bartolo & Booth, 2011; Di Bartolo, Hvorup, Locher, & Booth, 2011; see Brandts et al., 1989 for a theoretical treatment). Once the EMD unfolds, the domain–domain interactions no longer exist, and the TMD should unfold as if in isolation (Powl et al., 2012).

Sufficient experimental evidence has shown that upon thermal denaturation, the TMD loses tertiary structure whereupon the native helix–helix interactions are replaced by helix–lipids interactions. It is presumed that the TMD retains most of its *native* secondary structure (see reviews by

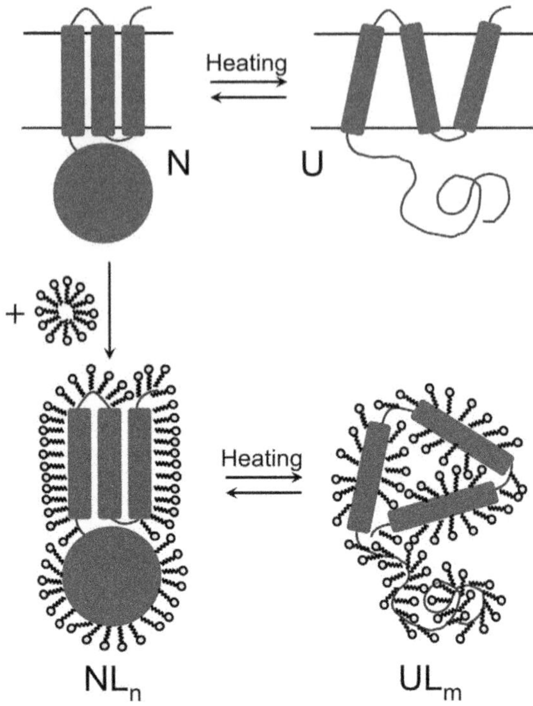

Figure 3 Detergent solubilization of an IMP with a transmembrane domain (TMD) and an EMD, and the thermal unfolding of the IMP in the lipid bilayer (top equilibrium) or in detergents (bottom equilibrium). N, the native state in the lipid bilayer; U, the thermally unfolded state in the lipid bilayer, the EMD is completely unfolded whereas the TMD primarily loses only the tertiary structure; NL_n, detergent solubilized IMP, with n detergent molecules bound; *Note*: the detergents bind the EMD too; UL_m, the unfolded state with m detergent molecules bound. Both unfolding processes are modeled as two-state to simplify the illustration.

Minetti & Remeta, 2006; Stanley & Fleming, 2008; Hong et al., 2009; Hong, 2014), because unfolding a helix inside the bilayer is energetically disfavored (Popot & Engelman, 1990; Ulmschneider, Doux, Killian, Smith, & Ulmschneider, 2010; White & Wimley, 1999; recently reviewed by Cymer, von Heijne, & White, 2015). Consequently, the unfolding enthalpy of the TMD is usually much lower than their soluble counterparts of the same size. The average specific unfolding enthalpy (h_u) of small globular proteins is 30–40 J/g and that of large soluble proteins is usually >20 J/g, whereas the h_u of some IMPs is <10 J/g. For example, the h_u of bacteriorhodopsin (bR), a single-domain helical IMP, is 15.8 J/g (Brouillette et al., 1987), but it has been estimated that at least 75% of the unfolding enthalpy comes from the dissociation of retinal. Haltia and

Freire (1995) have calculated that if only the extramembranous loops appreciably unfold then the h_u of bR is comparable to soluble proteins; this calculation is consistent with the limited loss of secondary structure observed by far-UV CD (Brouillette et al., 1987). Haltia and Freire (1995) performed similar calculations on a number of other IMPs and proposed that only the unfolding of extramembranous regions makes a significant contribution to the total unfolding enthalpy observed by DSC. Our observation from the DSC study on murine P-glycoprotein (Pgp) is consistent with this hypothesis (Bai et al., 2011, Yang et al., 2015, also see Section 6.2).

Upon detergent solubilization, the protein/lipid interactions in the TMDs are replaced by protein/detergent interactions, which are not ideal substitutes. Consequently, detergent solubilization often results in destabilization, when a comparison of thermal stability with the lipid embedded IMP can be made (Booth & Curnow, 2009; Brouillette et al., 1989; Galka et al., 2008; Heydenreich et al., 2015; Sedlák et al., 2014; Tol et al., 2013; Yang et al., 2015). Detergents may also disrupt the native quaternary structure of oligomeric membrane proteins (Brouillette et al., 1989; Galka et al., 2008). The protein–detergent complex (PDC) is commonly represented by a ring of detergent molecules surrounding a folded TMD with detergent hydrophobic tails in contact with the previously membrane-embedded regions (see, e.g., Delmar et al., 2015; Hong, 2014). However, if detergent is in excess (significantly above the CMC), which is always the case during detergent extraction of the IMP, then it will also bind to the EMD or the solvent-exposed extramembranous loops of the TMD (Fig. 3). In addition, some detergent molecules may penetrate into the protein core of the EMD resulting in a nonnative EMD structure. Collectively, our DSC and spectroscopic analysis of NBD1 unfolding provided support for this view (Yang et al., 2014; see Fig. 2). Upon thermal unfolding of the IMP in detergent, the transmembrane helices remain largely helical, but the unfolded protein tends to aggregate immediately, resulting in irreversible unfolding. A recent MD simulation by Neale et al. (2013) suggests that protein aggregation in detergents is mediated by a large network of aggregated detergents which hosts several embedded protein molecules.

Therefore, the instability of IMPs in detergents is due to several factors: the imperfect protection of the membrane-embedded regions, nonnative detergent/protein interactions in the TMD, detergent binding to and destabilization of the EMD, and irreversible unfolding due to aggregation. Among these factors, destabilization of the EMD can significantly influence the overall IMP stability when there is a substantial EMD structure and a concomitant extensive domain interface between the EMD and TMD.

With respect to DSC analysis, TMD thermal unfolding may not be observed because the contribution to ΔH_c from the TMD is often insignificant compared to the contribution from the EMD (Fodor et al., 2008; Haltia & Freire, 1995; Yoneda et al., 2013). Therefore, the stability of the EMD may be the only readout from the DSC experiment. However, this can also be the case when indirect methods such as intrinsic tryptophan fluorescence or far-UV CD are used to monitor unfolding, since little loss of secondary structure occurs in the TMD during unfolding and tryptophans may be insensitive to changes in environment during unfolding within the membrane. Based on both published and unpublished results, this seems to be the case for murine Pgp (Bai et al., 2011; Yang et al., 2015; see Section 6.2). The GPCR class seems to be an exception since, like the related bR, essentially the entire protein is membrane embedded. Nevertheless, thermal unfolding experiments have been successful in following the loss of native function and presumably tertiary structure, if not secondary structure. What sets these studies apart is the method for following thermal unfolding which, by and large, has been based on the temperature-dependent loss of ligand binding and, thus, have not directly monitored changes in either tertiary or secondary structure (Heydenreich et al., 2015; Serrano-Vega et al., 2008).

An important "take home" point from the above discussion is that thermal stability methods used to identify stabilizing detergents may only detect changes in EMD stability. This possibility was earlier raised by Tulumello and Deber (2012) who theorized that detergent interactions with the EMDs may account for detergent destabilization of IMPs based on their observation that transmembrane helices retain the same structure in both harsh and mild detergents. Moreover, for the purpose of downstream characterization, it may only be necessary to maintain or improve EMD stability (He et al., 2015). However, one should be aware that even when TMD unfolding cannot be directly monitored, changes in TMD stability can influence EMD stability via domain–domain interactions under certain conditions, and if these conditions are met, DSC will detect the changes in TMD stability via their effects on the EMD. For example, shown in Fig. 4 are simulated DSC curves (the theoretical basis for these was first introduced by Brandts et al., 1989) for a hypothetical protein with two interacting domains. Domain A has lower intrinsic stability than domain B. When the two DSC transitions overlap, an increase in the T_m of domain B (resulting from, e.g., stabilization of domain B by detergent or mutagenesis) results in an increase in the T_m of domain A. However, if the intrinsic stabilities of domains A and B are very different and the two transitions do not overlap, then a further increase in the stability of domain B has no effect

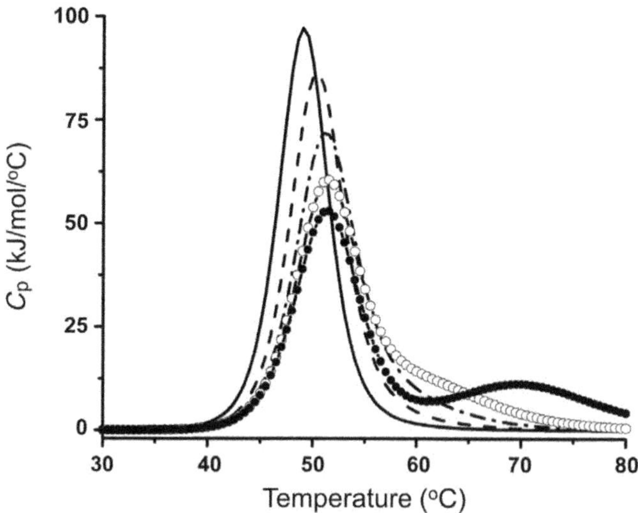

Figure 4 Simulated DSC curves of a protein with two interacting domains. The intrinsic unfolding parameters (i.e., in the absence of domain–domain interactions) are constant for domain A: $T_m^{A,\,Int} = 45\ °C$ and $\Delta H_c^{A,\,Int} = 418\ kJ/mol$; and for domain B: $T_m^{B,Int}$ varies from 45 to 70 °C and $\Delta H_c^{B,\,Int} = 209\ kJ/mol$. The free energy ($\Delta G_{int}$) of the domain–domain interaction is 8.36 kJ/mol. Solid line, $T_m^{B,\,Int} = 45\ °C$: the intrinsic stability of the two domains is the same, and due to domain interaction, both domains are stabilized and unfold cooperatively at 49 °C. Dashed line, $T_m^{B,\,Int} = 50\ °C$: the transition is slightly broadened, but both domains still unfold together at 50.5 °C. Dash dotted line, $T_m^{B,\,Int} = 55\ °C$: the transition is slightly asymmetrical with an apparent T_m of 51 °C. Open circles, $T_m^{B,\,Int} = 60\ °C$: a shoulder appears at the right side of the main transition that represents the unfolding of domain B, the T_m of domain A is 51.5 °C. Solid circles, $T_m^{B,\,Int} = 70°C$: the two domains unfold separately, domain A unfolds at 51.5 °C, and domain B at 70 °C. The theoretical basis of the equations used for this simulation can be found in Brandts et al. (1989).

on domain A. These simulations assume that when domain B is stabilized, there is no change in the A–B domain interface. If, however, the domain interface is strengthened when domain B is stabilized, then domain A will be stabilized by the increased domain–domain interaction even when the intrinsic stabilities are very different.

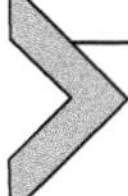

4. COMPLEMENTARY NONCALORIMETRY APPROACHES TO STUDY THERMAL UNFOLDING

Structural information provided by DSC is strictly inferred from other sources. If a structural model is available, e.g., via homology to an existing crystal structure, or even a prediction based on sequence analysis, one can begin to assign DSC transitions to the unfolding of probable domains or structural features. These suggested assignments can be tested with indirect thermal unfolding methods that monitor a specific structural feature. For example, far-UV CD monitors changes in secondary structure; near-UV CD and intrinsic tryptophan fluorescence, from primarily buried tryptophans, detect changes in tertiary structure. (A terrific review of readily accessible and commonly used methods can be found in Eftink (1995).) Putative domain assignments can be further reinforced by complementary studies on domains fragments produced by selected proteolytic cleavage such as described in Meijberg et al. (1998) who used CD to support the domain assignment of DSC transitions of lipid reconstituted mannitol permease, or overexpression, e.g., Powl et al., 2012 used CD to identify the unfolding of the EMD of the NaCHBac sodium channel in CYMAL-5 detergent.

An important criterion that can be used in support of two-state unfolding is the coincidence of all normalized thermal unfolding transitions, regardless of the detection method from which they were obtained. When comparing data derived from different thermal unfolding methods, it is absolutely essential to ensure the temperature scan rates are identical and that the temperature of the cuvette is monitored directly in the case of spectroscopic methods. When scan rate and temperature are carefully controlled, a comparison of different spectroscopic signals, to each other or to DSC, can be used to identify the existence of non-two-state unfolding, such as commonly observed for multidomain proteins. Spectroscopic thermal unfolding methods have been used frequently in DSC studies of membrane proteins to support unfolding mechanisms and/or domain assignments (examples can be found in Table 2).

5. CHOICE OF DETERGENTS

This topic has been covered in great detail in numerous places as cited in Section 1. To recapitulate briefly, the choice of detergents is ultimately determined by the goal of the study. For example, if the protein, similar to serum albumin or lipase, contains potential specific binding sites for amphipathic molecules (Kragh-Hansen et al., 2001; Mogensen, Sehgal, & Otzen, 2005; Nielsen, Borch, & Westh, 2000), then any type of detergent at a concentration below CMC and not in large excess of the protein may be suitable. Another example is using detergents during cell lysis to help solubilize proteins, because detergents have been shown to act as chemical chaperones to help stabilize partially unfolded proteins (Nath & Rao, 2001;

Rozema & Gellman, 1996) or keep proteins with low solubility from aggregating (Leibly et al., 2012). In these cases, uncharged detergents are preferred. If there is a need to completely remove the detergents once the protein is purified, detergents with higher CMC are easier to remove (Seddon, Curnow, & Booth, 2004). However, these detergents are more destabilizing than those with lower CMC (an example is shown in Section 6). Therefore, there usually is a trial and error process to discover the middle ground between stability and applicability. DSC can be an invaluable tool in this process.

As we have mentioned earlier, a common criterion for stability in detergent is whether the protein is monodisperse and remains so over the course of characterization. One caveat is that strongly denaturing detergents can yield a monodisperse preparation with similar secondary structure to that predicted for the native protein but is nevertheless devoid of native-like tertiary structure and function. Sometimes, functional assays may be absent or difficult to perform. In these cases, unfolding assays, especially DSC, may be the only way to determine whether the protein is folded or not. An example of Pgp denaturation by the anionic detergent LPG14 is shown in Section 6.2.

Other considerations when choosing detergents include the following:

5.1 Properties of the PDC

The presence of proteins may change the micellization properties of the detergent, especially when the protein binds to the detergent monomers. Our study of detergent interactions with NBD1 and lysozyme suggested that most nonionic and zwitterionic detergent *monomers* do not bind soluble proteins, whereas anionic detergent monomers do (Yang et al., 2014; see Section 6.1). Therefore, it is likely that for charged detergents, the presence of protein will lower the CMC by forming mixed PDCs or micelles. In the case of uncharged detergents, only micelles seem to bind soluble proteins. Upon binding, there is likely rearrangement in the detergent micelles to accommodate the surface properties of the protein and maximize hydrophobic interactions. Moreover, detergent molecules may cause changes in the protein's conformation to achieve better interactions, similar to an "induced fit" caused by stoichiometric ligands. However, because of the large number of detergents bound, the altered conformation is usually a less stable one. For both charged and uncharged detergents, the PDCs likely have different aggregation numbers, i.e., number of detergent molecules per micelle, compared to pure detergent micelles. The size of the PDCs can be determined by light scattering (Slotboom et al., 2008), ultracentrifugation, or size-exclusion chromatography (le Maire et al., 2008).

In Section 2, we have emphasized the necessity of knowing the concentrations of both the monomer and micelles in the pure detergents. To analyze the DSC data, it is not necessary to know the actual CMC in the presence of proteins or the number of detergent molecules in the PDCs, because using the thermodynamic cycle shown in Fig. 2B, the detergent is considered a ligand. Therefore, the concentrations of free and bound detergent monomers or micelles are dictated by the binding parameters, and knowing the total ligand concentrations is sufficient.

5.2 Working with Detergent Mixtures

The CMC, N_{agg}, and micelle composition of a binary mixed detergent system are determined by the micellization properties of the individual detergents (Vora, George, Hemangi, & Bahadur, 1999). Once these properties are experimentally determined, the mixed detergent system can be treated as a "single detergent" for DSC data analysis.

5.3 New Classes of Detergents and Nondetergent Alternatives

Traditional detergents are linear molecules with one head and one tail. Sometimes, the tail may contain monocyclic (e.g., CYMAL) or polycyclic (e.g., CHAPS) groups. Two classes of novel synthetic detergents that have gained popularity in recent years are the branch-chained detergents (Hong et al., 2010; Zhang, Tao, & Hong, 2011) and the cholesterol-like facial amphiphiles (Lee et al., 2013). Amphiphile is a term generally used to describe any structural class of detergents that deviates from the classical polar head-linear hydrophobic tail structure. Our experiences with these detergents (Yang et al., 2014, and unpublished results on NBD1) indicate that the principles remain the same. Readers are referred to the citations above for details on these detergents. Also see the review by Sadaf et al. (2015).

Nondetergent alternatives include amphipols (Kleinschmidt & Popot, 2014; Tribet, Audebert, & Popot, 1996) and lipopeptides (McGregor et al., 2003; Privé, 2009). Although these amphiphiles have been shown to improve IMP thermal stability by spectroscopic methods (amphipols reviewed by Kleinschmidt & Popot, 2014; lipopeptides in Wang et al., 2013), to our knowledge, there are no published DSC studies on proteins in these amphiphiles. Therefore, it is necessary to collect DSC scans on each new amphiphile without protein to determine if an amphiphile transition will interfere with the protein's unfolding signal.

6. CASE STUDIES

6.1 DSC of NBD1 from Human CFTR in Detergents

The goal of this study was to identify detergents suitable for the purification of full-length CFTR (our study laid the foundation for the study published by Hildebrandt et al., 2014, which identified better detergents for CFTR purification). We chose to study detergent effects on NBD1 (Yang et al., 2014) because this EMD plays a critical role in CFTR folding and stability (Protasevich et al., 2010 and references cited within; He et al., 2015). It is common for EMDs to autonomously fold in isolation, and it has been shown that the structures of these domains are generally similar to soluble proteins. Hence, studying NBD1 as an independent folding unit allowed us to distinguish detergent effects on the EMD from those on the TMD. Members from three detergent classes, anionic, nonionic, and zwitterionic, were chosen, which included those commonly used for CFTR purification and other popular detergents for IMP purification. Besides DSC, CD was used as a complementary method to monitor the change in secondary structure caused by the detergents.

Surprisingly, all detergents studied destabilize NBD1, but to different extents. An important observation was the effective harshness on NBD1 mirrors the general trend observed for their effect on IMPs, i.e., anionic > zwitterionic > nonionic. For example, Galka et al. (2008) studied the *E. coli* glycerol facilitator in three different detergents, SDS (anionic), LPC14 (zwitterionic), and DDM (nonionic). They found that in DDM, the protein retains its native tetrameric structure, whereas in SDS, there is more monomer. In LPC14, the protein remains tetrameric, but the thermal unfolding is less cooperative than in DDM. In our NBD1 study, we included more detergents and found the effects of all detergents from the same class were similar with just one or two exceptions. All anionic detergents, along with OG (the nonionic detergent with the highest CMC) and FC-14 (the harshest zwitterionic detergent), denature NBD1, as indicated by the complete loss of the DSC transition at certain detergent concentrations (see Fig. 1); whereas all nonionic detergents, except OG, destabilize but do not denature. The destabilizing effects of nonionic detergents plateau at high detergent concentrations (Fig. 5 shows the effect of DDM as an example), which suggests saturation of binding to both the N and

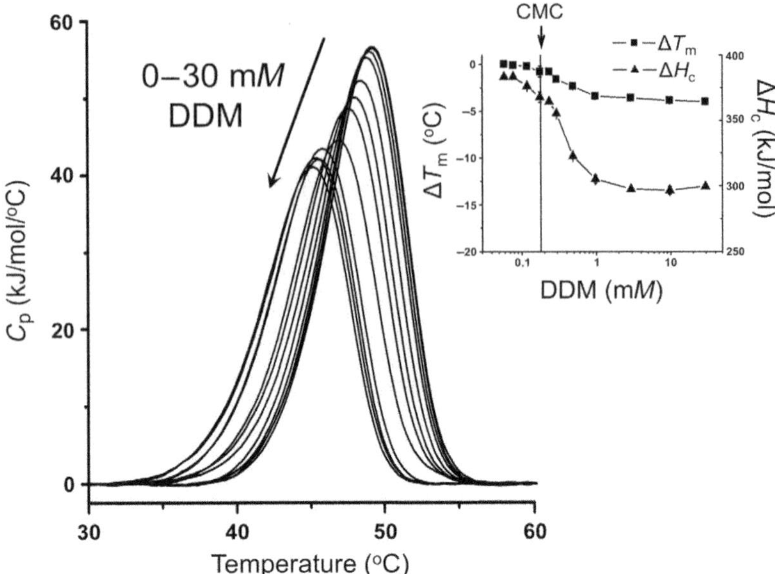

Figure 5 DSC of NBD1 from human CFTR in the presence of increasing amount of DDM from 0 to 30 m*M*. 0.5 mg/ml NBD1 in 20 m*M* HEPES, pH 7.5, 150 m*M* NaCl, 1 m*M* TCEP, 10% glycerol and 10% ethylene glycol, 20 μ*M* ATP, 3 m*M* $MgCl_2$ with DDM added at 0, 0.059, 0.085, 0.12, 0.18, 0.24, 0.30, 0.49, 0.98, 2.9, 9.8, and 30 m*M*. DSC scan rate was 2 °C/min. Inset: ΔH_c and change in T_m (ΔT_m) as a function of DDM concentration. *Reproduced with permission from Yang et al. (2014) with change in the unit of the* Y*-axis.*

U states (model in Fig. 2B). The effect of zwitterionic detergents is more diverse, likely due to their structural diversity, but most of them are expected to denature NBD1 only at high concentrations that are seldom used in protein extraction and purification, i.e., most zwitterionic detergents are more denaturing than nonionic detergents, but they do not completely denature under normal IMP solubilization conditions.

Information on the secondary structure obtained by the CD measurements provided further insights into the detergent effect. For all the denaturing detergents, except OG, CD indicated that the detergent-denatured protein contains a high content of nonnative helicity. On the other hand, no change in secondary structure was observed even in the presence of saturating concentrations of nonionic detergents. Therefore, harsher detergents affect NBD1 secondary structure, whereas milder detergents apparently only affect tertiary structure.

All the above conclusions were drawn based on the qualitative features of detergent-dependent changes in the DSC curves. Further analysis using the thermodynamic model shown in Fig. 2B to globally fit the DSC curves provided the binding stoichiometry, affinity, and enthalpy of each detergent. Because the unfolding of NBD1 is irreversible, approximation to two-state unfolding was applied, and as a result, the binding parameters obtained were not true thermodynamic parameters. However, these apparent binding parameters provided insights into the correlation between the detergent CMC and harshness.

Usually, detergents with a relatively small CMC are milder than those with a large CMC. Otzen proposed that because the CMC represents the concentration of the monomeric detergent in solution, monomers are likely the species that interact with the protein and cause destabilization (Otzen, 2011). However, our thermodynamic modeling suggests that nonionic or zwitterionic detergent monomers do not bind to the native state. It is the micelles that bind the protein and cause destabilization. We compared the apparent binding parameters of three maltosides with increasing tail lengths, DM (C10), UDM (C11), and DDM (C12) and a corresponding decrease in CMC. We found the extent of destabilization is determined by the number of detergent molecules bound per protein, and this number is inversely proportional to the micelle size. Therefore, it appears that DDM is the mildest among the three because its micelles are the largest, not because its CMC is the smallest. Additional experiments are required to confirm our hypothesis.

Another important finding of this study is the detergent effects on NBD1 correlate well with the observed purification yield of full-length CFTR and its ability to bind ATP, suggesting the importance of keeping these EMDs stable and folded during the course of membrane protein extraction and purification. This finding laid the foundation for the discovery of better detergents for the purification of CFTR (Hildebrandt et al., 2014). Due to the low expression level and low purification yield, we have not obtained enough purified CFTR for DSC experiments to directly correlate with the NBD1 DSC data. We therefore conducted DSC experiments on the homolog murine Pgp. Details are in the following section.

6.2 DSC of Murine Pgp in Detergents

Pgp is a prototypical ATP-binding cassette transporter that shares high sequence similarity with CFTR. Crystal structures of Pgp have been used for homology modeling of CFTR. In DSC, we observed an excellent correlation between the detergent effects on Pgp and their effects on CFTR NBD1. As shown in Fig. 6, the top panel contains the DSC curves of

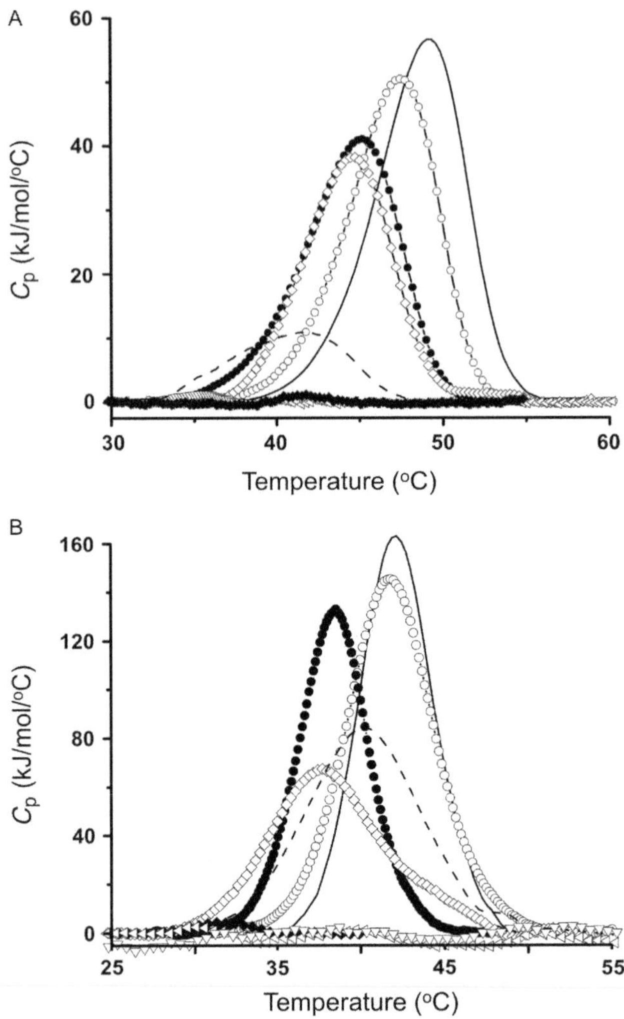

Figure 6 Effect of detergents on the DSC transitions of CFTR NBD1 and full-length murine Pgp. (A) 0.5 mg/ml CFTR NBD1 in 20 m*M* HEPES, pH 7.5, 150 m*M* NaCl, 1 m*M* TCEP, 10% glycerol and 10% ethylene glycol, 20 μ*M* ATP, 3 m*M* $MgCl_2$ with detergent as indicated. Black solid line: in the absence of detergent; open circles: 10 m*M* DMNG; solid circles: 18 m*M* DDM; open diamonds: 41 m*M* CHAPS; dashed line: 1 m*M* LPG14; solid diamonds: 6.3 m*M* LPG14; open down triangles: 40 m*M* OG; open left triangles: 10 m*M* FC-14. (B) 0.6 mg/ml murine Pgp in 20 m*M* HEPES, pH 7.5, 50 m*M* NaCl, 10% glycerol, 0.1 m*M* TCEP, and 0.5 m*M* DDM with added detergents as indicated. Black solid line: without added detergent; open circles: 10 m*M* DMNG; solid circles: in total of 18 m*M* DDM; open diamonds: 41 m*M* CHAPS; dashed line: 1 m*M* LPG14; solid diamonds: 4.2 m*M* LPG14; open down triangles: 40 m*M* OG; open left triangles: 13 m*M* FC-14. DSC scan rate was 2 °C/min for all. *Panel (A) Data from Yang et al. (2014), replotted. Panel (B) Data from Yang et al. (2015), manuscript in preparation.*

NBD1 in the presence of several different detergents (data from Yang et al., 2014, replotted), and the bottom panel contains the DSC curves of Pgp in the presence of the same detergents at approximately the same concentrations (data from Yang et al., 2015). The extent of destabilization caused by the nondenaturing detergents was CHAPS > DDM > DNMG, which was the same for both NBD1 and Pgp. Detergents that denatured NBD1, i.e., OG, FC-14, or LPG14, also denatured Pgp. The data reinforce the idea that the effect of detergents on the thermal stability of the EMDs makes a major contribution to their effect on the entire IMP.

7. EXPERIMENTAL PROCEDURES

7.1 Sample Preparation

Step 1. Prepare detergent stock solutions. Detergents usually come as powder and are stored at −20 °C. Since many detergents are hygroscopic, before weighing the powder out to prepare stock solution, let the bottle warm up to RT before opening. Make a concentrated stock detergent (e.g., 100–200 m*M*) in water, aliquot, and overlay with nitrogen or other inert gas to prevent oxidation and then store at −20 °C. Note that it is better to think in terms of molar concentration than %w/v, because detergents have different MW and the %w/v concentrations are not directly comparable among different detergents.

Step 2. Determine detergent CMC and N_{agg} in the DSC buffers. We routinely use the procedure described in Jumpertz et al. (2011) to determine the CMC. This assay monitors the large increase in the quantum yield of a fluorescent dye, Hoechst 33342, when it is sequestered in a hydrophobic environment such as inside a detergent micelle. Briefly, a set of detergent solutions at various [detergent], both below and above its CMC, are prepared. Also included in the solutions is Hoechst 33342 at a constant concentration. The [dye] needs to be optimized for the fluorimeter used, but a good starting concentration is between 5 and 10 μ*M*. The assay can be performed in cuvettes or in a 96-well assay plate (Yang et al., 2014). It is typical to observe a low fluorescence when the [detergent] is less than the CMC, and a large and steady increase in fluorescence when the [detergent] goes above the CMC. A simplified data analysis procedure involves linear regression of the points below and above the CMC, and the [detergent] at the intersection of the two lines is the CMC. Alternatively, nonlinear least square curve fitting using the equations in Jumpertz et al. (2011) can provide a slightly more accurate CMC.

The N_{agg} can be determined using the fluorescence quenching assay developed by Tummino and Gafni (1993). This assay monitors the decrease in the fluorescence of pyrene, when another dye, coumarin 153 (C153), is incorporated into the same micelle. The assay is usually carried out with a large excess of detergent micelles and a small amount of the dyes, typically 2 μM of pyrene and 0–20 μM of C153. The slope of the linear plot of $\mathrm{Ln}(F/F_o)$ versus $[\mathrm{C153}]_m$ equals the negative inverse of the [micelle], where F is the fluorescence of pyrene in the presence of C153 and F_o is that in the absence of C153. The $[\mathrm{C153}]_m$ used in the plot is not the total C153 concentration but that concentration which is partitioned into the micelles. Therefore, it is also necessary to determine the partition coefficient of C153 in the same detergent.

Step 3. Decide what protein and detergent concentrations to use. The sensitivity of the DSC varies among the models. Given a detection limit, the minimal required protein concentration can be calculated. For example, we have determined that the h_u of Pgp in DDM is approximately 6.5 J/g (Bai et al., 2011). The detection limit of our Cap-DSC instrument is 0.1 μW, i.e., 0.1 μJ/s. In order to observe a good signal-to-noise ratio, the signal should be three times the detection limit, which is 0.3 μJ/s. The temperature range from the beginning to the end of the unfolding transition is usually less than 20 °C. If the scan rate is 1 °C/min, then the base of the unfolding peak is less than 20 min, which is 1200 s. The area of the unfolding peak can be approximated by the area of a triangle with the same base and height as the bell-shaped unfolding peak. Therefore, the minimum detectable heat is 0.3 μJ/s × 1200 s × 0.5 (the formula for the area of a triangle), which is 180 μJ. The cell volume of the Cap-DSC is 0.13 ml. In order to generate 180 μJ of heat, the amount of Pgp needed is 180 μJ/(0.13 ml × 6.5 J/g) = 0.22 g/l (i.e., mg/ml). For membrane proteins with unknown enthalpy of unfolding, we recommend using twice the concentration, which is approximately 0.5 mg/ml. The h_u of soluble proteins is at least two times that of Pgp, so a good starting concentration for a soluble protein is 0.25 mg/ml. It is important that the protein concentration is accurately determined because the final thermodynamic parameters are highly dependent on the user's input of protein concentration during data analysis. Methods for determining protein concentration have been reviewed by Kelly et al. (2005), and possible interference by detergents was covered by Linke (2009). It is equally important that if the protein is not at least 95% pure, then the fraction of protein should be known with some reliability for purposes of calculating its concentration. In no case should the purity be less than 80–90%, because abundant contaminating proteins may contribute to the unfolding signal. As for the detergent concentration, depending on the application, you may want to limit it to below the CMC unless the effect of saturation is being investigated. In any case, the molar concentrations of both the monomer and micelle will be needed based on the following formula:

If $[\text{Detergent}]_{\text{total}} < \text{CMC}$, then $[\text{Monomer}] = [\text{Detergent}]_{\text{total}}$ and $[\text{Micelle}] = 0$,

If $[\text{Detergent}]_{\text{total}} \geq \text{CMC}$, then $[\text{Monomer}] \approx \text{CMC}$, and $[\text{Micelle}] = ([\text{Detergent}]_{\text{total}} - \text{CMC})/\text{N}_{\text{agg}}$.

Step 4. Prepare the DSC and reference samples. For IMPs, if the protein is purified in detergent(s) other than that intended for DSC, then it is necessary to perform detergent exchange. See Linke (2009) for methods of detergent removal and exchange. The most effective method for detergent exchange is washing extensively with the target detergent when the IMP is immobilized on an affinity or ion-exchange resin.

7.2 Conducting the DSC Experiments

The DSC profile of a protein that unfolds irreversibly may be scan-rate dependent and/or protein-concentration dependent. In order to determine the true effect of the detergents, it is necessary to keep the protein concentration and scan rate constant when the detergent concentration is varied. Reversibility can be crudely assessed by rescanning of the same sample that has been heated and cooled. Finer detail can be obtained by scanning partially through the DSC transition followed by a cooling cycle, and rescanning. It is also important to conduct the proper control experiments. A scan of both DSC cells filled with the detergent in buffer is necessary, because it represents the instrument baseline, and will be subtracted from the protein scan during the first step of data analysis. Also, a possible contribution by the detergent to the DSC transition(s) should be ruled out with a "detergent scan" obtained by filling the sample cell with the same detergent buffer mixture that is contained in the protein sample, while the reference cell contains the detergent-free buffer. Even if the detergent does not exhibit a transition, it is still necessary to keep the detergent concentration the same in the protein sample and in the reference sample. A large slope in the protein scan usually indicates a mismatch in the buffering components between the two samples, and the mismatch in detergent concentration is the most likely cause. An effective way to prevent this mismatch is to use appropriately sized filters when concentrating proteins in detergent, so that the pure detergent micelles can pass through the filter and not be concentrated.

7.3 DSC Data Analysis

7.3.1 *Basic Data Analysis with Software Built-In Models*

An outline of the steps involved in the basic analysis can be found in Cooper et al. (2000) and Johnson (2005). The instrument manuals provided by the DSC manufacturers usually contain a detailed guide for basic analysis as well. Basic data analysis involves data transformation to obtain the molar heat capacity (C_p) curve, and curve fitting with the built-in models in the instrument-provided software package. Data transformation usually includes the subtraction of the buffer scan from the protein scan and a mathematical operation to convert the raw C_p (usually in the unit of mJ/°C) into molar C_p (in the unit of kJ/mol). As we have mentioned earlier, using the correct protein concentration is critical in this step. For curve fitting, a sigmoidal or cubic baseline is usually subtracted from the molar C_p curve to floor the pre-transition and post-transition baselines to zero, because most built-in models assume that the ΔC_p (change in the protein's heat capacity upon unfolding) is zero. The ΔH_c is obtained by integrating the baseline-subtracted C_p curve. The ΔH_v is obtained by fitting the curve to the chosen model. The simplest model is the two-state model. If the two-state model does not provide an adequate fit, then more sophisticated models, either provided by the software or written by the users, will be needed (e.g., Protasevich et al., 2010; Yang et al., 2014).

7.3.2 *Curve Fitting with More Sophisticated Models*

Using the thermodynamic cycle shown in Fig. 2B, a set of DSC curves obtained in the presence of varying detergent concentration can be globally fitted to obtain the detergent/protein-binding parameters. A mathematical implementation of the model can be found in the Supplemental Materials of our NBD1-detergent interaction paper (Yang et al., 2014).

7.4 Avoiding Potential Pitfalls and Data Quality Assessment

It is useful to summarize here the key issues to keep in mind when conducting the DSC experiment that were raised in this section. A common experimental problem is not all DSC transitions result from unfolding or conformational changes of the target protein. This is one reason to conduct thermal unfolding experiments with complementary spectroscopic techniques that unambiguously detect a signal from protein (see Section 4 and Table 2). However, if the target protein is not sufficiently pure, then a transition identified as derived from protein may be due to a contaminating protein. A nonprotein transition may come from the detergent, which can micellize or demicellize as the temperature changes, with concomitant heat changes detected by DSC; conducting DSC of the detergent mixture without protein can be helpful in this case. The heat from a cooperative unfolding transition must not only be detectable (ΔH_c, see Section 7.1), but the transition must also be sufficiently sharp to be detected (ΔH_v). If it is unclear whether there is a cooperative transition or not, increasing the protein concentration will improve the signal-to-noise ratio. The average h_u of soluble proteins or IMPs listed in Section 3 should be used to estimate the expected ΔH_c. If the experimental ΔH_c is much lower than expected, then the protein may only partially unfold through the transition detected, or the protein sample may contain some denatured material that does not participate in the unfolding transition. These possibilities cannot be sorted out without conducting other, noncalorimetric, experiments to assess the protein's native folded state. It is also important to keep in mind that detergent solubilization of an IMP from the native membrane may not only result in destabilization but may, in fact, denature one or more domains. If it is not possible to assess the folded or functional state of the protein before solubilization, there is no way to unambiguously rule out this possibility. Stabilization or renaturation may or may not be realized by reconstitution back into a lipid environment once the protein has been solubilized.

In terms of data quality, a reliable and accurate protein concentration is of paramount importance. A key advantage of DSC comes from the ability to infer an unfolding mechanism for a given transition through a comparison of its calorimetric and van't Hoff enthalpies. For instance, $\Delta H_c > \Delta H_v$ suggests multiple domains may unfold within a single transition, while $\Delta H_c < \Delta H_v$ could be interpreted as oligomer unfolding. If the protein concentration is not known with confidence, it is easy to see how this comparison would be misleading, at the least. A caveat to keep in mind is that thermal denaturation of IMPs is irreversible and commonly leads to aggregation. If the aggregation event is kinetically fast compared to the instrument scan rate, then the unfolding transition will be artifactually sharpened, leading to an overestimation of ΔH_v and frequently an underestimation of ΔH_c. Aggregation may also be accompanied by heat evolution that is detected in the DSC experiment. This renders the determination of T_m problematic or impossible, depending on the severity of the heat effect. Adjustments in scan rate may improve the situation, as well as changes in the buffer and detergent constituents. If the scan rate is changed, it is crucial that all complementary unfolding methods use the same scan rate for appropriate comparisons.

8. CONCLUDING REMARKS

Here, we have provided evidence and reasons why DSC can be a useful and novel tool for exploring the structure of membrane proteins, even when only small amounts of pure material are available. While indirect methods can be used to assess thermal stability, the added information provided by DSC makes it an ideal characterization method. Moreover, the unique insights gained on the cooperativity and communication between structural domains is not duplicated by any other technique, even high-resolution crystallography. Potentially new insights into the molecular interactions between detergents and proteins are given here that can guide the experimentalist, as well as practical advice to conduct DSC experiments in the presence of detergent.

ACKNOWLEDGMENT

This work was supported, in part, by the Cystic Fibrosis Foundation Therapeutics, Inc.

REFERENCES

Alexandrov, A. I., Mileni, M., Chien, E. Y., Hanson, M. A., & Stevens, R. C. (2008). Microscale fluorescent thermal stability assay for membrane proteins. *Structure*, *16*, 351–359.

Anteneodo, C., Rodahl, A. M., Meiering, E., Heynen, M. L., Sennisterra, G. A., & Lepock, J. R. (1994). Interaction of dibucaine with the transmembrane domain of the Ca(2+)-ATPase of sarcoplasmic reticulum. *Biochemistry*, *33*(40), 12283–12290.

Bai, J., Swartz, D. J., Protasevich, I. I., Brouillette, C. G., Harrell, P. M., Hildebrandt, E., et al. (2011). A gene optimization strategy that enhances production of fully functional P-glycoprotein in Pichia pastoris. *PLoS One*, *6*(8), e22577.

Berr, S. S., & Jones, R. R. M. (1989). Small-angle neutron scattering from aqueous solutions of sodium perfluorooctanoate above the critical micelle concentration. *Physical Chemistry*, *93*(6), 2555–2558.

Bhaskara, R. M., & Srinivasan, N. (2011). Stability of domain structures in multi-domain proteins. *Scientific Reports*, *1*(40), 1–9.

Booth, P. J., & Curnow, P. (2009). Folding scene investigation: Membrane proteins. *Current Opinion in Structural Biology*, *19*, 8–13.

Brandts, J. F., Hu, C. Q., Lin, L. N., & Mas, M. T. (1989). A simple model for proteins with interacting domains. Applications to scanning calorimetry data. *Biochemical*, *28*(21), 8588–8596.

Broecker, J., & Keller, S. (2013). Impact of urea on detergent micelle properties. *Langmuir*, *29*, 8502–8510.

Brouillette, C. G., Muccio, D. D., & Finney, T. K. (1987). pH dependence of bacteriorhodopsin thermal unfolding. *Biochemistry, 26,* 7431–7438.

Brouillette, C. G., McMichens, R. B., Stern, L. J., & Khorana, H. G. (1989). Structure and thermal stability of monomeric bacteriorhodopsin in mixed phospholipid/detergent micelles. *Proteins, 5*(1), 38–46.

Chamani, J. (2010). Energetic domains analysis of bovine a-lactalbumin upon interaction with copper and dodecyl trimethylammonium bromide. *Journal of Molecular Structure, 979,* 227–234.

Chou, D. K., Krishnamurthy, R., Randolph, T. W., Carpenter, J. F., & Manning, M. C. (2005). Effects of Tween 20 and Tween 80 on the stability of Albutropin during agitation. *Pharmaceutical Sciences, 94*(6), 1368–1381.

Chu, Z., & Feng, Y. (2012). Empirical correlations between Krafft temperature and tail length for amidosulfobetaine surfactants in the presence of inorganic salt. *Langmuir, 28*(2), 1175–1181.

Cooper, A., Nutley, M. A., & Wadood, A. (2000). Differential scanning microcalorimetry. In S. E. Harding & B. S. Chowdhry (Eds.), *Protein-ligand interactions: Hydrodynamics and calorimetry* (pp. 287–318). Oxford, New York: Oxford University Press. See also, http://alan-cooper.org.uk/wp-content/uploads/2014/10/dsc.pdf.

Cymer, F., von Heijne, G., & White, S. H. (2015). Mechanisms of integral membrane protein insertion and folding. *Journal of Molecular Biology, 427*(5), 999–1022.

Delmar, J. A., Bolla, J. R., Su, C.-C., & Yu, E. W. (2015). Crystallization of membrane proteins by vapor diffusion. In A. K. Shukla (Ed.), *Methods in enzymology: Vol. 557. Membrane proteins—Engineering, purification, and crystallization* (pp. 363–392). Amsterdam: Elsevier. chapter 18; pp. 2–631.

Di Bartolo, N. D., Hvorup, R. N., Locher, K. P., & Booth, P. J. (2011). In vitro folding and assembly of the *Escherichia coli* ATP-binding cassette transporter, BtuCD. *The Journal of Biological Chemistry, 86,* 18807–18815.

Di Bartolo, N. D., & Booth, P. J. (2011). Unravelling the folding and stability of an ABC (ATP-binding cassette) transporter. *Biochemical Society Transactions, 39,* 751–760.

Eftink, M. R. (1995). Use of multiple spectroscopic methods to monitor equilibrium unfolding of proteins. In A. Hall, W. E. Balch, & C. J. Der (Eds.), *Methods in enzymology: Vol. 256* (pp. 487–512). Elsevier: Amsterdam.

Freire, E. (1995). Thermal denaturation methods in the study of protein folding. In M. L. Johnson & G. K. Ackers (Eds.), *Methods in enzymology: Vol. 259* (pp. 144–168). Amsterdam: Elsevier.

Fodor, E., Fedosova, N. U., Ferencz, C., Marsh, D., Pali, T., & Esmann, M. (2008). Stabilization of Na, K-ATPase by ionic interactions. *Biochimica et Biophysica Acta, 1778*(4), 835–843.

Galka, J. J., Baturin, S. J., Manley, D. M., Kehler, A. J., & O'Neil, J. D. (2008). Stability of the glycerol facilitator in detergent solutions. *Biochemistry, 47*(11), 3513–3524.

Garavito, R. M., & Ferguson-Miller, S. (2001). Detergents as tools in membrane biochemistry. *The Journal of Biological Chemistry, 276,* 32403–32406.

Gracia, C. A., Gómez-Barreiro, S., González-Pérez, A., Nimo, J., & Rodríguez, J. R. (2004). Static and dynamic light-scattering studies on micellar solutions of alkyldimethylbenzylammonium chlorides. *Journal of Colloid and Interface Science, 276*(2), 408–413.

Griko, Y. V., Rogov, V. V., & Privalov, P. L. (1992). Domains in lambda Cro repressor. A calorimetric study. *Biochemistry, 31*(50), 12701–12705.

Haltia, T., & Freire, E. (1995). Forces and factors that contribute to the structural stability of membrane proteins. *Biochimica et Biophysica Acta, 1241,* 295–322.

Haltia, T., Semo, N., Arrondo, J. L., Goñi, F. M., & Freire, E. (1994). Thermodynamic and structural stability of cytochrome c oxidase from Paracoccus denitrificans. *Biochemistry, 33*(32), 9731–9740.

Harris, N. J., & Booth, P. J. (2012). Folding and stability of membrane transport proteins in vitro. *Biochimica et Biophysica Acta, 1818*(4), 1055–1066.

He, L., Aleksandrov, A. A., An, J., Cui, L., Yang, Z., Brouillette, C. G., et al. (2015). Restoration of NBD1 thermal stability is necessary and sufficient to correct ΔF508 CFTR folding and assembly. *Journal of Molecular Biology, 427*(1), 106–120.

Heydenreich, F. M., Vuckovic, Z., Matkovic, M., & Veprintsev, D. B. (2015). Stabilization of G protein-coupled receptors by point mutations. *Frontiers in Pharmacology, 6*, 82.

Hildebrandt, E., Zhang, Q., Cant, N., Ding, H., Dai, Q., Peng, L., et al. (2014). A survey of detergents for the purification of stable, active human cystic fibrosis transmembrane conductance regulator (CFTR). *Biochimica et Biophysica Acta, 1838*(11), 2825–2837.

Hjelmeland, L. M. (1990). Solubilization of native membrane proteins. In M. P. Deutscher (Ed.), *Methods in enzymology: Vol. 182. Guide to protein purification* (pp. 253–264). Amsterdam: Elsevier.

Hong, H. (2014). Toward understanding driving forces in membrane protein folding. *Archives of Biochemistry and Biophysics, 564*, 297–313.

Hong, W. X., Baker, K. A., Ma, X., Stevens, R. C., Yeager, M., & Zhang, Q. (2010). Design, synthesis, and properties of branch-chained maltoside detergents for stabilization and crystallization of integral membrane proteins: Human connexin 26. *Langmuir, 26*(11), 8690–8696.

Hong, H., Joh, N. H., Bowie, J. U., & Tamm, L. K. (2009). Methods for measuring the thermodynamic stability of membrane proteins. In R. R. Burgess & M. P. Deutscher (Eds.), *Methods in enzymology: Vol. 455* (pp. 213–236). Amsterdam: Elsevier.

Jackson, W. M., Kostyla, J., Nordin, J. H., & Brandts, J. F. (1973). Calorimetric study of protein transitions in human erythrocyte ghosts. *Biochemistry, 12*, 3662–3667.

Jackson, M. B., & Sturtevant, J. M. (1978). Phase transitions of the purple membranes of Halobacterium halobium. *Biochemistry, 17*, 911–915.

Jakoby, W. B. (Ed.), (1984). *Methods in enzymology: Vol. 104. Part C: Enzyme purification and related techniques* (pp. 3–528). Amsterdam: Elsevier.

Johnson, D. M. (2005). *Differential scanning calorimetry: Theory and practice.* Northampton: Malvern Instruments. MicroCal, LLC Application Note, http://docslide.us/documents/faa-dsc-theory-and-practice.html.

Jumpertz, T., Tschapek, B., Infed, N., Smits, S. H., Ernst, R., & Schmitt, L. (2011). High-throughput evaluation of the critical micelle concentration of detergents. *Analytical Biochemistry, 408*, 64–70.

Kelly, S. M., Jess, T. J., & Price, N. C. (2005). How to study proteins by circular dichroism. *Biochimica et Biophysica Acta, 1751*, 119–139.

Kleinschmidt, J. H., & Popot, J. L. (2014). Folding and stability of integral membrane proteins in amphipols. *Archives of Biochemistry and Biophysics, 564*, 327–343.

Kragh-Hansen, U., Hellec, F., de Foresta, B., le Maire, M., & Mller, J. V. (2001). Detergents as probes of hydrophobic binding cavities in serum albumin and other water soluble proteins. *Biophysical Journal, 80*, 2898–2911.

Krell, T., Renauld-Mongénie, G., Nicolaï, M. C., Fraysse, S., Chevalier, M., Bérard, Y., et al. (2003). Insight into the structure and function of the transferrin receptor from Neisseria meningitidis using microcalorimetric techniques. *The Journal of Biological Chemistry, 278*(17), 14712–14722.

Krumova, S. B., Várkonyi, Z., Lambrev, P. H., Kovács, L., Todinova, S. J., Busheva, M. C., et al. (2014). Heat- and light-induced detachment of the light-harvesting antenna complexes of photosystem I in isolated stroma thylakoid membranes. *Journal of Photochemistry and Photobiology. B, Biology, 137*, 4–12.

le Maire, M., Arnou, B., Olesen, C., Georgin, D., Ebel, C., & Møller, J. V. (2008). Gel chromatography and analytical ultracentrifugation to determine the extent of detergent binding and aggregation, and Stokes radius of membrane proteins using sarcoplasmic reticulum Ca2+-ATPase as an example. *Nature Protocols*, *3*(11), 1782–1795.

Lee, S. C., Bennett, B. C., Hong, W. X., Fu, Y., Baker, K. A., Marcoux, J., et al. (2013). Steroid-based facial amphiphiles for stabilization and crystallization of membrane proteins. *Proceedings of the National Academy of Sciences of the United States of America*, *110*(13), E1203–E1211.

Leibly, D. J., Nguyen, T. N., Kao, L. T., Hewitt, S. N., Barrett, L. K., & Van Voorhis, W. C. (2012). Stabilizing additives added during cell lysis aid in the solubilization of recombinant proteins. *PLoS One*, 7(12), 1–14.

Lin, S.-H., & Guidotti, G. (2009). Purification of membrane proteins. In R. R. Burgess & M. P. Deutscher (Eds.), *Methods in enzymology: Vol. 463. Guide to protein purification* (2nd ed., pp. 619–629). Amsterdam: Elsevier. pp. 388–397.

Linke, D. (2009). Detergents: An overview. Guide to protein purification. In R. B. Richard & M. P. Deutscher (Eds.), *Methods in enzymology: Vol. 463* (2nd ed., pp. 1–851). Amsterdam: Elsevier.

Lundahl, P., Mascher, E., Kameyama, K., & Takagi, T. (1990). Water-soluble proteins do not bind octyl glucoside as judged by molecular sieve chromatographic techniques. *Journal of Chromatography*, *518*, 111–121.

Mao, D., & Wallace, B. A. (1984). Differential light scattering and absorption flattening optical effects are minimal in the circular dichroism spectra of small unilamellar vesicles. *Biochemistry*, *23*(12), 2667–2673.

McGregor, C. L., Chen, L., Pomroy, N. C., Hwang, P., Go, S., Chakrabartty, A., et al. (2003). Lipopeptide detergents designed for the structural study of membrane proteins. *Nature Biotechnology*, *21*(2), 171–176.

Mancusso, R., Karpowich, N. K., Czyzewski, B. K., & Wang, D. N. (2011). Simple screening method for improving membrane protein thermostability. *Methods*, *55*(4), 324–329.

Meijberg, W., Schuurman-Wolters, G. K., Boer, H., Scheek, R. M., & Robillard, G. T. (1998). The thermal stability and domain interactions of the mannitol permease of Escherichia coli. A differential scanning calorimetry study. *The Journal of Biological Chemistry*, *273*(33), 20785–20794.

Minetti, C. A., & Remeta, D. P. (2006). Energetics of membrane protein folding and stability. *Archives of Biochemistry and Biophysics*, *453*, 32–53.

Moon, C. P., & Fleming, K. G. (2011). Using tryptophan fluorescence to measure the stability of membrane proteins folded in liposomes. In M. L. Johnson, J. M. Holt, & G. K. Ackers (Eds.), *Methods in enzymology: Vol. 492* (pp. 189–211). Amsterdam: Elsevier.

Moraes, I., Evans, G., Sanchez-Wetherby, J., Newstead, S., & Stewart, P. D. (2014). Membrane protein structure determination—The next generation. *Biochimica et Biophysica Acta*, *1838*, 78–87.

Mogensen, J. E., Sehgal, P., & Otzen, D. E. (2005). Activation, inhibition, and destabilization of Thermomyces lanuginosus lipase by detergents. *Biochemistry*, *44*, 1719–1730.

Nath, D., & Rao, M. (2001). Artificial chaperone mediated refolding of xylanase from an alkalophilic thermophilic Bacillus sp. Implications for in vitro protein renaturation via a folding intermediate. *European Journal of Biochemistry*, *268*, 5471–5478.

Neale, C., Ghanei, H., Holyoake, J., Bishop, R. E., Privé, G. G., & Pomès, R. (2013). Detergent-mediated protein aggregation. *Chemistry and Physics of Lipids*, *169*, 72–84.

Neugebauer, J. M. (1990). Detergents: An overview. In M. P. Deutscher (Ed.), *Methods in enzymology: Vol. 182. Guide to protein purification* (pp. 1–894). Amsterdam: Elsevier.

Newstead, S., Ferrandon, S., & Iwata, S. (2008). Rationalizing alpa-helical membrane protein crystallization. *Protein Science*, *17*, 466–472.

Nielsen, A. D., Borch, K., & Westh, P. (2000). Thermochemistry of the specific binding of C12 surfactants to bovine serum albumin. *Biochimica et Biophysica Acta*, *1479*(1–2), 321–331.

Otzen, D. E., Sehgal, P., & Westh, P. (2009). Alpha-lactalbumin is unfolded by all classes of surfactants but by different mechanisms. *Journal of Colloid and Interface Science*, *329*, 273–283.

Otzen, D. E. (2011). Protein–surfactant interactions: A tale of many states. *Biochimica et Biophysica Acta*, *1814*, 562–591.

Piszczek, G., D'Auria, S., Staiano, M., Rossi, M., & Ginsburg, A. (2004). Conformational stability and domain coupling in D-glucose/D-galactose-binding protein from Escherichia coli. *The Biochemical Journal*, *381*(1), 97–103.

Popot, J. L., & Engelman, D. M. (1990). Membrane protein folding and oligomerization: The two-stage model. *Biochemistry*, *29*(17), 4031–4037.

Powl, A. M., Miles, A. J., & Wallace, B. A. (2012). Transmembrane and extramembrane contributions to membrane protein thermal stability: Studies with the NaChBac sodium channel. *Biochimica et Biophysica Acta*, *1818*(3), 889–895.

Privalov, P. L. (1979). Stability of proteins: Small globular proteins. *Advances in Protein Chemistry*, *33*, 167–241.

Privé, G. G. (2007). Detergents for the stabilization and crystallization of membrane proteins. *Methods*, *41*, 388–397.

Privé, G. G. (2009). Lipopeptide detergents for membrane protein studies. *Current Opinion in Structural Biology*, *19*(4), 379–385.

Protasevich, I., Yang, Z., Wang, C., Atwell, S., Zhao, X., Emtage, S., et al. (2010). Thermal unfolding studies show the disease causing F508del mutation in CFTR thermodynamically destabilizes nucleotide-binding domain 1. *Protein Science*, *19*(10), 1917–1931.

Reyes-Alcaraz, A., Martínez-Archundia, M., Ramon, E., & Garriga, P. (2011). Salt effects on the conformational stability of the visual G-protein-coupled receptor rhodopsin. *Biophysical Journal*, *101*(11), 2798–2806.

Roman, E. A., & Gonzalez Flecha, F. L. (2014). Kinetics and thermodynamics of membrane protein folding. *Biomolecules*, *4*, 354–373.

Roy, A. (2015). Membrane preparation and solubilization. In A. K. Shukla (Ed.), *Methods in enzymology: Vol. 557. Membrane proteins—Engineering, purification and crystallization* (pp. 2–631). Amsterdam: Elsevier. chapter 3.

Rozema, D., & Gellman, S. H. (1996). Artificial chaperone-assisted refolding of denatured-reduced lysozyme: Modulation of the competition between renaturation and aggregation. *Biochemistry*, *35*(49), 15760–15771.

Sadaf, A., Cho, K. H., Byrne, B., & Chae, P. S. (2015). Amphipathic agents for membrane protein study. In A. K. Shukla (Ed.), *Methods in enzymology: Vol. 557* (pp. 57–94). Amsterdam: Elsevier. chapter 4; pp. 2–631.

Seddon, A. M., Curnow, P., & Booth, P. J. (2004). Membrane proteins, lipids and detergents: Not just a soap opera. *Biochimica et Biophysica Acta*, *1666*(1-2), 105–117.

Sedlák, E., Varhač, R., Musatov, A., & Robinson, N. C. (2014). The kinetic stability of cytochrome C oxidase: Effect of bound phospholipid and dimerization. *Biophysical Journal*, *107*(12), 2941–2949.

Serrano-Vega, M. J., Magnani, F., Shibata, Y., & Tate, C. G. (2008). Conformational thermostabilization of the β1-adrenergic receptor in a detergent-resistant form. *Proceedings of the National Academy of Sciences of the United States of America*, *105*, 877–882.

Shukla, A. K. (Ed.), (2015a). *Methods in enzymology: Vol. 557. Membrane proteins—Engineering, purification and crystallization* (pp. 2–631). Amsterdam: Elsevier.

Shukla, A. K. (Ed.), (2015b). *Methods in enzymology: Vol. 556. Membrane proteins—Production and functional characterization*. Amsterdam: Elsevier.

Slotboom, D. J., Duurkens, R. H., Olieman, K., & Erkens, G. B. (2008). Static light scattering to characterize membrane proteins in detergent solution. *Methods*, *46*, 73–82.

Slowik, D., & Henderson, R. (2015). Benchmarking the stability of human detergent-solubilised voltage-gated sodium channels for structural studies using eel as a reference. *Biochimica et Biophysica Acta*, *1848*(7), 1545–1551.

Sonoda, Y., Newstead, S., Hu, N. J., Alguel, Y., Nji, E., Beis, K., et al. (2011). Benchmarking membrane protein detergent stability for improving throughput of high-resolution X-ray structures. *Structure*, *19*(1), 17–25.

Stanley, A. M., & Fleming, K. G. (2008). The process of folding proteins into membranes: Challenges and progress. *Archives of Biochemistry and Biophysics*, *469*(1), 46–66.

Stevens, R. (Ed.), (2011). *Methods: Vol. 55. Membrane protein technologies for structural biology* (pp. 271–420). Amsterdam: Elsevier.

Sturtevant, J. M. (1987). Biochemical applications of differential scanning calorimetry. *Annual Review of Physical Chemistry*, *38*, 463–488.

Tate, C. G. (2010). Practical considerations of membrane protein instability during purification and crystallisation. In I. Mus-Veteau (Ed.), *Methods in molecular biology: Vol. 601. Heterologous expression of membrane proteins* (pp. 187–203). Amsterdam: Elsevier.

Tol, M. B., Deluz, C., Hassaine, G., Graff, A., Stahlberg, H., & Vogel, H. (2013). Thermal unfolding of a mammalian pentameric ligand-gated ion channel proceeds at consecutive, distinct steps. *The Journal of Biological Chemistry*, *288*(8), 5756–5769.

Tribet, C., Audebert, R., & Popot, J. L. (1996). Amphipols: Polymers that keep membrane proteins soluble in aqueous solutions. *Proceedings of the National Academy of Sciences of the United States of America*, *93*(26), 15047–15050.

Tsujii, K., Saito, N., & Takeuchi, T. (1980). Krafft points of anionic surfactants and their mixtures with special attention to their applicability in hard water. *The Journal of Physical Chemistry*, *84*(18), 2287–2291.

Tulumello, D. V., & Deber, C. M. (2012). Efficiency of detergents at maintaining membrane protein structures in their biologically relevant forms. *Biochimica et Biophysica Acta*, *1818*, 1351–1358.

Tummino, P. J. And, & Gafni, A. (1993). Determination of the aggregation number of detergent micelles using steady-state fluorescence quenching. *Biophysical Journal*, *64*(5), 1580–1587.

Uchiyama, S., Ohshima, A., Yoshida, T., Ohkubo, T., & Kobayashi, Y. (2013). Thermodynamic assessment of domain-domain interactions and in vitro activities of mesophilic and thermophilic ribosome recycling factors. *Biopolymers*, *100*(4), 366–379.

Uddin, Y. M., & Schmidt-Krey, I. (2015). Inducing two-dimensional crystallization of membrane proteins by dialysis for electron crystallography. In A. K. Shukla (Ed.), *Methods in enzymology: Vol. 557. Membrane proteins—Engineering, purification and crystallization* (pp. 351–362). Amsterdam: Elsevier. chapter 17; pp. 2–631.

Ulmschneider, M. B., Doux, J. P., Killian, J. A., Smith, J. C., & Ulmschneider, J. P. (2010). Mechanism and kinetics of peptide partitioning into membranes from all-atom simulations of thermostable peptides. *Journal of the American Chemical Society*, *132*(10), 3452–3460.

Vajdos, F. F., Ultsch, M., Schaffer, M. L., Deshayes, K. D., Liu, J., Skelton, N. J., et al. (2001). Crystal structure of human insulin-like growth factor-1: Detergent binding inhibits binding protein interactions. *Biochemistry*, *40*(37), 11022–11029.

Vora, S., George, A., Hemangi, D., & Bahadur, P. (1999). Mixed micelles of some anionic-anionic, cationic-cationic, and ionic-nonionic surfactants in aqueous media. *Journal of Surfactants and Detergents*, *2*, 213–221.

Waldron, T. T., & Murphy, K. P. (2003). Stabilization of proteins by ligand binding: Application to drug screening and determination of unfolding energetics. *Biochemistry*, *42*, 5058–5064.

Wallace, B. A., & Mao, D. (1984). Circular dichroism analyses of membrane proteins: an examination of differential light scattering and absorption flattening effects in large membrane vesicles and membrane sheets. *Analytical Biochemistry*, *142*(2), 317–328.

Wang, X., Huang, G., Yu, D., Ge, B., Wang, J., Xu, F., et al. (2013). Solubilization and stabilization of isolated photosystem I complex with lipopeptide detergents. *PLoS One*, *8*(9), e76256.

White, S. H., & Wimley, W. C. (1999). Membrane protein folding and stability: Physical principles. *Annual Review of Biophysics and Biomolecular Structure*, *28*, 319–365.

Wiener, M. C. (2004). A pedestrian guide to membrane protein crystallization. *Methods*, *34*, 364–372.

Yang, Z. W., Tendian, S. W., Carson, W. M., Brouillette, W. J., DeLucas, L. J., & Brouillette, C. G. (2004). Dimethyl sulfoxide at 2.5% (v/v) alters the structural cooperativity and unfolding mechanism of dimeric bacterial NAD+ synthetase. *Protein Science*, *13*(3), 830–841.

Yang, Z., Wang, C., Zhou, Q., An, J., Hildebrandt, E., Aleksandrov, L. A., et al. (2014). Membrane protein stability can be compromised by detergent interactions with the extramembranous soluble domains. *Protein Science*, *23*(6), 769–789.

Yang, Z., Zhou, Q., Mok, L., Singh, A., Bai, J., Urbatsch, I., & Brouillette, C. (2015). Interactions and cooperativity between P-glycoprotein structural domains determined by thermal unfolding. Manuscript in preparation.

Yoneda, J. S., Rigos, C. F., & Ciancaglini, P. (2013). Addition of subunit γ, K+ ions, and lipid restores the thermal stability of solubilized Na, K-ATPase. *Archives of Biochemistry and Biophysics*, *530*(2), 93–100.

Zhang, Q., Tao, H., & Hong, W.-X. (2011). New amphiphiles for membrane protein structural biology. In R. Stevens (Ed.), *Membrane protein technologies for structural biology: Vol. 55* (pp. 271–420). Amsterdam: Elsevier.

CHAPTER FOURTEEN

Orthogonal Methods for Characterizing the Unfolding of Therapeutic Monoclonal Antibodies: Differential Scanning Calorimetry, Isothermal Chemical Denaturation, and Intrinsic Fluorescence with Concomitant Static Light Scattering

Deniz B. Temel, Pavel Landsman, Mark L. Brader[1]
Protein Pharmaceutical Development, Biogen, Cambridge, Massachusetts, USA
[1]Corresponding author: e-mail address: mark.brader@biogen.com

Contents

Methods in Enzymology, Volume 567
ISSN 0076-6879
http://dx.doi.org/10.1016/bs.mie.2015.08.029

Abstract

Evaluating prospective protein pharmaceutical stability from accelerated screening is a critical challenge in biotherapeutic discovery and development. Measurements of protein unfolding transitions are widely employed for comparing candidate molecules and formulations; however, the interrelationships between intrinsic protein conformational stability and pharmaceutical robustness are complex and thermal unfolding measurements can be misleading. Beyond the discovery phase of drug development, astute formulation design is one of the most crucial factors enabling the protein to resist damage to its higher order structure—initially from bioprocessing stresses, then from stresses encountered during its journey from the product manufacturing site to the bloodstream of the patient. Therapeutic monoclonal antibodies are multidomain proteins that represent a large and growing segment of the biotechnology pipeline. In this chapter, we describe how differential scanning calorimetry may be leveraged synergistically with isothermal chemical denaturation and intrinsic fluorescence with concomitant static light scattering to elucidate characteristics of mAb unfolding and aggregation that are helpful toward understanding and designing optimal pharmaceutical compositions for these molecules.

1. INTRODUCTION

Proteins are complex delicate molecules with a fragile higher order structure essential to biological function. In the quest to develop these molecules into stable therapeutic products, it is important to engineer and select molecules robust to the stresses of manufacturing, storage, shipping, and administration to the patient, which typically occurs as a subcutaneous injection or an intravenous infusion. Once the molecule has been selected for clinical development, its prospects for attaining maximal product stability reside heavily with the formulation scientist. Key therapeutic and product distinctions stem directly from the protective qualities of the formulation influencing attributes such as: a requirement for frozen storage versus refrigerated storage, the necessity of lyophilization for long-term storage stability versus the convenience of a liquid, hours of room temperature stability versus weeks or months of in-use stability, compatibility with pharmaceutical preservatives, and high concentration stability (100–250 mg/mL) with acceptably low viscosity. These attributes all connect with the protein formulator's ability to confer physical and chemical stability to the protein via manipulation of solution conditions using pharmaceutically acceptable additives.

Thermal unfolding techniques represent a precise and relatively convenient method of characterizing the ability of a protein to resist unfolding in response to thermal stress. This characteristic is influenced both by the protein sequence and by the specific solution conditions under which the protein is formulated. Thus, thermal unfolding techniques are applied by protein engineers in late discovery to help optimize and select the best molecule, as well as by formulation scientists in development who seek to identify optimal solution conditions for pharmaceutical stability. Differential scanning calorimetry (DSC) represents a highly effective method for comparing variants of a protein and for resolving effects of specific formulation variables on conformational stabilities of individual domains. During bioprocessing, storage, transport, and use, therapeutic proteins are exposed to many different interfacial stresses that cause partial protein unfolding to form aggregation-competent species. Therefore, maximizing conformational stability has become regarded as a highly effective strategy for reducing susceptibility to aggregation (Manning, Chou, Murphy, Payne, & Katayama, 2010).

Another trend within the biotechnology industry, driven primarily by the preeminence of mAb therapies, has been a distinct shift in product development emphasis toward subcutaneous injectables that require high concentration formulations in the ~100–250 mg/mL range to enable convenient delivery from a prefilled syringe-based device. A major consequence of this trend has been that material requirements for formulation development have increased by orders of magnitude thus encouraging development of screening techniques feasible with small protein quantities. To some extent, this has caused formulation scientists to spurn DSC in favor of techniques such as differential scanning fluorimetry (DSF) using extrinsic fluorescent dyes that can be run at high throughput on 96-well plates consuming much smaller quantities of protein. However, DSC can offer unique advantages that add significant value to the formulation development of mAbs, especially when used in concert with orthogonal unfolding techniques to enable a more detailed understanding of how formulation variables impact the unfolding and aggregation propensity of specific domains.

2. THERAPEUTIC MONOCLONAL ANTIBODIES

The therapeutic and diagnostic utility of antibodies (immunoglobulins) has made a tremendous impact on the biotechnology pipeline over the past 20 years. This stems from their safety, long-lasting pharmacokinetics, and versatility to generate a vast range of specificities of high affinity. The use of monoclonal antibodies (mAbs) for cancer therapy (e.g., hematological malignancies and solid tumors) and for inflammatory and immune diseases including rheumatoid arthritis, Crohn's disease, ulcerative colitis, juvenile arthritis, psoriasis, psoriatic arthritis, and others has achieved considerable success in recent years. Antibodies are relatively large (MW 150 kDa) Y-shaped proteins that are present in blood at 10–15 mg/mL. The structure of an immunoglobulin G (IgG) molecule is shown in Fig. 1. The complementarity-determining regions (CDRs) within the Fab (fragment antigen-binding) region bind to specific targets and cause antagonism or signaling. The Fc (fragment crystallizable) region of a mAb is composed of the hinge and constant heavy-chain domains (C_H2 and C_H3) and has other functions, such as complement fixation or binding to Fc receptors. Of the classes of antibodies, IgG1 is most commonly developed for therapeutic applications; however, other immunoglobulin types (e.g., IgG2, IgG4) and mAb-related products (e.g., bispecifics, Fc-fusion proteins, Fabs, and monobodies) are also being used therapeutically. In 2014, the US Food and Drug Administration approved 41 new molecular entities which included 12 novel biologics of which 5 were mAbs (Morrison, 2015). It can take 10–15 years to develop a new therapeutic mAb at an estimated average pre-tax industry cost per new prescription drug approval of $2.6 billion (inclusive of failures and capital costs) (Grabowski & Hansen, 2014).

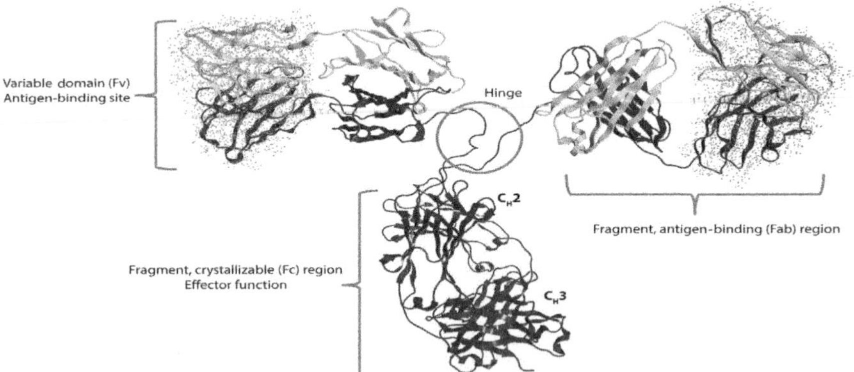

Figure 1 The IgG is a tetrameric molecule of molecular weight approximately 150 kDa, six Ig-fold domains, two heavy chains (blue) and two light chains (green), a single N-glycosylation site in each heavy chain and multiple intra- and interchain disulfide bonds. Fv regions that contain the CDRs are surrounded by dots. (See the color plate.)

Consequently, in addition to the pharmacology of the molecule, there is a lot riding on the ability of the biotech company to engineer and select a "developable" molecule then formulate it for optimal bioprocessing robustness, storage stability, and ease of administration. The term "developable" is increasingly being used to refer to physicochemical properties of a molecule that affect its suitability for large scale manufacture, formulation, and delivery in the most therapeutically and commercially preferred format.

Thermal unfolding-based methodologies have been employed extensively to screen and optimize conformational stability to aid optimal formulation of mAbs (Bhambhani et al., 2012; Goldberg, Bishop, Shah, & Sathish, 2011; He, Hogan, Latypov, Narhi, & Razinkov, 2010). DSC unfolding profiles have been reported for a large number of mAbs, both pipeline molecules and commercialized products. Garber and Demarest (2007) recorded DSC on 17 intact human(ized) antibodies under a common citrate buffer solution condition. Their work showed that stability variations between antibodies of a similar subclass are derived from the variable domains. However, the multidomain nature of mAbs adds complexity to interpretations of unfolding behavior. Three components (thermogram peaks) are generally observed in the DSC profile of a mAb. The peak with the lowest T_m usually corresponds to the C_H2 domain followed by the Fab then the C_H3 domain at highest temperature. However, this profile can show considerable variation, with sometimes two components apparent and sometimes four. The Fab transition peak is usually observed as the largest enthalpy transition (Fig. 2). A more detailed discussion on the deconvolution of mAb thermograms and assignment of peaks to specific domain unfolding events is provided in Section 5.

3. PERFORMING AND DETECTING PROTEIN UNFOLDING

Proteins are only marginally stable because the free energy change that results from folding into their native conformation is small. Protein unfolding can occur due to bioprocessing stresses and under conditions of pharmaceutical storage and delivery. This often leads to aggregation and loss of potency. These unfolding processes can be adequately modeled by gradual (gradient) unfolding induced under a controlled experimental setting. This is achieved by applying a gradient increase of a varied property. While there are several experimental approaches commonly used to induce protein unfolding, the current discussion will focus on the two of the most common:

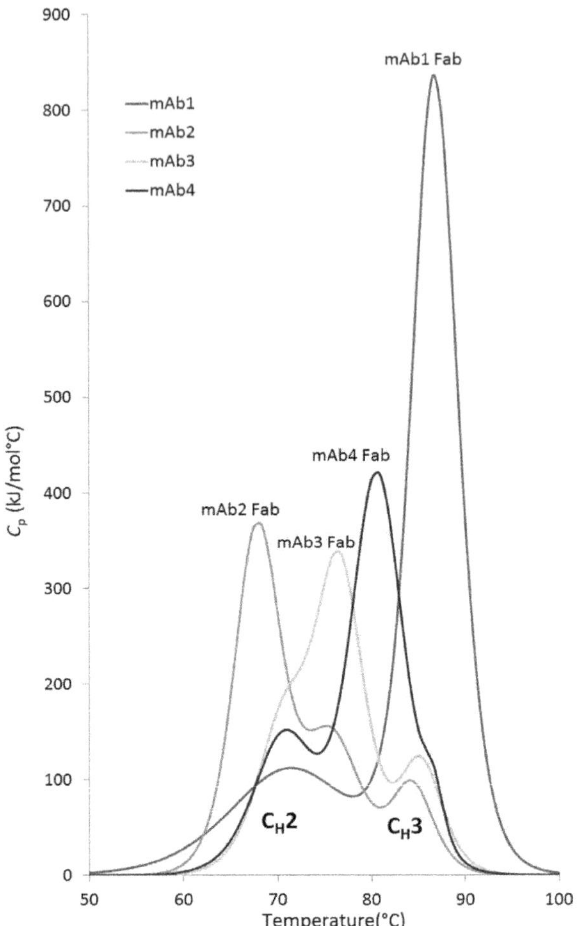

Figure 2 Thermal unfolding curves of human(ized) IgG1 antibodies. Note that the unfolding transitions of the Fab domains are highly variable. (See the color plate.)

(1) perturbation of protein native state by temperature increase "thermal unfolding," whereby the varied property is temperature of the sample, and (2) by titrating a chaotropic (denaturing) agent into the protein sample thereby producing "chemical unfolding," or "isothermal chemical denaturation" (ICD) with the denaturant concentration C as the varied property. In the simplest case, a two-state model is adopted for the unfolding reaction of a single-domain protein, whereas in the case of a multidomain protein, the stepwise independent unfolding of protein molecular fragments (domains) is assumed, following a mechanism:

$$\mathrm{N} \overset{\text{Perturbant}}{\Longleftrightarrow} \mathrm{U} \tag{1}$$

Reversible unfolding from the native protein state (N) to the unfolded state (U) is necessarily assumed for application of thermodynamic analysis to the unfolding reaction. During the unfolding experiment, an experimental observable, e.g., optical/spectroscopic property or heat is monitored during the controlled protein perturbation and yields different values for N and U states. The result of this experiment is usually plotted directly or as a derivative as depicted in Fig. 3. The output curve in Fig. 3A displays a typical sigmoidal shape representing the unfolding transition, with the left and right plateaus of the sigmoid corresponding to the maximally populated folded and unfolded states. The midpoint of the sigmoid (denoted as T_m or melting point on the temperature scale) is equivalent to the T-coordinate point at which the populations of folded and unfolded molecules are equal. It follows that the T_m point in the differential plot corresponds to the peak.

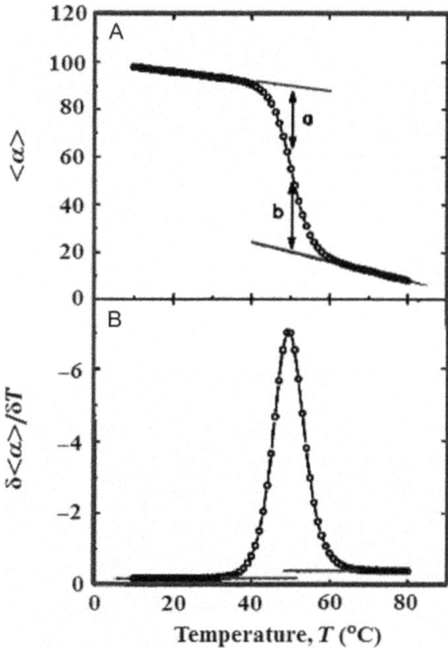

Figure 3 Simulated representation of two-state thermal unfolding (varied property is temperature T) as measured by averaged normalized output signal $\langle \alpha \rangle$. (A) Direct plot of $\langle \alpha \rangle$ versus T; (B) derivative plot. Signal baselines are shown as straight lines assuming linear dependence of α_N and α_U with T. The mole fraction of the U state is given as $a/(a+b)$. This simulation is with $\Delta H_U = 80{,}000$ cal/mol and $T_m = 50$ °C. *Adapted from Eftink (1995) with permission.*

Many proteins, mAbs included, are not single domains but are composed of multiple subunits. The Fab domain along with C_H2 and C_H3 domains of the Fc region may unfold independently in a noncooperative manner under the experimental conditions applied. Unfolding thus proceeds in a stepwise manner via several distinct intermediate states. This leads to the appearance of multiple peaks in the unfolding profile as evident from the thermograms of Fig. 2. In this case, the thermodynamic analysis implies a reversible unfolding process proceeding through several intermediates. The latter implication necessarily limits the speed of the thermal scan, in order to reach equilibration at each scan point and thus allow adequate application of thermodynamic calculations. Furthermore, it should be emphasized that in case of thermal unfolding, protein denaturation cannot in general be considered reversible, as under these conditions it is accompanied/followed by aggregation and possibly chemical degradation (in Eq. (2), A stands for the irreversibly degraded state):

$$N \xLeftrightarrow{\text{Perturbant}} U \xRightarrow{\text{Chemical degradation and aggregation}} A \tag{2}$$

In the case of a thermal unfolding experiment, this analysis remains adequate as long as the experimental scan rate is low enough (e.g., <1 °C/min for a DSC scan), in order to achieve equilibrium at every point of the scan (Lepock et al., 1992). However, thermodynamic models cannot be applied once the rate of the irreversible degradation process exceeds certain limits, which usually happens at the latter part of a thermal scan when intensive aggregation of unfolded protein occurs.

3.1 Differential Scanning Calorimetry

DSC is a classical thermally induced unfolding technique based on direct monitoring of heat effects accompanying the conformational changes as a protein unfolds (Freire, 1995). The resulting thermogram can be regarded as a sensitive biophysical signature of the protein and as such DSC is used extensively in formal biopharmaceutical comparability studies (Johnson, 2013; Wen et al., 2012). However, in this chapter, we focus on DSC in the context of formulation and developability screening rather than analytical comparability analysis. Briefly, DSC measures the heat capacity of the sample as a continuous function of temperature. In the case of a thermal unfolding experiment, DSC determines the heat uptake in the overall endothermic unfolding process that occurs in a protein solution externally heated at a uniform rate. The precisely controlled heating condition is applied to both the protein solution in the sample cell and simultaneously to a matching reference cell containing identical protein-free solution. Thereby in every heating cycle during the temperature ramp, the heat absorbed by the unfolding protein in the sample cell will form a temperature differential ($\Delta T > 0$) between sample and reference cell. The temperature difference between the two cells is detected by a sensor and a compensating thermal power is applied. This electrical power signal provides a direct measure of the heat capacity difference between the sample and the solvent. The two cells are then actively returned to the thermal equilibrium ($\Delta T' = 0$), prior to the next heating cycle.

In a standard temperature ramp experiment, the sample heating is performed at a constant rate, thus the output DSC signal can be instantaneously normalized from the initially measured calorimetric heat flux $J(t)$, where t is time, to the sample heat capacity $C_p(T)$, where T is temperature in °C. The latter is further normalized against protein molarity, to yield protein molar heat capacity plotted against temperature, $C_p(T)$, that is, the DSC normal thermogram. From Kirchhoff's law (Eq. 3), the molar enthalpy ΔH of the unfolding is calculated as area under the curve in the thermogram that in the simplest case (cooperatively unfolded one-domain protein) is depicted by Fig. 3B, with the $C_p(T)$ corresponding to the $\delta(\alpha_U)/\delta T)$ axis:

$$\Delta H_m = \int_{T_0}^{T_E} C_p \mathrm{d}T \tag{3}$$

where ΔH_m is overall enthalpy change at midpoint temperature T_m corresponding to the maxima of the peak. The lower and upper limits of integration T_0 and T_E, respectively, stand for start and end temperatures of the scan, while the integration is performed with the thermogram corrected for the instrumental baseline, which necessarily changes over the course of protein unfolding (as apparent from the horizontal lines of Fig. 3B corresponding to the baselines of the native and unfolded states). This change is caused by the difference in heat capacities, between the native and the unfolded states of the protein, which is denoted as ΔC_p. A routine nonlinear fit procedure performed by the standard DSC instrument software calculates T_m, ΔH_m, and ΔC_p values from the thermogram, assuming the two-state equilibrium model and given the baseline correction set by the user.

Now consider the Helmholtz equation $\Delta G_u = \Delta H_u - T\Delta S_u$ describing free energy change of the thermal unfolding. It then becomes, respectively

$$\Delta G_u = \int_{T_0}^{T_E} \Delta C_p dT - T \int_{T_0}^{T_E} \frac{\Delta C_p}{T} dT \tag{4}$$

For a monomeric protein that undergoes a cooperative unfolding transition this can be simplified to:

$$\Delta G_u = \Delta H_m + \Delta C_p (T - T_m) - T\left(\Delta S_m + \Delta C_p \ln (T/T_m)\right) \tag{5}$$

where T_m is used as a reference temperature (Doyle et al., 2013), and $\Delta S_m = \Delta H_m / T_m$.

While in this simplest case, thermal unfolding of a single-domain protein results in a single-peak thermogram routinely approximated with a two-state unfolding model, multidomain proteins, such as antibodies, usually reveal more than one peak in a thermogram. Several endothermic peaks related to sequential unfolding of various domains may be fully or partially resolved and require application of suitable multipeak fitting models.

3.2 Intrinsic Fluorescence

Intrinsic fluorescence (IF) originates from the tryptophan, tyrosine, and phenylalanine residues present in the protein sequence. These residues are usually monitored with an excitation wavelength in the range 260–280 nm and with emission spectra starting from 320 nm. On a per residue basis, the tryptophan fluorescence is the predominant contributor the total IF. Tryptophan emission can also be selectively excited at fixed wavelength 290–300 nm. Tryptophan (Trp) emission proves to be especially sensitive to the polarity of the local environment, making it suitable for reporting on the local-specific conformational changes as the protein unfolds. The usual observation related to the unfolding is for the fluorescence emission maximum to undergo a "red" shift (toward longer emission wavelength, from ~330 nm to above 350 nm). This corresponds to the increased exposure of the tryptophan groups to the solvent in the unfolded state. A mAb typically contains ~20 tryptophan residues which enables sensitive IF monitoring under the unfolding experimental conditions. With these multiple environment-sensitive intrinsic fluorophores spread throughout the mAb structure, the global monitoring of the unfolding process becomes possible.

Experimental outputs calculated from IF measurements (shifts, ratios, intensities): In order to analyze the thermal unfolding profile and calculate the inflection midpoints (T_m temperatures), a Boltzmann-type equation is usually solved with nonlinear fitting algorithms, as follows:

$$\alpha = \alpha_N + \frac{(\alpha_U - \alpha_N)}{1 + e^{\frac{T_m - T}{c}}} \tag{6}$$

Equation (6) represents a four-parametric fit where α is the output fluorescence parameter measured in the experiment as a function of the independent variable, T. The parameters α_N and α_U are baseline values of the output assumed in simplest case to be temperature-independent, T_m is melting temperature (half-transition) and C is the slope factor.

When IF is monitored as the output there are several options for data reduction:

1. The fluorescence intensity at a single fluorescence emission wavelength.
2. The difference between the fluorescence intensities at two different fluorescence emission wavelengths.
3. The ratio of the fluorescence intensities at two different fluorescence emission wavelengths.
4. The area under the curve for a specified fluorescence emission wavelength range.
5. The fluorescence emission wavelength change upon unfolding (i.e., the red shift).

The optimal observable will be the one that is most sensitive to the unfolding transition of the protein being studied. In most cases for mAbs, bispecifics, fusion proteins, and antibody drug conjugates, a good choice will be the ratio of the fluorescence intensity at the fluorescence emission wavelength peak for the unfolded protein to the fluorescence intensity at the fluorescence emission wavelength peak for the native protein.

3.3 Right-Angle Light Scattering

Right-angle light scattering represents an exquisitely sensitive and simple method for early detection of protein aggregation. The extreme sensitivity of the scattering intensity signal to the size of the scattering species follows from the foundations of light scattering theory (Yguerabide & Yguerabide, 1998). Rayleigh law postulates that in the ideal case, the intensity of light with wavelength λ scattered by a small spherical particle not absorbing light considerably at the given wavelength, at the angle θ related to the incidental light beam is proportional to the power six of the particle diameter (d), for $d \leq \lambda/20$,

$$I = I_0 \frac{8\pi^4 n_0^4 d^6}{r^2 \lambda_0^4} \left| \frac{m^2 - 1}{m^2 + 2} \right|^2 \left(1 + \cos^2\theta\right), \tag{7}$$

where I and I_0 are the intensities of the scattered and incidental light in arbitrary units, $m = n_p/n_0$, whereas n_p and n_0 are the respective refractive indices of the particle and the surrounding media, λ is the illuminating wavelength in vacuum. In the case of "right-angle" light scattering (RALS), $\theta = 90°$ and $\cos\theta = 0$, so the equation simplifies to:

$$I = I_0 \frac{8\pi^4 n_0^4 d^6}{r^2 \lambda_0^4} \left| \frac{m^2 - 1}{m^2 + 2} \right|^2, \tag{8}$$

which can be further multiplied by the particle concentration per volume unit, in case of dilute suspensions, to give total light scattering from the sample.

In the case of an antibody molecule irradiated by light of $\lambda \sim 200$ nm, aggregating particles are expected to exceed the $\sim\lambda/20$ Rayleigh model limitation. Therefore, Eq. (8) must be replaced by an even more complicated Mie formalism, accounting for the light absorption, turbidity, and multiple scattering centers of a larger particle (Yguerabide & Yguerabide, 1998). Nevertheless, the sharp increase of the scattered intensity at the very initial stages of protein self-association illustrated by the Rayleigh equation suggests that the light scattering output serves as an extremely sensitive indicator of self-association that may accompany protein unfolding. The experiment is sometimes referred to as differential scanning light scattering (DSLS) (Senisterra & Finerty, 2009). The corresponding profiles often display a sigmoidal shape resembling the Boltzmann equation (Eq. 9), which can be described empirically in the following form:

$$\mathrm{I} = \mathrm{I}_0 + \frac{I_{sat}}{1 + \mathrm{e}^{\frac{T_{agg} - T}{\mathrm{B}}}} \tag{9}$$

where I is output scattering intensity plotted against temperature T and reaching the saturation value I_{sat} at higher temperatures, from the scattering background I_0, while the term T_{agg} corresponds to the midpoint of the transition depicted by the sigmoid with B as a constant. Correlation of T_{agg} values with T_m obtained from spectroscopic profiles and DSC thermograms have been demonstrated (Senisterra & Finerty, 2009). Another useful parameter that can be obtained from the analysis of the DSLS profile is the aggregation onset T_{on}, which is often calculated from a sigmoid or its differential plot as the temperature corresponding to a specified signal increase above the background, e.g., $10\% \times (I_{sat} - I_0)$. Thus, DSLS can be used as an indirect reporter of an unfolding process where other techniques (such as CD or fluorimetry) may not be suitable due to the accompanying aggregation (Senisterra & Finerty, 2009).

3.4 Isothermal Chemical Denaturation

In the case of an ICD experiment, the free energy change for each domain (i) to unfold becomes:

$$\Delta G_{Di} = \Delta G_{0i} - m_i[\text{Denaturant}] \tag{10}$$

Where ΔG_{Di} is the measured Gibbs energy of denaturation and ΔG_{0i} is the Gibbs energy at zero denaturant concentration. The ΔG_0 value determined from the raw data analysis defines the stability of the protein or domain under the specific conditions in which the measurements are performed (including all solvents, excipients, ligands, etc. present in the solution except the denaturant.) For a protein with n cooperative domains, the overall fractional degree of denaturation is proportional to the sum of the degree of denaturation of each domain as shown by Eq. (11), or in terms of the denaturation equilibrium constant as shown by Eq. (12).

$$F_D = \sum_{i=1}^{n} f_i F_{Di} \tag{11}$$

$$F_D = \sum_{i=1}^{n} f_i \frac{K_i}{(1 + K_i)}. \tag{12}$$

The constant f_i accounts for the fact that not all domains are equal in size or contribute the same to the observable used to monitor the denaturation process (Fig. 4).

3.5 Differential Scanning Fluorimetry

The changes in protein conformation and aggregation state that occur during a thermal ramp can also be monitored by extrinsic fluorescence using a reactive fluorescent dye that undergoes a change in emission intensity and spectral profile when it binds to the unfolding protein or aggregates thereof.

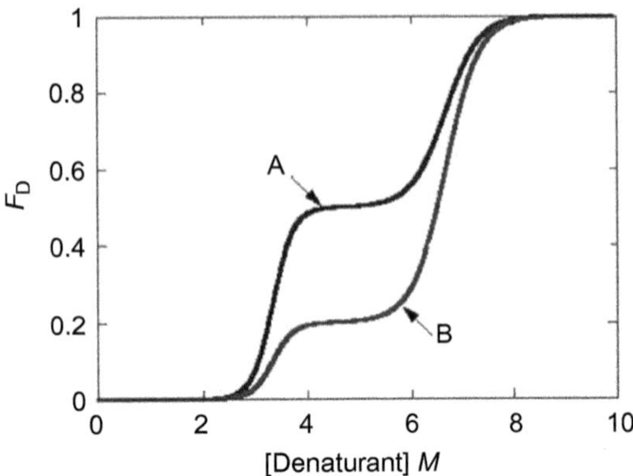

Figure 4 Simulated chemical denaturation experiments for two two-domain proteins. The two chemical denaturation experiments are characterized by identical thermodynamic parameters, except that in the (A) blue curve (dark gray in the print version) each domain contributes 0.5 ($f_1 = f_2 = 0.5$) and in the (B) red curve (light gray in the print version) the first domain contributes 0.2 ($f_1 = 0.2$; $f_2 = 0.8$). Since $\Sigma f_i = 1$, for two transitions it is sufficient to specify f_1 since $f_2 = 1 - f_1$.

This is the basis of the technique often referred to as DSF. It has been described extensively in the literature applied to mAb formulation screening (Razinkov, Treuheit, & Becker, 2013; Samra & He, 2012). External protein-reactive probes may vary in their binding specificities. For example, Sypro Orange, Nile Red, and 8-anilinonaphthalenesulfonate bind to hydrophobic regions of the protein whereas Thioflavin T binds preferentially to "beta-pleated stack" and related motifs formed in misfolded, refolded, or aggregated protein species formed in the course of the unfolding. The so-called "molecular rotor" microviscosity probe, DCVJ, has also been applied in DSF monitoring (Hawe, Filipe, & Jiskoot, 2010). Their extreme sensitivity, structural specificity, and versatility make extrinsic probes useful for high-throughput screening.

However, several disadvantages exist with the use of extrinsic probes. The fluorescence properties of these molecules are usually affected by polysorbates which is unfortunate because most protein formulations contain polysorbates as a stabilizer against interfacial damage. Another issue is that the binding of the external probe molecule can interfere with the unfolding of the protein and is, therefore, usually restricted to determining the start of unfolding. Yet another difficulty with emission monitoring in a classical fluorimetric setting of this experiment is light scattered from aggregates formed by unfolded protein species at elevated temperatures, especially at later stages of the thermal ramp (Garidel, Hegyi, Bassarab, & Weichel,

2008). A promising new variation of DSF that overcomes this problem is based on IF detection of thermal unfolding. The Prometheus NT.48 instrument by NanoTemper Technologies (South San Francisco, USA) uses a capillary-based system (instead of cuvettes) in conjunction with a UV detection system designed specifically to monitor shifts in the fluorescence emission of tryptophan. It can be applied over a wide concentration range (~5 μg to 150 mg/mL), requires very small protein quantities and can accommodate 48 samples per run. This instrument produces high-resolution thermal unfolding curves that allow detection and analysis of the transitions associated with a multidomain protein. This capability is illustrated by the data of Fig. 5.

4. EXPERIMENTAL

4.1 Differential Scanning Calorimetry

We have used a MicroCal Auto-Capillary-DSC instrument for determination of T_m values. In a multiwell DSC experiment, each sample measurement requires a pair of wells. The first is for buffer reference and the second for sample measurement. It is important to match the buffer in the reference and protein cell meticulously. Our suggestion is to either extensively dialyze or use a centrifugal filter unit to achieve an exact buffer match. Dilute the protein in the buffer of interest then concentrate. Repeat three times and save the last flow-through buffer to run as the buffer reference to achieve an exact buffer match. It is recommended to include reference standard runs between each set of samples and at the end of the run to check the stability of the instrument and reproducibility of the data. Running three buffer samples and using the last buffer run for baseline subtraction is advised. Protein samples may be run as triplicates to estimate standard deviation.

Protein concentrations can be in the range 0.5–2 mg/mL. Each buffer and sample well must contain at least 400 μL of solution for analysis. Both sample and buffer should be degassed under a vacuum for about 15 min. It is essential to avoid introducing bubbles into the system as they will adversely affect the baseline. A suitable temperature range for unfolding of an antibody is 20–100 °C. Almost all proteins will unfold at this final temperature. A key parameter that needs to be considered carefully is the thermal scan rate. The relationship between scan rate and resolution of thermograms can be appreciated from Eq. (13).

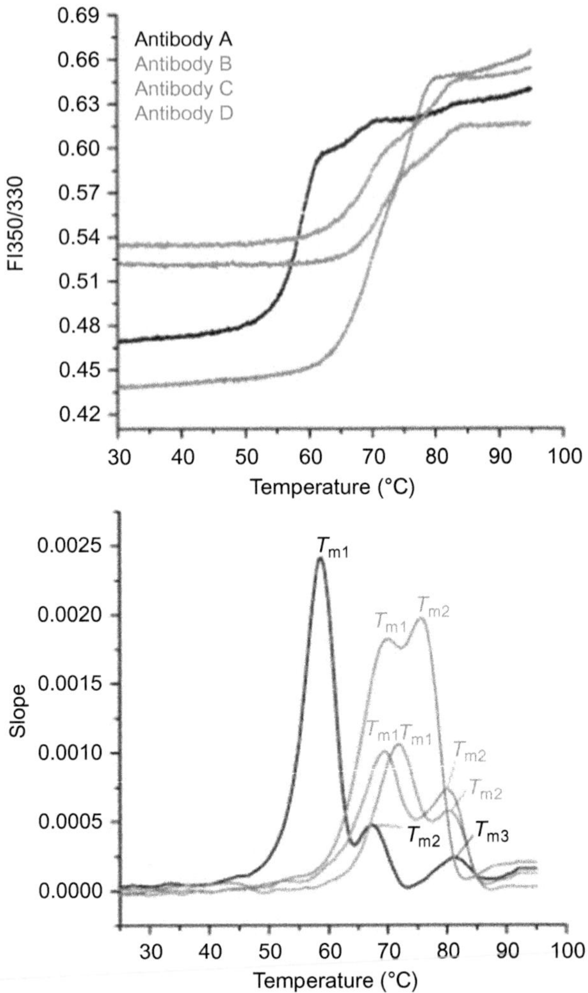

Figure 5 High-resolution intrinsic fluorescence data from Prometheus NT.48 thermal unfolding experiments on four therapeutic mAbs. Plots of the fluorescence ratio (FI350/330) and the corresponding first derivative are shown resolving the unfolding of distinct antibody domains. *Unpublished data provided courtesy of NanoTemper Technologies Inc.* (See the color plate.)

$$C_p = \left(\frac{dH}{dT}\right) = \left(\frac{dH}{dt}\right)\left(\frac{dt}{dT}\right) \tag{13}$$

where $\left(\frac{dH}{dt}\right)$ is the shift in the baseline of thermogram and $\left(\frac{dt}{dT}\right)$ is inverse of scan rate. Since the heat capacity of the molecule will not change under a specific buffer condition, a lower scan rate will increase the shift in the baseline. This will lead to increased resolution of transitions. Typical scan rates can range from 60 to 120 °C/h. Usually 90 °C/h (1.5 °C/min) is a suitable rate to obtain adequately high signal-to-noise ratio with mAbs.

The MicroCal analysis package is integrated into OriginLab software. A good baseline is important for reliable data analysis. Reference baselines, generated from buffer versus buffer scans, are subtracted from the protein scans before data analysis. Data are then normalized with respect to concentration- and scan rate, followed by subtraction of native state baselines. Deconvolution is performed using a Gaussian model by manually selecting peak centers as an initial approximation for iterations of a least squares fitting algorithm. T_m (and when available, ΔH and ΔC_p) values may be obtained from the software calculations. Often the protein precipitates at the end of the run preventing reliable estimation of the unfolded baseline. Consequently, for formulation screening the only unfolding parameter that can be obtained routinely is T_m.

4.2 Thermal Unfolding with Multimodal Detection

To characterize the thermal unfolding of mAbs using IF and light scattering concomitantly, we have used the Optim1000 instrument (Unchained Labs, Pleasanton, CA). This instrument requires only 9 μL of sample which is introduced to the instrument via microcuvette arrays that can accommodate 48 samples per run. In the case of mAbs, the sample concentration can range from about 1–100 mg/mL. It is important to avoid introducing bubbles during cuvette filling because these will interfere with the light scattering signal. We suggest a start temperature of 35 °C with an end temperature 95 °C and a step size of 0.5 °C min^{-1} (to enable direct comparison with DSC data keep the ramp step size and scan rate the same as in the DSC experiment). We recommend running samples in triplicate from which standard deviations can be calculated. Fluorescence signal tuning is one of the key considerations affecting the signal-to-noise ratio. Whenever possible, we suggest using similar concentration samples within each run.

The Optim1000 software is embedded within the Igor Pro 6.34A suite. The software provides various options to visualize and analyze the data. The raw data are examined for outliers, which can arise from air bubbles or extraneous large particles that cause sharp spikes in the data. With clean data established, sigmoidal fitting is performed and the derivative curve calculated. Onset (Fig. 6) and transition (Fig. 7) temperatures can be calculated from various measurables including light scattering at 266 nm, ratio of fluorescence intensity at 350 nm to at 330 nm (Fl350/330) or barycentric mean

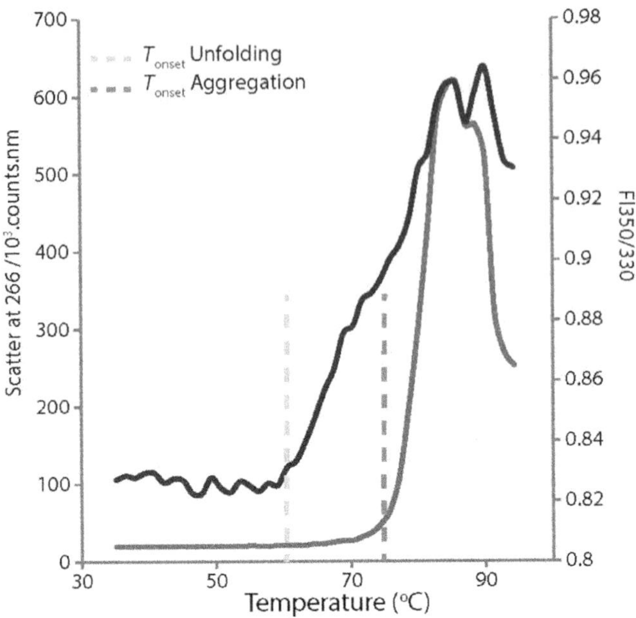

Figure 6 Onset temperatures of unfolding (red) and aggregation (blue) from intrinsic fluorescence and light scattering respectively for an IgG4 mAb. (See the color plate.)

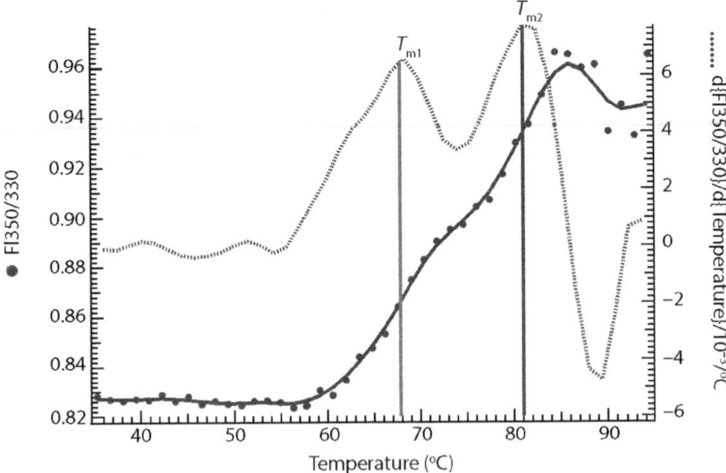

Figure 7 Melting temperature domain transitions. Melting temperatures are identified from the first derivative of fluorescence ratio curve.

of fluorescence which corresponds to the change in the center of mass of the wavelength of the fluorescence emission.

4.3 Isothermal Chemical Denaturation

ICD experiments use chemical denaturants to unfold proteins by steadily increasing the concentration of the denaturant. This is achieved by combining a "formulation solution" with a "denaturant solution" which is identical to the formulation solution except that it contains denaturant at high concentration (e.g., 10 *M* urea). The most commonly used denaturants are urea and guanidinium chloride (GdmCl). On occasions when stronger denaturants are required *N,N*-dimethylurea or guanidinium thiocyanate may be utilized. These denaturants are available commercially as high purity reagents. Their solubilities are given in Table 1. Guanidinium solutions are stable for months at room temperature, but urea solutions readily decompose to cyanate and ammonium ions in a process accelerated at higher pH values. The cyanate ions can react with amino groups on proteins. Consequently, urea stock solutions should not be used for longer than 2 days.

We have used the AVIA Model 2304 ICD instrument which performs automated preparation and incubation of denaturation solutions. Stock formulation and denaturant solutions are most readily prepared by first preparing a formulated buffer solution at a higher concentration than required for the formulation and denaturant solutions. An aliquot of the high concentration formulated buffer is then diluted into the denaturant to form the denaturant stock solution. A second aliquot is diluted with water to the same final concentration to form the formulated buffer solution. The instrument dispenses a fixed aliquot of protein stock solution into each well separately then combines calculated volumes of formulated buffer and denaturant solutions to create the denaturant gradient series at a fixed protein concentration in the formulation of interest. The stock solutions are most conveniently prepared by weight according to Table 2. These values will yield denaturant solutions at the indicated denaturant molarity, and the denaturant and formulation solutions will have identical buffer concentrations. An important point to emphasize is that once the formulation and denaturant buffers have been

Table 1 Solubility of Denaturants

Property	Urea	*N,N*-Dimethylurea	GdmCl	GdmSCN
Molecular weight (g/mol)	60.056	88.11	95.533	118.16
Solubility (25 °C)	10.5 *M*	8.7 *M*	8.5 *M*	6.6 *M*

Table 2 Denaturant Solution Preparation Guide

Denaturant Solution	Required Multiple of Final Buffer Strength	Denaturant Grams of Denaturant Per Gram of Buffer	Formulation Grams of Water Per Gram of Buffer
10 *M* Urea[a]	1.830	1.103	0.830
8 *M* Urea	1.570	0.755	0.570
6 *M* Urea	1.374	0.495	0.374
8 *M* GdmCl	2.380	1.816	1.380
6 *M* GdmCl[b]	1.766	1.009	0.766
8 *M* Dimethylurea	2.520	1.774	1.520
6 *M* GdmSCN	2.136	1.515	1.136

[a]Using urea, it is most convenient to prepare denaturant solutions at 10 *M*. The ICD-2304 instrument will automatically prepare the correct denaturant concentration for each point in the denaturation curve.
[b]Similarly, when using GdmCl it is usually most convenient to prepare the denaturant solution at 6 *M*.

prepared they must be carefully adjusted to precisely the same pH value because the denaturant may change the pH of buffer solution significantly.

For example, in order to prepare 10 *M* urea denaturant buffer and corresponding formulation buffer:

1. Make a 1.83 × concentrated buffer stock.
2. Denaturant: For every 1.103 g of urea, add 1.000 g of 1.83 × concentrated buffer stock.
3. Formulated buffer for every 1.000 g of 1.83 × concentrated buffer stock, add 0.830 g of water.

The concentration of denaturant in the denaturant solutions can be accurately determined using a temperature-controlled refractometer and the following equations (Pace, 1986).

$$[\text{Urea}] = 117.66 \times \Delta N + 29.753 \times (\Delta N)^2 + 185.56 \times (\Delta N)^3 \quad (14)$$

$$[\text{GdmCl}] = 57.147 \times \Delta N + 38.68 \times (\Delta N)^2 - 91.60 \times (\Delta N)^3 \quad (15)$$

where ΔN is the difference between the refractive index of the denaturant solution and buffer.

There are experimental considerations that influence the choice of urea or GdmCl as denaturant. Urea may be preferred for formulation screening studies because, unlike GdmCl, it does not change the ionic strength of the solution. However, GdmCl is a stronger denaturant and will generally result in more complete unfolding. The number of observable unfolding transitions depends on the effectiveness of the denaturant and the stability of the protein. This is exemplified in Fig. 8 where the fluorescence signal clearly shows that in the unfolded state of the urea denaturation (A), the baseline has not reached saturation. However, the urea denaturation data, although incomplete, can still be very accurate for the transition that is unfolded. Even in the example shown in Fig. 8, the partially unfolded baseline for the observed transition could easily be fit and useful information for ΔG_1 could be obtained as a function of intrinsic changes to the protein or formulation.

It is important to optimize working protein concentration so as not to over saturate the fluorescence detector. The user can tune the working protein solution readily by inspecting the 100% native and 100% denatured fluorescence emission spectra. It is important to verify that the native and denatured profiles exhibit good signal-to-noise and are distinct from one another.

To analyze the ICD curve, the native and unfolded baselines are determined initially by the software or the user and held constant while the remaining parameters (f_i: the fraction of the total signal attributed to the transition, ΔG_{0i}: Gibbs free energy of the transition and m_i: slope of the transition) are determined by nonlinear least squares fitting to the data. Higher order models require three additional parameters per added transition. The baseline selection for native and denatured states is extremely important for rigorous determination of ΔG values and the analyst should pay close

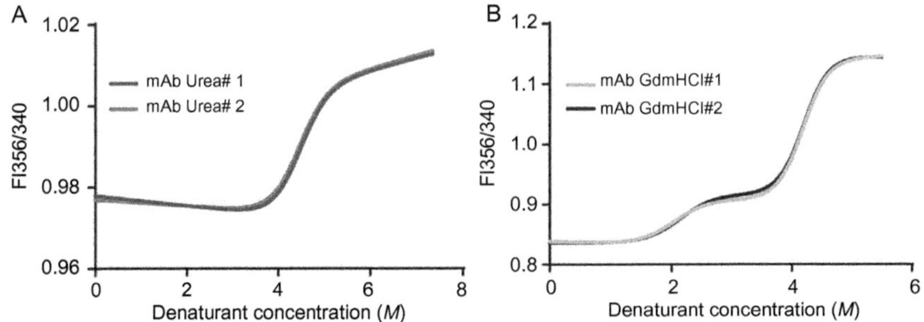

Figure 8 ICD profiles (duplicates) of an IgG1 mAb using (A) urea and (B) GdmCl as denaturants. Fluorescence 356/340 signal clearly shows that in the unfolded state of the urea denaturation, baseline did not reach a plateau. However with GdmCl denaturation, the baseline is stable for both native and unfolded states. Two two-state transitions are observed with GdmCl denaturation whereas only one for urea.

attention to these. An important function to be aware of is the "Protein Groups and Replicates" function. If fluorescence data for replicates of each sample are collected, ΔG values may be calculated from the fitting using an average "m" value. This procedure generally yields the highest level of reproducibility of results. Similarly, if replicate data are available and a two two-state state fitting model is being used, we also suggest fitting using the average F value of the replicates.

5. DATA INTERPRETATION

5.1 Identifying the Fab and Fc Transitions

It is not always straightforward to unambiguously assign the transitions in the DSC thermogram to specific domains. Generally, the largest enthalpy transition arises from the Fab unfolding. Often this is easily distinguished by eye from the C_H2 and C_H3 transitions, but not always (Ionescu, Vlasak, Price, & Kirchmeier, 2008). Therefore, digestion of mAb into Fab and Fc fragments can be helpful to enable examining these fragments separately by DSC. Ionescu et al. have shown how the DSC unfolding transitions of the isolated Fab and Fc fragments of an IgG1 mAb corresponded closely with those in the intact molecule. We recommend using Immunoglobulin G-degrading enzyme of *S. pyogenes* (IdeS). This enzyme is a cysteine protease which has high degree of specificity toward human IgG without degrading IgM, IgA, IgD, and IgEs (von Pawel-Rammingen, Johansson, & Bjorck, 2002). IdeS (FabRICATOR® Genovis, Sweden) cleaves IgG at the hinge region separating $F(ab')_2$ and Fc (An, Zhang, Mueller, Shameem, & Chen, 2014). Since digestion of the enzyme is highly specific to the single site, over digestion does not occur with increased incubation time.

The commercially available FabRICATOR® kit allows researchers to digest the IgG and purify its $F(ab')_2$ and Fc fragments. The enzyme is covalently immobilized on highly cross-linked agarose beads. This resin is incubated with the IgG sample at room temperature for around 15 min (time may be adjusted depending on the protein). To separate the resulting fragments an affinity column supplied with the kit is used. This comprises a multispecies Fc affinity matrix with llama antibody fragments cross-linked to agarose beads. The flow-through after digestion contains the fragmented species. The $F(ab')_2$ and Fc fragments can then be purified using an affinity column (CaptureSelect® Life Technologies, USA). The previously digested IgG sample is incubated in this column. The Fc fragment will bind to the column and the $F(ab')_2$ fragment will flow-through. The Fc fragment can subsequently be eluted by passing low pH buffer through the column followed by immediate neutralization of the flow-through. After isolation of $F(ab')_2$ and Fc fragments, DSC can be run to identify the melting temperatures corresponding to these domains.

The glycans present in mAbs have a significant effect on thermal stability. This effect can be leveraged to identify the C_H2 domain with which the glycan is associated. Glycan removal from the C_H2 domain of an IgG1 mAb usually produces a significant shift to lower temperature of the first transition, with negligible effect on the second transition. Deglycosylation of the mAb Fc can be achieved readily with an enzymatic digest (Ionescu et al., 2008).

5.2 Identifying the Aggregation-Prone Domain

By overlaying the thermal ramp data detected using DSC, IF and light scattering outputs, it is possible to gain insight into the relationship between heat-induced domain unfolding and aggregation. Figure 9 shows DSC data superimposed with the corresponding IF (as intensity ratio, Fl350/330) and light scattering (as intensity) profiles. The three unfolding transitions of the DSC thermogram are clearly evident as inflections in the IF profile. It is evident from the light scattering profile that under the conditions of this experiment, unfolding of the first domain (C_H2) does not result in aggregation, whereas

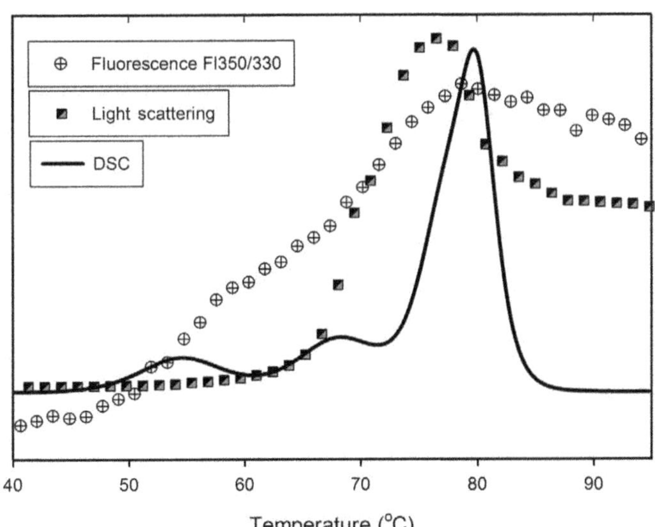

Figure 9 Overlaid DSC, intrinsic fluorescence (ratio Fl350/330) and light scattering (intensity) profiles of an IgG4 mAb (the respective signals are graphically fit in the same *Y* scale).

there is a major increase in light scattering associated with the second domain that unfolds (C_H3). By identifying the aggregation-prone domain in this manner it becomes possible to evaluate the effect of formulation conditions or sequence variants on the aggregation propensity of this domain. It has been shown that formulation conditions can shift the aggregation-prone domain from the first domain to unfold to the second one (Brader et al., 2015) which may be indicative of an improved formulation.

5.3 Assessing Cooperativity of Unfolding

The single peak of the Fab transition arises as a consequence of the domains of the Fab region unfolding in a cooperative manner as if a single domain. However, formulation conditions can alter the cooperativity of domain unfolding, as shown in Fig. 10 for a mAb in citrate buffer versus a

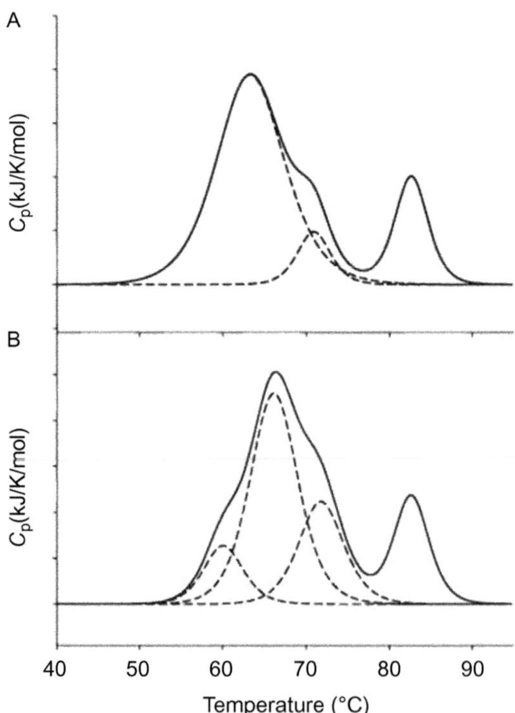

Figure 10 Formulation conditions can alter the cooperativity of domain unfolding. In this example, the first T_m of this mAb occurs at a lower value (60.1 °C) in its optimized formulation (A) than it does in a citrate screening buffer (63.3 °C) (B). However, the improved stability of the optimized formulation is reflected in the large (~3 °C) positive shift in the Fab transition for the optimized formulation relative to the screening buffer.

pharmaceutically optimized formulation. In the former case, three components are apparent versus four components for the latter. The C_H3 transition at 82.6 °C remained invariant. An important implication of this example is that techniques that focus only on characterizing the temperature of the first unfolding transition can be oblivious to the enhanced stability conferred by the formulation condition on the rest of the molecule. Recent results have suggested that the first DSC unfolding transition is in fact a poor general indicator of improved formulation stability and that the Fab transition appears to be a more meaningful correlate (Brader et al., 2015). Indeed the Fab transition of the mAb shown in Fig. 10 has shifted significantly from 63.3 to 66.2 °C which likely reflects the enhanced overall pharmaceutical stability of that formulation. The number of transitions, and hence the corresponding model for the multidomain unfolding, is determined based on the best mathematical fit to the experimental thermogram. Details of experiment and thermogram deconvolution are provided in Section 4.1.

5.4 Unusual Transition Phenomena

The effect of unfolding a globular protein is to cause exposure of the hydrophobic core to the solvent. Because the tryptophan groups are hydrophobic, protein unfolding usually enhances their solvent exposure which results in a characteristic red shift. However, occasionally mAbs do not follow this pattern as shown by the example in Fig. 11. The IF of this mAb appears very different from what is usually observed. Between about 40 and 60 °C, there

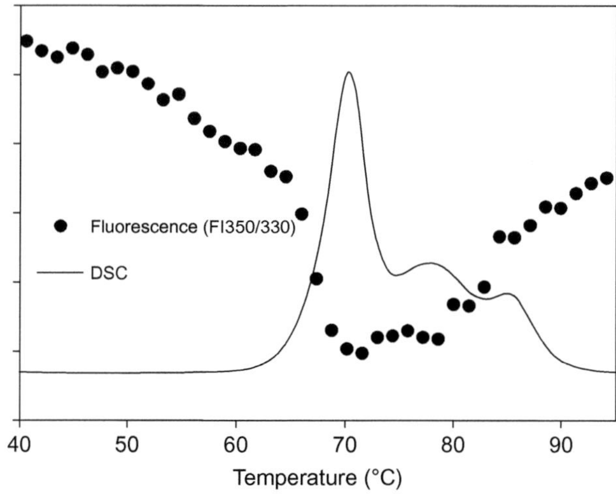

Figure 11 Unusual fluorescence behavior is sometimes observed, as for this IgG1 mAb.

is a decrease in the Fl350/330 ratio that does not correspond to an observable transition in the DSC thermogram. This suggests that there is a relatively low temperature-induced effect on the protein that does not cause a measurable heat change by DSC. This may be a subtle conformational change that affects the tryptophan groups or a solvent rearrangement that has a significant spectroscopic effect but a negligible thermal effect. Between about 65 and 70 °C, there is a significant sharp decrease in the Fl350/330 ratio that corresponds approximately to the Fab transition by DSC. Beyond the Fab unfolding, the usual red shifted fluorescence change is observed. The IF signal is dominated by tryptophan but also contains contributions from the tyrosines and phenylalanines. Consequently, the observed signal is a weighted average corresponding to the spectral microenvironments of all the fluorophores in the sequence. It is recognized that thermally induced unfolding is likely more complicated than the simple two-state model of folded and unfolded protein and the occurrence of unusual IF profiles such as that of Fig. 11 seems to support this. It is unclear exactly why the IF change is sometimes so different. The observation of spectral changes over the 40–60 °C temperature range is unusual and could be considered undesirable. However, it is important to note that the molecule giving rise to this behavior is a glycosylated IgG1 mAb which is a common subtype for therapeutic mAbs and furthermore, this particular molecule has been shown to be highly stable as a liquid formulation drug product. Therefore, the observation of an "anomalous" spectroscopic profile upon heating should not in and of itself disqualify the molecule from consideration as a viable development candidate.

5.5 Using ICD for Formulation Screening and Analysis

In principle, both thermal and chemical denaturation techniques may be used to determine ΔG as a measure of a protein's intrinsic conformational stability. However, in practice a limitation of thermal unfolding techniques can be the irreversibility of denaturation due to aggregation and precipitation invalidating thermodynamic analysis. Formulation development may be significantly aided by using chemical denaturation as a reliable technique for measuring ΔG and its response to pH, ionic strength, excipients, and protein concentration. Because it is possible to screen a large number of samples using a few mg of protein, ICD is an attractive method for extensive screening of pharmaceutical excipients and formulation design space. Stabilizing formulation conditions will increase ΔG whereas destabilizing conditions will decrease it. An example of just how dramatically the ICD profile of a mAb can vary with pH is shown in Fig. 12 (Table 3). The quantity of denatured protein present in each formulation is directly related to ΔG.

Additional insights into native protein self-association and denatured state protein aggregation may be obtained by measuring changes in ΔG as a function of protein concentration. In the absence of aggregation, the equilibrium of monomeric protein between the native and denatured states is expected to be independent of the protein concentration. In our simple two two-state model, it is possible that the native state may undergo

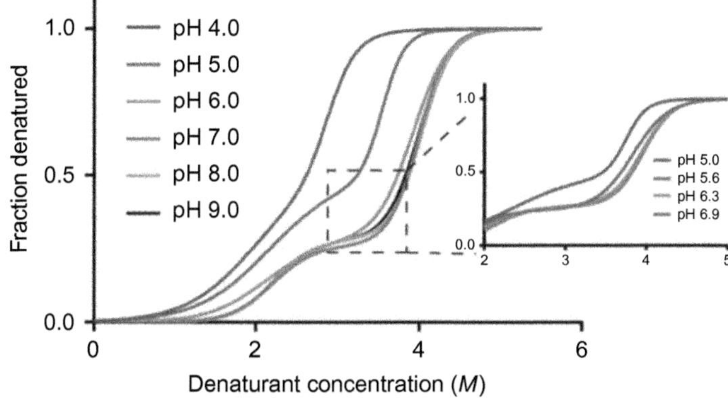

Figure 12 The effect of pH on the ICD profile of an IgG1 mAb. Inspection of the expanded region suggests a formulation pH in the range 6.3–6.9 would maximize folded stability. (See the color plate.)

Table 3 ICD Parameters Obtained from Analysis of the Denaturation Profiles Presented in Fig. 12

pH	ΔG_1 (kJ/mol)	m_1 (kJ/mol·M)	ΔG_2 (kJ/mol)	m_2 (kJ/mol·M)
4.00	11.02 ± 0.51	8.20	25.4 ± 0.65	9.25
5.00	16.6 ± 0.28	8.87	39.1 ± 0.40	11.3
5.60	17.2 ± 0.17	8.66	49.2 ± 0.48	12.8
6.30	18.4 ± 0.35	8.70	57.4 ± 1.00	14.4
7.00	21.6 ± 0.17	9.96	54.0 ± 0.60	13.6
9.00	24.0 ± 0.44	10.8	52.4 ± 0.09	13.1

self-association (i.e., aggregation without unfolding) and/or that the denatured state may undergo aggregation. If either of these processes occur, then the equilibrium process is no longer strictly monomeric and there will be a concentration dependence of ΔG. Because native state self-association stabilizes the protein, and the associated state is proportionately more populated at higher concentration, ΔG will increase with protein concentration. Conversely, if the denatured state undergoes aggregation ΔG will decrease with protein concentration. Using ICD to gain insight into how the protein responds to formulation conditions and protein concentration can thus help point to specific compositional strategies to improve formulation stability.

5.6 Selecting the Best Molecule and Formulation

It has been stated repeatedly in the pharmaceutical sciences literature that formulations with the highest T_m values are expected to produce the lowest aggregation rates upon long-term storage. The assertion that thermal stability measured by DSC correlates with long-term pharmaceutical stability data also continues to persist as a dogma in the literature (Blake-Haskins, Lin, Wu, Perkins, & Spitznagel, 2014; Wen, Hymes, & Narhi, 2004). It is our experience, and others, that often the best formulations do not have the highest T_m values. Excipients such as arginine HCl and polysorbates can be excellent stabilizers for mAb formulations; however, these excipients usually decrease T_m, indicating that they confer stability via mechanisms other than by increasing the conformational stability of the native state. The complexities associated with predicting stability at low temperature from T_m and other stability measurements at high temperature have been summarized (Brader et al., 2015; Brummitt, Nesta, & Roberts, 2011; Hawe et al., 2012).

The unfolding techniques described herein can be leveraged to gain insight into the conformational stability of the mAb and its response to sequence modification and solution conditions. These data can contribute significantly toward a detailed understanding of the solution behavior of the molecule thereby creating a sound underpinning for subsequent formulation design and/or molecule selection. It is important to emphasize that while intrinsic protein conformational stability is clearly a major element of pharmaceutical stability and robustness, it is not the only element. The degradation pathways for mAbs are numerous and complex. Colloidal stability and solubility are also very important and become increasingly so in higher concentration formulations. Although multiple physical and chemical mechanisms of degradation can be enhanced via unfolded or partially unfolded states, this is not true of all mAb degradation pathways. Consequently, optimizing formulation conditions may involve a trade-off between maximizing conformational stability and controlling another degradation pathway.

In summary, thermal shifts associated with the Fab domains appear to provide a better indication of enhanced overall pharmaceutical stability than those associated with the first unfolding transition (Brader et al., 2015). ICD provides direct access to ΔG and can provide unique insight into excipient effects on the respective stabilities of the native and unfolded states. Thermal unfolding detected by IF offers speed of analysis and high throughput with extremely small protein quantities. Incorporation of light scattering detection can reveal the aggregation-prone domain and excipient effects upon it. MAbs possessing a wide range of T_m values and thermal unfolding profiles can be formulated into developable and manufacturable products (Brader et al., 2015). The quest of finding the best formulation can be aided significantly by a thorough characterization of unfolding behavior using orthogonal methods such as those described here.

ACKNOWLEDGMENTS

We thank Richard K. Brown (Avia Biosystems), Nicole Bouley Ford (NanoTemper Technologies Inc.), and John M. Lindsay (Genovis) for valuable input. We thank Julie Wei for a critical reading and helpful comments on the draft manuscript.

REFERENCES

An, Y., Zhang, Y., Mueller, H. M., Shameem, M., & Chen, X. (2014). A new tool for monoclonal antibody analysis: Application of IdeS proteolysis in IgG domain-specific characterization. *MAbs*, *6*(4), 879–893. http://dx.doi.org/10.4161/mabs.28762.

Bhambhani, A., Kissmann, J. M., Joshi, S. B., Volkin, D. B., Kashi, R. S., & Middaugh, C. R. (2012). Formulation design and high-throughput excipient selection based on structural integrity and conformational stability of dilute and highly concentrated IgG1 monoclonal antibody solutions. *Journal of Pharmaceutical Sciences*, *101*(3), 1120–1135. http://dx.doi.org/10.1002/jps.23008.

Blake-Haskins, A. W., Lin, Y.-H., Wu, Z., Perkins, M. D., & Spitznagel, T. M. (2014). *Biophysical and structural characterization needed prior to proof of concept*. Hoboken, NJ: John Wiley and Sons Inc.

Brader, M. L., Estey, T., Bai, S. J., Alston, R. W., Lucas, K. K., Lantz, S., et al. (2015). Examination of thermal unfolding and aggregation profiles of a series of developable therapeutic monoclonal antibodies. *Molecular Pharmaceutics*, *12*(4), 1005–1017. http://dx.doi.org/10.1021/mp400666b.

Brummitt, R. K., Nesta, D. P., & Roberts, C. J. (2011). Predicting accelerated aggregation rates for monoclonal antibody formulations, and challenges for low-temperature predictions. *Journal of Pharmaceutical Sciences*, *100*(10), 4234–4243. http://dx.doi.org/10.1002/jps.22633.

Doyle, C. M., Rumfledt, J. A., Broom, H. R., Broom, A., Stathopulos, P. B., Vassall, K. A., et al. (2013). Energetics of oligomeric protein folding and association. *Archives of Biochemistry and Biophysics*, *531*, 20.

Eftink, M. R. (1995). Use of multiple spectroscopic methods to monitor equilibrium unfolding of proteins. *Methods in Enzymology*, *259*, 487–512.

Freire, E. (1995). Thermal denaturation methods in the study of protein folding. *Methods in Enzymology*, *259*, 144–168.

Garber, E., & Demarest, S. J. (2007). A broad range of Fab stabilities within a host of therapeutic IgGs. *Biochemical and Biophysical Research Communications*, *355*(3), 751–757. http://dx.doi.org/10.1016/j.bbrc.2007.02.042.

Garidel, P., Hegyi, M., Bassarab, S., & Weichel, M. (2008). A rapid, sensitive and economical assessment of monoclonal antibody conformational stability by intrinsic tryptophan fluorescence spectroscopy. *Biotechnology Journal*, *3*(9–10), 1201–1211. http://dx.doi.org/10.1002/biot.200800091.

Goldberg, D. S., Bishop, S. M., Shah, A. U., & Sathish, H. A. (2011). Formulation development of therapeutic monoclonal antibodies using high-throughput fluorescence and static light scattering techniques: Role of conformational and colloidal stability. *Journal of Pharmaceutical Sciences*, *100*(4), 1306–1315. http://dx.doi.org/10.1002/jps.22371.

Grabowski, H. G., & Hansen, R. W. (2014). Cost to develop and win marketing approval for a new drug is $2.6 billion [Press release].

Hawe, A., Filipe, V., & Jiskoot, W. (2010). Fluorescent molecular rotors as dyes to characterize polysorbate-containing IgG formulations. *Pharmaceutical Research*, *27*(2), 314–326. http://dx.doi.org/10.1007/s11095-009-0020-2.

Hawe, A., Wiggenhorn, M., van de Weert, M., Garbe, J. H. O., Mahler, H. C., & Jiskoot, W. (2012). Forced degradation of therapeutic proteins. *Journal of Pharmaceutical Sciences*, *101*(3), 895–913. http://dx.doi.org/10.1002/jps.22812.

He, F., Hogan, S., Latypov, R. F., Narhi, L. O., & Razinkov, V. I. (2010). High throughput thermostability screening of monoclonal antibody formulations. *Journal of Pharmaceutical Sciences*, *99*(4), 1707–1720. http://dx.doi.org/10.1002/jps.21955.

Ionescu, R. M., Vlasak, J., Price, C., & Kirchmeier, M. (2008). Contribution of variable domains to the stability of humanized IgG1 monoclonal antibodies. *Journal of Pharmaceutical Sciences*, *97*(4), 1414–1426. http://dx.doi.org/10.1002/jps.21104.

Johnson, C. M. (2013). Differential scanning calorimetry as a tool for protein folding and stability. *Archives of Biochemistry and Biophysics*, *531*(1–2), 100–109.

Lepock, J. R., Ritchie, K. P., Kolios, M. C., Rodahl, A. M., Heinz, K. A., & Kruuv, J. (1992). Influence of transition rates and scan rate on kinetic simulations of differential scanning calorimetry profiles of reversible and irreversible protein denaturation. *Biochemistry*, *31*(50), 12706–12712. http://dx.doi.org/10.1021/bi00165a023.

Manning, M. C., Chou, D. K., Murphy, B. M., Payne, R. W., & Katayama, D. S. (2010). Stability of protein pharmaceuticals: An update. *Pharmaceutical Research*, *27*(4), 544–575. http://dx.doi.org/10.1007/s11095-009-0045-6.

Morrison, C. (2015). Fresh from the biotech pipeline-2014. *Nature Biotechnology*, *33*(2), 125–128.

Pace, C. N. (1986). Determination and analysis of urea and guanidine hydrochloride denaturation curves. *Methods in Enzymology*, *131*, 266–280.

Razinkov, V. I., Treuheit, M. J., & Becker, G. W. (2013). Methods of high throughput biophysical characterization in biopharmaceutical development. *Current Drug Discovery Technologies*, *10*(1), 59–70.

Samra, H. S., & He, F. (2012). Advancements in high throughput biophysical technologies: Applications for characterization and screening during early formulation development of monoclonal antibodies. *Molecular Pharmaceutics*, *9*(4), 696–707. http://dx.doi.org/10.1021/mp200404c.

Senisterra, G. A., & Finerty, P. J., Jr. (2009). High throughput methods of assessing protein stability and aggregation. *Molecular Biosystems*, *5*(3), 217–223. http://dx.doi.org/10.1039/b814377c.

von Pawel-Rammingen, U., Johansson, B. P., & Bjorck, L. (2002). IdeS, a novel streptococcal cysteine proteinase with unique specificity for immunoglobulin G. *The EMBO Journal*, *21*(7), 1607–1615. http://dx.doi.org/10.1093/emboj/21.7.1607.

Wen, J., Arthur, K., Chemmalil, L., Muzammil, S., Gabrielson, J., & Jiang, Y. (2012). Applications of differential scanning calorimetry for thermal stability analysis of proteins: Qualification of DSC. *Journal of Pharmaceutical Sciences*, *101*(3), 955–964. http://dx.doi.org/10.1002/jps.22820.

Wen, J., Hymes, K., & Narhi, L. O. (2004). Studying antibodies and Fc related proteins by DSC. *Protein Science*, *13*, 236.

Yguerabide, J., & Yguerabide, E. E. (1998). Light-scattering submicroscopic particles as highly fluorescent analogs and their use as tracer labels in clinical and biological applications—I. Theory. *Analytical Biochemistry*, *262*(2), 137–156. http://dx.doi.org/10.1006/abio.1998.2759.

CHAPTER FIFTEEN

IATC, DSC, and PPC Analysis of Reversible and Multistate Structural Transition of Cytochrome *c*

Shun-ichi Kidokoro[1], Shigeyoshi Nakamura
Department of Bioengineering, Nagaoka University of Technology, Nagaoka, Japan
[1]Corresponding author: e-mail address: kidokoro@nagaokaut.ac.jp

Contents

Abstract

Development of precise calorimeters has enabled us to monitor the structural transition of biomolecules by calorimetry to characterize the thermodynamic property changes accompanying three-dimensional structure change. We developed isothermal acid-titration calorimetry to evaluate the pH dependence of protein enthalpy, and demonstrated the thermodynamic transition between the native and molten globule (MG) states of cytochrome *c* with very small enthalpy change (~20 kJ/mol) by this method. The double deconvolution method with precise differential scanning calorimetry has revealed the MG state as an equilibrium intermediate state of the reversible thermal

Methods in Enzymology, Volume 567
ISSN 0076-6879
http://dx.doi.org/10.1016/bs.mie.2015.08.002

transition of the protein, and pressure perturbation calorimetry has succeeded in determining its volumetric properties. These examples strongly indicate the importance of a precise calorimetry and analysis model in the field of protein research.

1. INTRODUCTION

For the development of precise, easy-to-use, commercially available calorimeters in the past three decades, the thermodynamic properties of many kinds of proteins have been clarified by calorimetry. In this chapter, the analytic models of isothermal acid-titration calorimetry (IATC), differential scanning calorimetry (DSC), and pressure perturbation calorimetry (PPC) are discussed using a reversible and multistate transition of the three-dimensional structure of cytochrome *c* as an interesting example. As the structural transition is unimolecular, in this chapter we do not discuss nonisomeric systems such as those that include a self-dissociation/association process.

1.1 Calorimetry and Thermodynamic Properties of Reversible Structural Transitions

Gibbs free energy controls almost all of the molecular basis of enzymes, such as folding/unfolding of the three-dimensional structure, stability, ligand (activator/inhibitor) binding, protein–protein interaction, substrate specificity, and catalytic activity. Therefore, measurement of the free energy function is very important in order to understand and/or design the physical properties of enzymes. In many cases, however, there is no direct measurement for free energy; instead, it is usually estimated by analyzing experimental observations and fitting it with an appropriate model. In this case, the reliability of the calculated free energy depends largely on the validity of the model.

Calorimetry is the measurement of the heat accompanying a chemical process. For a process at constant pressure, the enthalpy change, ΔH, can be measured directly through the observed heat, q, using Eq. (1), derived from the first thermodynamic law:

$$\Delta H = q \text{ (in the process of constant pressure)} \tag{1}$$

If the process is reversible, the observed heat corresponds to the entropy change, ΔS, as described in Eq. (2) by the second law:

$$\Delta S = \frac{q}{T} \text{ (in the reversible process)} \tag{2}$$

where T is the temperature. These two relationships between thermodynamic functions and heat make calorimetry a unique method for evaluating Gibbs free energy, G, because G is composed of enthalpy and entropy as in Eq. (3):

$$G = H - TS \tag{3}$$

DSC data of reversible transitions enable us to determine the Gibbs free energy as a function of temperature with no model or assumption (Kidokoro & Wada, 1987). In the case of IATC, only enthalpy can be determined by Eq. (1) because the titration or mixing process is not reversible even if the structural transition itself is reversible. Therefore, one must select the appropriate model to determine the entropy and free energy changes. The issue on model selection and validation is not special for IATC but common for ITC analysis of molecular interaction (Nakamura, Koga, Shibuya, Seo, & Kidokoro, 2013).

As pressure is not constant in the case of PPC, Eq. (1) does not hold; entropy can be determined instead using Eq. (2) because the pressure perturbation is so small that the process induced by the pressure change is usually reversible. In this case, the pressure dependence of the entropy under constant temperature, which corresponds to the temperature dependence of the volume under constant pressure, can be directly evaluated without a model. We still need an appropriate model with which to analyze the temperature dependence of the PPC data.

It is worth noting that it is not the absolute value but the relative difference in energies that has real physical meaning for Gibbs free energy and enthalpy, and the equilibrium between the two states should be determined by the standard Gibbs free energy change, $\Delta G°(T, \text{pH})$, under the condition of constant pressure. Its temperature and pH dependence are described as Eq. (4) or Eqs. (5) and (6), respectively:

$$\frac{\partial \Delta G°}{\partial T} = -\Delta S° \tag{4}$$

$$\frac{\partial}{\partial T}\left(\frac{\Delta G°}{T}\right) = -\frac{\Delta H}{T^2} \tag{5}$$

$$\frac{\partial \Delta G°}{\partial \text{pH}} = (\ln 10) RT\Delta\nu \tag{6}$$

where R is the gas constant and $\Delta\nu$ is the change of the proton binding number (Kidokoro, Miki, & Wada, 1990). As these thermodynamic functions are also functions of temperature and pH, these dependences are described as follows:

$$\frac{\partial \Delta S^\circ}{\partial T} = \frac{\Delta C_\mathrm{p}}{T} \tag{7}$$

$$\frac{\partial \Delta S^\circ}{\partial \mathrm{pH}} = -(\ln 10)R\left(\Delta\nu + T\frac{\partial \Delta\nu}{\partial T}\right) \tag{8}$$

$$\frac{\partial \Delta H}{\partial T} = \Delta C_\mathrm{p} \tag{9}$$

$$\frac{\partial \Delta H}{\partial \mathrm{pH}} = -(\ln 10)RT^2\frac{\partial \Delta\nu}{\partial T} \tag{10}$$

When we assume the pH dependence of the enthalpy change, $\partial\Delta H/\partial\mathrm{pH}$ is a constant, $\Delta H'$, and the heat capacity change becomes a linear function of temperature as described in Eq. (11):

$$\Delta C_\mathrm{p}(T) = \Delta C_\mathrm{p}'(T - T_0) + \Delta C_{\mathrm{p},0} \tag{11}$$

where $\Delta C_\mathrm{p}'$ is a constant representing the temperature dependence of the heat capacity change and a constant, $\Delta C_{\mathrm{p},0}$ is defined by Eq. (12):

$$\Delta C_{\mathrm{p},0} = \Delta C_\mathrm{p}(T_0) \tag{12}$$

The standard Gibbs free energy change can be described as a function of temperature and pH in Eq. (13):

$$\begin{aligned}\Delta G^\circ(T,\mathrm{pH}) = {} & \Delta G_0^\circ\frac{T}{T_0} - [\Delta H_0 + \Delta H'(\mathrm{pH}-\mathrm{pH}_0)]\left(\frac{T}{T_0}-1\right)\\ & -\Delta C_{\mathrm{p},0}\left[T\ln\left(\frac{T}{T_0}\right) - (T-T_0)\right]\\ & -\frac{\Delta C_\mathrm{p}'}{2}\left[T^2 - T_0^2 - 2T_0T\ln\left(\frac{T}{T_0}\right)\right]\\ & + (\ln 10)RT\int_{\mathrm{pH}_0}^{\mathrm{pH}}\Delta\nu(T_0,\mathrm{pH})\,\mathrm{dpH}\end{aligned} \tag{13}$$

where ΔG_0° and ΔH_0 are the standard Gibbs free energy change and enthalpy change at the reference condition, $T = T_0, \mathrm{pH} = \mathrm{pH}_0$, as defined by Eqs. (14) and (15), respectively:

$$\Delta G_0^\circ = \Delta G^\circ(T_0, \mathrm{pH}_0) \tag{14}$$

$$\Delta H_0 = \Delta H(T_0, \mathrm{pH}_0) \tag{15}$$

The enthalpy change and the change in the number of bound protons are functions of temperature and pH as described by Eqs. (16) and (17), respectively:

$$\Delta H(T,\text{pH})=\frac{\Delta C_\text{p}'}{2}(T-T_0)^2+\Delta C_{\text{p},0}(T-T_0)+\Delta H_0+\Delta H'(\text{pH}-\text{pH}_0) \tag{16}$$

$$\Delta\nu(T,\text{pH})=\Delta\nu(T_0,\text{pH})-\frac{\Delta H'(T-T_0)}{(\ln 10)RT_0T} \tag{17}$$

When we consider n thermodynamic states, $(n-1)$ standard Gibbs free energy changes, $\Delta G^\circ_{0i}(T,\text{pH})$ $(i=1,2,\ldots,n-1)$, determine the equilibrium and the mole fractions of the native state, $f_N(T,\text{pH})$, and those of other $(n-1)$ states. $f_i(T,\text{pH})$, as in Eq. (18):

$$f_N(T,\text{pH})=\left[1+\sum_{i=1}^{n-1}\exp\left(-\frac{\Delta G^\circ_{0i}}{RT}\right)\right]^{-1}$$
$$f_i(T,\text{pH})=\frac{\sum_{i=1}^{n-1}\exp\left(-\frac{\Delta G^\circ_{0i}}{RT}\right)}{1+\sum_{j=1}^{n-1}\exp\left(-\frac{\Delta G^\circ_{0j}}{RT}\right)} \tag{18}$$

1.2 Molten Globule State of Cytochrome *c*

Several globular proteins are reported to exhibit an equilibrium molten globule (MG) state, which is a compact denatured state with a native-like secondary structure but a largely disordered tertiary structure (Arai & Kuwajima, 2000; Ohgushi & Wada, 1983; Ptitsyn, 1995). The MG state of cytochrome *c* at low pH and high anion concentration was one of the first identified examples of the MG state (Ohgushi & Wada, 1983) (we designated it the MG1 state).

While the MG state of α-lactoalbumin shows noncooperative thermal unfolding (Dolgikh et al., 1985), the MG state of cytochrome *c* shows a cooperative thermal transition (Hamada, Kidokoro, Fukada, Takahashi, & Goto, 1994; Nakamura, Baba, & Kidokoro, 2007; Nakamura, Seki, Katoh, & Kidokoro, 2011; Potekhin & Pfeil, 1989). Although the MG1-to-D transition of horse cytochrome *c* was initially described as a two-state transition based on DSC and isothermal titration calorimetry (ITC) data (Hamada et al., 1994; Potekhin & Pfeil, 1989), the thermal MG1-to-D transition under an acidic condition was confirmed with the three-state transition including an intermediate state (we designated it the MG2 state) by circular dichroism (CD) (Kuroda, Kidokoro, & Wada, 1992) and calorimetric studies (Nakamura et al., 2011).

In order to show that the MG state is a thermodynamic state that is distinguishable from the native state, we used IATC, which can evaluate the pH dependence of enthalpy directly, because it allowed us to observe the pH transition from the N to MG1 state under a high salt concentration by decreasing the pH to acidic values. In Section 2, the IATC method is reviewed and is applied to observe the pH transition. A pH transition with rather small molar enthalpy change, 20 kJ/mol, was clearly observed through this method.

From the direct observation of molar enthalpy change from the N-to-D transition of cytochrome *c* by IATC, a rather significant difference, about 100 kJ/mol, was found in the molar enthalpy change between the native (N) and denatured (D) states when a two-state model was used for DSC analysis. We have shown that a three-state analysis is needed to explain the precise DSC data, and the molar enthalpy change based on the model agrees with the enthalpy of IATC (Nakamura et al., 2007). In Section 3, the double deconvolution (DD) method and nonlinear least-squares fitting with a multistate model are described in detail.

The temperature dependence of the partial molar volume of a protein molecule can be obtained directly by PPC. If we obtain or estimate the partial molar volume at a temperature, for example, at room temperature, the partial molar volume can be obtained as a function of temperature. As the observed volume function, $V(T)$, is the average of those of the thermodynamic states, the molar volume of each state can be separately determined if we obtain the mole fraction of each state as a function of temperature. From the case of the three-state transition of cytochrome *c* in a weak acid pH, the partial molar volume of the intermediate, an MG state, can be determined as described in Section 4.

2. ISOTHERMAL ACID-TITRATION CALORIMETRY

Because protein molecules are composed of acidic and basic amino acids and have amino and carboxyl termini, protons are general ligands for proteins. Proton binding and/or release significantly affects the entropy and enthalpy change for the reaction.

As indicated by Eq. (1), the pH dependence of enthalpy could be observed in theory by titrating the acid or base to the protein solution. The dilution enthalpy of acid or base into water is not negligible, but it can be corrected for if appropriate reference experiments are performed. The neutralization heat is generally too big to correct precisely, and the neutralization heat is usually observed by titrating the base into the protein solution due to proton release from the protein, while no OH^- ion is released when protons are titrated to the protein. We have developed IATC using an ITC device and pH measurement.

2.1 Measurement of pH Dependence of Protein Enthalpy

The detailed experimental procedure for IATC was described in our paper (Nakamura & Kidokoro, 2004). In brief, lyophilized protein powder was dissolved in 20 m*M* NaCl solution and the protein solution was dialyzed against the same solvent. The pH of the solution was adjusted to around pH 7 with 50 m*M* NaOH. Protein solutions with concentrations of about 0.5 mg/mL were found to be sufficient to obtain the pH profile in a MicroCal MCS-ITC (Malvern Instruments Ltd., Worcestershire, UK), at temperatures ranging from 15 to 65 °C. HCl solutions (50–500 m*M*) containing 20 m*M* NaCl were added for the acid solution of titration. In the ITC measurement, each 0.5–15 μL acid solution was titrated into the protein solution in a 1.4 mL cell, and the induced heat from each injection was evaluated. The dilution heat was monitored by performing the same titration into a NaCl solution in the absence of cytochrome *c*.

The pH of the solution after each acid titration was monitored outside the calorimeter using a pH meter fitted with a glass electrode, calibrated with standard buffers at each temperature. The solutions were identical to those used in the ITC. In the pH measurement of the study, the initial volume of the protein solution was 10 mL and the injection volumes were determined in order to reproduce the ratio corresponding to each injection of the ITC measurement. The titration vessel should be kept at a constant temperature. As the calorimeter cells were made from Hastelloy, which is known to be damaged when placed in strong acid solution, the lowest pH of the sample was confirmed to be above pH 1.5.

IATC has been used to measure the pH function of enthalpy of bovine ribonuclease A at several temperatures, as shown in Fig. 1. In this case, the pH profile at 15 °C can be considered the native state, and the profile can be fit to an exponential function as in the figure. The protonation enthalpy of histidine residues, whose pK_a is around neutral pH, is negative and relatively significant. Thus, the pH profile in the region from neutral to weak acidity seems reasonable. On the other hand, the protonation enthalpy of carboxyl

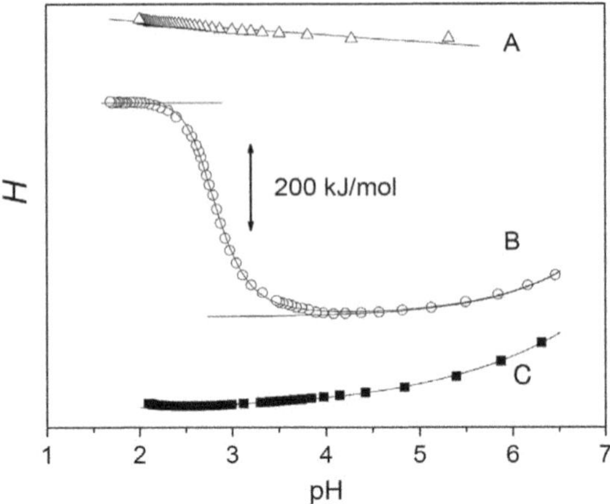

Figure 1 The enthalpy of bovine ribonuclease A was determined as a function of pH at 15 (C), 40 (B), and 65 °C (A). The solid lines were determined by least-squares fitting with an exponential function (15 °C), a linear function (65 °C), and a theoretical curve calculated by a two-state transition model (40 °C), respectively. *Nakamura and Kidokoro (2004). Reprinted with permission from Elsevier.*

groups of aspartate and glutamate is so small that the pH dependence at acidic pH is negligible. As seen in the pH profile at 65 °C, representing the denatured state, the enthalpy shows little pH dependence. At 40 °C, the pH dependence of molar enthalpy shows a clear pH transition from the native to the denatured state by acid titration.

Figure 2 shows the pH functions of molar enthalpy of horse cytochrome *c* in 500 m*M* KCl at several temperatures in the range of 16–25 °C evaluated by IATC. The protein concentration was 0.5 mg/mL and the titration was conducted via injections of 2, 5, and 10 μL of a 20–400 m*M* HCl solution in 500 m*M* KCl. The reference titration for dilution heat was performed with a 500 m*M* KCl solution. The temperature of the titration vessel for pH measurement was kept constant at the same temperature as that of ITC measurement. Although the molar enthalpy change was very small, around 20 kJ/mol in this case, a distinct pH transition was observed.

2.2 Analysis with Two-State Model and Validation of the Model

As described in Section 1.1, while ITC can measure molar enthalpy change, it does not determine the molar entropy change directly. An appropriate model, therefore, is necessary to determine the entropy and Gibbs energy change accompanying the structural transition based on the IATC data. Since the results of the analysis depend on the model, it is important to check the validity of the model used for the analysis.

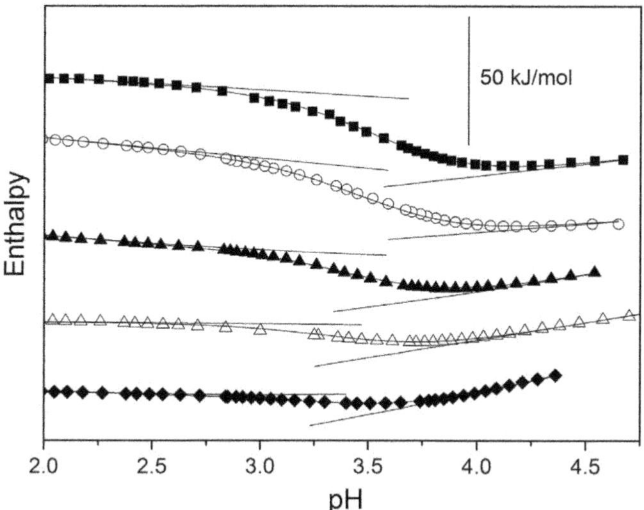

Figure 2 Enthalpy functions of pH evaluated by isothermal acid-titration calorimetry (IATC). Enthalpy functions of horse cytochrome *c* in the transition from the native (N) state to the MG1-state at 16 (filled diamonds), 17.5 (open triangles), 20 (filled triangles), 22.5 (open circles), and 25 °C (filled squares) in 500 m*M* KCl. Solid lines show the theoretical fitting curves and baselines of the N and MG1 states calculated by the global fitting method. *Nakamura et al. (2011). Reprinted with permission from the American Chemical Society.*

The simplest model for a protein structural transition is a two-state model including only the N and D states. This means that the sum of the mole fractions of N and D states can be considered to be one; in other words, this sum in any other thermodynamic state is very small to be negligible. The equilibrium between the states is described as Eq. (19)

$$\mathrm{N} \rightleftarrows \mathrm{D} \tag{19}$$

and determined by the equilibrium constant, $K(T, \mathrm{pH})$, or the standard Gibbs free energy change, $\Delta G^{\circ}(T, \mathrm{pH})$, under constant pressure (see Eq. 13) as Eq. (20):

$$K(T,\mathrm{pH}) = \frac{f_{\mathrm{D}}(T,\mathrm{pH})}{f_{\mathrm{N}}(T,\mathrm{pH})} = \exp\left(-\frac{\Delta G^{\circ}(T,\mathrm{pH})}{RT}\right) \tag{20}$$

where f_{N} and f_{D} are the mole fractions of the N and D states, respectively.

When we analyze the IATC data at a temperature, T, Eqs. (13) and (16) can be simplified as Eqs. (21) and (22), respectively, by setting $T_0 = T$:

$$\Delta G^{\circ}(T_0, \mathrm{pH}) = \Delta G_0^{\circ} + (\ln 10)RT_0\int_{\mathrm{pH}_0}^{\mathrm{pH}} \Delta\nu(T_0, \mathrm{pH})\mathrm{dpH} \tag{21}$$

$$\Delta H(T_0, \mathrm{pH}) = \Delta H_0 + \Delta H'(\mathrm{pH} - \mathrm{pH}_0) \tag{22}$$

In order to analyze the IATC data, we must assume the pH dependence of the molar enthalpy of both the N-state, $H_N(T_0, \mathrm{pH})$, and the proton binding number difference, $\Delta\nu(T_0, \mathrm{pH})$. The following pH function was proposed and found to work well for several targets (Nakamura & Kidokoro, 2004; Nakamura & Kidokoro, 2005; Nakamura et al., 2011):

$$H_N(T_0, \mathrm{pH}) = \Delta H_{p,N}\{\exp[A(\mathrm{pH} - \mathrm{pH}_0)] - 1\} + H_{N,0} \tag{23}$$

$$\Delta\nu(T_0, \mathrm{pH}) = \Delta\nu_0 \exp[-k(\mathrm{pH} - \mathrm{pH}_0)] \tag{24}$$

where $\Delta H_{p,N}$, A, and k are the constants for describing the pH dependence, $H_{N,0}$ is defined by Eq. (25), and $\Delta\nu_0$ is defined by Eq. (26):

$$H_{N,0} = H_N(T_0, \mathrm{pH}_0) \tag{25}$$

$$\Delta\nu_0 = \Delta\nu(T_0, \mathrm{pH}_0) \tag{26}$$

When we use Eq. (24) for the proton binding number change, the pH dependence of standard Gibbs free energy change, Eq. (21), can be described simply as Eq. (27):

$$\Delta G^{\circ}(T_0, \mathrm{pH}) = \Delta G_0^{\circ} + (\ln 10)RT_0\frac{\Delta\nu_0}{k}\{1 - \exp[-k(\mathrm{pH} - \mathrm{pH}_0)]\} \tag{27}$$

It is worth noting that pH_0 can in theory be set at any pH. In IATC data analysis, the midpoint pH, pH_d, at which the standard Gibbs free energy change becomes zero at $T = T_0$, is convenient to use. When we set $\mathrm{pH}_0 = \mathrm{pH}_d$, ΔG_0° should be zero and Eqs. (13) and (27) are described as Eqs. (28) and (29), respectively:

$$\begin{aligned}\Delta G^{\circ}(T, \mathrm{pH}) = &-[\Delta H_0 + \Delta H'(\mathrm{pH} - \mathrm{pH}_d)]\left(\frac{T}{T_0} - 1\right)\\ &-\Delta C_{p,0}\left[T \ln\left(\frac{T}{T_0}\right) - (T - T_0)\right]\\ &-\frac{\Delta C_p'}{2}\left[T^2 - T_0^2 - 2T_0 T \ln\left(\frac{T}{T_0}\right)\right]\\ &+ (\ln 10)RT\frac{\Delta\nu_0}{k}\{1 - \exp[-k(\mathrm{pH} - \mathrm{pH}_d)]\}\end{aligned} \tag{28}$$

$$\Delta G^{\circ}(T_0, \mathrm{pH}) = (\ln 10)RT_0\frac{\Delta\nu_0}{k}\{1 - \exp[-k(\mathrm{pH} - \mathrm{pH_d})]\} \quad (29)$$

For the analysis of pH transition, the pH dependence of the protein enthalpy, $H(T_0, \mathrm{pH})$, then becomes Eq. (30):

$$\begin{aligned} H(T_0, \mathrm{pH}) &= H_\mathrm{N}(T_0, \mathrm{pH}) + \Delta H(T_0, \mathrm{pH}) f_\mathrm{D}(T_0, \mathrm{pH}) \\ &= H_\mathrm{N}(T_0, \mathrm{pH}) + \Delta H(T_0, \mathrm{pH})\frac{\exp\left(-\frac{\Delta G^{\circ}(T_0, \mathrm{pH})}{RT}\right)}{1 + \exp\left(-\frac{\Delta G^{\circ}(T_0, \mathrm{pH})}{RT}\right)} \end{aligned} \quad (30)$$

The theoretical curves in Fig. 1 were calculated using Eq. (30), whose eight parameters, $\Delta H_{\mathrm{p,N}}$, A, $H_{\mathrm{N},0}$, $\Delta\nu_0$, k, ΔH_0, $\Delta H'$, and $\mathrm{pH_d}$, were optimized to fit the experimental data by the nonlinear least-squares method. The transition enthalpy, ΔH_0, at 40 °C (312 ± 3 kJ/mol) coincided very well with the transition enthalpy observed by DSC (Nakamura & Kidokoro, 2004). The pH dependence of enthalpy change, $\Delta H'$, was found to be very small, namely 0 to −3 kJ/mol, in this case.

When the IATC is performed at several different temperatures, the temperature dependence of the fitting parameters can be obtained. From the temperature dependence of midpoint pH, $\mathrm{pH_d}$, and the change in proton binding number, the van't Hoff enthalpy can be evaluated based on Eq. (31) (Nakamura & Kidokoro, 2004). If the enthalpy coincides with the directly observed calorimetric enthalpy, ΔH_0, it strongly supports the validity of the model assumed for the analysis:

$$\Delta H_{\mathrm{vH}}(T_0, \mathrm{pH_d}) = (\ln 10)RT_0^2\Delta\nu_0\left.\frac{\mathrm{dpH_d}}{\mathrm{d}T}\right|_{T=T_0} \quad (31)$$

For example, in the case of the acid transition of bovine ribonuclease A whose IATC data are shown in Fig. 1, the temperature dependence of $\mathrm{pH_d}$ was evaluated as $0.071 \pm 0.002\ \mathrm{K}^{-1}$ at 40 °C, as shown in Fig. 3, and the van't Hoff enthalpy was calculated to be 320 ± 10 kJ/mol by Eq. (31). This value agreed well with the calorimetric enthalpy obtained by IATC at 40 °C, 312 ± 3 kJ/mol (Nakamura & Kidokoro, 2004). Therefore, the validity of the two-state model was confirmed for the transition.

Global analysis of IATC data at several different temperatures is another possible way to validate the model. In Fig. 2, the enthalpy functions at all

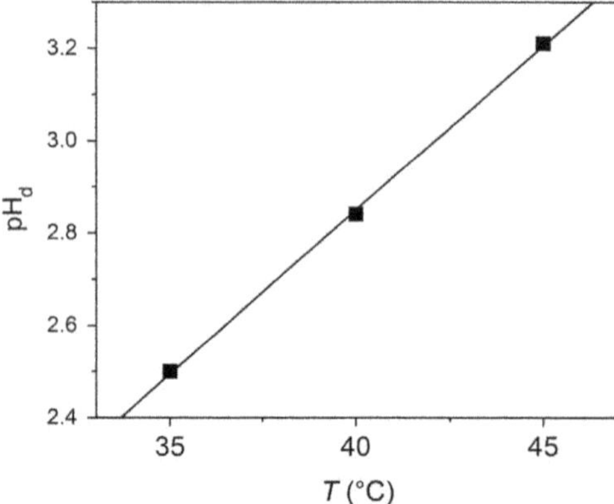

Figure 3 The temperature dependence of the midpoint pH, pH_d, of bovine ribonuclease A by IATC at several temperatures. The symbols show the parameters at each temperature, and the solid line was the best-fitted linear function to evaluate dpH_d/dT as $0.071 \pm 0.002\ K^{-1}$. *Nakamura and Kidokoro (2004). Reprinted with permission from Elsevier.*

temperatures were analyzed by a two-state global fitting method. In the global analysis, we use Eq. (28), and the parameters in this equation were globally optimized (Nakamura & Kidokoro, 2005). The fitting parameter for the temperature dependence of the heat capacity change, $\Delta C_p'$, can be fixed to zero to fit the experimental data within error in Fig. 2. This indicated that the set of thermodynamic parameters for the two-state model were consistent with all the IATC data at various temperatures, thus supporting the validity of the model.

2.3 Transition Enthalpy of Cytochrome *c*

As DSC has been used for the calorimetric evaluation of the structural transition of proteins, the structural transition should have a positive enthalpy change because the temperature dependence of the equilibrium constant is caused by the molar enthalpy change, as in Eq. (32):

$$\frac{\partial \ln K}{\partial T} = \frac{\partial}{\partial T}\left(-\frac{\Delta G^\circ}{RT}\right) = \frac{\Delta H}{RT^2} \tag{32}$$

When the molar enthalpy change is small, the thermal transition will become very broad and the transition cannot be distinguished from the temperature dependence of the heat capacity of the native state. For that reason, most molar enthalpies reported by DSC have been above 100 kJ/mol.

Because the pH dependence of equilibrium constants is derived not from the enthalpy change but from the proton binding number difference as

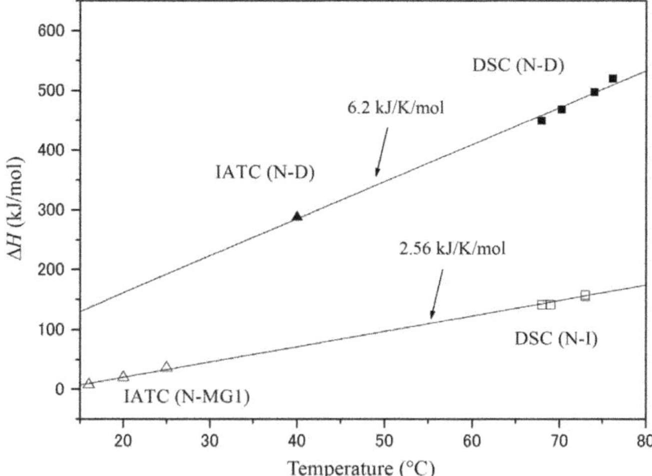

Figure 4 Temperature dependence of the enthalpy change in the structure transition of horse cytochrome *c* examined by DSC and IATC. Filled squares show the ΔH values of the N-to-D transition of horse cytochrome *c* at pH 4 examined by DSC. The filled triangle shows the ΔH of the N-to-D transition examined by IATC. Empty squares show the ΔH values of the N-to-I transition of horse cytochrome *c* at pH 4 examined by DSC (Nakamura et al., 2007). Empty triangles show the ΔH values of the N-to-MG1 transition of horse cytochrome *c* at pH 2.5 in 500 m*M* KCl examined by IATC (Nakamura et al., 2011). Solid lines show the linear fitting curves of IATC and DSC results. *Nakamura et al. (2011). Reprinted with permission from the American Chemical Society.*

shown in Eq. (33), IATC can be used to observe the acid transition even when the molar enthalpy change is very small:

$$\frac{\partial \ln K}{\partial \mathrm{pH}} = \frac{\partial}{\partial \mathrm{pH}}\left(-\frac{\Delta G^{\circ}}{RT}\right) = -(\ln 10)\Delta\nu \tag{33}$$

Actually, IATC can successfully detect the native to MG state transition of cytochrome *c* with a molar enthalpy change of only 20 kJ/mol, as shown in Fig. 4, and these values are consistent with the enthalpies by DSC as shown in Section 3.

3. DIFFERENTIAL SCANNING CALORIMETRY

3.1 Check for Equilibrium and Concentration Dependence

When the thermal transition of the protein is reversible and the folding/unfolding equilibrium can be reached rapidly enough to show no scan-rate dependence, DSC can determine the Gibbs free energy of the protein directly, as described in Section 1.1. Although the scan-rate dependence is rarely checked in the literature, it is worth noting that reversibility is a necessary but not sufficient condition to indicate that the reaction is in equilibrium. If the thermal transition is irreversible and has apparent scanning rate dependence, a kinetic model should be used to obtain correct information on the protein stability.

Another important experimental check is the dependence of the DSC data on protein concentration. When the quaternary structure of the protein changes and the dissociation or association of the subunits accompanies the thermal transition, the transition should show concentration dependence. For example, the apparent Gibbs free energy change of the transition depends on the concentration of the protein, c, as $(m_D - m_N)RT \ln c$ (Kidokoro, Uedaira, & Wada, 1988) where m_D and m_N are the stoichiometry coefficients of the denatured and the native states, respectively. The n-times difference of the concentration therefore affects the midpoint temperature, ΔT_m, as in Eq. (34):

$$\Delta T_m = \frac{(m_D - m_N)RT_m^2 \ln n}{\Delta H(T_m)} \tag{34}$$

When m_D, m_N, n, T_m, and $\Delta H(T_m)$ are 2, 1, 2, 320 K, and 1000 kJ/mol, respectively, for a typical dimer to monomer thermal transition with a double or half protein concentration change, ΔT_m is expected to be 0.6 K. This is usually sufficient for the recent precise DSC apparatus to detect.

3.2 Double Deconvolution Method and Two-State Analysis

For the reversible and unimolecular thermal transition, the mole fractions of the native and denatured states, f_N and f_D, can be easily calculated from the Gibbs free energy of the system, N-state, and D-state, $G(T)$, $G_N(T)$, and $G_D(T)$, with Eq. (35):

$$\begin{aligned} f_N(T) &= \exp[-G_N(T) + G(T)] \\ f_D(T) &= \exp[-G_D(T) + G(T)] \end{aligned} \tag{35}$$

The midpoint temperature, T_m, is easily determined as the temperature at which both mole fractions become equal:

$$f_N(T_m) = f_D(T_m) \tag{36}$$

Figure 5A shows, for example, the sum of the two mole fractions in Eq. (35) for the thermal N-to-D transition of horse cytochrome c at pH 4.1. This strongly indicates the existence of a stable intermediate state. Using

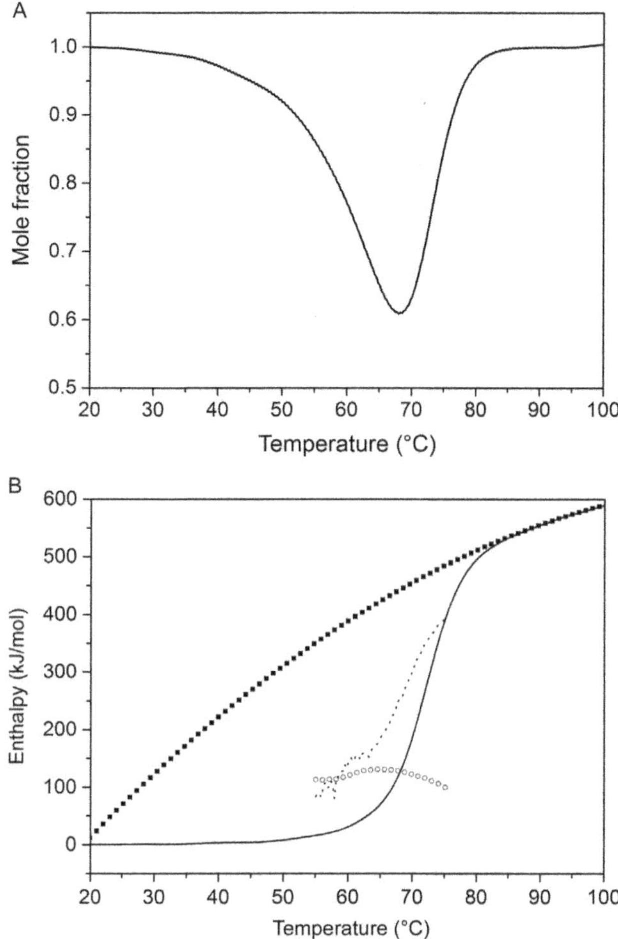

Figure 5 (A) The sum of the mole fraction of the native and denatured states of cytochrome *c* in the thermal transition at pH 4.1 calculated by the double deconvolution (DD) method (Kidokoro & Wada, 1987). (B) The temperature dependence of the enthalpies of cytochrome *c* determined by the DD method. The solid lines show the ΔH_N, which is defined as $H_{obs} - H_N$, where H_{obs} is the observed enthalpy and H_N is the enthalpy of the N state of cytochrome *c*. The filled squares show the calorimetric enthalpy, ΔH_{cal}, which is defined as $H_D - H_N$, where H_D is the enthalpy of the D state of cytochrome *c*. The open circles show the ΔH_{NI}, which is defined as $H_I - H_N$, where H_I is the enthalpy of the intermediate state on the thermal N-to-D transition. The dotted lines show the van't Hoff enthalpy. *Nakamura et al. (2007). Reprinted with permission from Elsevier.*

the mole fractions, the molar enthalpy of the intermediate state can be calculated as shown in Fig. 5B. This DD method requires no assumptions regarding the number of thermodynamic states (Kidokoro & Wada, 1987). The results are typically superior to those of the traditional two-state analysis, and they completely cover the information provided by the traditional analysis, as discussed below. Thus, the DD method should be used instead of the two-state analysis.

Two-state analysis is used to determine the midpoint temperature, T_m, and two kinds of enthalpies; the calorimetric enthalpy, ΔH_{cal}, and the van't Hoff enthalpy, ΔH_{vH}. Although these enthalpies can be defined as functions of temperature as seen in Fig. 5B, only the values at the midpoint temperature are usually used to check the validity of the model. Of course, the T_m and ΔH_{vH} of the two-state analysis are not correct in the case of $\Delta H_{cal} \neq \Delta H_{vH}$.

3.3 Multistate Model Fitting

Thermodynamic parameters determined by the DD method can be further refined by nonlinear least-squares fitting of the parameters to experimental data. When we analyze the DSC data at a constant pH, Eqs. (13) and (16) can be simplified as Eqs. (37) and (38), respectively, by setting $pH_0 = pH$:

$$\Delta G^\circ(T, pH_0) = \Delta G_0^\circ \frac{T}{T_0} - \Delta H_0\left(\frac{T}{T_0} - 1\right) - \Delta C_{p,0}\left[T \ln\left(\frac{T}{T_0}\right) - (T - T_0)\right] - \frac{\Delta C_p'}{2}\left[T^2 - T_0^2 - 2T_0 T \ln\left(\frac{T}{T_0}\right)\right] \tag{37}$$

$$\Delta H(T, pH_0) = \frac{\Delta C_p'}{2}(T - T_0)^2 + \Delta C_{p,0}(T - T_0) + \Delta H_0 \tag{38}$$

When we consider n thermodynamic states, $(n-1)$ standard Gibbs free energy changes, $\Delta G_{0i}^\circ(T, pH_0)$ $(i = 1, 2, \ldots, n-1)$, determines the equilibrium and the mole fractions of the native state, $f_N(T, pH_0)$ as well as those of other $(n-1)$ states, $f_i(T, pH_0)$, as in Eq. (18). Using these functions, the heat capacity can be calculated as shown in Eq. (39), where $f_0 = f_N$ and the heat capacity of the native state, $C_N(T)$, is assumed to be a linear function of temperature:

$$C_p(T) = C_N(T) + \sum_{i=1}^{n-1} \Delta C_{0i}(T) f_i(T) + \frac{\sum_{i,j=0}^{n-1} \left(\Delta H_{0j}(T) - \Delta H_{0i}(T)\right)^2 f_i(T) f_j(T)}{2RT^2} \tag{39}$$

Figure 6 shows the fitting results with a three-state model. The model well describes a thermal transition under weak acidic pH condition.

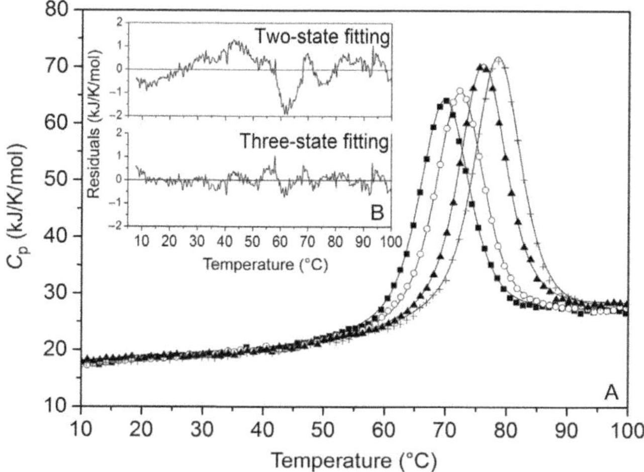

Figure 6 (A) The temperature dependence of the partial molar heat capacity of horse cytochrome *c* at pH 3.8 (closed squares), 4.1 (open circles), 4.5 (closed triangles), and 5.0 (crosses). The solid lines show the theoretical fitting curves determined by the analysis with the three-state transition model. (B) The residuals of two-state (upper graph) and three-state (lower graph) fittings. *Nakamura et al. (2007). Reprinted with permission from Elsevier.*

Although the data under this condition were traditionally analyzed by the two-state model, the clear systematic deviation of the fitting function from the data is observed as shown in Fig. 6B.

Additionally, the N-to-D enthalpy change, ΔH_{ND}, was found to be underestimated by about 100 kJ/mol because the heat capacity of the native state was inappropriately fitted in the case of the two-state model (Nakamura et al., 2007). Therefore, this example clearly indicates the importance of both precise calorimetry and the use of the appropriate model for the analysis. If the noise is large, we cannot recognize the deviation between the theoretical and experimental data, and if we use an incorrect model, the thermodynamic parameters can systematically deviate from the real values.

By CD spectral and solution X-ray scattering analyses, the intermediate state above was found to have the structural property of the MG state (Nakamura et al., 2007, 2011). Considering that cytochrome *c* was traditionally reported to be a typical small globular protein that was proven to show two-state thermal transition by DSC, the recent precise DSC measurement is expected to reveal several other examples that possess the stable MG state under natural solution conditions.

4. PRESSURE PERTURBATION CALORIMETRY WITH DSC ANALYSIS

In the case of the reversible process, the entropy change can be obtained directly with Eq. (2). The pressure response under constant temperature and pH is described in Eq. (40) using one of the Maxwell relations:

$$\frac{1}{T}\left(\frac{\mathrm{d}'q}{\mathrm{d}p}\right)_{T,\mathrm{pH}} = \left(\frac{\partial S}{\partial p}\right)_{T,\mathrm{pH}} = -\left(\frac{\partial V}{\partial T}\right)_{p,\mathrm{pH}} \tag{40}$$

PPC was then used to evaluate the temperature dependence of the partial molar volume of proteins (Lin, Brandts, Brandts, & Plotnikov, 2002). The 0.1–0.5 MPa pressure, limited by the pressure allowance of the calorimeter cell, is used for the pressure perturbation and is so small that the response is completely reversible in many cases. By measuring the response heat at several different temperatures, the partial molar volume can be obtained. If the thermal transition of the protein is reversible and in equilibrium, the volumetric characterization of the protein accompanying the thermal transition can be evaluated by PPC.

4.1 Multistate Fitting of PPC

When we consider n thermodynamic states for the thermal transition of the protein, the mole fractions of the native state, $f_{\mathrm{N}}(T, \mathrm{pH}_0)$, and those of other $(n-1)$ states, $f_i(T, \mathrm{pH}_0)$, are determined as shown in Eq. (18). The temperature dependence of the partial molar volume is described as in Eq. (41) (Nakamura et al., 2012):

$$\left(\frac{\partial V}{\partial T}\right)_{p,\mathrm{pH}} = \left(\frac{\partial V_{\mathrm{N}}}{\partial T}\right)_{p,\mathrm{pH}} + \sum_{i=1}^{n-1}\left(\frac{\partial \Delta V_{0i}}{\partial T}\right)_{p,\mathrm{pH}} f_i(T) + \frac{\sum_{i,j=0}^{n-1}(\Delta V_{0j} - \Delta V_{0i})(\Delta H_{0j} - \Delta H_{0i}) f_i(T) f_j(T)}{2RT^2} \tag{41}$$

where V_{N} is the partial molar volume of the native state and ΔV_{0i} is the change in the partial molar volume from the native state to the ith state.

Almost all the thermodynamic parameters in Eq. (39) except the parameters for $C_{\mathrm{N}}(T)$ are included in Eq. (41), and in Eq. (42) additionally the volumetric parameters for $\left(\frac{\partial V_{\mathrm{N}}}{\partial T}\right)_{p,\mathrm{pH}}$ and $\Delta V_{0i}(T)$ are needed to describe the partial molar volume of the protein. We have proposed the use of the exponential function of temperature for the temperature dependence of V_N as in Eq. (42) and that of the linear function for $\Delta V_{0i}(T)$ as in Eq. (43), respectively

$$\left(\frac{\partial V_N}{\partial T}\right)_{p,\mathrm{pH}} = b_1\left(e^{-b_2 T - T_0} - 1\right) + s_0 \tag{42}$$

$$\Delta V_{0i} = s_i(T - T_0) + v_i \tag{43}$$

where b_1, b_2, s_i, and v_i are the fitting parameters.

Figure 7 shows an example of the three-state model fitting of cytochrome *c*. In this analysis, all the fitting parameters other than the volumetric

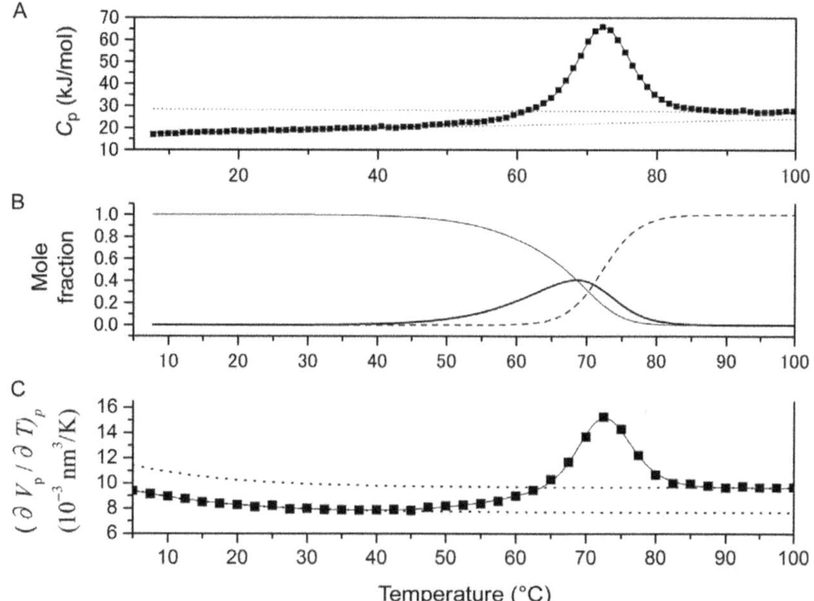

Figure 7 (A) The heat capacity function of 1 mg/mL cytochrome *c* at pH 4.1 in 50 m*M* sodium acetate/acetic acid buffer evaluated by DSC measurement. Filled squares show the experimental data. Solid lines indicate the theoretical fitting curve calculated by three-state transition analysis. Dotted lines illustrate the baselines of the N and D states calculated by three-state transition analysis. (B) The temperature dependence of the mole fraction of each state, i.e., the N (thin line), I (thick line), and D (broken line) states, in the thermal transition of 1 mg/mL cytochrome *c* at pH 4.1. (C) The temperature dependence of $(\partial V_p/\partial T)_p$ of 15 mg/mL cytochrome *c* at pH 4.1 in 50 m*M* sodium acetate/acetic acid buffer. Solid lines show the theoretical fitting curves calculated by three-state transition analysis using DSC parameters. Dotted lines illustrate the baselines of the N and D states calculated by three-state transition analysis using DSC parameters. *Nakamura et al. (2012). Reprinted with permission from the American Chemical Society.*

Table 1 Thermodynamic Parameters of the Thermal Transition of Cytochrome *c* at pH 4.1 (Nakamura et al., 2012)

	Transition	T_m (°C)	ΔV^a (10^{-3} nm^3)	$\left(\frac{\partial \Delta V_p}{\partial T}\right)_p$ (10^{-3} nm^3/K)	$\Delta\alpha_p$ (10^{-4} K^{-1})
Cytochrome *c*[b]	N-I	75 ± 30	100 ± 200	−1 ± 20	−0.5 ± 10
	N-D	72 ± 4	90 ± 10	2.1 ± 0.2	1.3 ± 0.1
Cytochrome *c*[c]	N-I	69 ± 1	23.3–26.9	1.07–1.41	0.71–0.93
	N-D	70.2 ± 0.3	69.8–72.1	1.76–1.79	1.16–1.18

[a]Evaluated at T_m.
[b]Evaluated by an analysis where all parameters are free.
[c]Evaluated by an analysis with transition parameters determined by DSC.

parameters were determined by DSC data as described in Section 3.3, and only the volumetric parameters are fitted with PPC data. The fitting parameters are listed in Table 1. When we use both DSC and PPC data, all the parameters can be determined with the three-state model. If all the parameters are determined only with PPC, however, the estimation errors for not only the volumetric parameters but also the midpoint temperature become very large, as the table shows. Even in the case of a simpler two-state transition, the volumetric parameters can be determined precisely by using DSC data together (Nakamura et al., 2012).

5. SUMMARY

For rapid and reversible thermal transition, the pH dependence of the enthalpy, the temperature dependence of Gibbs free energy, and that of partial specific volume can be directly obtained by IATC, DSC, and PPC. An appropriate model is necessary to determine the Gibbs free energy change for IATC analysis, and the validity of the model can be checked by analyzing the temperature dependence of the IATC data. The pH transition of N-to-MG1 state of cytochrome *c* under high salt condition was successfully analyzed by a two-state model even with small enthalpy change about 20 kJ/mol. In DSC analysis, on the other hand, the Gibbs free energy change can be directly determined by DD method without assuming the number of the thermodynamic states. While the N-to-D transition of the protein was approximated to be two-state traditionally, the MG1 state was observed as a stable intermediate in the transition. The mole fractions of the three states determined by DSC were found to be important to determine the volumetric parameters of these states by PPC analysis.

ACKNOWLEDGMENT

This work was supported by JSPS KAKENHI (grant no. 25400426).

REFERENCES

Arai, M., & Kuwajima, K. (2000). Role of the molten globule state in protein folding. *Advances in Protein Chemistry*, *53*, 209–282.

Dolgikh, D. A., Abaturov, L. V., Bolotina, I. A., Brazhnikov, E. V., Bychkova, V. E., Gilmanshin, R. I., et al. (1985). Compact state of a protein molecule with pronounced small-scale mobility: Bovine alpha-lactalbumin. *European Biophysics Journal*, *13*, 109–121.

Hamada, D., Kidokoro, S., Fukada, H., Takahashi, K., & Goto, Y. (1994). Salt-induced formation of the molten globule state of cytochrome *c* studied by isothermal titration calorimetry. *Proceedings of the National Academy of Sciences of the United States of America*, *91*, 10325–10329.

Kidokoro, S., Miki, Y., & Wada, A. (1990). Physical and biological stability of globular proteins. In M. Hatano (Ed.), *Protein structural analysis, folding and design* (pp. 75–91). Tokyo: Japan Scientific Societies Press & Elsevier.

Kidokoro, S., Uedaira, H., & Wada, A. (1988). Determination of thermodynamic functions from scanning calorimetry data. II. For the system that includes self-dissociation/association process. *Biopolymers*, *27*, 271–297.

Kidokoro, S., & Wada, A. (1987). Determination of thermodynamic functions from scanning calorimetry data. *Biopolymers*, *26*, 213–229.

Kuroda, Y., Kidokoro, S., & Wada, A. (1992). Thermodynamic characterization of cytochrome *c* at low pH. Observation of the molten globule state and of the cold denaturation process. *Journal of Molecular Biology*, *223*, 1139–1153.

Lin, L. N., Brandts, J. F., Brandts, J. M., & Plotnikov, V. (2002). Determination of the volumetric properties of proteins and other solutes using pressure perturbation calorimetry. *Analytical Biochemistry*, *302*, 144–160.

Nakamura, S., Baba, T., & Kidokoro, S. (2007). A molten globule-like intermediate state detected in the thermal transition of cytochrome *c* under low salt concentration. *Biophysical Chemistry*, *127*, 103–112.

Nakamura, S., & Kidokoro, S. (2004). Isothermal acid-titration calorimetry for evaluating the pH dependence of protein stability. *Biophysical Chemistry*, *229*, 229–249.

Nakamura, S., & Kidokoro, S. (2005). Direct observation of the enthalpy change accompanying the native to molten-globule transition of cytochrome *c* by using isothermal acid-titration calorimetry. *Biophysical Chemistry*, *113*, 161–168.

Nakamura, S., & Kidokoro, S. (2012). Volumetric properties of the molten globule state of cytochrome *c* on the thermal three-state transition evaluated by pressure perturbation calorimetry. *The Journal of Physical Chemistry. B*, *116*, 1927–1932.

Nakamura, S., Koga, S., Shibuya, N., Seo, K., & Kidokoro, S. (2013). A new multi-binding model for isothermal titration calorimetry analysis of the interaction between adenosine 5′-triphosphate and magnesium ion. *Thermochimica Acta*, *563*, 82–89.

Nakamura, S., Seki, Y., Katoh, E., & Kidokoro, S. (2011). Thermodynamic and structural properties of the acid molten globule state of horse cytochrome *c*. *Biochemistry*, *50*, 3116–3126.

Ohgushi, M., & Wada, A. (1983). 'Molten-globule state': A compact form of globular proteins with mobile side-chains. *FEBS Letters*, *164*, 21–24.
Potekhin, S., & Pfeil, W. (1989). Microcalorimetric studies of conformational transitions of ferricytochrome c in acidic solution. *Biophysical Chemistry*, *34*, 55–62.
Ptitsyn, O. B. (1995). Molten globule and protein folding. *Advances in Protein Chemistry*, *47*, 83–229.

CHAPTER SIXTEEN

The Complementarity of the Loop to the Stem in DNA Pseudoknots Gives Rise to Local TAT Base-Triplets

Calliste Reiling-Steffensmeier, Luis A. Marky[1]
Department of Pharmaceutical Sciences, University of Nebraska Medical Center, Omaha, Nebraska, USA
[1]Corresponding author: e-mail address: lmarky@unmc.edu

Contents

Abstract

Pseudoknots belong to an RNA structural motif that has significant roles in the biological function of RNA. An example is ribosomal frameshifting; in this mechanism, the formation of a local triplex changes the reading frame that allows for differences in the translation of mRNAs. In this work, we have used a combination of temperature-dependent UV spectroscopy and differential scanning calorimetry (DSC) to determine the unfolding thermodynamics of a set of DNA pseudoknots with the following sequence: d(TCTCTT_nAAAAAAAAGAGAT_5TTTTTTT), where "T_n" is a thymine loop with $n=5$ (*PsK-5*), 7 (*PsK-7*), 9 (*PsK-9*), or 11 (*PsK-11*). All four oligonucleotides form intramolecular pseudoknots, and the increase in the length of this loop yielded more stable pseudoknots due to higher transition temperatures and higher unfolding enthalpies. This indicates formation of one and three TAT/TAT stacks in *PsK-9* and *PsK-11*,

Methods in Enzymology, Volume 567
ISSN 0076-6879
http://dx.doi.org/10.1016/bs.mie.2015.07.029

respectively. We have flipped one AT for a TA base pair in the core stem of these pseudoknots, preventing in this way the formation of these base-triplet stacks. The DSC curves of these pseudoknots yielded lower unfolding enthalpies, confirming the formation of a local triplex in *PsK-9* and *PsK-11*. Furthermore, we have investigated the reaction of *PsK-5* and *PsK-9* with their partially complementary strands: directly by isothermal titration calorimetry and indirectly by creating a Hess cycle with the DSC data. Relative to the *PsK-5* reaction, *PsK-9* reacts with its complementary strand with less favorable free energy and enthalpy contributions; this indicates *PsK-9* is more stable and more compact due to the formation of a local triplex.

1. INTRODUCTION

The formation and prediction of secondary structures such as hairpins, triplexes, and pseudoknots are dependent on sequence; for example, a single-stranded DNA molecule is able to form an intramolecular structure if it has a sequence that is partially complementary to itself. It is essential to understand how this folding takes place to fully understand its biological function. A complete physical description of a particular DNA complex depends on the contributions from base pairing, base stacking, as well as ion binding and hydration. The formation and biological functions of these structures are well documented (Bock, Griffin, Latham, Vermaas, & Toole, 1992; Crooke, 1999; Firulli, Maibenco, & Kinniburgh, 1994; Fox, 1990; Gehring, Leroy, & Gueron, 1993; Han & Hurley, 2000; Helene, 1991, 1994; Juliano, Astriab-Fisher, & Falke, 2001; Kankia & Marky, 2001; Kaushik, Suehl, & Marky, 2007; Lee, Khutsishvili, & Marky, 2010; Mills et al., 2002; Rando et al., 1995; Rich, 1993; Soto, Loo, & Marky, 2002; Wang, Krawczyk, Bischofberger, Swaminathan, & Bolton, 1993). In particular, pseudoknots are an interesting and diverse RNA structural motif. This is due to the variation in their stem sequence, the fluctuation of loop length and sequence, as well as their interactions among themselves. Pseudoknots have been found to be involved in a variety of biological functions including telomerase (Huard & Autexier, 2002; Theimer, Blois, & Feigon, 2005), riboswitches (Souliere, Altman, & Micura, 2013), ribosomal frameshifting (Chen, Chang, Chou, Bustamante, & Tinoco, 2009; Huang, Yang, Wang, Cheng, & Du, 2014), self-splicing introns (Adams, Stahley, Kosek, Wang, & Strobel, 2004), and forming the catalytic core of various ribozymes (Isambert & Siggia, 2000; Ke, Zhou, Ding, Cate, & Doudna, 2004; Rastogi, Beattie, Olive, & Collins, 1996). In ribosomal frameshifting, a common mechanism found in viruses, a pseudoknot and slippery sequence are involved to change the reading frame allowing for different mRNAs to be translated (Huang et al., 2014). The high efficiency of frameshifting could be due to the formation of a local base-triplet stack within the pseudoknot to increase the stability of the terminal stem and its interaction with the ribosome (Chen et al., 2009).

Biological regulation can be accomplished by the targeting of pseudoknots with complementary nucleic acid oligonucleotides (ODNs) (Beal & Dervan, 1991; Folini, Pennati, & Zaffaroni, 2002; Khutsishvili et al., 2014; Lee, Olsen, Waters, Sukup, & Marky, 2008; Zahler, Williamson, Cech, & Prescott, 1991). ODNs, as drugs, are able to discriminate targets that differ by a single base, which presents an exquisite selectivity that can be used to control the expression of genes (Beal & Dervan, 1991; Crooke, 1999; Helene, 1991, 1994). There are three main approaches for the use of ODNs as modulators of gene expression: the antisense, antigene, and small interfering RNA strategies (Beal & Dervan, 1991). In the antisense strategy, an ODN binds to messenger RNA, forming a DNA/RNA hybrid duplex that inhibits translation by blocking the assembly of the translation machinery or by inducing an RNase H-mediated cleavage of their mRNA target (Crooke, 1999). In the antigene strategy, an ODN binds to the major groove of a DNA duplex, forming a triple helix (Mahato, Cheng, & Guntaka, 2005) that inhibits transcription, by competing with the binding of proteins that activate the transcriptional machinery (Helene, 1991; Soyfer & Potaman, 1996).

From a thermodynamic point of view, successful control of gene expression depends on the effective binding of a DNA oligonucleotide sequence to its target with tight affinity and specificity. This is provided by using a long sequence of 15–20 bases in length when targeting genes (Crooke, 1999); strong specificity is conferred by hydrogen bonding in the formation of Watson–Crick and/or Hoogsteen base pairs, while high affinity is provided by the large negative free energy upon formation of a duplex or triplex products, thereby competing efficiently with the proteins involved in transcription or translation. In the successful targeting of nucleic acid secondary structures with complementary strands, the strand must be able to invade and disrupt the secondary structure forming a large number of base-pair stacks in the duplex products.

Currently, our laboratory is interested in both the putative structure and overall physical properties for the folding (and unfolding) of nucleic acid stem–loop motifs. Our current understanding has been enhanced by thermodynamic investigations of the helix–coil transitions of model oligonucleotide compounds of known sequence, on both the stability and structure of DNA and RNA (Breslauer, Frank, Blocker, & Marky, 1986; Rentzeperis,

Kupke, & Marky, 1994; SantaLucia, Allawi, & Seneviratne, 1996; Sugimoto et al., 1995; Xia et al., 1998). Our laboratory is primarily focused on understanding the folding/unfolding of single-stranded DNA oligomers that are designed specifically to adopt intramolecular structures (Lee et al., 2011, 2008; Marky et al., 2007). Intramolecular complexes unfold with transition temperatures higher than their bimolecular counterparts due to a lower entropy penalty (Lee et al., 2011); this allows for the investigation of their physical properties over a wider temperature range.

In this work, we have determined the unfolding thermodynamics of a set of DNA pseudoknots as a function of the number of thymines in one of their loops. The increase in the length of this loop yielded pseudoknots with higher transition temperatures and higher unfolding enthalpies, indicating the formation of local TAT/TAT stacks in *PsK-9* and *PsK-11*. This observation is further confirmed in two different ways: (1) by flipping one AT for a TA base pair in the core stem of these pseudoknots, which prevents the formation of base-triplet stacks. We obtained lower unfolding enthalpies with these modified pseudoknots, and (2) by investigating the reaction of *PsK*-5 and *PsK-9* with their partially complementary strands. The reaction of *PsK-9* with its complementary strand takes place with lower favorable free energy and enthalpy contributions; this indicates *PsK-9* is more stable and more compact due to the formation of a local triplex.

2. MATERIALS AND METHODS

2.1 Materials

All oligonucleotides were synthesized by the Integrated DNA Technologies (IDT) (Coralville, IA), HPLC purified, and desalted by column chromatography using G-10 Sephadex exclusion chromatography. The sequences of oligonucleotides used in this work and their designation are shown in Table 1. The concentrations of the oligomer solutions were determined at 260 nm and 90 °C using an Aviv Spectrophotometer Model 14DS UV–Vis using the molar extinction coefficients shown in the last column of Table 1. These values were obtained by extrapolation of the tabulated values for dimers and monomeric bases (Borer, 1975; Cantor, Warshaw, & Shapiro, 1970) at 25–90 °C using procedures reported previously (Marky, Blumenfeld, Kozlowski, & Breslauer, 1983; Marky et al., 2007). The extinction coefficients of the duplexes (not shown in this table) are simply calculated by averaging the molar extinction coefficients of its component complementary single strands. Inorganic salts from Sigma were

Table 1 Sequences, Designations, and Molar Extinction Coefficients

Sequence	Name	ε_{260} (mM^{-1} cm^{-1})
d(TCTCTT_5AAAAAAAAGAGAT_5TTTTTTT)	*PsK-5*	331
d(TCTCTT_7AAAAAAAAGAGAT_5TTTTTTT)	*PsK-7*	344
d(TCTCTT_9AAAAAAAAGAGAT_5TTTTTTT)	*PsK-9*	367
d(TCTCTT_{11}AAAAAAAAGAGAT_5TTTTTTT)	*PsK-11*	376
d(CTTTTTTTTAAAAAAGAGA)	*CS-5*	181
d(CTTTTTTTTAAAAAAAAAAGAGA)	*CS-9*	237
d(TCTCTT_7AAAAATAAGAGAT_5TATTTTT)	*PsK-9 6TA*	363
d(TCTCTT_{11}AAAAATAAGAGAT_5TATTTTT)	*PsK-11 6TA*	374

reagent grade and used without further purification. Typical measurements were made in appropriate buffer solutions: 10 m*M* sodium phosphate with 100 m*M* NaCl at pH 7.0. All oligonucleotide solutions were prepared by dissolving the dry and desalted ODNs in buffer.

2.2 UV Melting Curves

Absorbance versus temperature profiles were measured at 260 nm with a thermoelectrically controlled Aviv Spectrophotometer Model 14DS UV–Vis (Lakewood, NJ). The temperature was scanned at a heating rate of 0.6 °C/min, and shape analysis of the melting curves yielded transition temperatures, T_Ms (Marky & Breslauer, 1987). The transition molecularity for the unfolding of a particular complex was obtained by monitoring T_M as a function of the strand concentration. Intramolecular complexes show a T_M independence on strand concentration, while the T_M of intermolecular complexes does depend on strand concentration (Marky et al., 2007).

2.3 Differential Scanning Calorimetry

The total heat required for the unfolding of each oligonucleotide (pseudoknot, single strand, or duplex product) was measured with a VP-DSC differential scanning calorimeter from Microcal (Northampton, MA). Standard thermodynamic profiles and T_Ms are obtained from the differential scanning calorimetry (DSC) experiments using the following relationships (Marky & Breslauer, 1987; Marky et al., 2007): $\Delta H_{cal} = \int \Delta C_p(T) dT$, $\Delta S_{cal} = \int \Delta C_p(T)/T dT$, and the Gibbs equation, $\Delta G^{\circ}_{(T)} = \Delta H_{cal} - T\Delta S_{cal}$, where ΔC_p is the anomalous heat capacity of the ODN solution during the unfolding process, ΔH_{cal} and ΔS_{cal} are the unfolding enthalpy and entropy, respectively, assumed to be temperature independent, and $\Delta G^{\circ}_{(T)}$ is the free energy at temperature T, normally 5 °C.

2.4 Isothermal Titration Calorimetry

The heat for the reaction of a pseudoknot with its complementary strand was measured directly by isothermal titration calorimetry (ITC) using the ITC_{200} from GE Microcal (Northampton, MA). A 40-μL syringe was used to inject the titrant; mixing was effected by stirring this syringe at 1000 rpm. Typically, we used five to seven injections of 2 μL of pseudoknot solution with at least twofold lower concentration than the solution of the complementary strand in the cell, and over a time of 4–8 min between injections. The reaction heat of each injection is measured by integration of the area of the injection curve, corrected for the dilution heat of the titrant, and normalized by the moles of titrant added to yield the reaction enthalpy, ΔH_{ITC} (Khutsishvili et al., 2014; Lee et al., 2011, 2008; Wiseman, Williston, Brandts, & Lin, 1989). All titrations were designed to obtain mainly the ΔH_{ITC} for each targeting reaction, by averaging at least five injections, which correspond to the formation of duplex products. To determine the free energy, ΔG_{ITC}, for each targeting reaction, we use the following relationship, $\Delta G_{ITC} = \Delta G_{HC}$ $(\Delta H_{ITC}/\Delta H_{HC})$ (Khutsishvili et al., 2014; Lee et al., 2011, 2008), while the Gibbs equation is used to determine the $T\Delta S_{ITC}$ parameter, where T is the temperature of the ITC experiments.

2.5 Overall Experimental Approach

We are investigating the thermodynamic stability of a set of DNA pseudoknots (Fig. 1A) with a variable loop length, which is complementary to the stem portion of the pseudoknot. Initially, we use UV melting techniques to follow the temperature unfolding of each molecule. We test if this unfolding takes place intramolecularly, by following the dependence of the transition temperature, T_M, on strand concentration. Then, we use DSC to obtain thermodynamic profiles for the unfolding of each molecule. The comparison of the resulting thermodynamic data yielded the loop contributions as its length increases. The sequence of the core stem of two pseudoknots, *PsK-9* and *PsK-11*, was changed by flipping one AT base pair of the homopurine/homopyrimidine stretch ($^{5'}$-A_7/T_7 to $^{5'}$-A_5TA/TAT_5) on the right of this stem (Fig. 2A), preventing in this way the formation

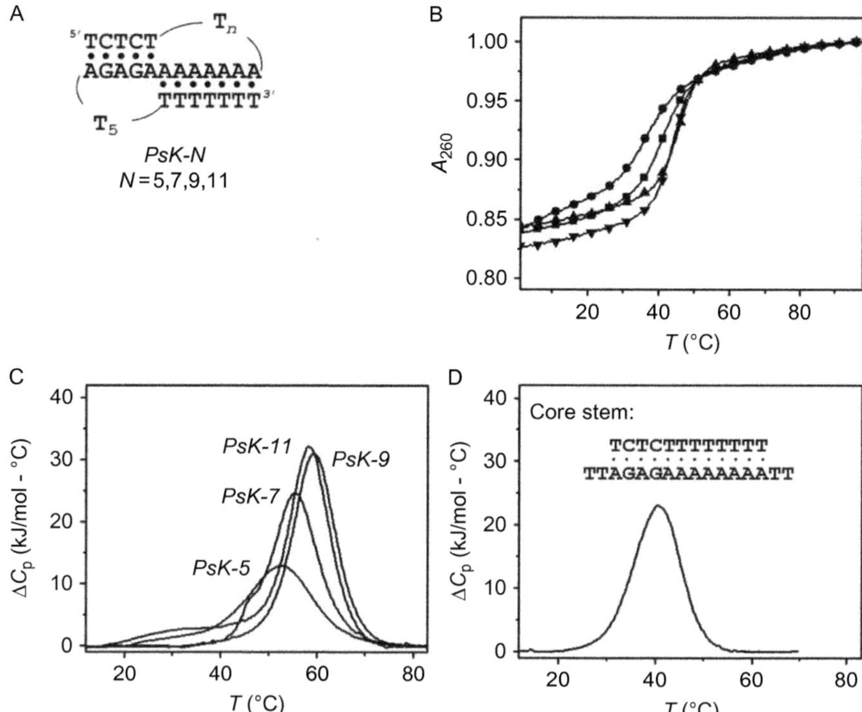

Figure 1 Unfolding of pseudoknots. (A) Cartoon of DNA Pseudoknots, (B) UV melting curves (•, *PsK-5*; ■, *PsK-7*; ▲, *PsK-9*; ▼, *PsK-11*), (C) DSC unfolding of pseudoknots, concentrations used: 126 μ*M* (*PsK-5*), 73 μ*M* (*PsK-7*), 120 μ*M* (*PsK-9*), and 53 μ*M* (*PsK-11*), (D) DSC unfolding of the core stem duplex with concentration, 120 μ*M*. All experiments were carried out in 10 m*M* NaPi, 100 m*M* NaCl at pH 7.0.

of base-triplet stacks. The lower magnitude of the unfolding enthalpies would tell us the absence of a local triplex. We use ITC and DSC (Khutsishvili et al., 2014; Lee et al., 2011, 2008) to investigate the reaction of two pseudoknots, *PsK*-5 and *PsK-9*, with their partially complementary strands, the comparison of their resulting thermodynamic data confirms whether or not a triplex is formed in the pseudoknot with the longer loop.

3. RESULTS AND DISCUSSION

3.1 Unfolding of Pseudoknots with Increasing Loop Length

The melting curves for pseudoknots with loops of five to nine thymines show monophasic transitions with T_Ms ranging from 53 to 59 °C

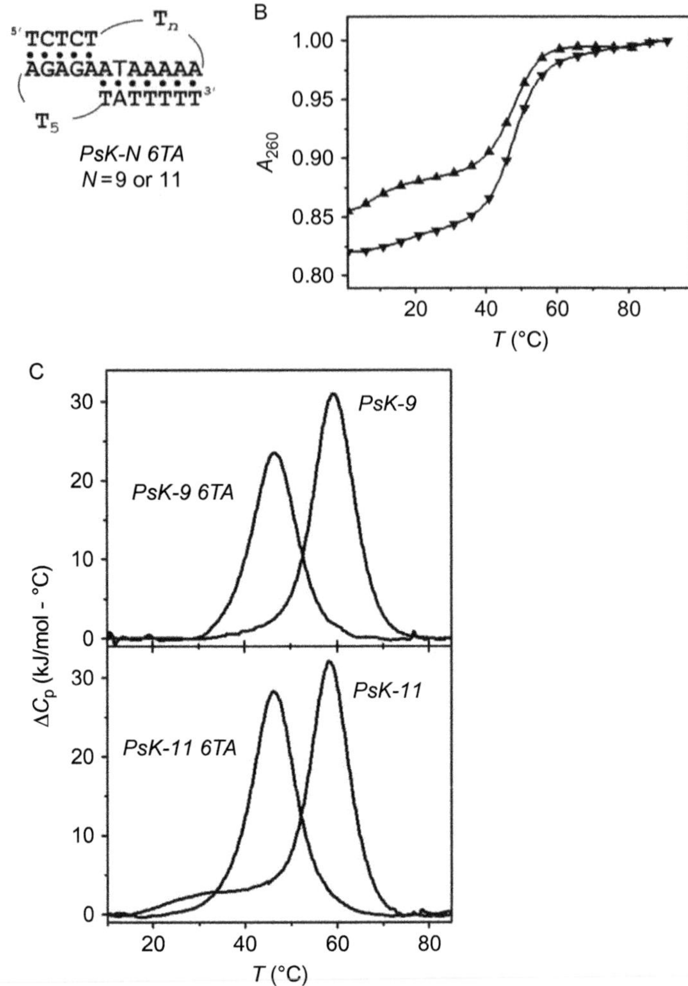

Figure 2 Unfolding of pseudoknots with a flipped AT base pair in the core stem. (A) Cartoon of modified pseudoknots, (B) UV melting curves of pseudoknots (▲, *PsK-9-6TA*; ▼, *PsK-11-6TA*), (C) DSC curves of *PsK-9* (120 μ*M*) and *PsK-9-6TA* (26 μ*M*) (top); *PsK-11* (52 μ*M*) and *PsK-11-6TA* (43 μ*M*) (bottom). All experiments were carried out in 10 m*M* NaPi, 100 m*M* NaCl at pH 7.0.

(Fig. 1A and B). This equates to an increase in 1 °C per thymine in the loop. However, when reaching a loop length of 11 thymines, there is a biphasic transition with T_Ms of 32 and 59 °C. These pseudoknots also have similar hyperchromicities of 15–17% at 260 nm. The main observation is that there is an increase in the T_M as the loop length of the pseudoknots increases, which is the opposite effect that is normally seen with increasing the loop length of straight hairpin loops (Reiling, Khutsishvili, Huang, & Marky, 2015; Rentzeperis, 1994). This can be rationalized by the helical stems becoming less accessible to the solvent due to the loop thymines constituting the ceiling of the nearby major groove, thereby making the pseudoknots more temperature stable. We follow the T_M as a function of strand concentration to determine the transition molecularity. Each pseudoknot forms intramolecularly due to similar T_Ms over a 10-fold increase in strand concentration (data not shown).

The DSC unfolding for these pseudoknots is shown in Fig. 1C, most of the pseudoknots unfold through asymmetric monophasic transitions due to the similar T_Ms of the two adjacent stems; however, *PsK-11* unfolds through a biphasic transition. The increase in the length of the right-side loop (Fig. 1A) from 5 to 11 thymines yielded total ΔH_{cal}s of 251 (*PsK-5*), 361 (*PsK-7*), 366 (*PsK-9*), and 460 kJ/mol (*PsK-11*) (Table 2). The predicted enthalpies from N–N parameters are 365 (SantaLucia, 1998) and 379 kJ/mol (Breslauer et al., 1986) for the duplex stem of this set of pseudoknots; however, we obtained experimentally a ΔH_{cal} of 290 kJ/mol in 116 m*M* Na^+ for this stem duplex with two dangling thymines at each end (TCTCTTTTTTTT/TTAAAAAAAGAGATT), in order to prevent fraying (Senior, Jones, & Breslauer, 1988; Fig. 1D). We use a value of 273 instead of the 290 kJ/mol for the unfolding of the core stem of the pseudoknots; this assumes that the middle AA/TT of each pseudoknot is considered as partially stacked. *PsK-5* unfolds with a ΔH_{cal} of 251 kJ/mol, 22 kJ/mol lower than the core stem, due to the constrained loops inducing unstacking of the stem and/or fraying at the 3′ end of the AT base pairs. It is important to emphasize that the loop is complementary to the

Table 2 Thermodynamic Unfolding Profiles for DNA Pseudoknots

Molecule		T_M (°C)	ΔH_{cal} (kJ/mol)	$T\Delta S_{cal}$ (kJ/mol)	ΔG°_{5} (kJ/mol)
PsK-5		52.8	251	216	35
PsK-7		56.2	361	306	55
PsK-9		59.3	366	307	59
PsK-11	1st	31.6	110	98	12
	2nd	58.7	350	293	57
Total			460	391	69

All experiments were carried out in 10 m*M* phosphate buffer at pH 7.0 and 100 m*M* NaCl. Experimental errors are as follows: T_M (±0.5 °C), ΔH (±5%), $T\Delta S$ (±5%), $\Delta G^{\circ}_{(5)}$ (±7%).

corresponding stem, nearby major groove of this stem, and the increase in the length of the loop on the right side (Fig. 1) yielded higher enthalpy terms. Relative to the control core stem ($\Delta H_{cal} = 273$ kJ/mol), we obtained enthalpy increases of 88 (*PsK-7*), 93 (*PsK-9*), and 187 kJ/mol (*PsK-11*), respectively. These enthalpies can be attributed to improving base-pair stacking of the stem, closing the fraying of the ends, and the local formation of TAT base triplets. We have estimated an enthalpy of 21 kJ/mol for the first two contributions, yielding excess enthalpies of 67 (*PsK-7*), 72 (*PsK-9*), and 166 kJ/mol (*PsK-11*). The enthalpy of a TAT/TAT base-triplet stack has been determined to be equal to 100 kJ/mol (Soto et al., 2002), and the enthalpy of the AA/TT base-pair stack is considered to be ~36 kJ/mol (Alessi, 1995; Breslauer et al., 1986; SantaLucia, 1998). This means that an enthalpy of 64 kJ/mol is needed to form a single base-triplet stack from the addition of the third strand (thymine loop) to complement the A_7/T_7 duplex stem. By dividing the excess enthalpy of each pseudoknot by 64 kJ/mol, we estimate that *PsK-7* and *PsK-9* form one TAT base-triplet stack, while *PsK-11* forms ~2.5 TAT/TAT base-triplet stacks.

3.2 Unfolding of Pseudoknots with a Flipped AT Base Pair in the Core Stem

To confirm the formation of base triplets in *PsK-9* and *PsK-11*, the sequence of the core stem of these pseudoknots (Fig. 1A) was changed by flipping one AT base pair of the homopurine/homopyrimidine stretch ($^{5'}$-A_7/T_7 to $^{5'}$-A_5TA/TAT_5) on the right side of this stem, yielding the *PsK-9-6TA* and *PsK-11-6TA* pseudoknots, shown in Fig. 2A. This disrupts a triplex helix because the shift of the third strand thymine to the other side of the major groove by ~10 Ås (Li, Fan, Zhang, Marky, & Gold, 2003) causes a triplex destabilization that spans five base triplets (Shikiya & Marky, 2005). The melting curves of *PsK-9-6TA* and *PsK-11-6TA* are monophasic with T_Ms around 46 °C and similar hyperchromicities of 16–17% at 260 nm (Fig. 2B). This indicates that the flip of an AT base pair causes a 13 °C decrease in their T_M relative to *PsK-9* and *PsK-11*. This can be rationalized in terms of a higher exposure to the solvent of both the loop and stem bases. Furthermore, the T_Ms remain the same for each pseudoknot over a 10-fold increase in strand concentration, indicating their intramolecular formation (data not shown).

PsK-9-6TA and *PsK-11-6TA* unfold through monophasic transitions due to the similar T_Ms of the left and right sequences in their stems (Fig. 2C). The increase in the length of the right-side loop from 9 to

Table 3 Thermodynamic Unfolding Profiles of Pseudoknots with a Flipped AT Base Pair in the Core Stem

Molecule		T_M (°C)	ΔH_{cal} (kJ/mol)	$T\Delta S_{cal}$ (kJ/mol)	ΔG°_{5} (kJ/mol)
PsK-9		59.3	366	307	59
PsK-9 6TA		46.6	302	262	40
PsK-11	1st	31.6	110	98	12
	2nd	58.7	350	293	57
Total			460	391	69
PsK-11 6TA		46.3	361	314	47

All experiments were carried out in 10 m*M* phosphate buffer at pH 7.0 and 100 m*M* NaCl. Experimental errors are as follows: T_M (±0.5 °C), ΔH (±5%), $T\Delta S$ (±5%), $\Delta G^{\circ}_{(5)}$ (±7%).

11 thymines yielded total ΔH_{cal}s of 302 (*PsK-9-6TA*) and 361 kJ/mol (*PsK-11-6TA*), Table 3. These are significant drops, 64 and 99 kJ/mol, relative to the enthalpies of the wild-type pseudoknots. The decrease in enthalpy can be attributed to the disruption of the base-triplet stacks due to the instability cause by the AT flipped base pair in this stem. Since the destacking of a third strand thymine reduces the total enthalpy by 64 kJ/mol (Soto et al., 2002), we can confirm that *PsK-9* is forming one TAT/TAT stack. Similar estimation indicates that *PsK-11* forms 1.5 TAT/TAT stacks, lower than suggested earlier. This apparent contradiction may be explained in terms of additional stacking contributions in the 11-thymine loop of *PsK-11-6TA* and lower enthalpy of its core stem (replacement of two AA/TT base-pair stacks for one AT/AT and one TA/TA base-pair stacks) and higher flexibility of this pseudoknot.

3.3 The Reactions of *PsK-5* and *PsK-9* with Complementary Strands Are Favorable

The reactions of *PsK-5* and *PsK-9* with their partially complementary strands were investigated directly by ITC, Fig. 3A, to further confirm if formation of a triplex helix is occurring when the loop length is greater than five thymines. The heat for each reaction was measured by ITC under unsaturated conditions, using ODN concentrations and temperatures that guaranteed 100% formation of the final duplex products (Fig. 3A). The ITC titrations are shown in Fig. 3B; the shape of these curves shows that the initial enthalpies are more exothermic, −134 and −138 kJ/mol, and

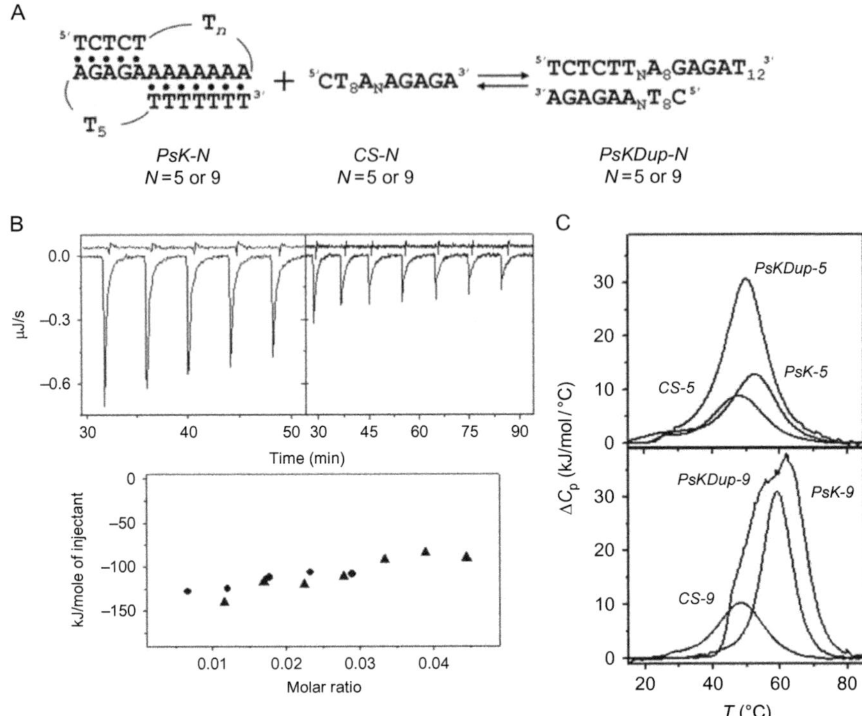

Figure 3 Thermodynamic data of the reaction of pseudoknots with partially complementary strands. (A) Cartoon of each reaction, (B) ITC data for the formation of duplex products, concentrations used: 65 μ*M* (*PsK-5*), 123 μ*M* (*CS-5*), 67 μ*M* (*PsK-9*), and 128 μ*M* (*CS-9*); and (C) DSC curves of reactants and products, concentrations used: 126 μ*M* (*PsK-5*), 182 μ*M* (*CS-5*), 39 μ*M* (*PsKDup-5*), 120 μ*M* (*PsK-9*), 136 μ*M* (*CS-9*), and 25 μ*M* (*PsKDup-9*). All experiments were carried out in 10 m*M* NaPi, 100 m*M* NaCl at pH 7.0.

gradually reaching a plateau at −114 and −89 kJ/mol, respectively. The average heat of all injections for each reaction yielded ΔH_{ITC} values of −118 (*PsK-5*) and −99 kJ/mol (*PsK-9*); the net exothermicity of these enthalpy values corresponds to a complete override of the endothermic heat contributions (disruption of the base-pair stacks of both pseudoknot and hairpin loop) by the exothermic heat contributions (formation of base-pair stacks of the duplex product). An extra exothermic/endothermic term should be included due to hydration changes from the participating reaction species, which may be accounted for the variability of the reaction enthalpies in these titrations.

We also investigated these two reactions indirectly using DSC. Figure 3C shows the DSC thermograms for the reactants (pseudoknot

Table 4 Thermodynamic Unfolding Profiles for the Reactants and Products

		T_M (°C)	ΔH_{cal} (kJ/mol)	$T\Delta S_{cal}$ (kJ/mol)	$\Delta G^\circ_{(5)}$ (kJ/mol)
Reaction 1					
PsK-5		52.8	251	216	35
CS-5		47.9	161	139	22
PsKDup-5		49.6	565	485	80
Reaction 2					
PsK-9		59.3	366	307	59
CS-9		48.6	203	175	28
PsKDup-9	1st	47.7	248	215	33
	2nd	58.3	427	358	69
Total			675	573	102

All experiments were done in 10 m*M* sodium phosphate buffer and 100 m*M* NaCl at pH 7.0. Experimental errors are as follows: T_M (±0.5 °C), ΔH (±5%), $T\Delta S$ (±5%), $\Delta G^\circ_{(5)}$ (±7%).

and complementary strand) and duplex products for each reaction and Table 4 shows the unfolding thermodynamic data. The DSC of each pseudoknot was discussed in an earlier section. Each complementary strand shows small DSC peaks with T_Ms and ΔHs of 47.9 °C, 161 kJ/mol (*CS-5*) and 48.6 °C, 203 kJ/mol (*CS-9*). UV melts as a function of strand concentration (data not shown) showed that their T_Ms remain constant; we conclude that each strand is forming a hairpin loop with a dangling end. Dividing their enthalpy values by the enthalpy of a DNA base-pair stack, ~36 kJ/mol, *CS-5* is forming five base-pair stacks in its stem, while *CS-9* is forming six base-pair stacks.

The UV melts of the duplex products show a slight increase in their T_M as the strand concentration is increased (data not shown). These T_M dependences are similar to the unfolding of DNA polymers, indicating that the melting of these duplexes is approaching polymer behavior. The DSC curves of each duplex product (Fig. 3C) show that *PsKDup-5* unfolds in an apparent monophasic transition with a T_M of 49.6 °C and ΔH_{cal} of 565 kJ/mol, while *PsKDup-9* unfolds in a biphasic transition with T_Ms of 47.7 and 58.3 °C and total ΔH_{cal} of 675 kJ/mol. Each DSC profile corresponds to the unfolding of the duplex followed by the folding and sequential unfolding of the corresponding pseudoknot and hairpin (single strands). For instance, the monophasic unfolding of *PsKDup-5* is due to the similar T_Ms of the structures formed by the reactants and product of this reaction, which are within 5 °C. On the other hand, *PsKDup-9* shows a biphasic transition (Fig. 3C), and the first transition corresponds to the unfolding of the duplex into partially folded *PsK-9* and *CS-9*, followed by the simultaneous unfolding of *CS-9* and *PsK-9*. The higher ΔH term of *PsKDup-9*, by 110 kJ/mol, when compared with *PsKDup-5* corresponds to the formation of three additional base-pair stacks.

We created Hess cycles with the unfolding data shown in Table 5 to generate indirectly thermodynamic profiles for each reaction; i.e., we added the thermodynamic profiles of the pseudoknot and single strand (hairpin) and subtracted the thermodynamic profiles of the duplex. The resulting data are shown in the last two entries of Table 5. This exercise yielded ΔG°_{HC} and ΔH_{HC} of −21 and −153 kJ/mol (*PsK-5*) and −14 and −105 kJ/mol (*PsK-9*), respectively. Both reactions are favorable and enthalpy driven. However, the targeting of *PsK-9* is less favorable, which is consistent with the higher stability of this pseudoknot. Furthermore, we obtained unfavorable $T\Delta S_{HC}$ terms of −132 (*PsK-5*) and −90 kJ/mol (*PsK-9*), which correspond to the net uptake of ions and water molecules by the duplex products of each reaction, since the conformational entropy change is considered similar for each reaction. Furthermore, the ΔH_{ITC}s of −118 (*PsK-5*) and −99 kJ/mol (*PsK*-9) obtained directly in the ITC experiments are in good agreement with the ΔH_{HC} values of −153 and −105 kJ/mol, respectively, obtained indirectly by the Hess cycles from the DSC data.

To obtain the free energy terms at the temperature of the ITC titrations, ΔG°_{ITC} at 5 °C, the ΔG°_{HC} values are multiplied by a temperature factor ($=\Delta H_{ITC}/\Delta H_{HC}$), which assumes ΔH_{HC}s to be independent of temperature, i.e., $\Delta Cp=0$. The $T\Delta S_{ITC}$ parameters were calculated using the Gibbs equation. The overall results are shown in Table 4. We obtained favorable

Table 5 Standard Thermodynamic Profiles of the Targeting Reactions

	ITC		Hess Cycle		
	ΔH_{ITC} (kJ/mol)	ΔG°_{ITC} (kJ/mol)	ΔH_{HC} (kJ/mol)	ΔG°_{HC} (kJ/mol)	$T\Delta S_{HC}$ (kJ/mol)
Reaction 1	−118	−16	−153	−21	−132
Reaction 2	−99	−13	−105	−15	−90

All experiments were done in 10 m*M* sodium phosphate buffer and 100 m*M* NaCl at pH 7.0. Experimental errors are as follows: ΔH_{ITC} (±5%), ΔG°_{ITC} (±7%), ΔH_{HC} (±10%), $T\Delta S_{HC}$ (±10%), ΔG°_{HC} (±14%).

ΔG°_{ITC} contributions for each reaction, and each complementary strand is able to invade and disrupt the pseudoknot structure. However, the reaction with *PsK-9* is less favorable in spite of forming a more stable duplex, by 3 kJ/mol. This result is consistent with its higher stability and the formation of a local triplex.

4. CONCLUSIONS

In this chapter, we used a set of DNA oligonucleotides to model the folding of pseudoknots, which is an important motif in the biology of RNA. We used DNA pseudoknots containing a variable number of thymines in one of the loops, complementary to the core sequence of the duplex stem. Our biological question is if this particular set of pseudoknots can illustrate the formation of local base triplets and base-triplet stacks, explaining to some extent the high efficiency in ribosomal frameshifting. Our experimental approach is to use a combination of temperature-dependent UV spectroscopy and calorimetry techniques (DSC and ITC) to determine the unfolding and targeting thermodynamics of these pseudoknots. The first observation is that DNA strands are able to form intramolecular pseudoknots; their transition temperatures remain the same with the 10-fold increase in strand concentration. This means that DNA strands are flexible and can mimic to some extent the tertiary structures formed by RNA molecules, which is important considering the cheaper cost of DNA strands. The unfolding DSC data showed that the increase in the length of the thymine loop yielded higher transition temperatures and higher unfolding enthalpies; this is contrary to the loop effects of straight hairpin loops and indicates that additional interactions are occurring within the pseudoknot as the loop length increases. Specifically, pseudoknots with longer thymine loops are thermodynamically more stable because the sequence complementarity of this loop with the stem (A_7T_7) allows them to form Hoogsteen base pairs. In this particular system, the magnitude of the excess enthalpies indicates formation of one and three TAT/TAT stacks in *PsK-9* and *PsK-11*, respectively. To confirm this observation, one AT base pair was flipped for a TA base pair in the A_7T_7 stem of these pseudoknots, which prevents the formation of base triplets. This resulted in modified pseudoknots with lower unfolding enthalpies, confirming the formation of a local triplex in *PsK-9* and *PsK-11*. Another way is to investigate the reaction of *PsK*-5 and *PsK-9* with their partially complementary strands using ITC. The reaction of *PsK-9* takes place with less favorable free energy and enthalpy contributions, indicating that *PsK-9* is more stable and more compact. This further confirms the formation of a local triplex because if a triplex was not forming, we would expect to see more favorable free energy and enthalpic contributions with the formation of additional base pairs in the formed duplex from the loop thymines. Triplex formation only takes place when the loop (third strand) is complementary to the homopurine–homopyrimidine stem of these pseudoknots.

Overall, this investigation confirms the flexibility of DNA oligonucleotides being able to form pseudoknots; therefore, they can be used to mimic known RNA secondary structures. In particular, we were able to show the formation of a local triplex helix that explains ribosomal frameshifting, which allows for differences in the translation of mRNAs.

"Practical considerations for the proper execution and analysis of calorimetric experiments with nucleic acids":

In this chapter, we have used DSC and ITC calorimetric techniques. The main experimental protocol to execute appropriate DSC and ITC experiments is to start with clean cells and to fill the calorimetric cells without air bubbles, which significantly impact the quality of the heat capacity measurements. It really takes some practice to master the later procedure:

- The cells can be cleaned with a mild detergent solution and rinse with water several times, and a final ethanol rinse. Flushing a strong stream of nitrogen then dries out the cells completely.
- To avoid air bubbles, the buffer or solution should be degassed and the cells filled appropriately, by using a syringe filled up with this solution without air bubbles.
- In the filling of the cells, the syringe needle should touch the bottom of the cell without scoring, to fill up its volume halfway and then do three to four injections of 0.1 mL solution to dislodge the air bubbles, if any. This procedure (pushes) can be repeated several times until confident that no air bubbles are present. The beginner investigator should practice the above procedure with water–water or buffer–buffer scans until he/she feels confident the cells are air bubbles free, by obtaining reproducible heating scans.
- The impact of dirty cells or bubbles can sometimes be observed in the ITC data. Look for stray peaks that do not coincide with normal injection peaks or single peaks that are split into two.

In the execution of oligonucleotide DSC scans, it is best to use concentrations that will produce heats in excess of 15 μJ. It is best to obtain three to four buffer–buffer scans and three to four oligomer solution versus buffer scans:

- All scans should be reproducible; one exception is the first oligomer solution versus buffer scan that deviates from the others because of the need of the cell walls to be wetted.
- Relative to the buffer versus buffer scan, the oligomer solution versus buffer scan should come out below; otherwise, it indicates the presence of air bubbles in the sample cell. The actual displacement corresponds to the exclusion of water molecules from the sample cell by the presence of solute.
- Good practice is to stay with the DSC instrument until the first scan is complete and to set up the instrument for more scans using the appropriate temperature interval and heating rates. For nucleic acid oligonucleotides, this heating rate can be set from 0.7 to 1 °C/min.
- In the analysis of DSC scans, the buffer versus buffer scan is subtracted from the oligomer solution versus buffer scan. Baselines are then drawn by extrapolating a straight line from the pre- and the posttransition baselines. This is required because heat capacity effects cannot be measured in the region of the DNA folding transition. The reason is that heat capacity differences of the helix and coil states of these molecules are within experimental error of the DSC measurement.
- To check for transition molecularity, dilute the oligomer solution in the sample cell and rescan it. Unimolecular transitions are concentration independent, whereas dilution yields lower T_Ms for higher order reactions.

The study of pseudoknots by ITC is a special case due to their partial complementarity and multistate folding. In our experience, it is best to load the binding partner with the least base stacking in the syringe. In many cases, these particular experiments can only measure reaction enthalpies (due to *c* values outside the suitable range) using solution concentrations with a 1 (pseudoknot in the syringe):2 ratio. It is best to use injections producing total heats in excess of 10 μJ.

ACKNOWLEDGMENTS

This work was supported by Grant MCB-1122029 from the National Science Foundation and GAANN Grant P200A120231 (C.R.) from the U.S. Department of Education.

REFERENCES

Adams, P. L., Stahley, M. R., Kosek, A. B., Wang, J., & Strobel, S. A. (2004). Crystal structure of a self-splicing group I intron with both exons. *Nature, 430*, 45–50.

Alessi, K. (1995). *Thermodynamic contributions of deoxyoligonucleotides containing A-T base-pairs.* Ph.D. dissertation, New York University.

Beal, P. A., & Dervan, P. B. (1991). Second structural motif for recognition of DNA by oligonucleotide-directed triple-helix formation. *Science, 251*, 1360–1363.
Bock, L. C., Griffin, L. C., Latham, J. A., Vermaas, E. H., & Toole, J. J. (1992). Selection of single-stranded DNA molecules that bind and inhibit human thrombin. *Nature, 355*, 564–566.
Borer, P. N. (1975). Optical properties of nucleic acids, absorptions and circular dichroism spectra. In G. D. Fasman (Ed.), *Handbook of biochemistry and molecular biology* (3rd ed., pp. 589–595). Cleveland, OH, USA: CRC Press.
Breslauer, K. J., Frank, R., Blocker, H., & Marky, L. A. (1986). Predicting DNA duplex stability from the base sequence. *Proceedings of the National Academy of Sciences of the United States of America, 83*, 3746–3750.
Cantor, C. R., Warshaw, M. M., & Shapiro, H. (1970). Oligonucleotide interactions. 3. Circular dichroism studies of the conformation of deoxyoligonucleotides. *Biopolymers, 9*, 1059–1077.
Chen, G., Chang, K. Y., Chou, M. Y., Bustamante, C., & Tinoco, I., Jr. (2009). Triplex structures in an RNA pseudoknot enhance mechanical stability and increase efficiency of -1 ribosomal frameshifting. *PNAS, 106*, 12706–12711.
Crooke, S. T. (1999). Molecular mechanisms of action of antisense drugs. *Biochimica et Biophysica Acta, 1489*, 31–43.
Firulli, A. B., Maibenco, D. C., & Kinniburgh, A. J. (1994). Triplex forming ability of a c-myc promoter element predicts promoter strength. *Archives of Biochemistry and Biophysics, 310*, 236–242.
Folini, M., Pennati, M., & Zaffaroni, N. (2002). Targeting human telomerase by antisense oligonucleotides and ribozymes. *Current Medicinal Chemistry. Anti-Cancer Agents, 5*, 605–612.
Fox, K. R. (1990). Long (dA)n × (dT)n tracts can form intramolecular triplexes under superhelical stress. *Nucleic Acids Research, 18*, 5387–5391.
Gehring, K., Leroy, J.-L., & Gueron, A. (1993). A tetrameric, DNA structure with protonated cytosine.cytosine base pairs. *Nature, 363*, 561–565.
Han, H., & Hurley, L. H. (2000). A potential target for anti-cancer drug design. *Trends in Pharmacological Sciences, 21*, 136–142.
Helene, C. (1991). Rational design of sequence-specific oncogene inhibitors based on antisense and antigene oligonucleotides. *European Journal of Cancer, 27*, 1466–1471.
Helene, C. (1994). Control of oncogene expression by antisense nucleic acids. *European Journal of Cancer, 30A*, 1721–1726.
Huang, X., Yang, Y., Wang, G., Cheng, Q., & Du, Z. (2014). Highly conserved RNA pseudoknots at gag-pol junction of HIV-1 suggest a novel mechanism for -1 ribosomal frameshifting. *RNA, 20*, 587–593.
Huard, S., & Autexier, C. (2002). Targeting human telomerase in cancer therapy. *Current Medicinal Chemistry. Anti-Cancer Agents, 2*, 577–587.
Isambert, H., & Siggia, E. D. (2000). Modeling RNA folding paths with pseudoknots: Application to hepatitis delta virus ribozyme. *Proceedings of the National Academy of Sciences of the United States of America, 97*, 6515–6520.
Juliano, R. L., Astriab-Fisher, A., & Falke, D. (2001). Macromolecular therapeutics: Emerging strategies for drug discovery in the postgenome era. *Molecular Interventions, 1*, 40–53.
Kankia, B. I., & Marky, L. A. (2001). Folding of the thrombin aptamer into a G-quadruplex with Sr^{2+}: Stability, heat, and hydration. *Journal of the American Chemical Society, 123*, 10799–10804.
Kaushik, M., Suehl, N., & Marky, L. A. (2007). Calorimetric unfolding of the bimolecular and i-motif complexes of the human telomere complementary strand, d(C3TA2)4. *Biophysical Chemistry, 126*, 154–164.
Ke, A., Zhou, K., Ding, F., Cate, J. H., & Doudna, J. A. (2004). A conformational switch controls hepatitis delta virus ribozyme catalysis. *Nature, 429*, 201–205.

Khutsishvili, I., Johnson, S. E., Reiling, C., Prislan, I., Lee, H. T., & Marky, L. A. (2014). Interaction of DNA intramolecular structures with their complementary strands: A thermodynamic approach for the control of gene expression. In V. A. Erdmann, et al. (Ed.), *Chemical biology of nucleic acids* (pp. 367–383). Berlin Heidelberg: Springer-Verlag. Chapter 20.

Lee, H.-T., Carr, C., Siebler, H., Waters, L., Khutsishvili, I., Iseka, F., et al. (2011). A thermodynamic approach for the targeting of nucleic acid structures using their complementary single strands. *Methods in Enzymology*, *492*, 1–26.

Lee, H.-T., Khutsishvili, I., & Marky, L. A. (2010). DNA complexes containing joined triplex and duplex motifs: Melting behavior of intramolecular and bimolecular complexes with similar sequences. *The Journal of Physical Chemistry. B*, *114*, 541–548.

Lee, H.-T., Olsen, C. M., Waters, L., Sukup, H., & Marky, L. A. (2008). Thermodynamic contributions of the reactions of DNA intramolecular structures with their complementary strands. *Biochimie*, *90*, 1052–1063.

Li, J.-S., Fan, Y.-H., Zhang, Y., Marky, L. A., & Gold, B. (2003). Design of triple helix forming C-glycoside molecules. *Journal of the American Chemical Society*, *125*, 2084–2093.

Mahato, R. I., Cheng, K., & Guntaka, R. V. (2005). Modulation of gene expression by antisense and antigene oligodeoxynucleotides and small interfering RNA. *Expert Opinion on Drug Delivery*, *2*, 3–28.

Marky, L. A., Blumenfeld, K. S., Kozlowski, S., & Breslauer, K. J. (1983). Salt-dependent conformational transitions in the self-complementary deoxydodecanucleotide d(CGCGAATTCGCG): Evidence for hairpin formation. *Biopolymers*, *22*, 1247–1257.

Marky, L. A., & Breslauer, K. J. (1987). Calculating thermodynamic data for transitions of any molecularity from equilibrium melting curves. *Biopolymers*, *26*, 1601–1620.

Marky, L. A., Maiti, S., Olsen, C. M., Shikiya, R., Johnson, S. E., Kaushik, M., et al. (2007). Building blocks of nucleic acid nanostructures: unfolding thermodynamics of intramolecular DNA complexes. In V. Labhasetwar & D. Leslie-Pelecky (Eds.), *Biomedical applications of nanotechnology* (pp. 191–225): John Wiley & Sons, Inc.

Mills, M., Lacroix, L., Arimondo, P. B., Leroy, J.-L., Francois, J.-C., Klump, H., et al. (2002). Unusual DNA conformations: Implications for telomeres. *Current Medicinal Chemistry: Anti-Cancer Agents*, *2*, 627–644.

Rando, R. F., Ojwang, J., Elbaggari, A., Reyes, G. R., Tinder, R., McGrath, M. S., et al. (1995). Suppression of human immunodeficiency virus type 1 activity in vitro by oligonucleotides which form intramolecular tetrads. *The Journal of Biological Chemistry*, *270*, 1754–1760.

Rastogi, T., Beattie, T. L., Olive, J. E., & Collins, R. A. (1996). A long-range pseudoknot is required for activity of the Neurospora VS ribozyme. *The EMBO Journal*, *15*, 2820–2825.

Reiling, C., Khutsishvili, I., Huang, K., & Marky, L. A. (2015). Effect of loop length on DNA pseudoknot stability. *The Journal of Physical Chemistry. B*, *119*, 1939–1946.

Rentzeperis, D. (1994). *Thermodynamics and ligand interactions of DNA hairpins*. Ph.D. dissertation, New York University.

Rentzeperis, D., Kupke, D. W., & Marky, L. A. (1994). Differential hydration of dA.dT base pairs in parallel-stranded DNA relative to antiparallel DNA. *Biochemistry*, *33*, 9588–9591.

Rich, A. (1993). DNA comes in many forms. *Gene*, *135*, 99–109.

SantaLucia, J., Jr. (1998). A unified view of polymer, dumbbell, and oligonucleotide DNA nearest-neighbor thermodynamics. *Biochemistry*, *95*, 1460–1465.

SantaLucia, J., Jr., Allawi, H. T., & Seneviratne, P. A. (1996). Improved nearest-neighbor parameters for predicting DNA duplex stability. *Biochemistry*, *35*, 3555–3562.

Senior, M., Jones, R. A., & Breslauer, K. J. (1988). Influence of dangling thymidine residues on the stability and structure of two DNA duplexes. *Biochemistry*, *27*, 3879–3885.

Shikiya, R., & Marky, L. A. (2005). Calorimetric unfolding of intramolecular triplexes: Length dependence and incorporation of single AT–TA substitutions in the duplex domain. *The Journal of Physical Chemistry. B, 109*, 18177–18183.

Soto, A. M., Loo, J., & Marky, L. A. (2002). Energetic contributions for the formation of TAT/TAT, TAT/CGC+, and CGC+/CGC+ base triplet stacks. *Journal of the American Chemical Society, 124*, 14355–14363.

Souliere, M. F., Altman, R. B., & Micura, R. (2013). Tuning a riboswitch response through structural extension of a pseudoknot. *PNAS, 110*, 3256–3264.

Soyfer, V. N., & Potaman, V. N. (1996). *Triple-helical nucleic acids.* New York: Springer-Verlag.

Sugimoto, N., Nakano, S., Katoh, M., Matsumura, A., Nakamuta, H., Ohmichi, T., et al. (1995). Thermodynamic parameters to predict stability of RNA/DNA hybrid duplexes. *Biochemistry, 34*, 11211–11216.

Theimer, C. A., Blois, C. A., & Feigon, J. (2005). Structure of the human telomerase RNA pseudoknot reveals conserved tertiary interactions essential for function. *Molecular Cell, 17*, 671–682.

Wang, K. Y., Krawczyk, S. H., Bischofberger, N., Swaminathan, S., & Bolton, P. H. (1993). The tertiary structure of a DNA aptamer which binds to and inhibits thrombin determines activity. *Biochemistry, 32*, 11285–11292.

Wiseman, T., Williston, S., Brandts, J. F., & Lin, L.-N. (1989). Rapid measurement of binding constants and heats of binding using a new titration calorimeter. *Analytical Biochemistry, 179*, 131–137.

Xia, T., SantaLucia, J., Jr., Burkard, M. E., Kierzek, R., Schroeder, S. J., Jiao, X., et al. (1998). Thermodynamic parameters for an expanded nearest-neighbor model for formation of RNA duplexes with Watson-Crick base pairs. *Biochemistry, 37*, 14719–14735.

Zahler, A. M., Williamson, J. R., Cech, T. R., & Prescott, D. M. (1991). Inhibition of telomerase by G-quartet DNA structures. *Nature, 350*, 718–720.

SECTION III

Calorimetry Facilities

CHAPTER SEVENTEEN

A High-Throughput Biological Calorimetry Core: Steps to Startup, Run, and Maintain a Multiuser Facility

Neela H. Yennawar*, Julia A. Fecko*, Scott A. Showalter*[,†,‡,§], Philip C. Bevilacqua*[,†,‡,§,1]

*Huck Institutes of the Life Sciences, The Pennsylvania State University, University Park, Pennsylvania, USA
[†]Department of Chemistry, The Pennsylvania State University, University Park, Pennsylvania, USA
[‡]Department of Biochemistry and Molecular Biology, The Pennsylvania State University, University Park, Pennsylvania, USA
[§]Center for RNA Molecular Biology, The Pennsylvania State University, University Park, Pennsylvania, USA
[1]Corresponding author: e-mail address: pcb5@psu.edu

Contents

Methods in Enzymology, Volume 567
ISSN 0076-6879
http://dx.doi.org/10.1016/bs.mie.2015.07.024

Abstract

Many labs have conventional calorimeters where denaturation and binding experiments are setup and run one at a time. While these systems are highly informative to biopolymer folding and ligand interaction, they require considerable manual intervention for cleaning and setup. As such, the throughput for such setups is limited typically to a few runs a day. With a large number of experimental parameters to explore including different buffers, macromolecule concentrations, temperatures, ligands, mutants, controls, replicates, and instrument tests, the need for high-throughput automated calorimeters is on the rise. Lower sample volume requirements and reduced user intervention time compared to the manual instruments have improved turnover of calorimetry experiments in a high-throughput format where 25 or more runs can be conducted per day. The cost and efforts to maintain high-throughput equipment typically demands that these instruments be housed in a multiuser core facility. We describe here the steps taken to successfully start and run an automated biological calorimetry facility at Pennsylvania State University. Scientists from various departments at Penn State including Chemistry, Biochemistry and Molecular Biology, Bioengineering, Biology, Food Science, and Chemical Engineering are benefiting from this core facility. Samples studied include proteins, nucleic acids, sugars, lipids, synthetic polymers, small molecules, natural products, and virus capsids. This facility has led to higher throughput of data, which has been leveraged into grant support, attracting new faculty hire and has led to some exciting publications.

1. INTRODUCTION

Conventional manual single-usage calorimeters have been used extensively in research labs for decades and provided accurate thermodynamic data of many interesting biological systems. Excellent calorimetric equipment from Microcal (now with Malvern instruments) and TA instruments for both isothermal titration calorimetry (ITC) and differential scanning calorimetry (DSC) have seen a wide range of use for a variety of applications. Microcal offers several different ITC instruments, including the VP ITC and ITC200. The manual MicroCal VP ITC, while providing excellent signal to noise, has a large sample cell volume of 1400 μL, seven times greater than both the semiautomated MicroCal ITC200 and fully automated Auto-iTC200 models. The tradeoff with the MicroCal ITC200 is slightly higher noise (0.2 vs. 0.15 ncal/s for the Auto-iTC200 vs. the VP ITC); however, less time is spent on manual cleaning between experiments with the Auto-iTC200. With the many experimental parameters and variations in binding partners that can be studied in any biological or biological-like system, there is a need for low sample consumption, good sensitivity, and high throughput, as from any automated instrument. While pharmaceutical companies have applied automated calorimeters to screen for small-molecule drugs for some time now, the academic community has only started to use high-throughput calorimeters and we anticipate this usage will continue to grow.

The Automated Biological Calorimetry core facility at the Pennsylvania State University is a one-of-a-kind fee-for-service facility in the north-east region of the United States. This high-throughput state-of-the-art capability was envisioned in September 2011 and has been ably serving the needs of the scientific community at Penn State. It was set up with funds made available through a multiuser research instrumentation grant (MRI) from the National Science Foundation. The cost of establishing the facility is approximately $400,000–$500,000 direct costs, including costs of instruments and service contracts. Any costs of staff are in addition to this total. In recent years, service contracts have been dropped in favor of an annual preventative maintenance (PM). This dollar amount is an ideal range for funding from both NSF and NIH instrumental programs. In our case, administrative support including salaries for the facility staff and the 500-square-foot laboratory space for housing all the instrumentation were generously provided by the Huck Institutes of the Life Sciences at the Penn State (University Park campus). Dr. Neela Yennawar was appointed as the director and Julia Fecko as research technologist at the facility. An advisory committee was formed including Penn State faculty comprised of Profs. Philip Bevilacqua (Chair), Sarah Assman, Kenneth Keiler, and Scott Showalter as well as Dr. Nigel Deighton (Director of research instrumentation at the Huck Institutes of the Life Sciences). We placed the instruments within an existing facility for X-ray crystallography, since this facility is already adept at handling large amounts of pure protein and RNA. With these players in place the core facility was launched. We present here a model of starting and running a high-throughput biological calorimetry facility. Demand for such a high-throughput core facility currently exists in many large research institutions, and availability of such technologies could lead to higher success rates in obtaining biological research grants.

2. INSTRUMENTS AT THE FACILITY

A MicroCal Auto-iTC200 ITC instrument and a Microcal VP-cap DSC instrument were purchased from General Electric (Microcal wing of GE, which is now with Malvern instruments) each of which has 96-well formats. At the time, these were the only automated high-throughput calorimetric instrumentation in the market. TA instruments has added both an automated Affinity ITC Auto and the 96-well plate format Nano DSC Auto-sampler in recent years. Here, we provide an introduction to the calorimetry equipment at our core facility.

2.1 Isothermal Titration Calorimeter—Microcal Auto-iTC200

The MicroCal Auto-iTC200 ITC instrument allows high-throughput direct and label-free measurement of binding affinity and thermodynamics (Fig. 1). It measures the heat absorbed or generated when molecules interact. All filling, injection, and cell cleaning functions are fully automated, controlled, and operated through Origin® software that includes integrated experiment design wizards to assist in selecting experimental parameters. Up to four 96-well trays can be loaded and programmed to run without user intervention. The raw data obtained in this experiment are heat released or absorbed as a function of injection volume. The cell volume is 200 μL

Figure 1 Auto-iTC200 instrument from MicroCal. This instrument allows up to four 96-well trays to be loaded and programmed without user intervention. *Figure used with permission from Malvern.*

(sample volume to be loaded in the 96-well tray is 400 μL) and total injection volume is 40 μL, with each injection volume typically in the range of 0.7–2 μL (titration volume to be loaded in the 96-well tray is 120 μL). This is as opposed to a cell volume of 1400 μL and sample volume of 300 μL in the manual MicroCal VP ITC. Data analysis is also performed with Origin® software using several options of fitting models to calculate reaction stoichiometry, equilibrium association constant (inverted to derive equilibrium dissociation constant value), and enthalpy. Entropy is not a fitting parameter in the software, but rather a derived parameter calculated using the Gibbs free energy equation dependent on the known temperature and derived equilibrium association constant.

2.2 Differential Scanning Calorimeter—Microcal VP-cap DSC

The Microcal VP-capillary DSC instrument allows high-throughput measurement of the thermal stability of macromolecules through an automated cell loading and cleaning system (Fig. 2). It measures differences in heat required to raise the temperature of the sample cell versus the reference cell. Like the Auto-iTC, all filling, injection, and cell cleaning functions are fully automated, controlled, and operated through Origin® software that includes

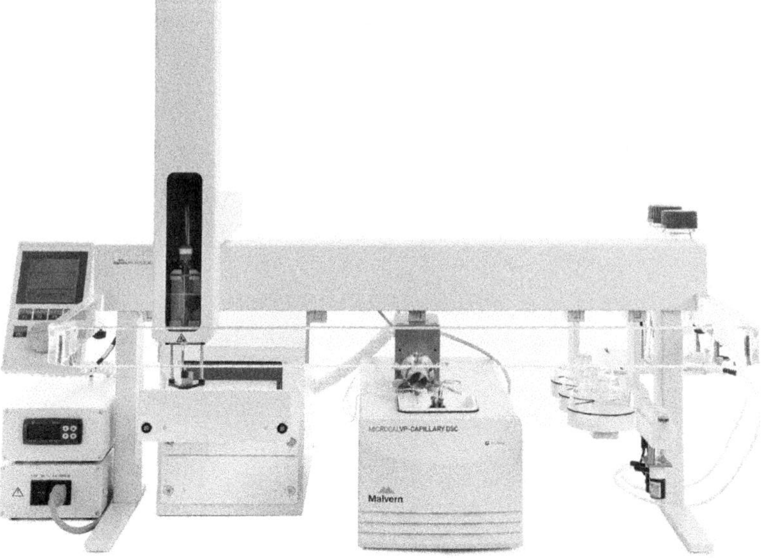

Figure 2 VP-cap DSC instrument from MicroCal. This instrument allows up to six 96-well trays to be loaded and programmed without user intervention. *Figure used with permission from Malvern.*

options to select experimental parameters. Up to six 96-well trays can be loaded and programmed to run without user intervention. The raw data obtained in this experiment are a plot of heat capacity difference as a function of temperature. The MicroCal VP-capillary DSC system has a sample cell and reference cell volume of 130 μL each (need to fill in 400 μL in the 96-well tray) as opposed to 500 μL in the manual MicroCal VP-DSC. Sample concentrations typically range from 0.1 to 10 mg/mL in both instrument setups. Data analysis is performed with the Origin® software and yields the macromolecule's melting transition midpoint temperature (T_m) as well as the enthalpy of folding, ΔH. The T_m is useful for characterizing the folding behavior of proteins, nucleic acids, micelles, and many other macromolecular systems, alone or as complexes. Furthermore, this information can optimize product shelf life, improve purification strategies, and evaluate protein constructs. In addition, affinities of small molecule to a protein target in a drug discovery pipeline can be ranked. Overall, DSC is very sensitive, easy-to-use, fast, and accurate for screening solution thermal behavior and interactions.

2.3 FT-IR Quantitation System

In order for users to obtain accurate figures for binding affinity, stoichiometry, enthalpy, and entropy, concentrations of the binding partners must be known precisely. To assist users to quantify their samples, the facility acquired a Millipore Direct Detect biomolecular quantitation system. Operation of the Direct Detect is based on infrared spectrometry and measures amide bond absorbance rather than aromatic residue absorbance for higher reproducibility across all proteins and peptides, with improved speed and accuracy over traditional biochemical assays. Because of the unique method of detection, this method also works in the presence of detergents and reducing agents commonly found in biological samples. As an added benefit, the method requires just 2 μL of sample per analysis. The Direct Detect can also be used with nucleic acids, although optical spectroscopy is typically used, and has proven to be a complementary instrument in the facility for both ITC and DSC runs. Alternately, nanodrop or nano-vue instruments can be used to determine biomolecular concentration.

2.4 Dynamic Light Scattering

Dynamic light scattering is a technique in which a beam of monochromatic laser light is directed through a biomolecule solution and fluctuations in scattered light intensity are analyzed. The experiment is noninvasive, requires only 12 μL of sample and can in a matter of minutes provide information about size and homogeneity of biomolecules. Sample characteristics like aggregation, folding, or conformation can be monitored as a function of various preparations, solvent conditions, temperature, or time. The facility houses a Viscotek 802 dynamic light-scattering instrument. The system is being used for prescreening samples under conditions in which the calorimetry experiment is being planned. The instrument can be operated between 4 and 60 °C. It is equipped with the OmniSIZE software that outputs information about the hydrodynamic radius (in the sample size range of 0.5 nm to 1 μm) along with percentage distribution of various components in the sample. Small traces of aggregation or oligomerization can be checked and corrected for by improving the buffer conditions or choosing the optimal temperature for the calorimetry experiment.

3. RUNNING OF THE FACILITY

In the first year of its operation, the calorimetry facility had users from 10 Penn State research labs across four departments—Profs. Graham Thomas (Biology/BMB), Stephen Benkovic (Chemistry), Philip Bevilacqua (Chemistry), Scott Showalter (Chemistry), Joshua Lambert (Food Science), Sarah Ades (BMB), Kenneth Keiler (BMB), Craig Cameron (BMB), Daniel Cosgrove (Biology), and Katsuhiko Murakami (BMB) and two outside (Boston college, Lebanon valley college) users as well. The operation has been smooth with an established workflow. A gradual growth in facility usage has been seen over the past 3 years, with more students receiving training annually. After initial training with sample preparation, instrument setup, and data analysis, students are able to directly operate and run the instruments independently with minimal support from the staff. The advisory committee has received positive feedback from the faculty regarding the usefulness of the facility. Having several faculty members keenly interested in the development of the facility is of paramount importance, especially in the first 2 years of setup and operation. The staff at GE (now Malvern) were knowledgeable and helpful with instrument maintenance issues, difficult-to-interpret experiments, and data analysis.

3.1 Steps Used in Setting Up the Core Facility

The following summarizes the steps and approximate timeframe needed in setting up the ITC and DSC capabilities at Penn State.

- Grouped eight faculty interested in calorimetry research and willing to participate and wrote a multiuser instrument and training grant—2 months.
- Gathered administrative support from Penn State for housing the instruments and staffing the core facility—1 month.
- Prepared shared instrumentation grant for NSF—2 months.
- Placed purchase orders to GE for the fully automated calorimeter instrumentation, Auto-iTC200, and VP-capDSC—2 months.
- Formed advisory committee consisting of four faculty members from the pool of interested faculty—2 weeks.
- Hired staff scientist at 50% effort to help run facility.
- Organized training workshops for staff and core facility lab users and advertised capabilities of the newly formed "Automated Biological Calorimtery Core" via open house, newsletters, Web site, and training opportunities—2 years.
- Currently operating the machine successfully without the annual service contract. Annual preventive maintenance is mandatory—ongoing.
- Providing user training every 3 months for all Penn State users—ongoing.

3.2 Role of Advisory Committee

The advisory committee meets with facility staff annually to discuss operations of the calorimetry facility. They have been active in helping run the facility smoothly and are in contact with the director on a monthly basis. The advisory committee plays an important role in setting policies and providing guidance for the facility. For example, it was the committee that considered the proposal to drop instrument service contracts and ultimately approved the decision. The advisory committee has also been actively involved in collaboration with the director to establish education, training, and outreach activities. The Showalter and Bevilacqua labs have also been involved in facility protocol establishment and periodically run tests, providing protein and RNA samples. Even after leaving Penn State, some of our students have continued to contribute ideas and suggestions that have helped to improve the usage of the calorimetry instrumentation.

3.3 Role of Facility Staff

The staff, which consists of the director and a staff scientist, train users, help design experiments and prepare samples, and address project-specific challenges. To increase awareness of the calorimetry equipment, the facility staff have met with faculty from various departments and presented in their group meetings. In order to highlight Penn State's new capability to a larger audience, they have also presented posters and talks at several national and international meetings and have advertised the facility through newsletters, open houses, and training sessions. Core facility usage has grown steadily over the past 4 years, with the Auto-iTC200 now running ~90% of the time including weekends, and the VP-cap DSC used 50% of the time.

3.4 Workflow

Users interested in microcalorimetry meet with the facility staff to discuss individual projects and optimal strategies for obtaining good quality calorimetric data. The staff become familiar with published literature on the project and the users are made aware of review articles and the basics of sample preparation (Baranauskiene, Petrikaite, Matuliene, & Matulis, 2009; Buurma & Haq, 2007; Chaires et al., 2015; Cooper, 1998; Feig, 2009; Ghai, Falconer, & Collins, 2012; Tellinghuisen, 2005; Velázquez-Campoy, Ohtaka, Nezami, Muzammil, & Freire, 2004; Zhao, Piszczek, & Schuck, 2015). Users can also learn more about the principles of calorimetry by taking a course in Biophysical Chemistry, Chem 540 offered at Penn State. The facility staff combines research literature with "the design of experiment" tool in the Origin® software to optimize starting conditions. AutoITC parameters include macromolecule and ligand concentrations, titration volumes, temperature, stirrer speed, and buffer choice, while cap-DSC parameters include temperature range, scan rate, and sample concentration. Once initial parameters are selected, users are trained on setting up their experiment using the 96-well plates, which are used in the fully automated equipment. In a 24-h period, depending on the chosen run parameters, ~8–10 titrations can be acquired on the Auto-iTC200 and 20–24 thermograms on the VP-cap DSC. This throughput enables users to optimize their run parameters and analyze their system quickly and efficiently. Some users find it useful to perform a dynamic light-scattering experiment to determine sample homogeneity. Others run a DSC experiment prior to an ITC run to determine the thermal denaturation midpoint so that the temperature-dependent ITC series can be designed with rational upper limits. Often the Auto-iTC200 is used during the pilot phase of experiments to rapidly screen multiple conditions and hone in on optimum experimental parameters. The staff assists in setting up the equipment and running the initial experiment as well as analyzing the data. After the results of the first set of runs, recommendations are made to improve data for ensuing experiments. Once a user is trained and feels comfortable running the equipment, they can schedule time to use the instrumentation independently and have more ownership and control of the outcome of the runs. Training in data analyses using Origin® and SEDPHAT software is also provided (Chaires et al., 2015; Zhao et al., 2015).

3.5 Outreach and Training

The facility has seen an increase in its user base through various advertising efforts. An annual open house is well attended by faculty, graduate students, and staff from various departments. Free user training sessions are offered on a quarterly basis where anyone from a diverse cross-section of research fields can learn about the theory of microcalorimetry and its applications, as well as gain practical hands-on experience. The facility also distributes a semiannual newsletter highlighting the latest news, publications, and ongoing research, which are posted on the Web site. Acknowledgments to the facility are made in publications resulting from core facility usage (Bastidas & Showalter, 2013; Khanna et al., 2015; Liu et al., 2013; Mukherjee et al., 2011; Patel, Blose, Sokoloski, Pollack, & Bevilacqua, 2012; Toroney, Hull, Sokoloski, & Bevilacqua, 2012; Warui et al., 2015), and highlighting these on the facility Web site is the best advertising possible.

The facility's Web site under the Huck Institutes of the Life Sciences details all the instrumentation and also features guidelines for successful sample preparation, applications, and testimonials (http://www.huck.psu.edu/content/instrumentation-facilities/automated-biological-calorimetry-facility). The Web site has been successful in reaching researchers outside the University and has recently generated use by industry.

The Automated Biological Calorimetry facility hosted a workshop featuring guest speakers including Prof. Andrew Feig from Wayne State University, technical talks from GE healthcare, and Penn State faculty, and student research presentations. The workshop offered informative presentations and technical discussions on automated ITC and DSC. It was well attended with participants from Penn State's College of Agricultural Sciences, and the Departments of Biochemistry and Molecular Biology, Chemistry, Biology, Chemical Engineering, and Physics. Many attendees provided positive feedback. The facility staff has presented poster sessions at several local, national, and international meetings. The director was invited to talk at the international calorimetry conference in Buzios, Brazil where she highlighted the collaborative crystallography and calorimetry work done with the Cosgrove lab (Georgelis, Yennawar, & Cosgrove, 2012).

3.6 Sample Preparation

Proper sample preparation is critical for achieving good microcalorimetry data for both the Auto-iTC200 and the VP-cap DSC. Details of sample preparation and experimental parameters are not very different for the automated and manual calorimeters and have been reviewed in detail in several earlier publications (Baranauskiene et al., 2009; Buurma & Haq, 2007; Chaires et al., 2015; Cooper, 1998; Feig, 2009; Ghai et al., 2012; Tellinghuisen, 2005; Velázquez-Campoy et al., 2004; Zhao et al., 2015). We provide a brief overview of these steps here.

1. Sample purification
2. Buffer matching
3. Design of experiments
4. Concentration determination
5. Buffer choice

1. Samples must be pure and free of degradation, and it is recommended that all molecules be gel or column purified before use. It is good practice to test macromolecular samples using dynamic light scattering in order to confirm homogeneity. The macromolecule should be tested for stability under the full range of experimental conditions. Precipitation will make the data unusable and could potentially damage the instruments via the fine tubing that is part of the liquid handling. Filtration may be necessary and degassing is recommended.
2. Buffer matching is crucial when preparing samples for both Auto-iTC200 and VP-cap DSC. Both syringe and cell contents must be very precisely buffer matched including all solvents and additives. Macromolecules should be dialyzed against their buffer to ensure exact salt matching of the sample. Small-molecule ligands should be dissolved into the dialysis flow-through, checked for pH changes, and matched. The dialysis buffer should also be used for the reference run in which ligand in buffer is titrated into buffer alone. The auto ITC equipment is so sensitive that even a slight mismatch of the buffer components between the macromolecule in the cell and the ligand in the syringe, even as low as 1%, can contribute substantial heat and mask the binding signal.

Any difference in pH between the buffer in the cell and syringe can have a similar detrimental effect. Particular attention should be given to lyophilized samples as there may be salts present prior to lyophilization that may not be accounted during the buffer matching. Some small-molecule ligands may need small percentage organic solvents like DMSO to dissolve. In this scenario, the same percentage of organic solvent should be added to the macromolecule as well.

3. An invaluable feature included in Auto-ITC200 software is the design of experiments simulation program written in Origin®, which allows users to optimize their starting concentrations. Through insights derived from prior research, using similar molecules or data from a previous run, one can estimate a stoichiometry (n), binding affinity (K_d), and enthalpy (ΔH) and enter them into the "design of experiment" aspect of the program. The software simulates the binding curve for the given concentrations of macromolecule and ligand. Initial experiments are performed by estimating the required sample concentration in the cell as $\sim 10 * K_d$ and concentration in syringe as $\sim$ cell concentration * no. of binding sites * 10.
4. Concentrations of the protein, DNA, RNA, or small molecule should be accurately known in order to obtain the best quantitative results. The core facility is equipped with the Millipore direct detect IR-based quantitation system as described above.
5. It is important to choose a buffer that has a low ΔH_{ion} (heat of ionization) thus not contributing excessive heat to the experiment. Examples of some buffers with low ΔH_{ion} include phosphate, acetate, formate, citrate, sulfate, cacodylate, and glycine (Baranauskiene et al., 2009; Buurma & Haq, 2007; Chaires et al., 2015; Cooper, 1998; Feig, 2009; Ghai et al., 2012; Tellinghuisen, 2005; Velázquez-Campoy et al., 2004; Zhao et al., 2015). Proper procedure for disposal of toxic and hazardous waste such as the cacodylate buffer should be followed. Tris buffer should be avoided for VP-cap DSC experiments as it has a high temperature dependence of its pK_a. Dithiothreitol should also be avoided, and tris(2-carboxyethyl)phosphine or beta-mercapto ethanol can be substituted as reductants instead.

3.7 Experimental Setup

3.7.1 For Auto-iTC200

Origin® software is used to control the instrument and for choosing the run parameters in the microcalorimeters. While the AutoITC can hold four

96-well trays, the VP-capDSC can hold six 96-well trays. We describe experimental setup of the Auto-iTC200, followed by the VP-cap DSC.

1. 96-well plate setup
2. Volumes
3. Auto sample setup
4. Experimental parameters

1. Once the samples are prepared, they are ready to be dispensed into the 96-well plate. Each experiment is comprised of a reference run as well as a sample run. The reference run mixes the ligand against the buffer to test if there is a reaction. The reference run will be subtracted from the sample run during the analysis stage, so it is essential that there be minimal or no reaction between the ligand and the buffer.

   Figure 3 illustrates a typical 96-well plate setup for the reference and sample runs. Four wells are utilized when setting up a run. The first well holds the sample, which will be dispensed into the cell. The second well holds the ligand, which will be titrated into the cell with the pipette. The third well holds buffer to be used as a prerinse. The prerinse will be dispensed into the cell prior to the macromolecule, held there for a minute and then discarded. This helps reduce noise during the sample run. A fourth well, which is empty at the start, can be used to save the sample after the titration if desired. With this setup, a total of 24 experiments can be run using a single 96-well plate. Typically, an entire plate is filled with replicates, as well as different samples, ligands, and buffers.

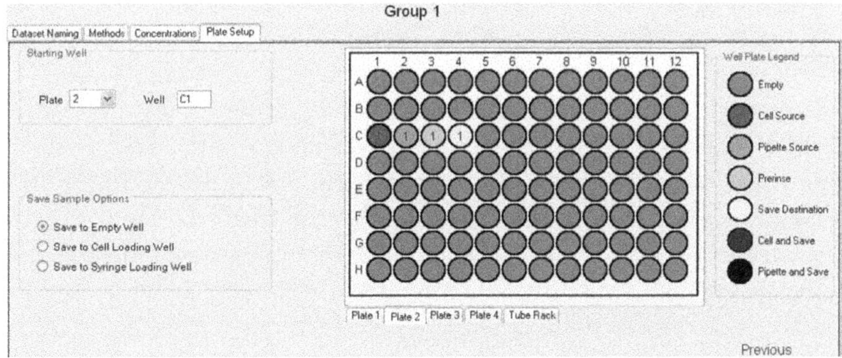

Figure 3 Plate setup for Auto-iTC200 experiment. Screen shot of the setup window for Auto-iTC200. Shown are four wells, which hold sample, ligand, and buffer, and collect sample, which comprise one experiment. All 96 wells can be filled to give 24 experiments. (See the color plate.)

2. The volume of the sample needed to be loaded in the 96-well tray—400 μL for the calorimeter cell, 120 μL for the titration syringe, and 400 μL for the prerinsing with matching buffer.
3. Once the plate has been filled without air bubbles and the adhesive lid has been secured, the plate is ready to be placed into the holding tray on the equipment. The Auto-iTC200 holds up to four plates and can be configured to run them in any desired order. The temperature of the tray can also be adjusted from 4 to 25 °C. The Auto-iTC200 is fully automated and can be set up to run continuously overnight. The operator simply has to enter in the number of samples, method of run, concentrations, and details of the plate setup.
4. The experimental parameters can also be customized for each individual experiment. They include total number of injections, cell temperature (value between 4 and 40 °C can be chosen), reference power, injection volume, spacing, and stir speed. An automated cleaning procedure is used between each run, with a more stringent cleaning used typically every 3–15 runs. These parameters can be fine-tuned in order to get a better-fitted binding curve.

3.7.2 For VP-cap DSC

1. 96-well plate setup
2. Volumes
3. Auto sample setup
4. Experimental parameters
1. The VP-cap DSC is fully automated so once the samples have been properly prepared they are ready to be dispensed into the 96-well plate. The VP-cap DSC is comprised of two cells, a reference cell and a sample cell. Each experiment consists of a reference run and a sample run. The reference run tests buffer against buffer to see if there is any heat, which will be subtracted from the sample run during the analysis stage. To set up the reference run, the first two wells of the plate will be used. The first well holds the buffer, which will be dispensed into the reference cell, while the second well also holds the buffer which will be dispensed into the sample cell. The sample run is similarly set up with the next two wells. The third well holds buffer for the reference cell, the fourth holds the sample for the sample cell.
2. The volume needed to be loaded in the 96-well tray for each of the reference cell and sample cell is 400 μL. An additional 200 mL of buffer is required to be used as a rinse between experiments.

3. Once the plate has been filled and the semipermeated lid has been secured to prevent evaporation, the plate is placed into the holding tray on the equipment. The DSC holds up to six plates and can be programmed to run the plates in any desired order. The temperature of the tray can be adjusted from 10 to 25 °C. The instrument is fully automated and can be set up to run continuously overnight. The operator simply has to enter in the number of samples, cleaning method, concentrations, and plate setup into the auto sampler.
4. The experimental scan parameters can also be customized for each individual experiment. They include starting temperature, final temperature (up to 120 °C), scan rate, prescan thermostat, postscan thermostat, and feedback mode. These parameters can all be finely adjusted in the DSC control tab.

The analysis of ITC binding curves and DSC thermograms are described in detail in the examples discussed in the following sections.

3.8 Maintaining the Facility and Cost Recovery

Having all the microcalorimetry equipment located at a central core facility is beneficial when it comes to maintaining the equipment. The research technologist takes ownership of the instrumentation and follows all cleaning protocols and oversees regular PM. The research technologist also knows the equipment well, which is advantageous when it comes to troubleshooting equipment issues. The following maintenance steps are followed diligently:

- A water install test is conducted on the Auto-iTC200 by using deionized, distilled water both in the cell as well as the syringe. Experimental parameters for the test usually are 19 injections, 2 μL injection volume, 5 μcal/s reference power, 120 s spacing, 750 rpm stir speed, high gain, and 25 °C cell temperature. Typically, a series of four to six water tests are run in a row to ensure that the baseline does not drift appreciably, that the injections are good (i.e., have a typical smooth shape), and that there is very slight heat and noise in the water–water injections. A good quality water–water injection profile is shown in Fig. 4. The mean energy per injection should be less than 1.5 μcal/s and the standard deviation less than 0.25 μcal/s. It is good practice to also incorporate several water–water runs in between the experiment schedule as a cross-check.
- The noise test is run on the Auto-iTC200 by using deionized, distilled water both in the cell and the syringe. There are no injections during the

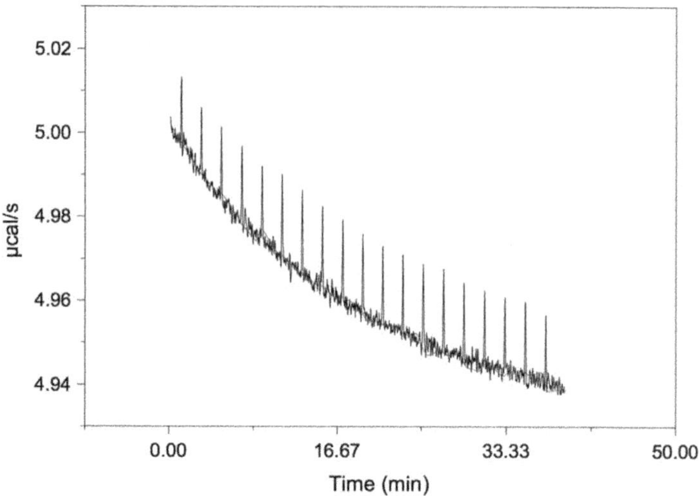

Figure 4 Water test result for Auto-iTC200. Test is done with titration of water against water and reference power set to 5 µcal/s. Shown here is a series of 19 injections of 2 µL each. An acceptable result is an upward deflection from heat of mixing of not more than 0.25 µcal/s, drift of a small fraction of the offset power with uniform and sharp injection peaks.

test and it runs for 20 min. The parameters for the test are 5 µcal/s reference power, 750 rpm stir speed, high gain, and 25 °C cell temperature. A good noise test run that "passes" the test will have only a slight drift in the baseline that is a small fraction of the offset power and uniform. The average RMS is not more than 1.5 ncal/s and the injections approach the baseline without significant shoulders or other protracted secondary features (Fig. 5). Excessive noise can indicate a dirty cell or a bent syringe. The noise test also will ensure the proper functioning of all electronics. Excessive noise during a sample run may indicate that the sample has some aggregation and needs to be filtered.

- An EDTA–$CaCl_2$ standard sample test is run on the Auto-iTC200 to ensure accurate heats and injections. 400 µL of 0.4 m*M* EDTA is used to fill the cell and 120 µL of 5 m*M* $CaCl_2$ is used in the titration syringe. The buffer used is 10 m*M* MES at pH 5.6. The test parameters are 19 injections, 2 µL injection volume, 150 s spacing, high gain, 750 rpm stir speed, 10 µcal/s reference power, and 25 °C cell temperature. Successful sample test results will have stoichiometry, binding affinity, and enthalpy within the limits listed for the standard (Griko, 1999; Fig. 6). In this instance, these values are 0.95, $1.26 \times 10^5\ M^{-1}$, and −4.27 kcal/mol (=−17.9 kJ/mol), respectively.

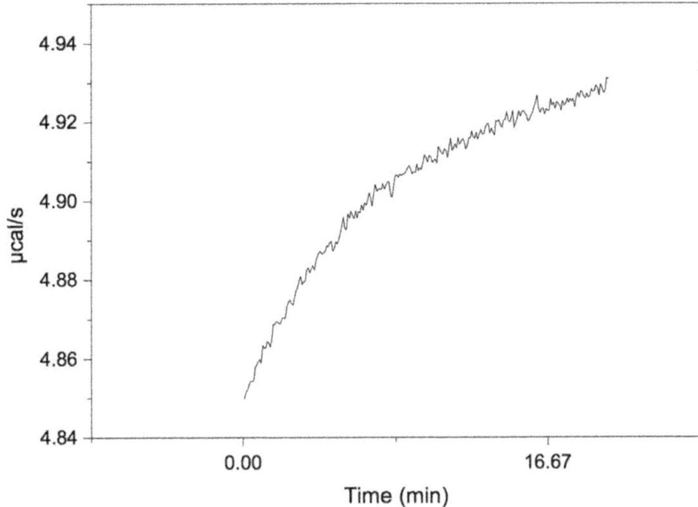

Figure 5 Noise test to ensure that the electronics in the instrument is working optimally.

- The Auto-iTC200 is set up to automatically clean the cell, cannulas, and syringe with 20% contrad70 between each sample run. An extensive cleaning protocol includes 30 min 20% contrad70 soak of the cell at 60 °C and is to be done between users.
- Cleaning cannula, syringe, and cell with 20% contrad70, rinsing with water, methanol, and nitrogen air blast between each experimental run.
- Changing water in the reference cell every 2 weeks.
- Maintenance log. A maintenance log is an essential tool to prevent downtime. Regular tests are routinely performed to ensure that the instrument is in the best working condition. Users book instrument time in advance and prepare samples accordingly. No service contract has been purchased in the year after the 3-year service contract expired and annual preventive maintenance has so far been sufficient for upkeep. Trained users are allowed to operate the machine and take ownership of their runs. Users also have key access to the instrument 24 h of the day, 7 days of the week.

The facility service fees (Table 1) collected have been sufficient to recover the cost of the preventive maintenance. No staff salaries are currently being recovered. The maintenance contract if purchased would increase the service fee to a level that may have an impact on usage. The cost recovery of the facility in the year 2014 is found in Table 1.

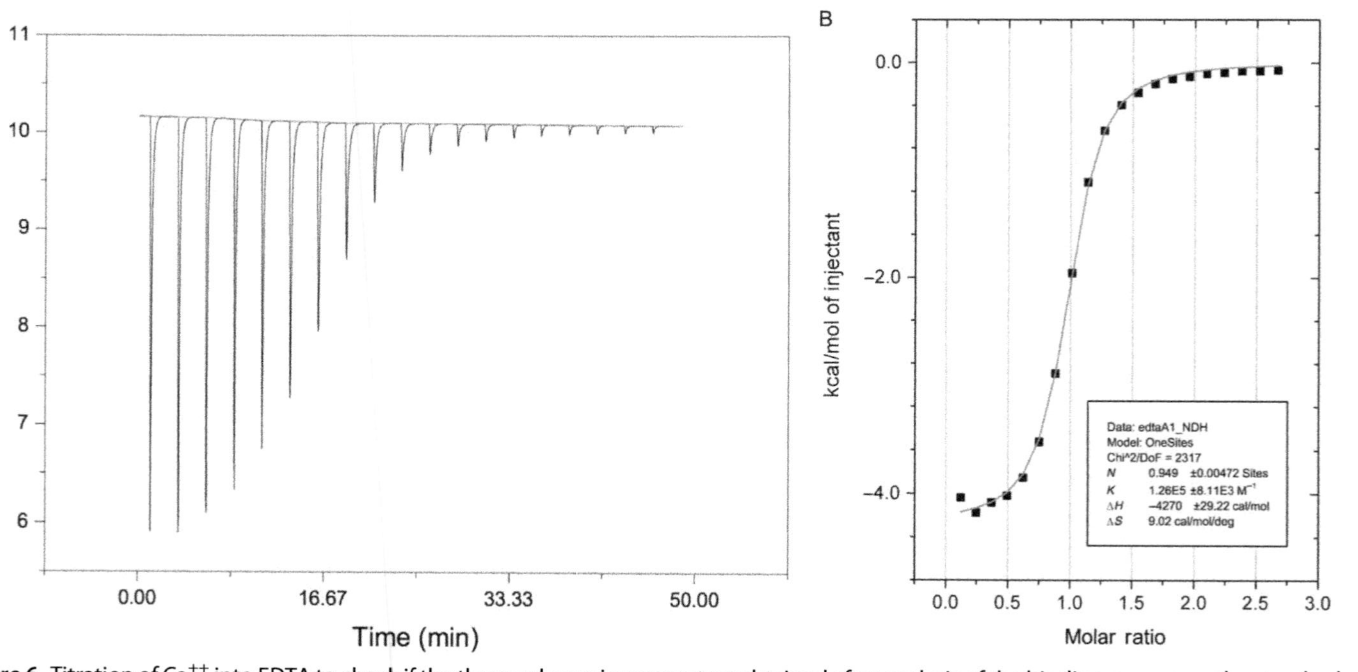

Figure 6 Titration of Ca^{++} into EDTA to check if the thermodynamic parameters obtained after analysis of the binding curve pass the standard test. (A) Injections of Ca^{2+}. (B) Integrated binding curve. Successful sample test results will have stoichiometry N, binding affinity K_a, and enthalpy ΔH within the limits listed for the standard, which are described in the text.

Table 1 Cost Recovery of Calorimetry Facility for Year 2014

Auto-iTC200 Expense Preventative Maintenance[a]	\$6700
ITC 106 days—internal (Penn State) experiments @ \$65 per day	\$6900
Difference	\$200

[a]The PM covers travel cost of service engineer, replacement of all tubing, pumps, cannula, checking for alignment, testing of all electronics, cleaning of cell and titration syringe, checking for leaks, and performing all standard tests such as temperature validation, differential power (*Y* axis) validation, water/water validation, noise validation, and Ca–EDTA validation.

4. SCIENCE FACILITATED

The Automated Biological Calorimetry Facility is employed by a wide spectrum of departments at the Penn State University Park campus including Agricultural Science, Biology, Biochemistry and Molecular Biology, Chemistry, Chemical Engineering, Food Science, and Engineering Science and Mechanics. In addition, Penn State Hershey College of Medicine, as well as some external research labs and universities, have used the facility. A number of exciting publications have ensued as a result of the capability (Bastidas & Showalter, 2013; Khanna et al., 2015; Liu et al., 2013; Mukherjee et al., 2011; Patel et al., 2012; Toroney et al., 2012; Warui et al., 2015). The speed of the Automated ITC and DSC combined with access to additional supporting instrumentation helps streamline research data collection considerably. The fully automated equipment enables users to set up a series of data acquisitions and also run experiments unattended overnight and during weekends for maximum efficiency. In addition, some of the Penn State faculty (Profs. Bevilacqua, Murakami, Showalter, Cameron) have included the facility as part of successful NIH and NSF grant applications.

4.1 Examples of Applications on the Auto-iTC200

The Bevilacqua lab from the Chemistry department has employed the high throughput and low sample requirements of the Auto-iTC200 in order to screen for NTP interactions with the protein kinase PKR showing that only ATP binds to the protein (Toroney et al., 2012; Fig. 7). These features also were essential in determination of the binding heat capacity for magnesium ion interactions with ATP and tRNA by rapidly performing a series of repeated titrations at different temperatures (Sokoloski, 2011). Both of these

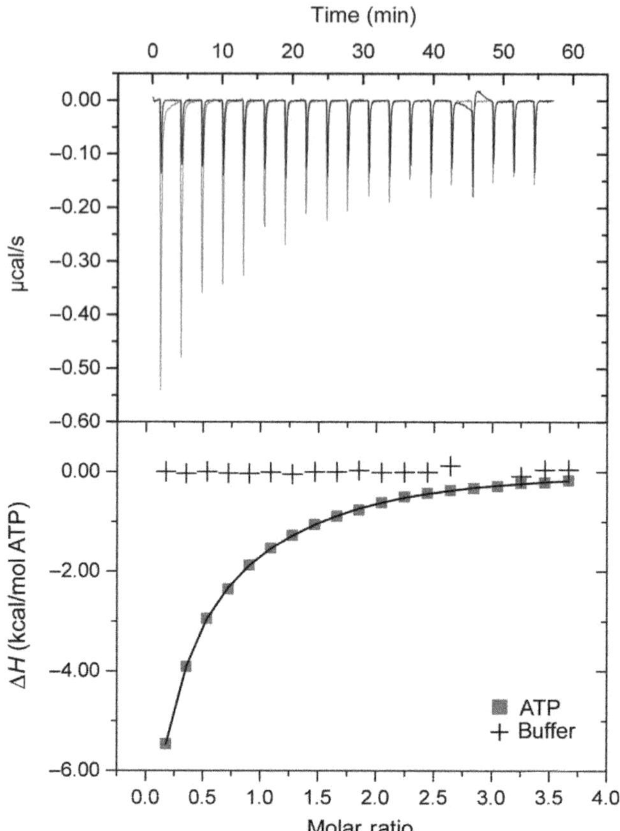

Figure 7 Sample output from Auto-iTC200. ITC titration curves for ATP binding to K296R mutant of PKR. Titration of ATP into buffer is included (black trace). ATP titration curve is fitted to a two-site binding model. The major contribution is with site 1 ($n = 1.32 \pm 0.09$) and a K_d of 19.6 ± 1.8 μM. *Figure adapted with permission from Toroney et al. (2012), copyright (2012) Cold Spring Harbor Press.*

studies benefited from the inherent advantages of the Auto-iTC200 as these series of measurements could be accomplished using the same amount of sample as a single titration in a traditional calorimeter and each series was carried out in a single day compared to a week or longer timeframe for acquiring the same data using a traditional calorimeter.

The Showalter lab from the Chemistry department applies biophysical chemistry techniques to understand the function of intrinsically disordered proteins and to define the features of protein–RNA interactions. They have had multiple users running hundreds of experiments on the Auto-iTC200 and have effectively utilized the automation for maximum throughput, which has enabled the investigation of DNA binding by the homeodomain transcription factor Pdx1, which elucidated the determinants of sequence selection that cause Pdx1 to select its target promoters against a nonspecific DNA background. This study generated the first testable hypotheses regarding the mechanism whereby Pdx1 binds with differential affinity among its multiple validated targets (Bastidas & Showalter, 2013). The Auto-iTC200 permitted the acquisition of up to 16 experimental titrations per 24 h period, even with water–water flushes included once per six experiments to provide extra cleaning. In a second study, binding between RNA and the protein TRBP2 was evaluated at different temperatures, in order to discover the thermodynamic driving forces behind target-RNA acquisition by TRBP. Due to the added time needed for temperature equilibration in these experiments, it was more typical to complete 10–12 experimental runs per 24 h period.

Other Auto-ITC applications include

- The Anantheswaran lab from the Food Science department studies the microwave processing and packaging of foods; as part of their study on diffusion of nisin through packaging materials they studied the binding interaction between nisin with food additives chitosan and alginate.
- The Benkovic Group from the Chemistry department is engaged in a variety of projects connected by the general theme of understanding enzyme catalysis at various levels. They used the Auto-iTC200 to study the binding between human PCNA and Polή.
- The Boehr lab's research in the Chemistry department is directed toward understanding catalytic mechanisms through the biophysical study of enzymes. The Auto-iTC200 was used in a study to assess the binding between a peptide and four different types of proteins. The user was able to set up multiple runs over several days to optimize conditions for each protein.
- The Cameron lab in the Biochemistry and Molecular Biology department has research interests including the role of RNA polymerases and RNA-binding proteins in viral infection and mitochondrial disease. The Auto-iTC200 was used in studying the binding reaction between the protein Bin1-SH3 and the intrinsically disordered domain from the protein NS5A, as well as interactions involving full-length NS5A. Similar studies were conducted with the cSrc-SH3 domain for comparison to the Bin1 data. In a very systematic study, the user made use of over 80 ITC experiments and also used the method of concatenation with continuous injections (Grossoehme, Akilesh, Guerinot, & Wilcox, 2006; Grossoehme, Spuches, & Wilcox, 2010) to reach saturation when concentrations were limited.

- The Elias lab from the Food Science department aims to understand the mechanistic basis of chemical food stability, especially as it relates to both product quality and health consequences in consumers. The Auto-iTC200 was used to study the interaction between gliadin (a major protein in gluten) with gallic acid and catechin, which required running 33 experiments using three different pH levels.
- The Golbeck lab at the Biochemistry and Molecular Biology department conducts research with emphasis on photosynthesis. Binding of menadione and quinone with proteins of the photosystem I reaction center was studied. Three different buffer concentrations were used in the preliminary phase to quickly determine best buffer conditions.
- The Thomas lab from the department of Biochemistry and Molecular Biology used the Auto-iTC200 to quantify the oligomerization-interaction between the tetramerization domains of α, β, and β (heavy) spectrins and several variants from the fruit fly genus *Drosophila* (Khanna et al., 2015).
- The Tien lab from the Biochemistry and Molecular Biology department focuses its research in three areas: characterization and biochemical analysis of cellulose synthesis in a variety of organisms, mechanism and regulation of fungal degradation of lignin and regulation of fungal degradation of lignin, and dissimilatory iron reduction. The Auto-iTC200 was used to study the interaction between the protein CCpax and cellulose.
- The Wood lab, which is affiliated with both the Biochemistry and Molecular Biology department and the Chemical Engineering department, studies the physiological relevance of bacterial toxin/antitoxin systems, the genetic basis of biofilm formation, and mechanisms to control biofilm formations for engineering applications. Binding affinity was studied between the protein Ygis and its small-molecule target, deoxycholate. In a second project, the Auto-iTC200 was used to study the interaction of five different metals with the protein RalR, and separately with DNA, using three different buffers. The user was able to perform more than 60 experiments over a 4-day period to find the best binding conditions.
- The Zydney lab from the Chemical Engineering department has research interests including bioseparations, artificial organs, and membrane processes. ITC was used to study the binding between DNA and various amino acids.

4.2 Examples of Applications on the VP-cap DSC

The Showalter lab from the Chemistry department utilizes the VP-cap DSC extensively in preparation for ITC, NMR spectroscopy, and many biochemical experiments. Critically, many of the Showalter laboratory's RNA-binding experiments must be performed at high RNA concentrations, where hairpin structures are often less thermodynamically stable than biologically irrelevant homodimers. The VP-cap DSC is used to establish conditions under which RNA samples retain their biologically relevant hairpin structures, which have distinct melting profiles from the duplex artifacts (Fig. 8). Additionally, many of the Showalter laboratory's ITC experiments are performed in temperature-dependent series that can only be designed effectively if the thermal denaturation midpoint temperature is known for all involved biological macromolecules. The VP-cap DSC provides a robust and time-efficient mechanism to generate this essential data because 10 runs or more can be collected in an overnight period, allowing many conditions to be efficiently screened. The screening capabilities of the VP-cap DSC thus make it a helpful companion to the Auto-iTC200.

Other VP-cap DSC applications include

- Fujirebio diagnostics Inc. tested the stability of six different antibodies in six different buffer formulations varying sucrose concentrations

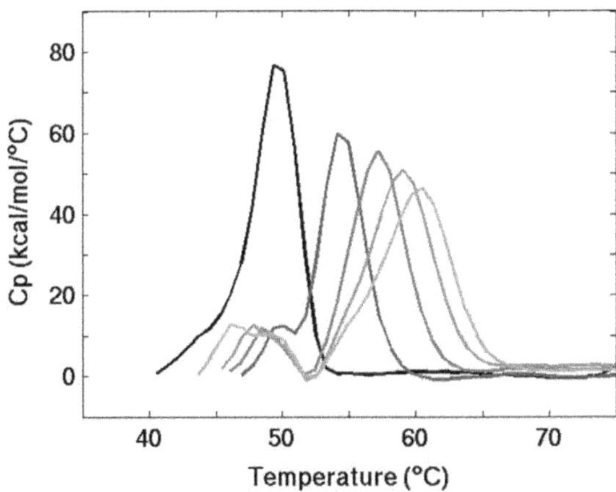

Figure 8 Sample output from VP-cap DSC. Screening for conditions that favor retention of RNA hairpin structure at high nucleic acid concentration is facilitated by the VP-cap DSC. In this experiment, 20 μ*M* RNA was screened for hairpin-to-duplex transitions as a function of total sodium chloride concentration (increasing from 40 m*M* through 200 m*M* in 40 m*M* increments, progressing from black to light gray).

and pH. All tests were done in duplicate and 96 tests were completed over a 5-day period.

- The Cremer group in the Chemistry department works at the crossroads of biological interfaces, nanomaterials, spectroscopy, and microfluidics. DSC was used to study the effect of ions on polyethylene oxide at varying Na_3PO_4 concentrations. They also compared the effect of solvation in H_2O versus D_2O on this interaction. A second study was conducted to measure the thermal melting temperature and enthalpy of unfolding seen for various proteins, the polymer PNiPAM-ddi, and elastin-like peptides in varying concentrations of TMAO.
- The Demirel lab from the Engineering Science and Mechanics department conducts research focusing on theory-driven functional materials synthesis and fabrication for designing novel engineering materials to produce next-generation materials for an array of fields including energy, biomedicine, and security/defense. They used the VP-cap DSC to determine the stability of a squid protein known to have some unusual material properties.
- The Harte lab from the Food Science department focuses on structure–function properties of milk proteins, with an emphasis on casein proteins. They used the VP-cap DSC to study how increasing concentrations of ethanol affect the stability of various proteins in cow's milk. Seventy eight experiments in total were conducted uninterrupted over a week with one easy setup.
- The Nixon lab from the Biochemistry and Molecular Biology department focuses on the functional basis of signal transduction and gene regulation by AAA+ ATPases in bacteria, degradation of lignocellulose by cellulases, and the synthesis of cellulose. They used the VP-cap DSC to study the folding and unfolding of several cellulases.

5. CONCLUSIONS

In this chapter, we described our high-throughput biological calorimetry facility including the instruments it houses, best practices for running the facility, and examples of the science facilitated. Having these calorimeters has inspired new science and collaborations across campus, which in turn has facilitated the writing of new grants. It has been a positive factor in new faculty recruitment. Automated calorimeters are fairly unusual on most college campuses; however, they offer many benefits to the campus and regional research community. Although the initial cost of acquisition is high, the costs of opening such a facility fit well into the funding levels and objective of most multiuser instrumentation grant programs. We hope to see an increase of high-throughput calorimeters in academic research and look forward to the new science they will generate.

ACKNOWLEDGMENTS

We acknowledge the National Science Foundation for the grant NSF-MRI award DBI-0922974 to Bevilacqua et al. and thank the Huck Institutes of the Life Sciences at Penn State for the continued administrative support. P. C. B. also acknowledges NIH (R01-GM095923). S. A. S. acknowledges NIH (R01-GM098451) and NSF (MCB-0953918). We thank Dr. Nigel Deighton for helpful comments on the chapter.

REFERENCES (*REPRESENTS THE BIBLIOGRAPHY RESULTING FROM THE EXISTENCE OF THE FACILITY)

Baranauskiene, L., Petrikaite, V., Matuliene, J., & Matulis, D. (2009). Titration calorimetry standards and the precision of isothermal titration calorimetry data. *International Journal of Molecular Sciences*, *10*, 2752–2762.

Bastidas, M., & Showalter, S. A. (2013). Thermodynamic and structural determinants of differential Pdx1 binding to elements from the insulin and IAPP promoters. *Journal of Molecular Biology*, *425*(18), 3360–3377.*

Buurma, N. J., & Haq, I. (2007). Advances in the analysis of isothermal titration calorimetry data for ligand–DNA interactions. *Methods*, *42*, 162–172.

Chaires, J. B., Hansen, L. D., Keller, S., Brautigam, C. A., Zhao, H., & Schuck, P. (2015). Biocalorimetry. *Methods*, *76*, 1–2.

Cooper, A. (1998). Microcalorimetry of protein-protein interactions. *Methods in Molecular Biology*, *88*, 11–22.

Feig, A. L. (2009). Studying RNA–RNA and RNA–protein interactions by isothermal titration calorimetry. *Methods in Enzymology*, *468*, 410–422.

Georgelis, N., Yennawar, N. H., & Cosgrove, D. J. (2012). Structural basis for entropy-driven cellulose binding by a type-A cellulose-binding module (CBM) and bacterial expansin. *Proceedings of the National Academy of Sciences of the United States of America*, *109*(37), 14830–14835.

Ghai, R., Falconer, R. J., & Collins, B. M. (2012). Applications of isothermal titration calorimetry in pure and applied research—Survey of the literature from 2010. *Journal of Molecular Recognition*, *25*, 32–52.

Griko, Y. V. (1999). Energetics of Ca(2+)-EDTA interactions: Calorimetric study. *Biophysical Chemistry*, *79*(2), 117–127.

Grossoehme, N. E., Akilesh, S., Guerinot, M. L., & Wilcox, D. E. (2006). Metal-binding thermodynamics of the histidine-rich sequence from the metal-transport protein IRT1 of Arabidopsis thaliana. *Inorganic Chemistry*, *45*(21), 8500–8508.

Grossoehme, N. E., Spuches, A. M., & Wilcox, D. E. (2010). Application of isothermal titration calorimetry in bioinorganic chemistry. *Journal of Biological Inorganic Chemistry*, *15*, 1183–1191.

Khanna, M. R., Mattie, F. J., Browder, K. C., Radyk, M. D., Crilly, S. E., Bakerink, K. J., et al. (2015). Spectrin tetramer formation is not required for viable development in Drosophila. *The Journal of Biological Chemistry*, *290*(2), 706–715.*

Liu, C. T., Hanoian, P., French, J. B., Pringle, T. H., Hammes-Schifferand, S., & Benkovic, S. J. (2013). Functional significance of evolving protein sequence in dihydrofolatereductase from bacteria to humans. *Proceedings of the National Academy of Sciences of the United States of America, 110*(25), 10159–10164.*

Mukherjee, S., Yakhnin, H., Kysela, D., Sokoloski, J., Babitzke, P., & Kearns, D. B. (2011). CsrA-FliW interaction governs flagellin homeostasis and a checkpoint on flagellar morphogenesis in Bacillus subtilis. *Molecular Microbiology, 82,* 447–461.*

Patel, S., Blose, J. M., Sokoloski, J. E., Pollack, L., & Bevilacqua, P. C. (2012). Specificity of the double-stranded RNA-binding domain from the RNA-activated protein kinase PKR for double-stranded RNA: Insights from thermodynamics and small-angle X-ray scattering. *Biochemistry, 51,* 9312–9322.*

Sokoloski, J. (2011). *Calorimetric studies of site-specific ligand binding by RNA.* PhD. Thesis. ProQuest Document ID: 1525026998.

Tellinghuisen, J. (2005). Optimizing experimental parameters in isothermal titration calorimetry. *The Journal of Physical Chemistry B, 109,* 20027–20035.

Toroney, R., Hull, C. M., Sokoloski, J. E., & Bevilacqua, P. C. (2012). Mechanistic characterization of the 5'-triphosphate-dependent activation of PKR: Lack of 5'-end nucleobase specificity, evidence for a distinct triphosphate binding site, and a critical role for the dsRBD. *RNA, 18,* 1862–1874.*

Velázquez-Campoy, A., Ohtaka, H., Nezami, A., Muzammil, S., & Freire, E. (2004). Isothermal titration calorimetry. *Current Protocols in Cell Biology, 17*(8), 1–24.

Warui, D., Pandelia, M. E., Rajakovich, L., Krebs, C., Bollinger, J. M., & Booker, S. (2015). Efficient delivery of long-chain fatty aldehydes from the nostoc punctiforme acyl–acyl carrier protein reductase to its cognate aldehyde deformylating oxygenase. *Biochemistry, 54*(4), 1006–1015.*

Zhao, H., Piszczek, G., & Schuck, P. (2015). SEDPHAT—A platform for global ITC analysis and global multi-method analysis of molecular interactions. *Methods, 76,* 137–148.

AUTHOR INDEX

Note: Page numbers followed by "*f*" indicate figures, and "*t*" indicate tables.

C

D

G

H

I

M

N

Q

R

SUBJECT INDEX

Note: Page numbers followed by "*f*" indicate figures, "*t*" indicate tables, and "*s*" indicate schemes.

D

E

J

K

L

M

U

V

Y

Z

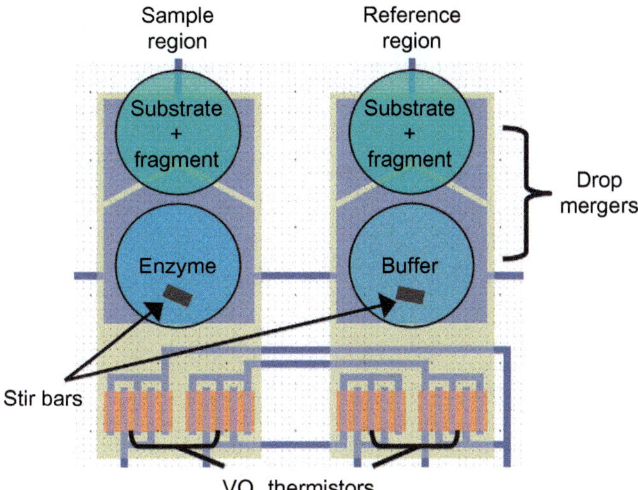

Michael I. Recht *et al.*, Figure 1 Schematic of a single enthalpy array detector. Sample and reference regions are designated based on the material (substrate (alone or with fragment), enzyme, or buffer) in the drops deposited on each region. PEGylated magnetic stir bars (400 μm × 200 μm × 15 μm) are manually placed on array drop merger elements using an *X–Y–Z* micromanipulator. A single stir bar is placed at the location where the enzyme solution or buffer solution will be deposited on the array. *Adapted with permission from Recht et al. (2009). Copyright, Elsevier, Inc.*

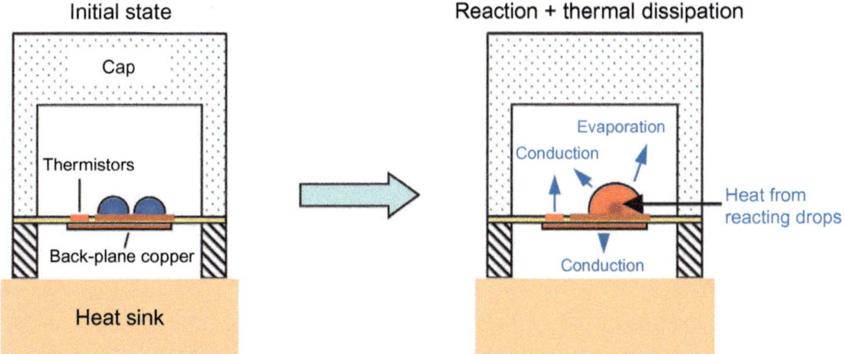

Michael I. Recht *et al.*, Figure 2 Side view of one detector of an array before and after drop merging. The drops and detector are isolated from the environment by a cap above and a heat sink below. Pogo pins through the cap provide electrical connections from the array to the electronics on a printed circuit board (not shown). The copper heat sink has a large mass × specific heat, which serves to lessen thermal fluctuations. The back-plane copper underneath the drops and thermistors evens out the temperature. Upon merging, drops heat (or cool for endothermic reactions) and the heat dissipates by conductive and evaporative heat transport.

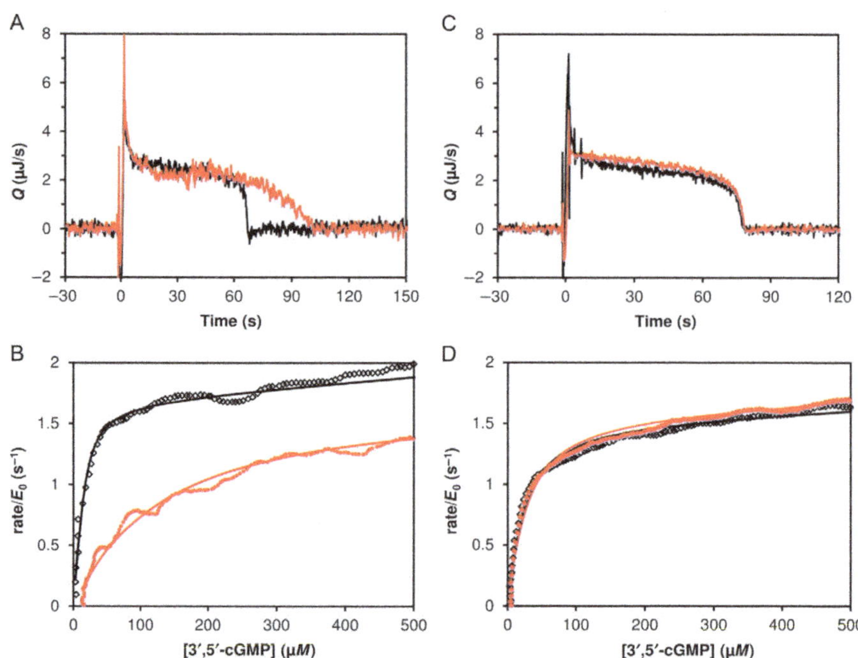

Michael I. Recht *et al.*, Figure 5 PDE10A hydrolysis of 3′,5′-cGMP in the absence (black, open) and presence (maroon, solid) of fragments. Reactions contained 15 μ*M* PDE10A and 2 m*M* 3′,5′-cGMP. Solid curves are fit of data to Eq. (2). (A) Plot of power (*Q*) versus time for the reaction in absence and presence of 2 m*M* fragment ZT0429. Note the signal returns to original levels as substrate is completely hydrolyzed and in particular more slowly than the reaction in the absence of inhibitor. This difference reflects a change in apparent K_M for 3′,5′-cGMP in the presence of ZT0429. (B) Rate versus remaining 3′,5′-cGMP concentration in the absence and presence of 2 m*M* ZT0429. Solid curves are fit of data to Eq. (2); K_I for ZT0429 is 300 μ*M*. (C) Plot of power (*Q*) versus time for the reaction in absence and presence of 2 m*M* fragment ZT0417. Note the signal for the reaction in the presence of fragment is nearly identical to the control reaction. (D) Rate versus remaining 3′,5′-cGMP concentration in the absence and presence of 2 m*M* ZT0417. Solid curves are fit of data to Eq. (2); K_I for ZT0417 is $\gg$2000 μ*M*.

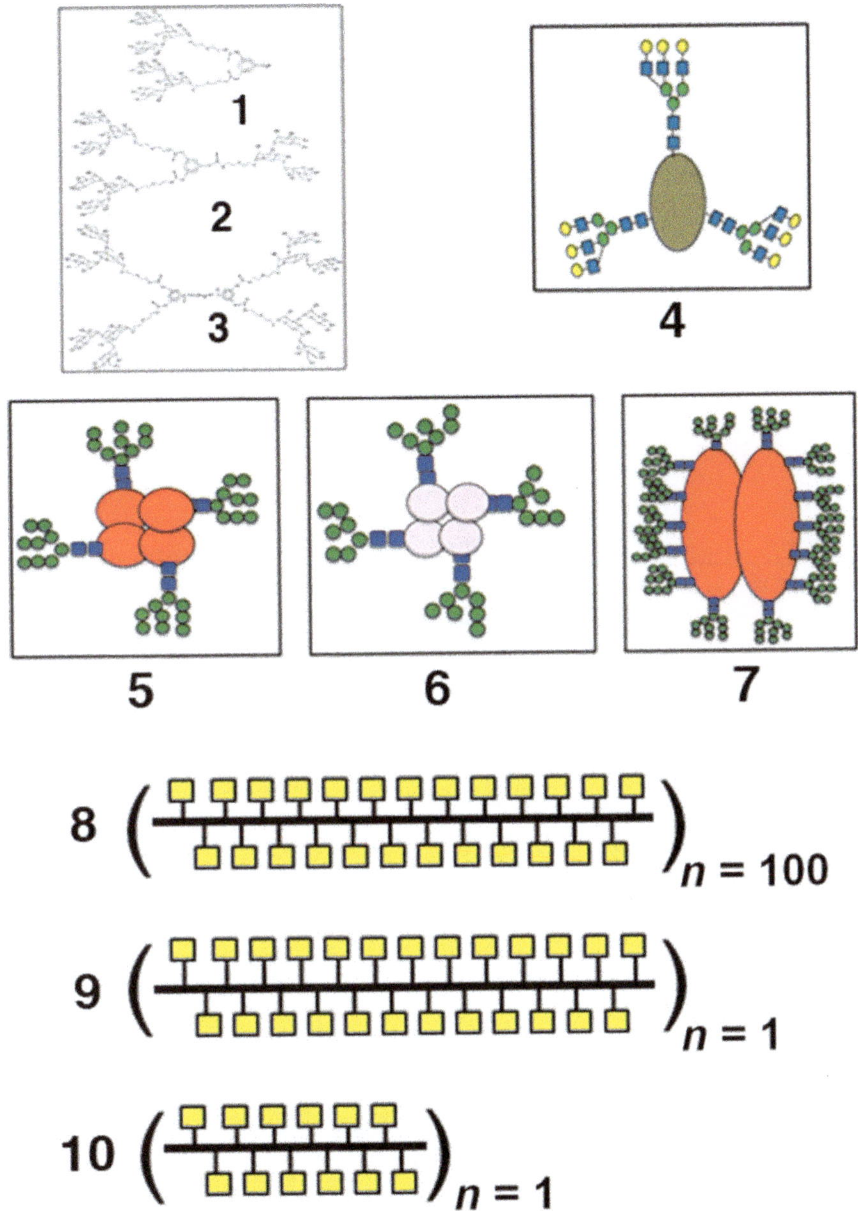

Tarun K. Dam *et al.*, Figure 1 Multivalent glycoconjugates used in the present studies: synthetic multivalent analogs (**1**–**3**); globular glycoproteins including asialofetuin (ASF) (**4**), SBA (soybean agglutinin) (**5**), avidin (**6**), and invertase (**7**); porcine submaxillary mucins (PSM) including the 100-repeat 81-residue polypeptide O-glycosylation domain

(*Continued*)

Tarun K. Dam *et al.*, Figure 1—Cont'd of PSM containing only peptide-linked-GalNAc residues (Tn-PSM) (**8**), the single 81-residue polypeptide O-glycosylation domain of PSM containing peptide-linked α-GalNAc residues (81-mer Tn-PSM) (**9**), and the 38/40-residue polypeptide(s) derived from the 81-residue polypeptide O-glycosylation domain of PSM containing peptide-linked α-GalNAc residues (38/40-mer Tn-PSM) (**10**). The number of glycan chains in **8** (Tn-PSM) is ~2300. The number of α-GalNAc residues in **9** (81-mer Tn-PSM) is ~23, while the number of α-GalNAc residues in **10** (38/40-mer Tn-PSM) is ~11–12.

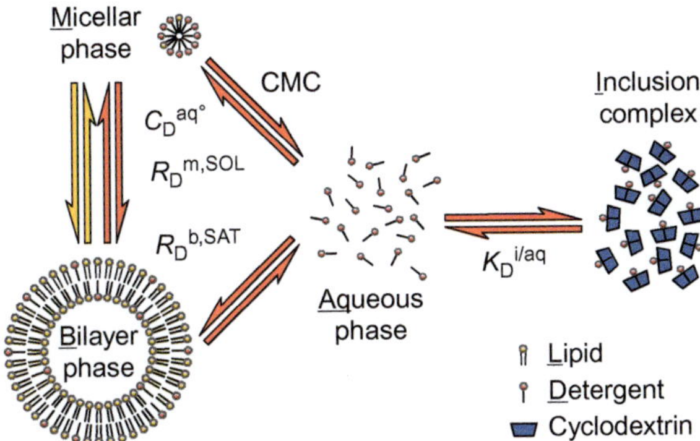

Martin Textor and Sandro Keller, Figure 1 Overview of chemical species, pseudophases, and linked equilibria involved in CD-mediated reconstitution and thermodynamic parameters required for its quantitative description.

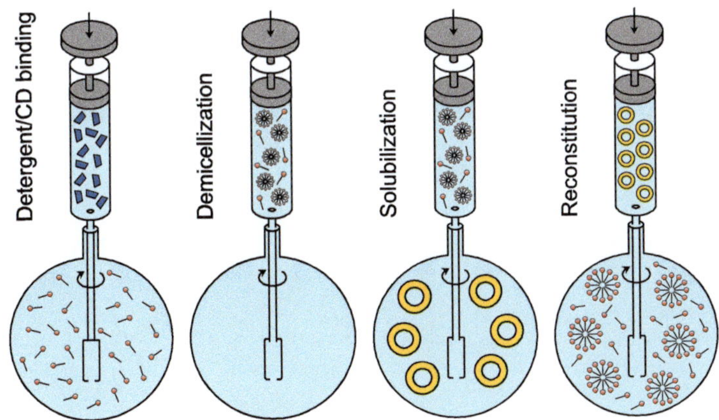

Martin Textor and Sandro Keller, Figure 2 Calorimetric experiments for obtaining thermodynamic parameters describing CD-driven reconstitution. Molecular species depicted include CD (trapezoids), lipid vesicles (circles), and detergent (sticks).

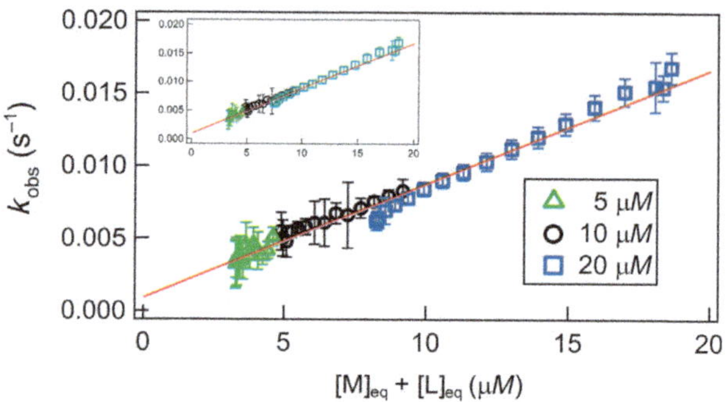

Kirk A. Vander Meulen *et al.*, Figure 4 Demonstration of consistency in the near-equilibrium approximation, using TT–RR experiments across a range of $[M]_{init}$ at the experimental condition 0.3 m*M* $MgCl_2$, 30 °C. Main figure: plot of best-fit k_{obs} against the total unbound RNA concentration ($[M]_{eq}+[L]_{eq}$) for a set of titrations in which the starting cell concentration of a macromolecule, M (RR in this case), was varied and titrated with its ligand, L (TT). For all experiments, the syringe [TT] is 171 μ*M*; green triangles, black circles, and blue squares reflect separate experiments in which $[M]_{init}$ is 5, 10, or 20 μ*M*, respectively. The red trendline is the result of a global, linear fit; results from the individual datasets are provided in Table 2. The inset figure plots the same data except that the plot for the 20 μ*M* dataset (cyan) employed a K_d that is decreased by 10% over its best-fit value.

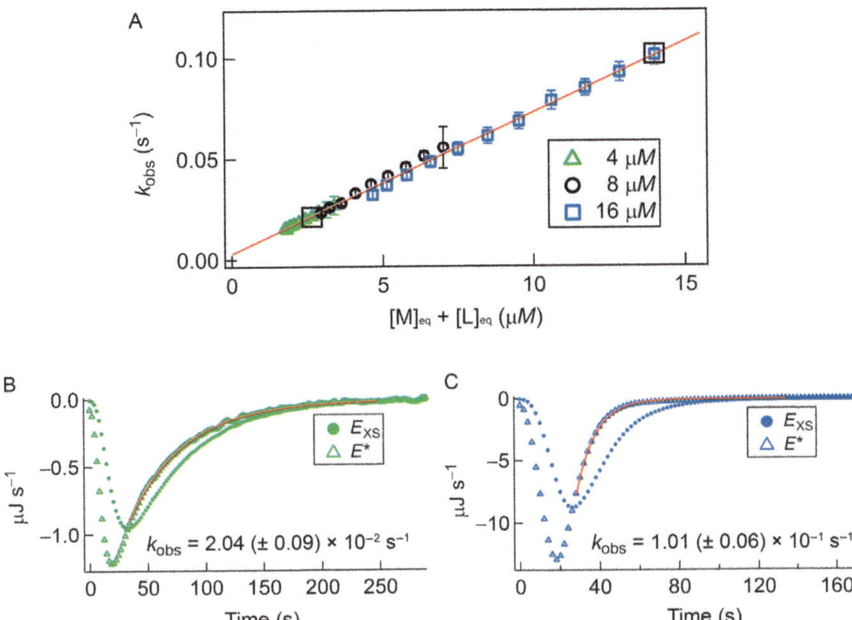

Kirk A. Vander Meulen *et al.*, Figure 5 Demonstration of consistency in the near-equilibrium approximation at the upper end of the calorimetry-accessible kinetics window, using TT–RR experiments across a range of $[M]_{init}$ at the experimental condition 200 m*M* KCl, 25 °C. (A) Plot of best-fit k_{obs} against unbound RNA for a set of TT–RR titrations in which $[M]_{init}$ was varied. For all experiments, the syringe ligand (TT) concentration is 240 μ*M*; green triangles, black circles, and blue squares reflect separate experiments in which the starting cell concentration of RR ($[M]_{init.}$) is 4, 8, or 16 μ*M*, respectively. The red trendline is the result of a global, linear fit; results from the individual datasets are provided in Table 3. Bottom panels: demonstration of increasing importance of deconvolution for fast association kinetics. Plots depict power traces before (E_{XS}, closed circles) and after (E^*, open triangles) deconvolution using the sixth injection from the 4 μ*M* dataset (B) and the second injection from the 16 μ*M* dataset (C) as examples. Red curves plot best-fit exponential functions and fit intervals, and associated composite rate constants (k_{obs}) are listed. The listed uncertainties reflect only injection-level sources (see Section 4.2.3). Boxed data points in (A) correspond to the injections displayed in (B) and (C).

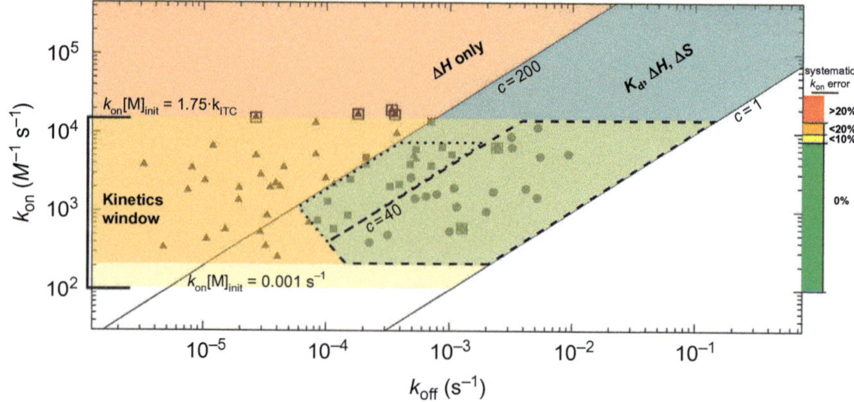

Kirk A. Vander Meulen *et al.*, Figure 8 Relationship between the kinetics (k_{on} and k_{off}) and the attainable thermodynamic and kinetic parameters by ITC, illustrated for a titration experiment conducted with $[M]_{init} = 10\ \mu M$ and instrument time constant of 12.5 s. The three main regions for consideration are denoted in blue, salmon, and yellow; corresponding overlapping regions are colored accordingly. The blue diagonal band represents the optimal thermodynamics regime for ITC in which $1 \leq c \leq 200$, and for which the full complement of thermodynamic information (K_d, ΔH, and subsequently ΔS) can be obtained. The salmon-colored region above the $c = 200$ line reflects the tight-binding regime for which, thermodynamically, only ΔH can be obtained. The horizontal yellow band represents the kinetics regime ($1 \times 10^2\ M^{-1}\ s^{-1} < k_{on} \leq 1.4 \times 10^4\ M^{-1}\ s^{-1}$). Within a subrange of the overlap of the kinetic regime with the optimal thermodynamic regime (largely consisting of $1 \leq c \leq 40$) is the near-equilibrium approximation regime, marked by dashed lines, for which this approximation is valid over the entirety of the titration. A dotted line encloses an adjacent region where, in our experience, the near-equilibrium approximation yields k_{on} values consistent with the general method. Closed symbols denote the regions within this map that we have examined using TT–RR, and the asterisk reflects the AdoMet–SET7/9 experiment. Symbol positions are "concentration-adjusted" such that they reflect the rate constants that would exhibit the experimentally observed kinetics if all experiments were conducted at $[M]_{init} = 10\ \mu M$. TT–RR closed circles and squares represent average kinetic parameters obtained at an experimental condition where k_{on} and the full complement of thermodynamic information were obtained (corresponding data are provided in Table 4). For data represented as circles ($c < 40$), k_{off} was also obtained independently of the K_d using the near-equilibrium approximation. Triangles represent conditions where we were able to obtain k_{on}, but only ΔH in terms of thermodynamics. For these experiments, a K_d extrapolated from the [salt]- or temperature dependence for TT–RR association was employed in the general equation; alternatively, for experiments in which $c > 200$, the tight-binding approximation could have been used to avoid this extrapolation requirement (additionally $\partial k_{on}/\partial K_d$ is rather negligible if $c > 200$ so precise knowledge of K_d is not required). The k_{off} for these datasets (as well as that of AdoMet–SET7/9), however, is known only as accurately as is the K_d, and thus, the associated uncertainty would be quite high in some cases. Note the truncation at the

(*Continued*)

Kirk A. Vander Meulen *et al.*, Figure 8—Cont'd lower left of the optimal thermodynamics regime: this denotes a region where slow kinetics prevented our accurate measurement of K_d. Boxed triangles represent data used as maximum rate constant test-cases in Fig. 7, and boxed circles were used as near-equilibrium test-cases in Figs. 4 and 5. The colored scale and labels on the right chart are inferred deconvolution-introduced errors: for $k_{on}[M]_{init} < 0.14$, the error should be less than 20% and we set this as the absolute upper limit for kinetics analysis.

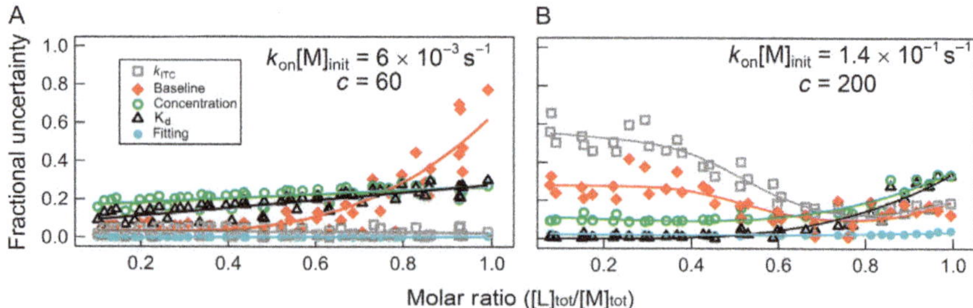

Kirk A. Vander Meulen *et al.*, Figure 9 Estimated fractional uncertainty for k_{on} measurements at the (A) lower and (B) upper end of accessibility. (A) Summary data of five experiments collected for TT–RR binding at 5 °C, 100 m*M* KCl, where the average forward rate constant is 600 $M^{-1}\ s^{-1}$. In each experiment, the starting cell concentration was 10 μ*M*, so $k_{on}[M]_{init}=0.006\ s^{-1}$. (B) Summary of data collected for the SET7/9 lysine methyltransferase, for which the average forward rate constant is 22,000 $M^{-1}\ s^{-1}$, and the average starting cell concentration was 6.4 μ*M* (average $k_{on}[M]_{init}=0.14\ s^{-1}$). Each symbol reflects the estimated contribution of each source of uncertainty to a single injection at that point in a titration. Averaging of all data points in a titration reduces the uncertainty from uncorrelated sources of error as discussed in Section 4.2.3.

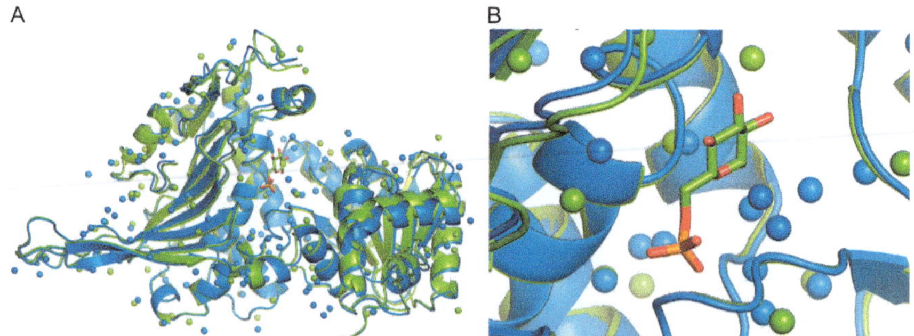

M. Lucia Bianconi, Figure 5 Comparison between the crystal structures of the apo- and G6P-bound enzyme. The crystal structures of the apo-enzyme (1H9B.pdb, in green) and the G6P-bound enzyme (1E77.pdb, in blue) were superimposed. Upon G6P binding, no significant structural change was observed, but the amount of water molecules in the solvation shell as well as in the active site significantly decreases. On the right, a zoom of the active site with a G6P molecule bound.

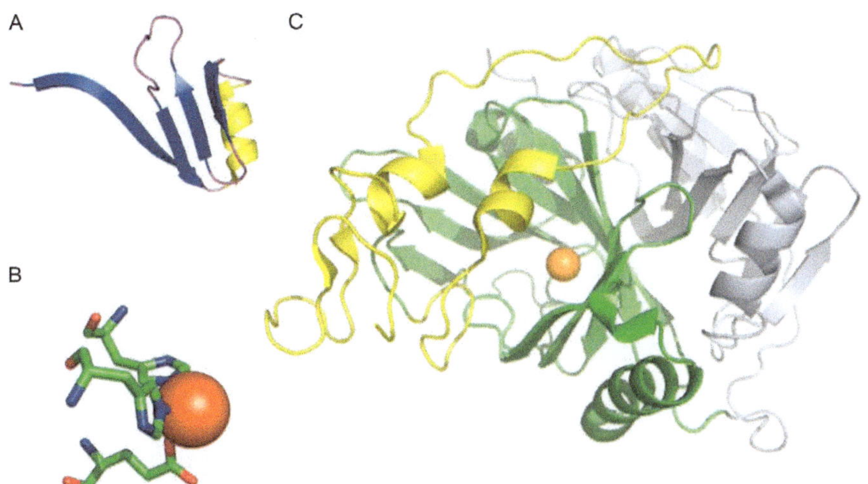

Kate L. Henderson *et al.*, Figure 2 The structure of HPCD. (A) The βαβββ motif, of which HPCD contains multiple copies to form the active site cavity. (B) The globular structure of the HPCD monomeric unit, which shows the N-terminal domain (blue), the C-terminal domain (green) which holds the 2H1C facial triad, and the lid domain (yellow). (C) The iron(II) center ligated to the 2H1C (green residues) of HPCD.

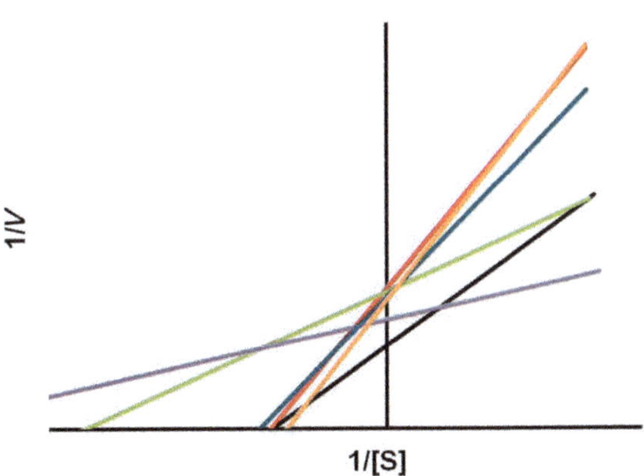

Kate L. Henderson *et al.*, Figure 5 Lineweaver–Burk plots of the steady-state kinetic data associated with HPCA ring opening by HPCD in different buffer systems at pH 7.2. Legend: MOPS (black), TRIS (red), ACES (green), PIPES (purple), phosphate (blue), and HEPES (orange).

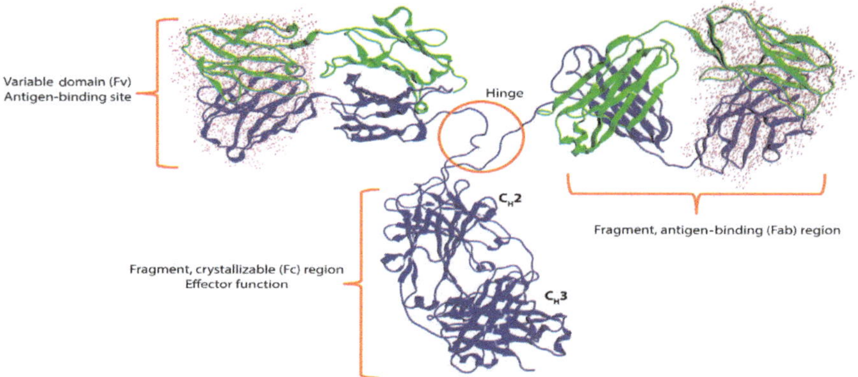

Deniz B. Temel *et al.*, Figure 1 The IgG is a tetrameric molecule of molecular weight approximately 150 kDa, six Ig-fold domains, two heavy chains (blue) and two light chains (green), a single N-glycosylation site in each heavy chain and multiple intra- and interchain disulfide bonds. Fv regions that contain the CDRs are surrounded by dots.

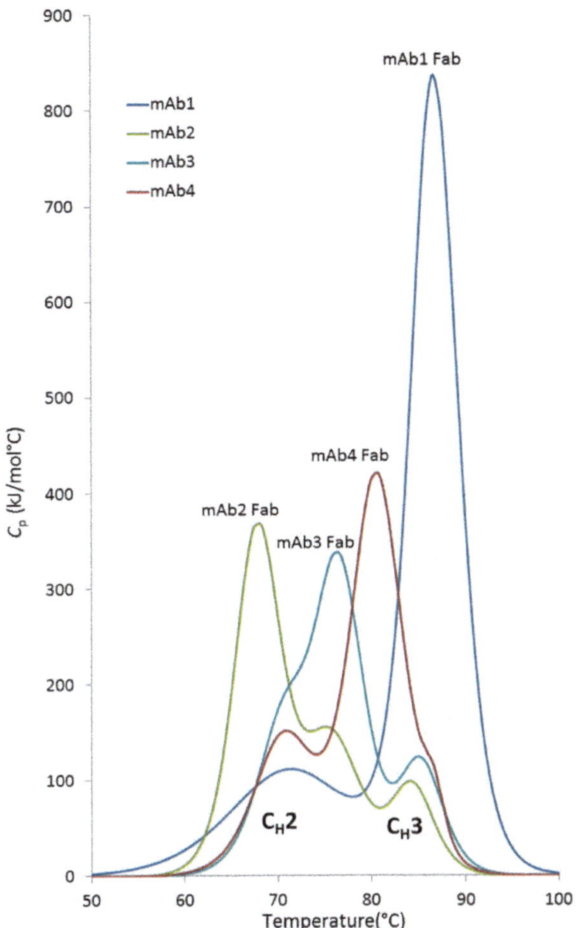

Deniz B. Temel *et al.*, Figure 2 Thermal unfolding curves of human(ized) IgG1 antibodies. Note that the unfolding transitions of the Fab domains are highly variable.

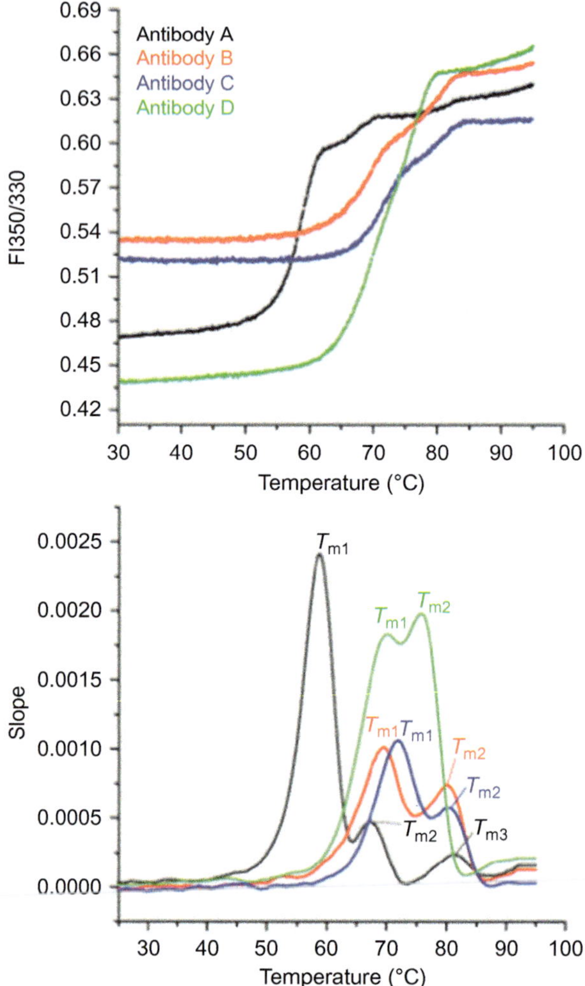

Deniz B. Temel *et al.*, Figure 5 High-resolution intrinsic fluorescence data from Prometheus NT.48 thermal unfolding experiments on four therapeutic mAbs. Plots of the fluorescence ratio (FI350/330) and the corresponding first derivative are shown resolving the unfolding of distinct antibody domains. *Unpublished data provided courtesy of NanoTemper Technologies Inc.*

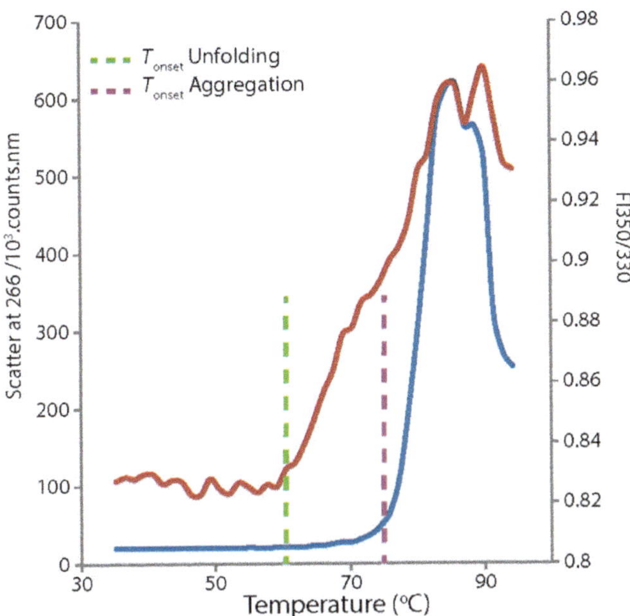

Deniz B. Temel *et al.*, Figure 6 Onset temperatures of unfolding (red) and aggregation (blue) from intrinsic fluorescence and light scattering respectively for an IgG4 mAb.

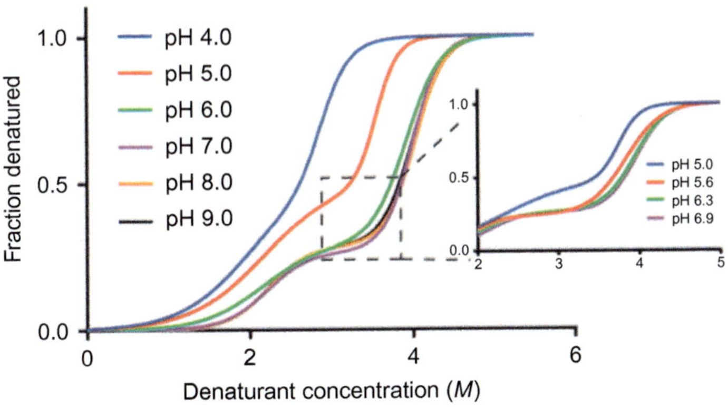

Deniz B. Temel *et al.*, Figure 12 The effect of pH on the ICD profile of an IgG1 mAb. Inspection of the expanded region suggests a formulation pH in the range 6.3–6.9 would maximize folded stability.

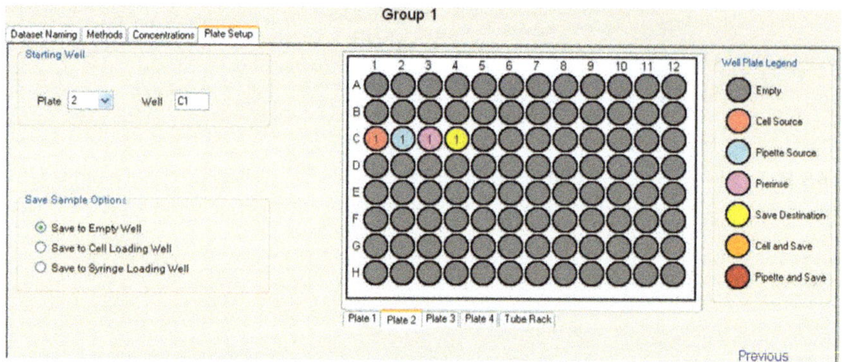

Neela H. Yennawar *et al.*, Figure 3 Plate setup for Auto-iTC200 experiment. Screen shot of the setup window for Auto-iTC200. Shown are four wells, which hold sample, ligand, and buffer, and collect sample, which comprise one experiment. All 96 wells can be filled to give 24 experiments.